Flowchart of the Process of Statistical Hypothesis Testing: Two-Sample Situations

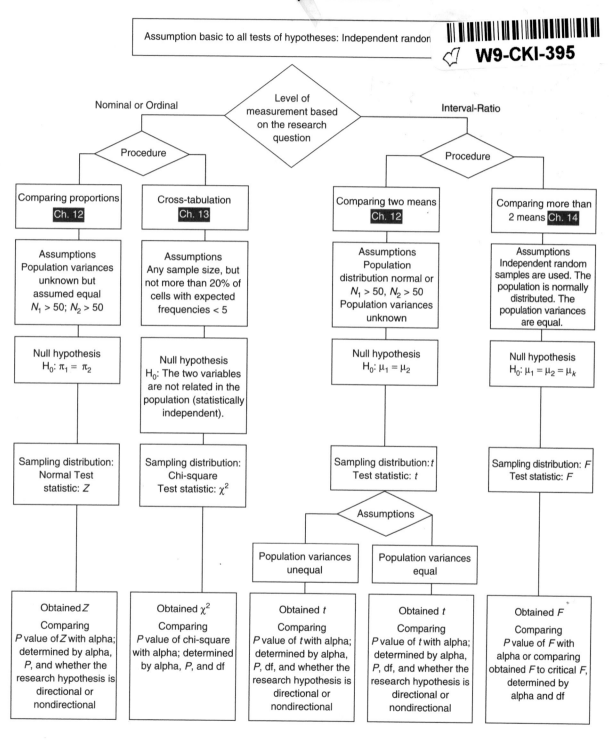

SOCIAL
STATISTICS
for a
DIVERSE SOCIETY
4th Edition

Chava Frankfort-Nachmias
University of Wisconsin

Anna Leon-Guerrero
Pacific Lutheran University

SOCIAL
STATISTICS
for a
DIVERSE SOCIETY
4th Edition

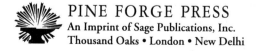

PINE FORGE PRESS
An Imprint of Sage Publications, Inc.
Thousand Oaks • London • New Delhi

For information, address:

Pine Forge Press
A Sage Publications Company
2455 Teller Road
Thousand Oaks, California 91320
(805) 499–4224
E-mail: sales@pfp.sagepub.com

Sage Publications Ltd
1 Oliver's Yard
55 City Road
London EC1Y 1SP
United Kingdom

Sage Publications India Pvt. Ltd.
M-32, Market
Greater Kailash I
New Delhi 110 048 India

Printed in the United States of America on acid-free paper

Library of Congress Cataloging-in-Publication Data

Frankfort-Nachmias, Chava.
Social statistics for a diverse society / Chava Frankfort-Nachmias, Anna Leon-Guerrero.— 4th ed.
   p. cm.
Includes bibliographical references and index.
ISBN 1-4129-1517-1 (cloth)
1. Social sciences—Statistical methods. 2. Statistics. I. Leon-Guerrero, Anna. II. Title.
HA29.N25 2006
519.5—dc22

                         2005008726

05 06 07 08 09 10 9 8 7 6 5 4 3 2

Acquiring Editor:	Jerry Westby
Editorial Assistant:	Laura Shigemitsu
Associate Editor:	Katja Werlich Fried
Production Editor:	Sanford Robinson
Copy Editor:	David Kaplan, Print Matters, Inc.
Typesetter:	C&M Digitals (P) Ltd.
Indexer:	John Hulse
Cover Designer:	Michelle Kenny

To all our friends

ABOUT THE AUTHORS

Chava Frankfort-Nachmias is Associate Professor of Sociology and Director of the Center for Jewish Studies at the University of Wisconsin, Milwaukee, where she teaches courses in research methods, statistics, and gender. She is the author of *Research Methods in the Social Sciences* (with David Nachmias), co-editor of *Sappho in the Holy Land* (with Erella Shadmi) and author of numerous publications on ethnicity and development, urban revitalization, science and gender, and women in Israel. She was the recipient of the University of Wisconsin System teaching improvement grant on integrating race, ethnicity, and gender into the social statistics and research methods curriculum.

Anna Leon-Guerrero is Associate Professor of Sociology at Pacific Lutheran University, Washington. She received her Ph.D. in sociology from the University of California, Los Angeles. She teaches courses in statistics, social theory, and social problems. Her areas of research and publications include job retention and social welfare policy and social service program evaluation. She is the author of *Social Problems: Community, Policy and Social Action.*

CONTENTS

PREFACE

Y ou may be reading this introduction on the first day or sometime during the first week of your statistics class. You probably have some questions about statistics and concerns about what your course will be like. Math, formulas, calculations? Yes, those will be part of your learning experience. However, there is more.

Throughout our text, we emphasize the relevance of statistics in our daily and professional lives. In fact, statistics is such a part of our lives that its importance and uses are often overlooked. How Americans feel about a variety of political and social topics—safety in schools, gun control, abortion, affirmative action, or our president—are measured by surveys and polls and reported daily by the news media. Consider how news programs are already predicting the Republican and Democratic front runners for the next presidential election based on poll results. The latest from a health care study on women was just reported on a morning talk show. And that outfit you just purchased—it didn't go unnoticed. The study of consumer trends, specifically focusing on teens and young adults, helps determine commercial programming, product advertising and placement, and ultimately, consumer spending.

Statistics is not just a part of our lives in the form of news bits or information. And it isn't just numbers either. Throughout this book, we encourage you to move beyond being just a consumer of statistics and begin to recognize and utilize the many ways that statistics can increase our understanding of our world. As social scientists, we know that statistics can be a valuable set of tools to help us analyze and understand the differences in our American society and the world. We use statistics to track demographic trends, to assess differences among groups in society, and to make an impact on social policy and social change. Statistics can help us gain insight into real-life problems that affect our lives.

◉ TEACHING AND LEARNING GOALS

The following three teaching and learning goals continue to be the guiding principles of our book, as they were in the third edition.

The first goal is to introduce you to social statistics and demonstrate its value. Although most of you will not use statistics in your own student research, you will be expected to read and interpret statistical information presented by others in professional and scholarly publications, in the workplace, and in the popular media. This book will help you understand the concepts behind the statistics so that you will be able to assess the circumstances in which certain statistics should and should not be used.

Our second goal is to demonstrate that substance and statistical techniques are truly related in social science research. A special quality of this book is its integration of statistical techniques with substantive issues of particular relevance in the social sciences. Your learning will not be limited to statistical calculations and formulas. Rather, you will become proficient in statistical techniques while learning about social differences and inequality through numerous substantive examples and real-world data applications. Because the world we live in is characterized by a growing diversity—where personal and social realities are increasingly shaped by race, class, gender, and other categories of experience—this book teaches you basic statistics while incorporating social science research related to the dynamic interplay of social variables.

Many of you may lack substantial math background, and some of you may suffer from the "math anxiety syndrome." This anxiety often leads to a less-than-optimal learning environment, with students trying to memorize every detail of a statistical procedure rather than attempting to understand the general concept involved. Hence, our third goal is to address math anxiety by using straightforward prose to explain statistical concepts and by emphasizing intuition, logic, and common sense over rote memorization and derivation of formulas.

▣ DISTINCTIVE AND UPDATED FEATURES OF OUR BOOK

The three learning goals we emphasize are accomplished through a variety of specific and distinctive features throughout this book.

A Close Link Between the Practice of Statistics, Important Social Issues, and Real-World Examples   A special quality of this book is its integration of statistical technique with pressing social issues of particular concern to society and social science. We emphasize how the conduct of social science is the constant interplay between social concerns and methods of inquiry. In addition, the examples throughout the book—most taken from news stories, government reports, scholarly research, the National Opinion Research Center, General Social Survey and the International Social Survey Programme—are formulated to emphasize to students like you that we live in a world in which statistical arguments are common. Statistical concepts and procedures are illustrated with real data and research, providing a clear sense of how questions about important social issues can be studied with various statistical techniques.

A Focus on Diversity—U.S. and International   A strong emphasis on race, class, and gender as central substantive concepts is mindful of a trend in the social sciences toward integrating issues of diversity in the curriculum. This focus on the richness of social differences within our society and our global neighbors is manifested in the application of statistical tools to examine how race, class, gender, and other categories of experience shape our social work and explain social behavior. Examples from the International Social Survey Programme data set help expand our statistical focus beyond the United States to other nations.

Reading the Research Literature In your student career and in the workplace, you may be expected to read and interpret statistical information presented by others in professional and scholarly publications. The statistical analyses presented in these publications are a good deal more complex than most class and textbook presentations. To guide you in reading and interpreting research reports written by social scientists, most chapters include a section presenting excerpts of published research reports utilizing the statistical concepts under discussion.

Integration and Review Chapters Two special review chapters are included in your student CD. The first is a review of descriptive statistical methods (Chapters 2–8), and the second reviews inferential statistics (Chapters 10–14). These review chapters provide an overview of the interconnectedness of the statistical concepts in this book and help test your abilities to cumulatively apply the knowledge from previous chapters. Both chapters include flowcharts that summarize the systematic approach utilized in the selection of statistical techniques as well as exercises that require the use of several different procedures. As in our text chapters, each web chapter concludes with an SPSS demonstration, SPSS exercises and end of chapter exercises.

Tools to Promote Effective Study Each chapter closes with a list of main points and key terms discussed in that chapter. Boxed definitions of the key terms also appear in the body of the chapter, as do learning checks keyed to the most important points. Key terms are also clearly defined and explained in the index/glossary, another special feature in our book. Answers to all the odd-numbered problems in the text are included in the back of the book. Complete step-by-step solutions are in the manual for instructors, available from the publisher upon adoption of the text.

Emphasis on Computing SPSS® for Windows® is used throughout the book, although the use of computers is not required to learn from the text. Real data are used to motivate and make concrete the coverage of statistical topics. These data, from the General Social Survey and the International Social Survey Programme, are included in a disk packaged with every copy of the text. At the end of each chapter, we feature a demonstration of a related SPSS procedure, along with a set of exercises.

▣ HIGHLIGHTS OF THE FOURTH EDITION

We have made a number of important changes to this book in response to the valuable comments that we have received from the many instructors adopting the third edition and from other interested instructors (and their students).

- *Clearer and more concise presentation of topics* We have carefully edited the discussion of statistical procedures and concepts, reducing the redundancy of statistical procedures and clarifying examples, while at the same time preserving the book's easily understood style. For example, please refer to our revised discussion on the calculation of IQV in Chapter 5 and our briefer discussion on PRE measurements in Chapter 7.
- *Revisions to Chapter 8 "Regression and Correlation"* We have expanded our discussion in this chapter to include multiple regression.

- New *Chapter 14* Analysis of Variance. In this edition our last chapter is on analysis of variance (ANOVA). The chapter begins with a detailed computational example, highlights ANOVA applications to regression techniques, and reviews two examples from research literature.
- *Real-world examples and exercises* A hallmark of our first three editions was their extensive use of real data from a variety of sources for chapter illustrations and exercises. Throughout the fourth edition, we have updated the majority of exercises and examples based on General Social Survey, International Social Survey, or U.S. Census data.
- *SPSS version 13.0* Packaged with this text, on an optional basis, is SPSS Student Version 13.0. SPSS demonstrations and exercises have been updated, using version 13.0 format. Appendix E, How to Use a Statistical Package, has also been updated to highlight 13.0 features. Please telephone the publisher at (805) 499–4224, or access their Web site at www.pineforge.com to learn how to order the book packaged with the student version of SPSS v.13.0.
- *General Social Survey 2002 and International Social Survey Programme 2000* As a companion to the fourth edition's SPSS demonstrations and exercises, we have created four data sets. Those of you with the student version of SPSS 13.0 will work with two separate GSS files: GSS Module A and GSS Module B. The GSS2002PFP.SAV contains an expanded selection of variables and cases from the 2002 General Social Survey. The ISSP00PFP.SAV contains selection of variables and cases from the 2000 International Social Survey based on respondents from 38 countries. SPSS exercises at the end of each chapter utilize certain variables from all data modules. There is ample opportunity for instructors to develop their own SPSS exercises using these data.
- *Supplemental tools on important topics* The fourth edition's discussion of inferential statistics remains focused on Z, t, and chi-square. The Pine Forge Press Series in Research Methods and Statistics, of which this book is a part, includes additional supplementary volumes by Paul Allison on regression and by Robert Leik on analysis of variance. These supplements were written to closely coordinate with this text and are available from the publisher.

◙ ACKNOWLEDGMENTS

We are both grateful to Jerry Westby, Series Editor for Sage Publications, for his commitment to our book and for his invaluable assistance through the production process.

Many manuscript reviewers recruited by Pine Forge provided invaluable feedback. For their comments to the fourth edition, we thank Michelle Bata, University of Arizona; Michael Bourgeois, University of California, Santa Barbara; Karen Clarke, Manchester University; Juanita Firestone, University of Texas, San Antonio; Judith Gonyea, Boston University; Ted Greenstein, North Carolina State University; Andrew Jorgenson, University of California, Riverside; Suman Kakar, Florida International University; Jeffrey Kentor, University of Utah; Sally A. Raskoff, Los Angeles Valley College; Mark J. Schaefer, Louisiana State University; Paul von Hippel, Ohio State University; William E. Wagner, California State University, Bakersfield; N. Eugene Walls, University of Notre Dame; and James White, Clemson University.

We are grateful to Katja Fried and Sanford Robinson for guiding the book through the production process. We would also like to acknowledge Karen Wiley and the rest of the Pine Forge Press staff for their assistance and support throughout this project. This text would not have been possible without the editing support of Richard Rothschild, David Kaplan, Michael Bourgeois, and Jack DeWaard and instructional material support from Jack DeWaard and Billy Wagner. Our deepest thanks for your contributions.

Chava Frankfort-Nachmias would like to thank and acknowledge the following: I am intellectually indebted to Elizabeth Higginbotham and Lynn Weber Cannon for an instructive and inspiring SWS workshop on integrating race, class, and gender in the sociological curriculum. At that workshop the idea to work on this book began to emerge. I was also greatly influenced by the pioneering work of Margaret L. Andersen and Patricia Hill Collins, who developed an interdisciplinary and inclusive framework for transforming the curriculum.

My profound gratitude goes to friends and colleagues who have stood by me, cheered me on, and understood when I was unavailable for long periods due to the demands of this project. Indispensable assistance in preparing the fourth edition manuscript was provided at every step by my research assistant, Jack DeWaard. My work on this edition would not have been possible without his contributions. I would like to express my gratitude to Jack for his thoroughness and patience.

I am grateful to my students at the University of Wisconsin–Milwaukee, who taught me that even the most complex statistical ideas can be simplified. The ideas presented in this book are the products of many years of classroom testing. I thank my students for their patience and contributions.

Finally, I thank my partner, Marlene Stern, and my daughters, Anat and Talia, for their love, support, and faith in me.

Anna Leon-Guerrero expresses her thanks to the following: I wish to thank my statistics teaching assistants and students. My passion for and understanding of teaching statistics grow with each semester and class experience. I am grateful for the teaching and learning opportunities we have shared.

My work on this fourth edition could not have been possible without the contribution and efforts of Heather Ottum and Kathryn Irwin. In addition, I would like to express my gratitude to friends and colleagues for their encouragement and support throughout this project. My love and thanks to my husband, Brian Sullivan.

Chava Frankfort-Nachmias
University of Wisconsin–Milwaukee

Anna Leon-Guerrero
Pacific Lutheran University

The What and the Why of Statistics

Are you taking statistics because it is required in your major—not because you find it interesting? If so, you may be feeling intimidated because you know that statistics involves numbers and math. Perhaps you feel intimidated not only because you're uncomfortable with math, but also because you suspect that numbers and math don't leave room for human judgment or have any relevance to your own personal experience. In fact, you may even question the relevance of statistics to understanding people, social behavior, or society.

In this course, we will show you that statistics can be a lot more interesting and easy to understand than you may have been led to believe. In fact, as we draw upon your previous knowledge and experience and relate materials to interesting and important social issues, you'll begin to see that statistics is not just a course you have to take but a useful tool as well.

There are two major reasons why learning statistics may be of value to you. First, you are constantly exposed to statistics every day of your life. Marketing surveys, voting polls, and the findings of social research appear daily in newspapers and popular magazines. By learning statistics you will become a sharper consumer of statistical material. Second, as a major in the social sciences, you may be expected to read and interpret statistical information presented to you in the workplace. Even if conducting research is not a part of your job, you may still be expected to understand and learn from other people's research or to be able to write reports based on statistical analyses.

Just what is statistics anyway? You may associate the word with numbers that indicate birth rates, conviction rates, per-capita income, marriage and divorce rates, and so on. But the word statistics also refers to a set of procedures used by social scientists. They use these procedures to organize, summarize, and communicate information. Only information represented by numbers can be the subject of statistical analysis. Such information is called data; researchers use statistical procedures to analyze data to answer research questions and test theories. It is the latter usage—answering research questions and testing theories—that this textbook explores.

Statistics   A set of procedures used by social scientists to organize, summarize, and communicate information.

Data   Information represented by numbers, which can be the subject of statistical analysis.

▣ THE RESEARCH PROCESS

To give you a better idea of the role of statistics in social research, let's start by looking at the **research process**. We can think of the research process as a set of activities in which social scientists engage so they can answer questions, examine ideas, or test theories.

As illustrated in Figure 1.1, the research process consists of five stages:

1. Asking the research question

2. Formulating the hypotheses

3. Collecting data

4. Analyzing data

5. Evaluating the hypotheses

Each stage affects the *theory* and is affected by it as well. Statistics are most closely tied to the data analysis stage of the research process. As we will see in later chapters, statistical analysis of the data helps researchers test the validity and accuracy of their hypotheses.

Figure 1.1   The Research Process

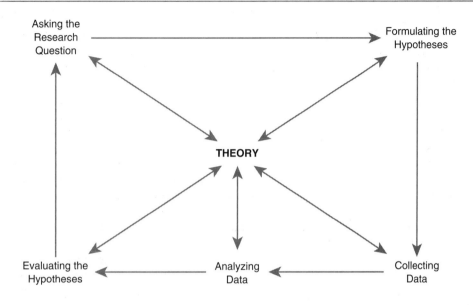

Research process   A set of activities in which social scientists engage to answer questions, examine ideas, or test theories.

⊡ ASKING RESEARCH QUESTIONS

The starting point for most research is asking a *research question*. Consider the following research questions taken from a number of social science journals:

"Does cost control influence the quality of health care?"

"Has sexual harassment become more widespread during the past decade?"

"Does social class influence voting behavior?"

"What factors influence the economic mobility of female workers?"

These are all questions that can be answered by conducting **empirical research**—research based on information that can be verified by using our direct experience. To answer research questions we cannot rely on reasoning, speculation, moral judgment, or subjective preference. For example, the questions "Is racial equality good for society?" and "Is an urban lifestyle better than a rural lifestyle?" cannot be answered empirically because the terms *good* and *better* are concerned with values, beliefs, or subjective preference and, therefore, cannot be independently verified. One way to study these questions is by defining *good* and *better* in terms that can be verified empirically. For example, we can define *good* in terms of economic growth and *better* in terms of psychological well-being. These questions could then be answered by conducting empirical research.

You may wonder how to come up with a research question. The first step is to pick a question that interests you. If you are not sure, look around! Ideas for research problems are all around you, from media sources to personal experience or your own intuition. Talk to other people, write down your own observations and ideas, or learn what other social scientists have written about.

Take, for instance, the issue of gender and work. As a college student about to enter the labor force, you may wonder about the similarities and differences between women's and men's work experiences and about job opportunities when you graduate. Here are some facts and observations based on research reports and our own personal experiences: In 2002 women who were employed full-time earned $530 per week on average; men who were employed full-time earned $680 per week on average.[1] Women's and men's work are also very different. Women continue to be the minority in many of the higher-ranking and higher-salaried positions in professional and managerial occupations. For example, in 2002 women made up 11 percent of engineers, 37 percent of physicians, 19.4 percent of dentists, and 20 percent of architects. In comparison, among all those employed as preschool and kindergarten teachers, 97.7 percent were women. Among all secretaries in 2002, 98.6 percent were women.[2] These observations may prompt us to ask research questions such as: Are women paid, on average, less than men for the same types of work? How much change has there been in women's work over time?

Empirical research   Research based on evidence that can be verified by using our direct experience.

✔ *Learning Check.*   *Identify one or two social science questions amenable to empirical research. You can almost bet that you will be required to do a research project sometime in your college career. Get a head start and start thinking about a good research question now.*

▣ THE ROLE OF THEORY

You may have noticed that each preceding research question was expressed in terms of a *relationship*. This relationship may be between two or more attributes of individuals or groups,

---

[1]U.S.Bureau of Census, 2003. *Statistical Abstract of the United States*, Table 641.
[2]U.S. Bureau of Census, 2003. *Statistical Abstract of the United States*, Table 615.

such as gender and income or gender segregation in the workplace and income disparity. The relationship between attributes or characteristics of individuals and groups lies at the heart of social scientific inquiry.

Most of us use the term *theory* quite casually to explain events and experiences in our daily life. We may have a "theory" about why our boss has been so nice to us lately or why we didn't do so well on our last history test. In a somewhat similar manner, social scientists attempt to explain the nature of social reality. Whereas our theories about events in our lives are commonsense explanations based on educated guesses and personal experience, to the social scientist a theory is a more precise explanation that is frequently tested by conducting research.

A **theory** is an explanation of the relationship between two or more observable attributes of individuals or groups. The theory attempts to establish a link between what we observe (the data) and our conceptual understanding of why certain phenomena are related to each other in a particular way. For instance, suppose we wanted to understand the reasons for the income disparity between men and women; we may wonder whether the types of jobs men and women have and the organizations in which they work have something to do with their wages.

One explanation for gender inequality in wages is *gender segregation in the workplace*— the fact that American men and women are concentrated in different kinds of jobs and occupations. For example, in 2000, of the approximately 66 million women in the labor force, one-third (30%) worked in only 10 of the 503 occupations listed by the census.[3]

What is the significance of gender segregation in the workplace? In our society, people's occupations and jobs are closely associated with their level of prestige, authority, and income. The jobs in which women and men are segregated are not only different but also unequal. Although the proportion of women in the labor force has markedly increased, women are still concentrated in occupations with low pay, low prestige, and few opportunities for promotion. Thus, gender segregation in the workplace is associated with unequal earnings, authority, and status. In particular, women's segregation into different jobs and occupations from those of men is the most immediate cause of the pay gap. Women receive lower pay than men do even when they have the same level of education, skills, and experience as men in comparable occupations.

Theory   An elaborate explanation of the relationship between two or more observable attributes of individuals or groups.

▣ FORMULATING THE HYPOTHESES

So far we have come up with a number of research questions about the income disparity between men and women in the workplace. We have also discussed a possible explanation— a theory—that helps us make sense of gender inequality in wages. Is that enough? Where do we go from here?

---

[3]Barbara Reskin and Irene Padavic, *Women and Men at Work* (Thousand Oaks, CA: Pine Forge Press, 2002), p. 65.

Our next step is to test some of the ideas suggested by the gender segregation theory. But this theory, even if it sounds reasonable and logical to us, is too general and does not contain enough specific information to be tested. Instead, theories suggest specific concrete predictions about the way that observable attributes of people or groups are interrelated in real life. These predictions, called **hypotheses**, are tentative answers to research problems. Hypotheses are tentative because they can be verified only after they have been tested empirically.[4] For example, one hypothesis we can derive from the gender segregation theory is that wages in occupations in which the majority of workers are female are lower than the wages in occupations in which the majority of workers are male.

Not all hypotheses are derived directly from theories. We can generate hypotheses in many ways—from theories, directly from observations, or from intuition. Probably the greatest source of hypotheses is the professional literature. A critical review of the professional literature will familiarize you with the current state of knowledge and with hypotheses that others have studied.

Hypothesis   A tentative answer to a research problem.

Let's restate our hypothesis:

Wages in occupations in which the majority of workers are female are lower than the wages in occupations in which the majority of workers are male.

Notice that this hypothesis is a statement of a relationship between two characteristics that vary: *wages* and *gender composition* of occupations. Such characteristics are called variables. A **variable** is a property of people or objects that takes on two or more values. For example, people can be classified into a number of *social class* categories, such as upper class, middle class, or working class. Similarly, people have different levels of education; therefore, *education* is a variable. *Family income* is a variable; it can take on values from zero to hundreds of thousands of dollars or more. *Wages* is a variable, with values from zero to thousands of dollars or more. Similarly, *gender composition* is a variable. The percentage of females (or males) in an occupation can vary from 0 to 100. (See Figure 1.2 for examples of some variables and their possible values.)

Variable   A property of people or objects that takes on two or more values.

Each variable must include categories that are both *exhaustive* and *mutually exclusive*. Exhaustiveness means that there should be enough categories composing the variables to classify every observation. For example, the common classification of the variable *marital status* into the categories "married," "single," "divorced," and "widowed" violates the requirement

---

[4]Chava Frankfort-Nachmias and David Nachmias, *Research Methods in the Social Sciences* (New York: Worth Publishers, 2000), p. 56.

Figure 1.2   Variables and Value Categories

Variable	Categories
Social class	Upper class Middle class Working class
Religion	Christian Jewish Muslim
Monthly income	$1,000 $2,500 $10,000 $15,000
Gender	Male Female

of exhaustiveness. As defined, it does not allow us to classify same-sex couples or heterosexual couples who are not legally married. (We can make every variable exhaustive by adding the category "other" to the list of categories. However, this practice is not recommended if it leads to the exclusion of categories that have theoretical significance or a substantial number of observations.)

Mutual exclusiveness means that there is only one category suitable for each observation. For example, we need to define *religion* in such a way that no one would be classified into more than one category. For instance, the categories "Protestant" and "Methodist" are not mutually exclusive because Methodists are also considered Protestant and, therefore, could be classified into both categories.

✔ *Learning Check.*   *Review the definitions of* exhaustive *and* mutually exclusive. *Now look at Figure 1.2. What other categories could be added to the variable* religion *in order to be exhaustive and mutually exclusive? What other categories could be added to* social class? *To* income?

Social scientists can choose which level of social life to focus their research on. They can focus on individuals or on groups of people such as families, organizations, and nations. These distinctions are referred to as **units of analysis**. A variable is a property of whatever the unit of analysis is for the study. Variables can be properties of individuals, of groups (such as the family or a social group), of organizations (such as a hospital or university), or of societies (such as a country or a nation). For example, in a study that looks at the relationship between individuals' level of education and their income, the variable *income* refers to the income level of an individual. On the other hand, a study that compares how differences in corporations' revenues relate to differences in the fringe benefits they provide to their employees uses the variable *revenue* as a characteristic of an organization (the corporation). The variables *wages* and *gender composition* in our example are characteristics of occupations. Figure 1.3 illustrates different units of analysis frequently employed by social scientists.

Figure 1.3   Examples of Units of Analysis

Individual as unit of analysis:

How old are you?
What are your political views?
What is your occupation?

Family as unit of analysis:

How many children are in the family?
Who does the housework?
How many wage earners are there?

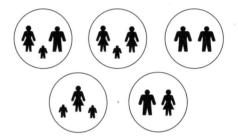

Organization as unit of analysis:

How many employees are there?
What is the gender composition?
Do you have a diversity office?

City as unit of analysis:

What was the crime rate last year?
What is the population density?
What type of government runs things?

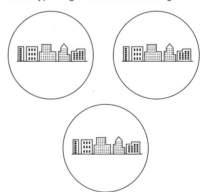

Unit of analysis   The level of social life on which social scientists focus. Examples of different levels are individuals and groups.

Independent and Dependent Variables: Causality

Hypotheses are usually stated in terms of a relationship between an *independent* and a *dependent variable.* The distinction between an independent and a dependent variable is an important one in the language of research. Social theories often intend to provide an explanation for social patterns or causal relations between variables. For example, according to the gender segregation theory, gender segregation in the workplace is the primary explanation (although certainly not the only one) of the male/female earning gap. Why should jobs where the majority of workers are female pay less than jobs that employ mostly men? One explanation is that "societies undervalue the work women do, regardless of what those tasks are, because women do them. . . . For example, our culture tends to devalue caring or nurturant work at least partly because it is done by women. This tendency accounts for child care workers' low rank in the pay hierarchy."[5]

In the language of research, the variable the researcher wants to explain (the "effect") is called the **dependent variable**. The variable that is expected to "cause" or account for the dependent variable is called the **independent variable**. Therefore, in our example, *gender composition of occupations* is the independent variable, and *wages* is the dependent variable.

Dependent variable   The variable to be explained (the "effect").

Independent variable   The variable expected to account for (the "cause" of) the dependent variable.

Cause-and-effect relationships between variables are *not* easy to infer in the social sciences. In order to establish that two variables are causally related, you need to meet three conditions: (1) the cause has to precede the effect in time, (2) there has to be an empirical relationship between the cause and the effect, and (3) this relationship cannot be explained by other factors.

Let's consider the decades-old debate about controlling crime through the use of prevention versus punishment. Some people argue that special counseling for youths at the first sign of trouble and strict controls on access to firearms would help reduce crime. Others argue that overhauling federal and state sentencing laws to stop early prison releases is the solution. Between 1993 and 1995, 24 states and the federal government adopted a "three strikes and you're out" policy enhancing sentencing for repeat offenders, usually imposing life prison terms on three-time felony offenders. Let's suppose that 10 years after the measure was introduced the crime rate declined in some of these states. Does the observation that the incidence of crime declined mean that the new measure caused this reduction? Not necessarily! Perhaps the rate of crime had been going down for other reasons, such as improvement in the economy, and the new measure had nothing to do with it. To demonstrate a cause-and-effect relationship, we would need to show three things: (1) the enactment of the "three strikes and you're out" measure was empirically associated with a decrease in crime; (2) the reduction of crime actually occurred *after* the enactment of this measure; and (3) the

---

[5]Reskin and Padavic, p. 144.

relationship between the reduction in crime and the "three strikes and you're out" policy is not due to the influence of another variable (for instance, the improvement of overall economic conditions).

Independent and Dependent Variables: Guidelines

Because of the limitations in inferring cause-and-effect relationships in the social sciences, be cautious about using the terms *cause* and *effect* when examining relationships between variables. However, using the terms *independent variable* and *dependent variable* is still appropriate even when this relationship is not articulated in terms of direct cause and effect. Here are a few guidelines that may help you to identify the independent and dependent variables:

1. The dependent variable is always the property you are trying to explain; it is always the object of the research.

2. The independent variable usually occurs earlier in time than the dependent variable.

3. The independent variable is often seen as influencing, directly or indirectly, the dependent variable.

The purpose of the research should help determine which is the independent variable and which is the dependent variable. In the real world, variables are neither dependent nor independent; they can be switched around depending on the research problem. A variable defined as independent in one research investigation may be a dependent variable in another.[6] For instance, *educational attainment* may be an independent variable in a study attempting to explain how education influences political attitudes. However, in an investigation of whether a person's level of education is influenced by the social status of his or her family of origin, *educational attainment* is the dependent variable. Some variables, such as race, age, and ethnicity, because they are primordial characteristics that cannot be explained by social scientists, are never considered dependent variables in a social science analysis.

✔ *Learning Check.*   *Identify the independent and dependent variables in the following hypotheses:*

- *Children who attended preschool day-care centers earn better grades in first grade than children who received home preschool care.*
- *People who attend church regularly are more likely to oppose abortion than people who do not attend church regularly.*
- *Elderly women are more likely to live alone than elderly men.*
- *Individuals with postgraduate education are more likely to have fewer children than those with less education.*

What are the independent and dependent variables in your hypothesis?

---

[6]Frankfort-Nachmias and Nachmias, p. 50.

▣ COLLECTING DATA

Once we have decided on the research question, the hypothesis, and the variables to be included in the study, we proceed to the next stage in the research cycle. This step includes measuring our variables and collecting the data. As researchers, we must decide how to measure the variables of interest to us, how to select the cases for our research, and what kind of data collection techniques we will be using. A wide variety of data collection techniques are available to us, from direct observations to survey research, experiments, or secondary sources. Similarly, we can construct numerous measuring instruments. These instruments can be as simple as a single question included in a questionnaire or as complex as a composite measure constructed through the combination of two or more questionnaire items. The choice of a particular data collection method or instrument to measure our variables depends on the study objective. For instance, suppose we decide to study how social class position is related to attitudes about abortion. Since attitudes about abortion are not directly observable, we need to collect data by asking a group of people questions about their attitudes and opinions. A suitable method of data collection for this project would be a *survey* that uses some kind of questionnaire or interview guide to elicit verbal reports from respondents. The questionnaire could include numerous questions designed to measure attitudes toward abortion, social class, and other variables relevant to the study.

How would we go about collecting data to test the hypothesis relating the gender composition of occupations to wages? We want to gather information on the proportion of men and women in different occupations and the average earnings for these occupations. This kind of information is routinely collected by the government and published in sources such as bulletins distributed by the U.S. Department of Labor's Bureau of Labor Statistics and the *Statistical Abstract of the United States*. The data obtained from these sources could then be analyzed and used to test our hypothesis.

Levels of Measurement

The statistical analysis of data involves many mathematical operations, from simple counting to addition and multiplication. However, not every operation can be used with every variable. The type of statistical operations we employ depends on how our variables are measured. For example, for the variable *gender*, we can use the number 1 to represent females and the number 2 to represent males. Similarly, 1 can also be used as a numerical code for the category "one child" in the variable *number of children*. Clearly, in the first example the number is an arbitrary symbol that does not correspond to the property "female," whereas in the second example the number 1 has a distinct numerical meaning that does correspond to the property "one child." The correspondence between the properties we measure and the numbers representing these properties determines the type of statistical operations we can use. The degree of correspondence also leads to different ways of measuring—that is, to distinct *levels of measurement*. In this section, we will discuss three levels of measurement: *nominal, ordinal,* and *interval-ratio.*

Nominal Level of Measurement   At the **nominal** level of measurement, numbers or other symbols are assigned to a set of categories for the purpose of naming, labeling, or classifying the observations. *Gender* is an example of a nominal-level variable. Using the numbers

1 and 2, for instance, we can classify our observations into the categories "females" and "males," with 1 representing females and 2 representing males. We could use any of a variety of symbols to represent the different categories of a nominal variable; however, when numbers are used to represent the different categories, we do not imply anything about the magnitude or quantitative difference between the categories. Because the different categories (for instance, males versus females) vary in the quality inherent in each but not in quantity, nominal variables are often called *qualitative*. Other examples of nominal-level variables are political party, religion, and race.

Ordinal Level of Measurement   Whenever we assign numbers to rank-ordered categories ranging from low to high, we have an **ordinal**-level variable. *Social class* is an example of an ordinal variable. We might classify individuals with respect to their social class status as "upper class," "middle class," or "working class." We can say that a person in the category "upper class" has a higher class position than a person in a "middle class" category (or that a "middle class" position is higher than a "working class" position), but we do not know the magnitude of the differences between the categories; that is, we don't know how much higher "upper class" is compared with "middle class."

Many attitudes we measure in the social sciences are ordinal-level variables. Take, for instance, the following question used to measure attitudes toward same-sex marriages: "Same-sex partners should have the right to marry each other." Respondents are asked to mark the number representing their degree of agreement or disagreement with this statement. One form in which a number might be made to correspond with the answers can be seen in Table 1.1. Although the differences between these numbers represent higher or lower degrees of agreement with same-sex marriage, the distance between any two of those numbers does not have a precise numerical meaning.

Interval-Ratio Level of Measurement   If the categories (or values) of a variable can be rank-ordered, and if the measurements for all the cases are expressed in the same units, then an **interval-ratio** level of measurement has been achieved. Examples of variables measured at the interval-ratio level are *age, income*, and *SAT scores*. With all these variables we can compare values not only in terms of which is larger or smaller, but also in terms of *how much* larger or smaller one is compared with another. In some discussions of levels of measurement you will see a distinction made between interval-ratio variables that have a natural zero point (where zero means the absence of the property) and those variables that have zero as an arbitrary point. For example, weight and length have a natural zero point, whereas temperature has an arbitrary zero point. Variables with a natural zero point are also called *ratio variables*.

Table 1.1   Ordinal Ranking Scale

Rank	*Value*
1	Strongly agree
2	Agree
3	Neither agree nor disagree
4	Disagree
5	Strongly disagree

In statistical practice, however, ratio variables are subjected to operations that treat them as interval and ignore their ratio properties. Therefore, no distinction between these two types is made in this text.

Nominal measurement   Numbers or other symbols are assigned to a set of categories for the purpose of naming, labeling, or classifying the observations.

Ordinal measurement   Numbers are assigned to rank-ordered categories ranging from low to high.

Interval-ratio measurement   Measurements for all cases are expressed in the same units.

Cumulative Property of Levels of Measurement   Variables that can be measured at the interval-ratio level of measurement can also be measured at the ordinal and nominal levels. As a rule, properties that can be measured at a higher level (interval-ratio is the highest) can also be measured at lower levels, but not vice versa. Let's take, for example, gender composition of occupations, the independent variable in our research example. Table 1.2 shows the percentage of women in four major occupational groups as reported in the 2003 *Statistical Abstract of the United States*.

Table 1.2   Gender Composition of Four Major Occupational Groups

Occupational Group	*% Women in Occupation*
Executive, administrative, and managerial	51
Administrative support	79
Transportation workers	10
Sales	49

Source: U.S. Bureau of the Census, *Statistical Abstract of the United States*, 2003, Table 675.

The variable *gender composition* (measured as the percentage of women in the occupational group) is an interval-ratio variable and, therefore, has the properties of nominal, ordinal, and interval-ratio measures. For example, we can say that the transportation group differs from the sales group (a nominal comparison), that sales occupations have more women than transportation occupations (an ordinal comparison), and that sales occupations have 39 percentage points more women (49 − 10) than transportation occupations (an interval-ratio comparison).

The types of comparisons possible at each level of measurement are summarized in Table 1.3 and Figure 1.4. Notice that differences can be established at each of the three levels, but only at the interval-ratio level can we establish the magnitude of the difference.

Table 1.3   Levels of Measurement and Possible Comparisons

Level	Different or Equivalent	Higher or Lower	How Much Higher
Nominal	Yes	No	No
Ordinal	Yes	Yes	No
Interval-ratio	Yes	Yes	Yes

✔ **Learning Check.**   *Make sure you understand these levels of measurement. As the course progresses, your instructor is likely to ask you what statistical procedure you would use to describe or analyze a set of data. To make the proper choice, you must know the level of measurement of the data.*

Figure 1.4   Levels of Measurement and Possible Comparisons: Education Measured on Nominal, Ordinal, and Interval-Ratio Levels

Possible Comparisons

Nominal Measurement

Difference or equivalence: These people have different types of education.

Graduated from public high school   Graduated from private high school   Graduated from military academy

Possible Comparisons

Ordinal Measurement

Ranking or ordering: One person is higher in education than another.

Holds a high school diploma   Holds a college diploma   Holds a Ph.D.

—◆?◆—
Distance Meaningless

Possible Comparisions

Interval-Ratio Measurement

How much higher or lower?

Has 8 years of education   Has 12 years of education   Has 16 years of education

4 years
Distance Meaningful

Levels of Measurement of Dichotomous Variables   A variable that has only two values is called a *dichotomous variable.* Several key social factors, such as gender, employment status, and marital status, are dichotomies; that is, you are male or female, employed or unemployed, married or not married. Such variables may seem to be measured at the nominal level: you fit in either one category or the other. No category is naturally higher or lower than the other, so they can't be ordered.

However, because there are only two possible values for a dichotomy, we can measure it at the ordinal or the interval-ratio level. For example, we can think of "femaleness" as the ordering principle for gender, so that "female" is higher and "male" is lower. Using "maleness" as the ordering principle, "female" is lower and "male" is higher. In either case, with only two classes, there is no way to get them out of order; therefore, gender could be considered at the ordinal level.

Dichotomous variables can also be considered to be interval-ratio level. Why is this? In measuring interval-ratio data, the size of the interval between the categories is *meaningful*: the distance between 4 and 7, for example, is the same as the distance between 11 and 14. But with a dichotomy, there is only one interval. Therefore, there is really no other distance to which we can compare it:

Mathematically, this gives the dichotomy more power than other nominal level variables (as you will notice later in the text).

For this reason, researchers often dichotomize some of their variables, turning a multicategory nominal variable into a dichotomy. For example, you may see race (originally divided into many categories) dichotomized into "white" and "nonwhite." Though this is substantively suspect, it may be the most logical statistical step to take.

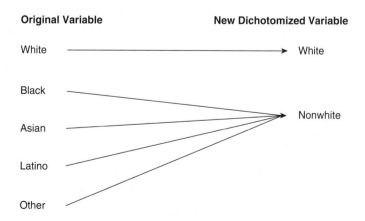

When you dichotomize a variable, be sure that the two categories capture a distinction that is important to your research question (for example, a comparison of the number of white versus nonwhite U.S. Senators).

Discrete and Continuous Variables

The statistical operations we can perform are also determined by whether the variables are continuous or discrete. *Discrete* variables have a minimum-sized unit of measurement, which cannot be subdivided. The number of children per family is an example of a discrete variable because the minimum unit is one child. A family may have two or three children, but not 2.5 children. The variable *wages* in our research example is a discrete variable because currency has a minimum unit (1 cent), which cannot be subdivided. One can have $101.21 or $101.22 but not $101.21843. Wages cannot differ by less than 1 cent—the minimum-sized unit.

Unlike discrete variables, *continuous* variables do not have a minimum-sized unit of measurement; their range of values can be subdivided into increasingly smaller fractional values. *Length* is an example of a continuous variable because there is no minimum unit of length. A particular object may be 12 inches long, it may be 12.5 inches long, or it may be 12.532011 inches long. Although we cannot always measure all possible length values with absolute accuracy, it is possible for objects to exist at an infinite number of lengths.[7] In principle, we can speak of a tenth of an inch, a ten-thousandth of an inch, or a ten-trillionth of an inch. The variable *gender composition of occupations* is a continuous variable because it is measured in proportions or percentages (for example, the percentage of women in medicine), which can be subdivided into smaller and smaller fractions.

This attribute of variables—whether they are continuous or discrete—affects subsequent research operations, particularly measurement procedures, data analysis, and methods of inference and generalization. However, keep in mind that, in practice, some discrete variables can be treated as if they were continuous, and vice versa.

✓ *Learning Check.* *Name three continuous and three discrete variables. Determine whether each of the variables in your hypothesis is continuous or discrete.*

▣ ANALYZING DATA AND EVALUATING THE HYPOTHESES

Following the data collection stage, researchers analyze their data and evaluate the hypotheses of the study. The data consist of codes and numbers used to represent our observations. In our example, each occupational group would be represented by two scores: (1) the percentage of women and (2) the average wage. If we had collected information on 100 occupations, we would end up with 200 scores, two per occupational group. However, the typical research project includes more variables; therefore, the amount of data the researcher confronts is considerably larger. We now must find a systematic way to organize these data,

---

[7]Ibid., p. 52.

analyze them, and use some set of procedures to decide what they mean. These last steps make up the *statistical analysis* stage, which is the main topic of this textbook. It is also at this point in the research cycle that statistical procedures will help us *evaluate* our research hypothesis and assess the theory from which the hypothesis was derived.

Descriptive and Inferential Statistics

Statistical procedures can be divided into two major categories: *descriptive statistics* and *inferential statistics*. Before we can discuss the difference between these two types of statistics, we need to understand the terms *population* and *sample*. A **population** is the total set of individuals, objects, groups, or events in which the researcher is interested. For example, if we were interested in looking at voting behavior in the last presidential election, we would probably define our population as all citizens who voted in the election. If we wanted to understand the employment patterns of Latinas in our state, we would include in our population all Latina women in our state who are in the labor force.

Although we are usually interested in a population, quite often, because of limited time and resources, it is impossible to study the entire population. Imagine interviewing all the citizens of the United States who voted in the last election, or even all the Latina women who are in the labor force in our state. Not only would that be very expensive and time consuming, but we would probably have a very hard time locating everyone! Fortunately, we can learn a lot about a population if we carefully select a subset from that population. A subset selected from a population is called a **sample.** Researchers usually collect their data from a sample and then generalize their observations to the larger population.

Population   The total set of individuals, objects, groups, or events in which the researcher is interested.

Sample   A relatively small subset selected from a population.

Descriptive statistics includes procedures that help us organize and describe data collected from either a sample or a population. Occasionally data are collected on an entire population, as in a census. **Inferential statistics**, on the other hand, is concerned with making predictions or inferences about a population from observations and analyses of a sample. For instance, the General Social Survey (GSS), from which numerous examples presented in this book are drawn, is conducted every year by the National Opinion Research Center (NORC) on a representative sample of about 2,800 respondents. The survey, which includes several hundred questions, is designed to provide social science researchers with a readily accessible database of socially relevant attitudes, behaviors, and attributes of a cross section of the U.S. adult population. NORC has verified that the composition of the GSS samples closely resembles census data. But because the data are based on a sample rather than on the entire population, the average for the sample does not equal the average of the population as a whole. For example, in the 2002 GSS, men and women were asked to report their total years of education. GSS researchers found the average to be 13.27 years. This average probably differs from the average of the population from which the GSS sample was drawn. The tools of statistical inference help determine the accuracy of the sample average obtained by the researchers.

Descriptive statistics Procedures that help us organize and describe data collected from either a sample or a population.

Inferential statistics The logic and procedures concerned with making predictions or inferences about a population from observations and analyses of a sample.

Evaluating the Hypotheses

At the completion of these descriptive and inferential procedures we can move to the next stage of the research process: the assessment and evaluation of our hypotheses and theories in light of the analyzed data. At this next stage new questions might be raised about unexpected trends in the data and about other variables that may have to be considered in addition to our original variables. For example, we may have found that the relationship between gender composition of occupations and earnings can be observed with respect to some groups of occupations but not others. Similarly, the relationship between these variables may apply for some racial/ethnic groups but not for others.

These findings provide evidence to help us decide how our data relate to the theoretical framework that guided our research. We may decide to revise our theory and hypothesis to take account of these later findings. Recent studies are modifying what we know about gender segregation in the workplace. These studies suggest that race as well as gender shape the occupational structure in the United States and help explain disparities in income. This reformulation of the theory calls for a modified hypothesis and new research, which starts the circular process of research all over again.

Statistics provides an important link between theory and research. As our example on gender segregation demonstrates, the application of statistical techniques is an indispensable part of the research process. The results of statistical analyses help us evaluate our hypotheses and theories, discover unanticipated patterns and trends, and provide the impetus for shaping and reformulating our theories. Nevertheless, the importance of statistics should not diminish the significance of the preceding phases of the research process. Nor does the use of statistics lessen the importance of our own judgment in the entire process. Statistical analysis is a relatively small part of the research process, and even the most rigorous statistical procedures cannot speak for themselves. If our research questions are poorly conceived or our data are flawed due to errors in our design and measurement procedures, our results will be useless.

▣ LOOKING AT SOCIAL DIFFERENCES

By the middle of this century, if current trends continue unchanged, the United States will no longer be a predominantly European society. Due mostly to renewed immigration and higher birth rates, the United States is being transformed into a "global society" in which nearly half the population will be of African, Asian, Latino, or Native-American ancestry.

Is the increasing diversity of American society relevant to social scientists? What impact will such diversity have on the research methodologies we employ?

In a diverse society stratified by race, ethnicity, class, and gender, less partial and distorted explanations of social relations tend to result when researchers, research participants, and the research process itself reflect that diversity. Such diversity shapes the research questions we ask, how we observe and interpret our findings, and the conclusions we draw.

How does a consciousness of social differences inform social statistics? How can issues of race, class, gender, and other categories of experience shape the way we approach statistics? A statistical approach that focuses on social differences uses statistical tools to examine how variables such as race, class, and gender, as well as other categories of experience such as age, religion, and sexual orientation, shape our social world and explain social behavior. Numerous statistical procedures can be applied to describe these processes, and we will begin to look at some of those options in the next chapter. For now, let's preview briefly some of the procedures that can be employed to analyze social differences.

In Chapter 2, we will learn how to organize information using descriptive techniques, such as frequency distributions, percentage distributions, ratios, and rates. These statistical tools can also be employed to learn about the characteristics and experiences of groups in our society that have not been as visible as other groups. For example, in a series of special reports published by the U.S. Census Bureau over the past few years, these descriptive statistical techniques have been used to describe the characteristics and experiences of Native Americans, Latinos, and the elderly in America.

In Chapter 3, we illustrate how graphic devices can highlight diversity. In particular, graphs help us to explore the differences and similarities among the many social groups coexisting within American society and emphasize the rapidly changing composition of the U.S. population. Using data published by the U.S. Census Bureau, we discuss various graphic devices that can be used to display differences and similarities among elderly Americans. For instance, by employing a simple graphic device called a bar chart, we depict variations in the living patterns of the elderly and show that in every age category elderly women are more likely than elderly men to live alone. Another graphic device, called a time series chart, shows changes over time in the percentages of divorced white American, African-American, and Latina women.

Whereas the similarities and commonalities in social experiences can be depicted using measures of central tendency (Chapter 4), the differences and diversity within social groups can be described using statistical measures of variation. For instance, we may want to analyze the changing age composition in the United States, or compare the degree of racial/ethnic or religious diversity in the 50 states. Measures such as the standard deviation and the index of qualitative variation (IQV) are calculated for these purposes. For example, using IQV, we can demonstrate that Maine is the least diverse state and Hawaii is the most diverse (Chapter 5).

In Chapters 6 through 8, we review several methods of bivariate analysis, which are especially suited for examining the association between different social behaviors and variables such as race, class, ethnicity, gender, and religion. We use these methods of analysis to show not only how each of these variables operates independently in shaping behavior, but also how they interlock in shaping our experience as individuals in society.[8]

---

[8]Patricia Hill Collins, "Toward a New Vision: Race, Class and Gender as Categories of Analysis and Connection" (Keynote address at Integrating Race and Gender Into the College Curriculum, a workshop sponsored by the Center for Research on Women, Memphis State University, Memphis, TN, 1989).

▣ A Closer Look 1.1
A Tale of Simple Arithmetic

How Culture May Influence How We Count

A second-grade schoolteacher posed this problem to the class: "There are four black-birds sitting in a tree. You take a slingshot and shoot one of them. How many are left?"

"Three," answered the seven-year-old European with certainty. "One subtracted from four leaves three."

"Zero," answered the seven-year-old African with equal certainty. "If you shoot one bird, the others will fly away."*

*Working Woman, January 1991, p. 45.

We will learn about inferential statistics in Chapters 10 through 14. Working with sample data, we examine the relationship among class, sex, or ethnicity and several social behaviors and attitudes. Inferential statistics, such as the *t*-test, chi-square, and *F* statistic, help us determine the error involved in using our samples to answer questions about the population from which they are drawn.

Finally, a word of caution about all statistical applications. Whichever model of social research you use—whether you follow a traditional one or integrate your analysis with qualitative data, whether you focus on social differences or any other aspect of social behavior—remember that any application of statistical procedures requires a basic understanding of the statistical concepts and techniques. This introductory text is intended to familiarize you with the range of descriptive and inferential statistics widely applied in the social sciences. Our emphasis on statistical techniques should not diminish the importance of human judgment and your awareness of the person-made quality of statistics. Only with this awareness can statistics become a useful tool for viewing social life.

▣ A Closer Look 1.2
Are You Anxious About Statistics?

Some of you are probably taking this introductory course in statistics with a great deal of suspicion and very little enthusiasm. The word **statistics** may make you anxious because you associate statistics with numbers, formulas, and abstract notations that seem inaccessible and complicated. It appears that statistics is not as integrated into the rest of your life as are other parts of the college curriculum.

Statistics is perhaps the most anxiety-provoking course in any social science curriculum. This anxiety often leads to a less than optimum learning environment, with students often trying to memorize every detail of a statistical procedure rather than trying to understand the general concept involved.

After many years of teaching statistics, we have learned that what underlies many of the difficulties students have in learning statistics is the belief that it involves mainly memorization of meaningless formulas.

There is no denying that statistics involves many strange symbols and unfamiliar terms. It is also true that you need to know some math to do statistics. But although the subject involves some mathematical computations, you will not be asked to know more than four basic operations: addition, subtraction, multiplication, and division. The language of statistics may appear difficult because these operations (and how they are combined) are written in a code that is unfamiliar to you. Those abstract notations are simply part of the language of statistics; much like learning any foreign language, you need to learn the alphabet before you can "speak the language." Once you understand the vocabulary and are able to translate the symbols and codes into terms that are familiar to you, you will feel more relaxed and begin to see how statistical techniques are just one more source of information.

The key to enjoying and feeling competent in statistics is to frame anything you do in a familiar language and in a context that is relevant and interesting. Therefore, you will find that this book emphasizes intuition, logic, and common sense over rote memorization and derivation of formulas. We have found that this approach reduces statistics anxiety for most students and improves learning.

Another strategy that will help you develop confidence in your ability to do statistics is working with other people. This book encourages collaboration in learning statistics as a strategy designed to help you overcome statistics anxiety. Over the years we have learned that students who are intimidated by statistics do not like to admit it or talk about it. This avoidance mechanism may be an obstacle to overcoming statistics anxiety. Talking about your feelings with other students will help you realize that you are not the only one who suffers from fears of inadequacy about statistics. This sharing process is at the heart of the treatment of statistics anxiety, not because it will help you realize that you are not the "dumbest" one in the class after all, but because talking to others in a "safe" group setting will help you take risks and trust your own intuition and judgment. Ultimately, your judgment and intuition lie at the heart of your ability to translate statistical symbols and concepts into a language that makes sense and to interpret data using newly acquired statistical tools.*

*This discussion is based on Sheila Tobias's pioneering work on mathematics anxiety. See especially Sheila Tobias, *Overcoming Math Anxiety* (New York: Norton, 1995), Chapters 2 and 8.

MAIN POINTS

- Statistics are procedures used by social scientists to organize, summarize, and communicate information. Only information represented by numbers can be the subject of statistical analysis.

- The research process is a set of activities in which social scientists engage to answer questions, examine ideas, or test theories. It consists of the following stages: asking the research question, formulating the hypotheses, collecting data, analyzing data, and evaluating the hypotheses.

- A theory is an elaborate explanation of the relationship between two or more observable attributes of individuals or groups.

- Theories offer specific, concrete predictions about the way observable attributes of people or groups would be interrelated in real life. These predictions, called hypotheses, are tentative answers to research problems.

- A variable is a property of people or objects that takes on two or more values. The variable the researcher wants to explain (the "effect") is called the dependent variable.

The variable that is expected to "cause" or account for the dependent variable is called the independent variable.

• Three conditions are required to establish causal relations: (1) the cause has to precede the effect in time, (2) there has to be an empirical relationship between the cause and the effect, and (3) this relationship cannot be explained by other factors.

• At the nominal level of measurement, numbers or other symbols are assigned to a set of categories to name, label, or classify the observations. At the ordinal level of measurement, categories can be rank-ordered from low to high (or vice versa). At the interval-ratio level of measurement, measurements for all cases are expressed in the same unit.

• A population is the total set of individuals, objects, groups, or events in which the researcher is interested. A sample is a relatively small subset selected from a population.

• Descriptive statistics includes procedures that help us organize and describe data collected from either a sample or a population. Inferential statistics is concerned with making predictions or inferences about a population from observations and analyses of a sample.

KEY TERMS

data
dependent variable
descriptive statistics
empirical research
hypothesis
independent variable
inferential statistics
interval-ratio measurement
nominal measurement

ordinal measurement
population
research process
sample
statistics
theory
unit of analysis
variable

ON YOUR OWN

Log on to the web-based student study site at http://www.pineforge.com/frankfort-nachmiasstudy4 for additional study questions, quizzes, web resources, and links to social science journal articles reflecting the statistics used in this chapter.

SPSS DEMONSTRATION

Introduction to Data Sets and Variables

We'll be using a set of computer data and exercises at the end of each chapter. All computer exercises are based on the program SPSS version 13 for Windows. There are two versions of the program: a standard version with no limits on the number of variables or cases, and a student version with a limit of 1,500 cases and 50 variables. Confirm with your instructor which SPSS version is available at your university.

Throughout this text, you'll be working with two data sets. The GSS2002PFP.SAV contains a selection of variables and cases from the 2002 General Social Survey. The GSS has been conducted annually since 1972. Conducted for the National Data Program for the Social Sciences at the National Opinion Research Center at the University of Chicago, the GSS was designed to provide social science researchers with a readily accessible database of socially relevant attitudes, behaviors, and attributes of a cross section of the U.S. population.

A total of 2,765 surveys were completed for the 2002 survey. We randomly selected a sample of 1,500 cases for our data sets. The data, obtained through a sampling design known as multistage probability sample, are representative of Americans 18 years of age or older. This means that the GSS data set allows us to estimate the characteristics, opinions, and behaviors of all noninstitutionalized, English-speaking, American adults in a given year.

The ISSP00PFP contains variables and cases from the International Social Survey Programme (ISSP), collected in 2000. The ISSP is a cross-national survey, represented by the GSS, conducted by NORC in the United States; the British Social Attitudes Survey (BSA), conducted by Social Community Planning Research (SCPR) in Great Britain; the Allgemeine Bevölkerungsumfrage der Sozialwissenschaften (ALLBUS), conducted by the Center for Survey Research and Methodology (ZUMA) in Germany; and the National Social Science Survey (NSSS), conducted by Austirlian National University (ANU) in Australia. In 2000, data was collected from 38 countries including the U.S. For 2000, the ISSP included questions on the state of the environment and environmental protection.

Those of you with the student version of SPSS 13.0 will work with two separate GSS files: GSS02PFP-A features gender and family issues, and GSS02PFP-B highlights race and government policy issues, along with the ISSP00PFP. The larger data set, GSS2002PFP, contains 146 variables, including all variables from modules A and B. The SPSS appendix on your data disk or CD explains the basic operation and procedures for SPSS for Windows Student Version. We strongly recommend that you refer to this appendix before beginning the SPSS exercises.

When you begin to use a data set, you should take the time to review your variables. What are the variables called? What do they measure? What do they mean? There are several ways to do this.

To review your data, you must first open the data file. Files are opened in SPSS by clicking on *File,* then *Open.* After switching directories and drives to the appropriate location of the files (which may be on a hard disk or on a CD), you select one data file and click on *OK.* This routine is the same each time you open a data file. SPSS automatically opens each data file in the SPSS Data Editor window labeled Data View. We'll use GSS02PFP-A.SAV for this demonstration.

One way to review the complete list of variables in a file is to click on the *Utilities* choice from the main menu, then on *Variables* in the list of submenu choices. A dialog box should open (as depicted in Figure 1.5). The SPSS variable names, which are limited to eight characters or less, are listed in the scroll box (left column). When a variable name is highlighted, the descriptive label for that variable is listed, along with any missing values and, if available, the value labels for each variable category. [As you use this feature, please note that sometimes SPSS mislabels the variable's measurement level. For example, for the variable MARITAL (marital status), SPSS identifies its measurement level as "ordinal." However, MARITAL is a nominal measurement. Always confirm that the reported SPSS measurement level is correct.] Variables are listed in alphabetical order.

A second way to review all variables is to click on *Utilities*, then on *Variables*. This choice tells SPSS to put the variable definition information in the Output–SPSS Viewer window. You can scroll up and down in the Output window to see the variables, and you can print the

Figure 1.5

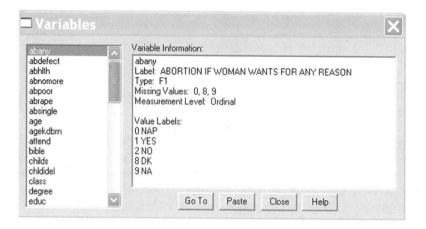

Figure 1.6

	abany	abdefect	abhlth	abnomore	abpoor	abrape	absingle
1	2	2	1	2	2	1	2
2	1	1	1	1	1	1	1
3	0	0	0	0	0	0	0
4	0	0	0	0	0	0	0
5	0	0	0	0	0	0	0
6	0	0	0	0	0	0	0
7	2	1	1	2	2	2	2
8	0	0	0	0	0	0	0
9	0	0	0	0	0	0	0
10	0	0	0	0	0	0	0

complete Output window to keep a printed version of the information. Variables are listed in the order in which they appear in the Data View window.

A third and final way to review all variables is through the Variable View window. Notice on the bottom of your screen that there are two tabs, one for *Data View* and the other for *Variable View*. Click on *Variable View* and you'll see all the variables listed in the order in which they appear in the Data View window (as depicted in Figure 1.6). Each column provides specific information about the variables. The columns labeled "Label" and "Values" provide the variable label (a brief label of what it's measuring) and value labels (for each variable category).

SPSS PROBLEM

Based on the *Utilities–Variables* option, review the variables from the GSS02PFP-A (or GSS02PFP-B). Can you identify three nominal and three ordinal variables? Based on the information in the dialog box or Variable View window, you should be able to identify the variable name, variable label, and category values. (Based on GSS2002PFP, can you identify three interval-ratio variables?)

CHAPTER EXERCISES

1. In your own words, explain the relationship of data (collecting and analyzing) to the research process. (Refer to Figure 1.1.)

2. Construct potential hypotheses or research questions to relate the variables in each of the following examples. Also, write a brief statement explaining why you believe there is a relationship between the variables as specified in your hypotheses.
   a. Gender and educational level
   b. Income and race
   c. The crime rate and the number of police in a city
   d. Life satisfaction and age
   e. A nation's military expenditures as a percentage of its gross domestic product (GDP) and that nation's overall level of security
   f. Care of elderly parents and race

3. Determine the level of measurement for each of the following variables.
   a. The number of people in your family
   b. Place of residence, classified as urban, suburban, or rural
   c. The percentage of university students who attended public high school
   d. The rating of the overall quality of a textbook, on a scale from "Excellent" to "Poor"
   e. The type of transportation a person takes to work (for example, bus, walk, car)
   f. The highest educational degree earned
   g. The U.S. unemployment rate
   h. The presidential candidate the respondent voted for in 2004

4. For each of the variables in Exercise 3 that you classified as interval-ratio, identify whether it is discrete or continuous.

5. Why do you think men and women, on average, do not earn the same amount of money? Develop your own theory to explain the difference. Use three independent variables in your theory, with annual income as your dependent variable. Construct hypotheses to link each independent variable with your dependent variable.

6. For each of the following examples, indicate whether it involves the use of descriptive or inferential statistics. Justify your answer.
   a. Estimating the number of unemployed people in the United States
   b. Asking all the students at a college their opinion about the quality of food at the cafeteria
   c. Determining the incidence of breast cancer among Asian women
   d. Conducting a study to determine the rating of the quality of a new automobile, gathered from 1,000 new buyers
   e. The average salaries of various categories of employees (for example, tellers and loan officers) at a large bank
   f. The change in the number of immigrants coming to the United States from Asian countries between 1995 to 2005

7. Identify three social problems or issues that can be investigated with statistics. (One example of a social problem is hate crimes.) Which one of the three issues would be the most difficult to study? Why? Which would be the easiest?

8. Construct measures of political participation at the nominal, ordinal, and interval-ratio levels. (*Hint:* You can use such behaviors as voting frequency or political party membership.) Discuss the advantages and disadvantages of each.

9. Variables can be measured according to more than one level of measurement. For the following variables, identify at least two levels of measurement. Is one level of measurement better than another? Explain.
   a. Individual age
   b. Annual income
   c. Religiosity
   d. Student performance
   e. Social class
   f. Attitude toward affirmative action

10. In a 1998 study Canadian researchers concluded, "Parental education plays a significant role in children's ability to match or improve upon their parents' educational attainment."[9]
   a. Identify the dependent and independent variables in this statement.
   b. Review the chapter discussion on cause-and-effect relationships (see pages 9–10). Are all three conditions met by the relationship identified in this study?
   c. What other variables may be involved in the relationship between the dependent and independent variables? Explain.

---

[9]*Canadian Social Trends* 49 (Summer 1998): 15.

Chapter 2

Organization of Information: Frequency Distributions

A s social researchers we often have to deal with very large amounts of data. For example, in a typical survey, by the completion of your data collection phase you will have accumulated thousands of individual responses represented by a jumble of numbers. To make sense out of these data you will have to organize and summarize them in some systematic fashion. The most basic method for organizing data is to classify the observations into a frequency distribution. A frequency distribution is a table that reports the number of observations that fall into each category of the variable we are analyzing. Constructing a frequency distribution is usually the first step in the statistical analysis of data.

Frequency distribution   A table reporting the number of observations falling into each category of the variable.

▣ FREQUENCY DISTRIBUTIONS

Let's begin with an example of a frequency distribution of a variable described in a study on Native Americans. When Columbus first encountered the original inhabitants of the Americas, people he later described as "Indios," nothing was known about their numbers, where they lived, or the characteristics of their social structure.[1] During the past 200 years we have gathered a wealth of information about other immigrant groups who settled in North America. In comparison, until a few years ago we knew little more about Native Americans than Columbus did 500 years ago.

Native Americans are not one group but many extremely diverse groups. Today there are about 200 different Native-American tribes characterized by distinct lifestyles and cultural practices. For the U.S. census, the term *American Indian* and *Alaska Native* is used to refer to people having origins in any of the original peoples of North and South America (including Central America). In its reports, the U.S. Census Bureau also refers to this group as "American Indian." The question of who is an American Indian is not a simple matter. Table 2.1 shows the frequency distribution of the variable identity categories of American Indian and Alaska Native populations for 2000. These categories are based on several different patterns of self-identified race. The first category includes persons who identify their race as American Indian and Alaska Native alone. The second category includes persons who report their race as American Indian and Alaska Native but include one or more other races (e.g., white, black or African American, or some other race).

Notice that the frequency distribution is organized in a table, which has a number (2.1) and a descriptive title. The title indicates the kind of data presented: "Categories of American-Indian and Alaska-Native Population." The table consists of two columns. The first column identifies the variable (identity) and its categories (American Indian and Alaska Native alone and American Indian and Alaska Native in combination with one or more other races). The second column, headed "Frequency (f)," tells the number of cases in each category as well as the total number of cases ($N = 4,119,301$). Notice also that the source of the table is clearly identified in a source note in the table. It shows that the table was adapted from a U.S. census report by S. Ogunwole and the data comes from the 2000 census. In general, the source of data for a table should appear as a source note in the table unless it is clear from the general discussion of the data.

What can you learn from the information presented in Table 2.1? The table shows that, as of 2000, approximately 4.1 million persons (4,119,301) reported that their race and/or ethnic ancestry was American Indian and Alaska Native. Out of this group, the majority, about 2.5 million persons, claim American Indian and Alaska Native ancestry alone for their racial background and the remaining 1,643,345 include persons who identify themselves as American Indian and Alaska Native in combination with one or more other races.

---

[1]C. Matthew Snipp, American Indians: The First of This Land. New York: Russell Sage Foundation, 1989, p. 1

Table 2.1   Frequency Distribution for Categories of American-Indian and Alaska-Native Population, 2000

Identity	Frequency (f)
American Indian and Alaska Native alone	2,475,956
American Indian and Alaska Native in combination with one or more other races	1,643,345
Total (N)	4,119,301

Source: Adapted from S. Ogunwole, 2002. *The American Indian and Alaska Native Population: 2000.* U.S. Bureau of Census, Census 2000 Brief. Based on 2000 census data.

✓ *Learning Check.*   *You will see frequency distributions throughout this book. Take the time to familiarize yourself with the parts in this basic example; they will get more complicated as we go on.*

▣ PROPORTIONS AND PERCENTAGES

Frequency distributions are helpful in presenting information in a compact form. However, when the number of cases is large, the frequencies may be difficult to grasp. To standardize these raw frequencies, we can translate them into relative frequencies—that is, proportions or percentages.

A proportion is a relative frequency obtained by dividing the frequency in each category by the total number of cases. To find a proportion *(P)*, divide the frequency *(f)* in each category by the total number of cases *(N)*:

$$P = \frac{f}{N} \qquad (2.1)$$

where
f = frequency
N = total number of cases

Thus, the proportion of American-Indian and Alaska-Native respondents who in 2000 identified themselves as "American Indian and Alaska Native alone" is

$$\frac{2,475,956}{4,119,301} = .60$$

The proportion who identified themselves as "American Indian and Alaska Native in combination with one or more other races" is

$$\frac{1,643,345}{4,119,301} = .40$$

Proportions should always sum to 1.00 (allowing for some rounding errors). Thus, in our example the sum of the two proportions is

$$.60 + .40 = 1.00$$

To determine a frequency from a proportion, we simply multiply the proportion by the total N:

$$f = P(N) \tag{2.2}$$

Thus, the frequency of Native-American respondents who in 2000 identified themselves as "American Indian and Alaska Native only" can be calculated as

$$.60(4,119,301) = 2,471,580.6$$

Note that the obtained frequency differs somewhat from the actual frequency of 2,475,956. This difference is due to rounding of the proportion. If we use the actual proportion instead of the rounded proportion, we obtain the correct frequency:

$$.60106217(4,119,301) = 2,475,956$$

✓ **Learning Check.**   *Compare group A with group B in Figure 2.1 and answer the following questions: Which group has the greater number of women? Which group has the larger proportion of women?*

We can also express frequencies as percentages. A **percentage** is a relative frequency obtained by dividing the frequency in each category by the total number of cases and multiplying by 100. In most statistical reports, frequencies are presented as percentages rather than proportions. Percentages express the size of the frequencies as if there were a total of 100 cases.

To calculate a percentage, simply multiply the proportion by 100:

$$\text{Percentage } (\%) = \frac{f}{N}(100) \tag{2.3}$$

or

$$\text{Percentage } (\%) = P(100) \tag{2.4}$$

Figure 2.1   Numbers and Proportions

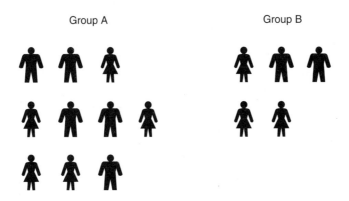

Thus, the percentage of respondents who identified themselves as "American Indian and Alaska Native alone" is

$$.60(100) = 60\%$$

The percentage who identified themselves as "American Indian and Alaska Native in combination with one or more other races" is

$$.40(100) = 40\%$$

Proportion   A relative frequency obtained by dividing the frequency in each category by the total number of cases.

Percentage   A relative frequency obtained by dividing the frequency in each category by the total number of cases and multiplying by 100.

✔ ***Learning Check.***   *Calculate the proportion of males and females in your statistics class. What proportion is female?*

▣ PERCENTAGE DISTRIBUTIONS

Percentages are usually displayed as percentage distributions. A percentage distribution is a table showing the percentage of observations falling into each category of the variable. For

Table 2.2   Frequency and Percentage Distributions for Categories of American Indian and Alaska Native Identity

Identity	Frequency (f)	Percentage
American Indian and Alaska Native alone	2,475,956	60.0
American Indian and Alaska Native in combination with one or more other races	1,643,345	40.0
Total (N)	4,119,301	100.0

Source: Adapted from S. Ogunwole, 2002. *The American Indian and Alaska Native Population: 2000.* U.S. Bureau of Census, Census 2000 Brief. Based on 2000 census data.

example, Table 2.2 presents the frequency distribution of categories of American Indian and Alaska Native identity (Table 2.1) along with the corresponding percentage distribution. Percentage distributions (or proportions) should always show the base *(N)* on which they were computed. Thus, in Table 2.2 the base on which the percentages were computed is $N = 4,119,301$.

Percentage distribution   A table showing the percentage of observations falling into each category of the variable.

◙ COMPARISONS

In Table 2.2 we illustrated that there are two different categories of American Indians. These distinctions among the American-Indian population raise important questions about our understanding of who should be considered in this racial category. For instance, population estimates vary considerably from 2.5 million American Indian and Alaska Native Americans, if we restrict our definition to only persons who identify their race as American Indian and Alaska Native, to almost 4.1 million persons if these identity categories are combined with one or more other races.

The decision to consider these groups separately or to pool them depends to a large extent on our research question. For instance, we know that over 880,000 persons who identify their race and ethnicity as American Indian were below the poverty line in 2001-2003 and over 1,000,000 did not have health insurance for the same time period.[2] Are these figures high or low? What do they tell us about the demographic characteristics, of American Indians?

To answer these questions and determine whether the two categories of American Indian identity have markedly different social characteristics we need to *compare* them. How do the numbers describing the poverty levels and unemployment rates of self-identified American

---

[2]DeNava-Walt, Carmen, Bernadette Proctor, and Robert J. Mills. 2004. *Income, Poverty, and Health Insurance Coverage in the United States.* U.S. Bureau of the Census, Current Population Reports, P60-226.

Indians compare with those numbers for American Indians of multiple ancestries or the population at large?

As students, as social scientists, and even as consumers, we are frequently faced with problems that call for some way to make a clear and valid comparison. For example, in 2002, 23 percent of children lived only with their mother.[3] Is this figure high or low? In 2000, 30.1 percent of those between 30 and 34 years of age had never been married.[4] Does this reflect a change in the American family? In each of these cases comparative information is required to answer the question and reach a conclusion.

These examples also illustrate several ways in which comparisons can be made. What we compare depends largely on the question we pose. Without a clearly formulated research question, it is difficult to decide which type of comparison to make. Several types of comparison are quite common in the social sciences. One type is the comparison between groups that have different characteristics—for example, comparisons between old and young, between white and Asian Pacific Islander, or as in our chapter example, between different categories of American-Indian identity. Sometimes we may be interested in looking at regional differences among groups or in comparing groups from different segments of society. You may have read news stories about contrasts in voting patterns between the North and the South, or the percentage of homeowners in central cities and suburbs. Also, we may be interested in comparing changes in the same group over time, such as the percentage change in foreign-born residents in the United States over the past decade or how the population has shifted from the cities toward the suburbs.

▣ STATISTICS IN PRACTICE: LABOR FORCE PARTICIPATION AMONG LATINOS

Very often we are interested in comparing two or more groups that differ in size. Percentages are especially useful for making such comparisons. For example, we know that differences in socioeconomic status mark divisions between populations, indicating differential access to economic opportunities. Labor participation is an important indicator of access to economic opportunities and is strongly associated with socioeconomic status. Table 2.3 shows the raw frequency distributions for the variable *labor force participation* for all three categories of Latino (or Hispanic) identity.

Which group has the highest relative number of persons who are not in the labor force? Because of the differences in the population sizes of the three groups, this is a difficult question to answer based on only the raw frequencies. To make a valid comparison we have to compare the percentage distributions for all three groups. These are presented in Table 2.4. Notice that the percentage distributions make it easier to identify differences between the groups. Compared with Puerto Ricans and Cubans, Mexicans have the highest percentage employed in the labor force (65% versus 56.4% and 51.8%). Among the three groups, Cubans have the lowest percentage (3.8% versus 5.8% and 5.3%) of persons who are unemployed.

---

[3]Fields, Jason. 2003. *Children's Living Arrangements and Characteristics, March 2002*. U.S. Bureau of the Census, Current Population Reports, P20-547.

[4]Fields, Jason and Lynne, Casper. 2001. *America's Families and Living Arrangements, 2000*. U.S. Bureau of the Census, Current Population Reports, P20-537.

Table 2.3   Employment Status of the Civilian Population, 2002

Employment Status	Mexican	Puerto Rican	Cuban
Employed	10,673,000	1,401,000	592,000
Unemployed	869,000	145,000	43,000
Not in the labor force	4,878,000	938,000	507,000

Source: U.S. Census Bureau. 2003 Statistical Abstract of the United States. Table 589.

Whenever one group is compared with another, the most meaningful conclusions can usually be drawn based on comparison of the relative frequency distributions. In fact, we are seldom interested in a single distribution. Most interesting questions in the social sciences are about differences between two or more groups.[5] The finding that the labor force participation patterns vary among different Latino groups raises doubt about whether Latinos can be legitimately regarded as a single, relatively homogeneous ethnic group. Further analyses could examine *why* differences in Latino identity are associated with differences in labor force participation patterns. Other variables that explain these differences could be identified. These kinds of questions can be answered using more complex multivariate statistical techniques that involve more than two variables. The comparison of percentage distributions is an important foundation for those more complex techniques.

✔ *Learning Check.* Examine Table 2.4 and answer the following questions: What is the percentage of Mexicans who are employed? What is the base (N) for this percentage? What is the percentage of Cubans who are not in the labor force? What is the base (N) for this percentage?

Table 2.4   Employment Status of the Civilian Population, 2002

Employment Status	Mexican	Puerto Rican	Cuban
Employed	65%	56.4%	51.8%
Unemployed	5.3%	5.8%	3.8%
Not in the labor force	29.7%	37.8%	44.4%
Total	100%	100%	100%
(N)	16,420,000	2,484,000	1,142,000

Source: U.S. Census Bureau. Statistical Abstract of the United States, 2003. Table 589.

---

[5]David Knoke and George W. Bohrnstedt, *Basic Social Statistics* (New York: Peacock Publishers, 1991), p. 25.

Before we continue, keep in mind that although we encourage you to begin thinking analytically about complex data, the basic procedures we'll review in the first 9 chapters only allow you to draw some tentative conclusions about differences between groups. To make valid comparisons you will need to consider the more complex techniques of sampling and statistical inference, which are discussed in the last five chapters. As you proceed through this book and master all the statistical concepts necessary for valid inference, you will be able to provide more complex interpretations.

▣ THE CONSTRUCTION OF FREQUENCY DISTRIBUTIONS

Up to now you have been introduced to the general concept of a frequency distribution. We saw that data can be expressed as raw frequencies, proportions, or percentages. We also saw how to use percentages to compare distributions in different groups.

In this section you will learn how to construct frequency distributions. Most often this can be done by your computer, but it is important to go through the process to understand how frequency distributions are actually put together.

For nominal and ordinal variables, constructing a frequency distribution is quite simple. Count and report the number of cases that fall into each category of the variable along with the total number of cases *(N)*. For the purpose of illustration, let's take a small random sample of 40 cases from our 2002 GSS sample and record their scores on the following variables: gender, a nominal level variable; degree, an ordinal measurement of education; and age and number of children, both interval-ratio level variables.

The gender of the respondents was recorded by the interviewer at the beginning of the interview. To measure degree, respondents were asked to indicate the highest degree completed: less than high school, high school, some college, bachelor's degree, and graduate degree. The first category represented the lowest level of education. Respondent age was calculated based on the respondent's birth year. The number of children was determined by the question "How many children have you ever had?" The answers given by our subsample of 40 respondents are displayed in Table 2.5. Note that each row in the table represents a respondent, whereas each column represents a variable. This format is conventional in the social sciences.

You can see that it is going to be difficult to make sense of these data just by eyeballing Table 2.5. How many of these 40 respondents are males? How many said they had a graduate degree? How many were older than 50 years of age? To answer these questions, we construct the frequency distributions for all four variables.

Frequency Distributions for Nominal Variables

Let's begin with the nominal variable, *gender.* First, we tally the number of males, then the number of females (the column of tallies has been included in Table 2.6 for the purpose of illustration). The tally results are then used to construct the frequency distribution presented in Table 2.6. The table has a title describing its content ("Frequency Distribution of the Variable Gender: GSS Subsample"). Its categories (male and female) and their associated frequencies are clearly listed; in addition, the total number of cases *(N)* is also reported. The Percentage column is the percentage distribution for this variable. To convert the Frequency column to percentages, simply divide each frequency by the total number of cases and

Table 2.5   A GSS Subsample of 40 Respondents

Gender of Respondent	Degree	Number of children	Age
M	Bachelor	1	43
F	High school	2	71
F	High school	0	71
M	High school	0	37
M	High school	0	28
F	High school	6	34
F	High school	4	69
F	Graduate	0	51
F	Bachelor	0	76
M	Graduate	2	48
M	Graduate	0	49
M	Less than high school	3	62
F	Less than high school	8	71
F	High school	1	32
F	High school	1	59
F	High school	1	71
M	High school	0	34
M	Bachelor	0	39
F	Bachelor	2	50
M	High school	3	82
F	High school	1	45
M	High school	0	22
M	High school	2	40
F	High school	2	46
M	High school	0	29
F	High school	1	75
F	High school	0	23
M	Bachelor	2	35
M	Bachelor	3	44
F	High school	3	47
M	High school	1	84
F	Graduate	1	45
F	Less than high school	3	24
F	Graduate	0	47
F	Less than high school	5	67
F	High school	1	21
F	High school	0	24
F	High school	3	49
F	High school	3	45
F	Graduate	3	37

Table 2.6   Frequency Distributions of the Variable Gender: GSS Subsample

Gender	Tallies	Frequency (f)	Percentage
Male	Ա1 Ա1 Ա1	15	37.5
Female	Ա1 Ա1 Ա1 Ա1 Ա1	25	62.5
Total (N)		40	100.0

multiply by 100. Percentage distributions are routinely added to almost any frequency table and are especially important if comparisons with other groups are to be considered. Immediately we can see that it is easier to read the information. There are 25 females and 15 males in this sample. Based on this frequency distribution, we can also conclude that the majority of sample respondents are female.

✓ ***Learning Check.*** *Construct a frequency and percentage distribution for males and females in your statistics class.*

Frequency Distributions for Ordinal Variables

To construct a frequency distribution for ordinal level variables, follow the same procedures outlined for nominal level variables. Table 2.7 presents the frequency distribution for the variable degree. The table shows that 60.0 percent, a majority, indicated that their highest degree was a high school degree.

The major difference between frequency distributions for nominal and ordinal variables is the order in which the categories are listed. The categories for nominal level variables do not have to be listed in any particular order. For example, we could list females first and males second without changing the nature of the distribution. Because the categories or values of ordinal variables are rank-ordered, however, they must be listed in a way that reflects their rank—from the lowest to the highest or from the highest to the lowest. Thus, the data on degree in Table 2.7 are presented in ascending order from "less than high school" (the lowest educational category) to "graduate" (the highest educational category).

Table 2.7   Frequency Distribution of the Variable Degree: GSS Subsample

Degree	Tallies	Frequency (f)	Percentage
Less than high school	||||	4	10.0
High school	Ա1 Ա1 Ա1 Ա1 ||||	24	60.0
Bachelor	Ա1 |	6	15.0
Graduate	Ա1 |	6	15.0
Total (N)		40	100.0

Figure 2.2   Forty Respondents from the GSS Subsample, Their Gender and Their Level of Degree (see Table 2.5)

✔ *Learning Check.* *Figures 2.2, 2.3, and 2.4 illustrate the gender and degree data in stages as presented in Tables 2.5, 2.6, and 2.7. To convince yourself that classifying the respondents by gender (Figure 2.3) and by degree (Figure 2.4) makes the job of counting much easier, turn to Figure 2.2 and answer these questions: How many men are in the group? How many women? How many said they completed a bachelor's degree? Now turn to Figure 2.3: How many men are in the group? Women? Finally, examine Figure 2.4: How many said they completed a bachelor's degree?*

Figure 2.3   Forty Respondents from the GSS Subsample, Classified by Gender (see Table 2.6)

Figure 2.4   Forty Respondents from the GSS Subsample, Classified by Gender and Degree

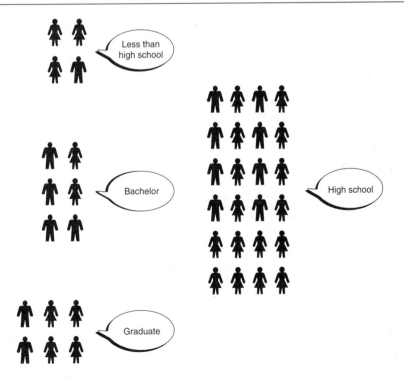

Frequency Distributions for Interval-Ratio Variables

We hope you agree by now that constructing frequency distributions for nominal and ordinal level variables is rather straightforward. Simply list the categories and count the number of observations that fall into each category. Building a frequency distribution for interval-ratio variables with relatively few values is also easy. For example, when constructing a frequency distribution for number of children, simply list the number of children and report the corresponding frequency, as shown in Table 2.8.

Table 2.8   Frequency Distribution of Variable Number of Children: GSS Subsample

Number of Children	Frequency (f)	Percentage
0	13	32.5
1	9	22.5
2	6	15.0
3	8	20.0
4	1	2.5
5	1	2.5
6	1	2.5
7+	1	2.5
Total (N)	40	100.0

Very often interval-ratio variables have a wide range of values, which makes simple frequency distributions very difficult to read. For example, take a look at the frequency distribution for the variable *age* in Table 2.9. The distribution contains age values ranging from 21 to 84 years. For a more concise picture, the large number of different scores could be reduced into a smaller number of groups, each containing a range of scores. Table 2.10 displays such a grouped frequency distribution of the data in Table 2.9. Each group, known as a *class interval,* now contains 10 possible scores instead of one. Thus, the ages of 21, 22, 23, 24, 28, and 29 all fall into a single class interval of 20–29. The second column of Table 2.10, Frequency, tells us the number of respondents that fall into each of the intervals—for example, that seven respondents fall into the class interval of 20–29. Having grouped the scores, we can clearly see that the biggest single age group is between 40 and 49 years (11 out of 40, or 27.5% of the sample). The percentage distribution we have added to Table 2.10 displays the relative frequency of each interval and emphasizes this pattern as well.

The decision as to how many groups to use and, therefore, how wide the intervals should be is usually up to the researcher and depends on what makes sense in terms of the purpose of the research. The rule of thumb is that an interval width should be large enough to avoid too many categories but not so large that significant differences between observations are concealed.[6] Obviously, the number of intervals depends on the width of each. For instance, if you are working with scores ranging from 10 to 60 and you establish an interval width of 10, you will have five intervals.

---

[6]Ibid., p. 41.

Table 2.9   Frequency Distribution of the Variable Age: GSS Subsample

Age of Respondent	Frequency (f)	Age of Respondent	Frequency (f)
21	1	59	1
22	1	62	1
23	1	67	1
24	2	69	1
28	1	71	4
29	1	75	1
32	1	76	1
34	2	82	1
35	1	84	1
37	2		
39	1		
40	1		
43	1		
44	1		
45	3		
46	1		
47	2		
48	1		
49	2		
50	1		
51	1		

Table 2.10   Grouped Frequency Distribution of the Variable Age: GSS Subsample

Age Category	Frequency (f)	Percentage
20–29	7	17.5
30–39	7	17.5
40–49	12	30.0
50–59	3	7.5
60–69	3	7.5
70–79	6	15.0
80–89	2	5.0
Total (N)	40	100.0

✔ **Learning Check.**   *Can you verify that Table 2.10 was constructed correctly? Use Table 2.9 to determine the frequency of cases that fall into the categories of Table 2.10.*

✔ **Learning Check.**   *If you are having trouble distinguishing between nominal, ordinal, and interval-ratio variables, go back to Chapter 1 and review the section on levels of measurement. The distinction between these three levels of measurement is important throughout the book.*

▣ CUMULATIVE DISTRIBUTIONS

Sometimes we may be interested in locating the relative position of a given score in a distribution. For example, we may be interested in finding out how many or what percentage of our sample was younger than 40 or older than 60. Frequency distributions can be presented in a cumulative fashion to answer such questions. A cumulative frequency distribution shows the frequencies at or below each category of the variable.

Cumulative frequencies are appropriate only for variables that are measured at an ordinal level or higher. They are obtained by adding to the frequency in each category the frequencies of all the categories below it.

Let's look at Table 2.11. It shows the cumulative frequencies based on the frequency distribution from Table 2.10. The cumulative frequency column, denoted by Cf, shows the number of persons at or below each interval. For example, you can see that 14 of the 40 respondents were 39 years old or younger, and 29 respondents were 59 years old or younger.

To construct a cumulative frequency distribution, start with the frequency in the lowest class interval (or with the lowest score, if the data are ungrouped) and add to it the frequencies in the next highest class interval. Continue adding the frequencies until you reach the last class interval. The cumulative frequency in the last class interval will be equal to the total number of cases (N). In Table 2.11 the frequency associated with the first class interval (20–29) is 7. The cumulative frequency associated with this interval is also 7, since there are no cases below this class interval. The frequency for the second class interval is 7. The cumulative frequency for this interval is $7 + 7 = 14$. To obtain the cumulative frequency of 26 for the third interval we add its frequency (12) to the cumulative frequency associated with the second class interval (14). Continue this process until you reach the last class interval. Therefore, the cumulative frequency for the last interval is equal to 40, the total number of cases (N).

We can also construct a cumulative percentage distribution ($C\%$), which has wider applications than the cumulative frequency distribution (Cf). A cumulative percentage distribution shows the percentage at or below each category (class interval or score) of the variable. A cumulative percentage distribution is constructed using the same procedure as for a cumulative frequency distribution except that the percentages—rather than the raw frequencies—for each category are added to the total percentages for all the previous categories.

Table 2.11 Grouped Frequency Distribution and Cumulative Frequency for the Variable Age: GSS Subsample

Age Category	Frequency (f)	Cf
20–29	7	7
30–39	7	14
40–49	12	26
50–59	3	29
60–69	3	32
70–79	6	38
80–89	2	40
Total (N)	40	

▣ A Closer Look 2.1
Real Limits, Stated Limits, and Midpoints of Intervals

The intervals presented in Table 2.10 constitute the categories of the variable age that we used to classify the survey's respondents. In Chapter 1 we noted that our variables need to be both exhaustive and mutually exclusive. These principles apply to the intervals here as well. This means that each of the 40 respondents can be classified into one and only one category. In addition, we should be able to classify all the possible scores.

In our example, these requirements are met: Each observation score fits into only one interval, and there is an appropriate category to classify each individual score as recorded in Table 2.10. However, if you looked closely at Table 2.10 you may have noticed that there is actually a gap of 1 year between adjacent intervals. A gap could create a problem with scores that have fractional values. Though age is conventionally rounded down, let's suppose for a moment that respondent's age had been reported with more precision. Where would you classify a woman who was 49.25 years old? Notice that her age would actually fall between the intervals 40–49 and 50–59! To avoid this potential problem, use the real limits shown in the following table rather than the stated limits listed in Table 2.10.

Real limits extend the upper and lower limits of the intervals by .5. For instance, the real limits for the interval 40–49 are 39.5–49.5; the real limits for the interval 50–59 are 49.5–59.5; and so on. [Scores that fall exactly at the upper real limit or the lower real limit of the interval (for example, 59.5 or 49.5) are usually rounded to the closest even number. The number 59.5 would be rounded to 60 and would thus be included in the interval 59.5–69.5.] In the following table, we include both the stated limits and real limits for the grouped frequency distribution of respondent's age. So where would you classify a respondent who was 49.25 years old? (Answer: in the interval 39.5–49.5.) How about 19.9? (In the interval 19.5–29.5.)

The midpoint is a single number that represents the entire interval. A midpoint is calculated by adding the lower and upper real limits of the interval and dividing by 2. The midpoint of the interval 19.5–29.5, for instance, is $(19.5 + 29.5) \div 2 = 24.5$. The midpoint for all the intervals of the table are displayed in the third column.

Even though grouped frequency distributions are very helpful in summarizing information, remember that they are only a summary and therefore involve a considerable loss of detail. Since most researchers and students have access to computers, grouped frequencies are used only when the raw data are not available. Most of the statistical procedures described in later chapters are based on the raw scores.

Respondent's Age		Respondent's Age	
Stated Limits	*Real Limits*	*Midpoint*	*Frequency (f)*
20–29	19.5–29.5	24.5	7
30–39	29.5–39.5	34.5	7
40–49	39.5–49.5	44.5	12
50–59	49.5–59.5	54.5	3
60–69	59.5–69.5	64.5	3
70–79	69.5–79.5	74.5	6
80–89	79.5–89.5	84.5	2
Total (*N*)			40

Table 2.12 Grouped Frequency Distribution and Cumulative
Percentages for the Variable Age: GSS Subsample

Age Category	Frequency (f)	Percentage (%)	C%
20–29	7	17.5	17.5
30–39	7	17.5	35.0
40–49	12	30.0	65.0
50–59	3	7.5	72.5
60–69	3	7.5	80.0
70–79	6	15.0	95.0
80–89	2	5.0	100.0
Total (N)	40	100.0	

Table 2.13 Relative Family Income: White Men Versus White Women

	White Men		White Women	
	%	C%	%	C%
Far below average	3	3	8	8
Below average	22	25	23	31
Average	48	73	52	83
Above average	25	98	15	98
Far above average	2	100	2	100
Total	100		100	
(N)	415			

Source: GSS Survey 2002.

In Table 2.12 we have added the cumulative percentage distribution to the frequency and percentage distributions shown in Table 2.10. The cumulative percentage distribution shows, for example, that 35 percent of the sample was younger than 40 years of age—that is, 39 years or younger.

Like the percentage distributions described earlier, cumulative percentage distributions are especially useful when you want to compare differences between groups. For an example of how cumulative percentages are used in a comparison, we have used the 2002 GSS data to contrast the opinions of white women and white men about their family income compared with other American families. Respondents were asked the following question: "Compared with American families in general, would you say your family income is far below average, below average, average, above average, or far above average?"

The percentage distribution and the cumulative percentage distribution for white women and white men are shown in Table 2.13. The cumulative percentage distributions suggest that relatively more white women than white men consider their family income to be average or lower. Whereas only 73 percent of the white males consider their family income average or lower (where lower includes the categories "below average" and "far below average"),

83 percent of the women ranked their income as average or lower. What might explain these differences? Gender discrimination might play a role in the income gap between white women and white men. These data prompt many other questions about the role that gender (and race) may play in income inequality. For instance, are these differences consistent across other racial groups? What would the differences be if we compared white men to black men? White women to Latina women?

Cumulative frequency distribution   A distribution showing the frequency at or below each category (class interval or score) of the variable.

Cumulative percentage distribution   A distribution showing the percentage at or below each category (class interval or score) of the variable.

▣ RATES

Terms such as *birth rate, unemployment rate,* and *marriage rate* are often used by social scientists and demographers and then quoted in the popular media to describe population trends. But what exactly are rates, and how are they constructed? A rate is a number obtained by dividing the number of actual occurrences in a given time period by the number of possible occurrences. For example, to determine the poverty rate for 2003, the U.S. Census Bureau took the number of men and women in poverty in 2003 (actual occurrences) and divided it by the total population in 2003 (possible occurrences). The rate for 2003 can be expressed as:

Poverty Rate, 2003 = (Number of people in poverty in 2003/Total population in 2003)

Since 35,869,000 people were living below the federal poverty line in 2003 and the number for the total population was 287,699,000, the poverty rate for 2003 can be expressed as:

Poverty Rate, 2003 = 35,869,000/287,699,000 = .125

The poverty rate in 2003 as reported by the U.S. Census Bureau was 12.5 percent (.125 × 100). Rates are often expressed as rates per thousand or hundred thousand to eliminate decimal points and make the number easier to interpret. For example, to express the poverty rate per thousand we multiply it by 1,000: .125 × 1000 = 125. This means that for every 1,000 people, 125 were living below the federal poverty line according to the U.S. Census Bureau definition.

The preceding poverty rate can be referred to as a *crude rate* because it is based on the total population. Rates can be calculated on the general population or on a more narrowly defined select group. For instance, poverty rates are often given for the number of people who are 18 years old or younger—highlighting how our young are vulnerable to poverty. The poverty rate for those 18 years or younger is:

Poverty Rate (18 years or younger, 2003) = 12,866,000/72,999,000 = .176 × 1,000 = 176

We could even take a look at the poverty rate for older Americans:

Poverty Rate (65 years or older, 2003) = 3,552,000/34,659,000 = .102 × 1,000 = 102

Rate A number obtained by dividing the number of actual occurrences in a given time period by the number of possible occurrences.

✔ **Learning Check.** *Law enforcement agencies routinely record crime rates (the number of crimes committed relative to the size of a population), arrest rates (the number of arrests made relative to the number of crimes reported), and conviction rates (the number of convictions relative to the number of cases tried). Can you think of some other variables that could be expressed as rates?*

▣ STATISTICS IN PRACTICE: MARRIAGE AND DIVORCE RATES OVER TIME

How can we examine the shifting marriage and divorce rates? Have marriage rates really declined? Have divorce rates risen? Like percentages, rates are useful in making comparisons between different groups and over time. The crude marriage rate of 9.8 for 1990 might be difficult to interpret by itself and will not answer our question of whether or not marriage rates have really changed. To illustrate how rates have changed over time, let's look at Table 2.14, which reports the marriage rates and divorce rates since 1970. The table shows that over the past two decades marriage has declined, whereas the divorce rate in 2000 was roughly the same as it was in 1995 but much higher than it was in 1970.

✔ **Learning Check.** *Make sure you understand how to read tables. Can you explain how we reached the preceding conclusions based on the information in Table 2.14?*

▣ READING THE RESEARCH LITERATURE[8]: STATISTICAL TABLES

Statistical tables that display frequency distributions or other kinds of statistical information are found in virtually every book, article, or newspaper report that makes any use of statistics.

---

[8]The idea of "Reading the Research Literature" sections that appear in most chapters was inspired by Joseph F. Healey, *Statistics: A Tool for Social Research*, 5th ed. (Belmont, CA: Wadsworth, 1999).

Table 2.14   Marriage and Divorce Rates per 1,000

Year	Marriage Rate	Divorce Rate
1970	10.6	3.7
1975	10.0	4.8
1980	10.6	5.2
1985	10.1	5.0
1990	9.8	4.7
1995	8.9	4.4
2000	8.5	4.2

Source: Data from U.S. Bureau of the Census, *Statistical Abstract of the United States, 2003*, Table 83.

However, the inclusion of statistical tables in a report or an article doesn't necessarily mean that the research is more scientific or convincing. You will always have to ask what the tables are saying and judge whether the information is relevant or accurately presented and analyzed. Most statistical tables presented in the social science literature are a good deal more complex than those we describe in this chapter. The same information can sometimes be organized in many different ways, and because of space limitations the researcher may present the information with minimum detail.

In this section we present some guidelines for how to read and interpret statistical tables displaying frequency distributions. The purpose is to help you see that some of the techniques described in this chapter are actually used in a meaningful way. Remember that it takes time and practice to develop the skill of reading tables. Even experienced researchers sometimes make mistakes when interpreting tables. So take the time to study the tables presented here; do the chapter exercises; and you will find that reading, interpreting, and understanding tables will become easier in time.

Basic Principles

The first step in reading any statistical table is to understand what the researcher is trying to tell you. There must be a reason for including the information, and usually the researcher tells you what it is. Begin your inspection of the table by reading its title. It usually describes the central contents of the table. Check for any source notes to the table. These tell the source of the data or the table and any additional information the author considers important. Next, examine the column and row headings and subheadings. These identify the variables, their categories, and the kind of statistics presented, such as raw frequencies or percentages. The main body of the table includes the appropriate statistics (frequencies, percentages, rates, and so on) for each variable or group as defined by each heading and subheading.

Table 2.15 was taken from an article written by Professors Eric Fong and Kumiko Shibuya about the residential patterns of different racial and ethnic groups. In their study, the researchers attempted to compare different home ownership patterns and suburbanization patterns among whites, blacks, five major Asian groups, and three major Hispanic groups. Data from the 1990 1% Public Use Microdata Sample (PUMS) based on 15 Primary Metropolitan Statistical Areas (PMSAs) and Metropolitan Statistical Areas (MSAs) are used for their analysis.

Table 2.15   Percentage Distribution of Housing Tenure Status by Residential Location of Major Racial and Ethnic Groups, 1990. Source: 1990 U.S. 1% PUMS Data

	Suburban Owner	Central City Owner	Suburban Renter	Central City Renter	Total
Whites	**50.4**	**16.8**	**17.1**	**15.7**	**100.0**
Blacks	**14.6**	**27.2**	**13.9**	**44.3**	**100.0**
Asians	**34.2**	**20.6**	**14.7**	**30.5**	**100.0**
Chinese	33.7	26.1	9.9	30.3	100.0
Japanese	39.2	20.8	17.8	22.2	100.0
Koreans	29.8	13.8	17.1	39.3	100.0
Vietnamese	31.0	11.8	24.3	32.9	100.0
Filipino	38.1	21.5	15.8	24.6	100.0
Hispanics	**22.9**	**15.3**	**20.8**	**40.9**	**100.0**
Cubans	36.3	16.6	16.2	30.9	100.0
Mexicans	26.0	15.5	26.8	31.8	100.0
Puerto Ricans	7.1	14.2	8.4	70.3	100.0
Total	41.7	18.5	16.8	23.0	100.0

Source: Adapted from Eric Fong and Kumiko Shibuya, "Suburbanization and Home Ownership: The Spatial Assimilation Process in U.S. Metropolitan Cities." *Sociological Perspectives 43*, no 1, (2000): 143.

In Table 2.15, the researchers display the percentages for housing status by location for each racial/ethnic group for 1990. Notice that the frequency (*f*) for each category is not reported. Although the table is quite simple, it is important to examine it carefully, including its title and headings, to make sure you understand what the information means.

✓ *Learning Check.*   *Inspect Table 2.15 and answer the following questions:*

- *What is the source of this table?*
- *How many variables are presented? What are their names?*
- *What is represented by the numbers presented in the first column? In the third column?*

What do the authors tell us about the table?:

"Table 1, which shows tenure status cross-classified by housing location, reveals four distinctive patterns. Column 1, which presents the proportions of suburban home owners, shows remarkable variations among groups. Although 42 percent of the total population of these 15 metropolitan areas have been able to own a house in the suburbs, the figures suggest that it is atypical for some groups. Whites have the highest share: about 50 percent are suburban home owners. Blacks and Puerto Ricans, on the other hand, show an extremely low level of suburban home ownership. Asians in general have higher suburban home ownership rates (34%) than Hispanics (23%). There are, however, substantial variations within these two groups, ranging from 39 percent for Japanese to only 7 percent for Puerto Ricans.

"Home ownership rates in the central city (Column 2) show a different picture. Blacks have the highest rate: about 27 percent of blacks own homes in central city areas. In fact, two-thirds of black home owners are found in those areas. Their higher home ownership rates in the central city may be explained by the fact that the cost of owning a house in the suburbs is higher. It is also possible that blacks are barred from the suburban areas because of discrimination in the housing market. Asian ethnic groups vary substantially in home ownership rates in the central city. The rate ranges from 12 percent for Vietnamese to 26 percent for Chinese. A higher rate of Chinese home owners in central cities may reflect the fact that Chinatowns have traditionally been located there. However, the rates among Hispanic ethnic groups are similar to one another. [They range from 14 percent for Puerto to Ricans to 17 percent for Cubans.]

"The results in column 3 suggest a pattern for suburban renters that is distinct from home ownership patterns in either the suburbs or the central city. The results indicate that whites and Hispanics have the highest proportions of suburban renters. Although the percentage of Asian suburban renters is low on average, some specific Asian ethnic groups, such as Vietnamese, have higher rates than whites. The relatively high number of Vietnamese renters may be related to the results of the dispersion by refugee settlement programs. The higher rate of suburban renters among these Asian groups may be related to the absence of their ethnic communities in central cities. . . .

"The results in column 4 reveal another unique pattern for central city renters. Among all groups, blacks and Puerto Ricans have the highest rates in this category. About 70 percent of Puerto Ricans and 44 percent of blacks are central city renters. Asians have a moderate percentage of central city renters, ranging from a low of 22 percent for Japanese to a high of 39 percent for Koreans. Whites have the lowest percentage of central city renters: about 16 percent.

"Overall, these data show the complicated nature of the residential distribution patterns of racial/ethnic groups in contemporary cities. The results suggest four distinctive pictures of tenure status in suburbs and central cities, which would be undetected if suburbanization or home ownership were studied separately. These distinctive patterns, however, may simply reflect the differences in the socioeconomic resources and acculturation levels of each group in both locations."[9]

For a more detailed analysis of the relationships between these variables, you need to consider some of the more complex techniques of bivariate analysis and statistical inference. We consider these more advanced techniques beginning in Chapter 6.

Tables with a Different Format

Tables can sometimes present data for only a subset of the sample. For example, Table 2.16, based on 2000 census data for white Americans, shows percentages for selected variables. However, only partial information on each of the variables is included, and therefore the percentages do not add up to 100 percent. For instance, 83.6 percent of white Americans

---

[9]Eric Fong and Kumiko Shibuya, "Urbanization and Home Ownership: The Spatial Assimilation Process in U.S. Metropolitan Areas," *Sociological Perspectives 43*, no.1 (2000):137–157.

Table 2.16   Selected Economic and Social Indicators for White Americans, 2000

Indicators	Percentage
Percent high school graduate or higher	83.6
Population 21 to 64 years, with a disability	17.3
Percent foreign born	6.3
Percent using public transportation to work	3.1
Percent in management, professional, and related occupations	35.6

Source: U.S. Census Bureau, *Statistical Abstract of the United States,* 2003, Table 40.

Table 2.17   Selected Economic and Social Indicators for White and Black Americans, 2000

Indicators	Percentage of Whites	Percentage of Blacks
Percent high school graduate or higher	83.6	72.3
Population 21 to 64 years, with a disability	17.3	27.7
Percent foreign born	6.3	6.1
Percent using public transportation to work	3.1	12.2
Percent in management, professional, and related occupations	35.6	25.2

have had a high school degree or higher. The remaining 16.4 who had less than a high school education are omitted from the table. Similarly, 3.1 percent used public transportation for work in 2000, omitting the 96.9 percent who used other transportation (including their own car or carpooling) to work.

Although the data displayed in Table 2.16 provide useful information, we are usually interested in answering questions that go beyond a simple description of how the variables are distributed. Most research usually goes on to make comparisons between groups or to compare one group at different times. For instance, to put the information on white Americans presented in Table 2.16 into a more meaningful context, we may want to compare it with that of other groups of Americans. Such a comparison allows us to answer questions such as how "high" is a percent with a high school degree or higher, and is the 3.1 percent using public transportation "high" or "low"?

Take a look at Table 2.17. It includes the information from Table 2.16, plus corresponding information on black Americans. Note the difference in the proportion of blacks and whites Americans (72.3% compared with 83.6%) with a high school degree or higher, the higher percentage of blacks with a disability (27.7 versus 17.3), and the notably higher rates of using public transportation to work (about four times as high).

◙ CONCLUSION

In the introduction to this chapter we told you that constructing a frequency distribution is usually the first step in the statistical analysis of data; we hope that by now you agree that constructing a basic frequency or percentage distribution is a fairly straightforward task. As you have seen in the examples in this chapter, distribution tables help researchers to organize, summarize, display, and describe data. Trends within groups and differences or similarities between groups can be identified using a simple distribution table.

In the chapters that follow, you will find that frequency distribution tables provide the basic information for graphically displaying data and calculating measures of central tendency and variability. In other words, you will see frequency and percentage distributions again and again, so make sure you have confidence in your ability to construct and read distribution tables before you proceed to the next chapters.

MAIN POINTS

• The most basic method for organizing data is to classify the observations into a frequency distribution—a table that reports the number of observations that fall into each category of the variable being analyzed.

• Constructing a frequency distribution is usually the first step in the statistical analysis of data.

• To obtain a frequency distribution for nominal and ordinal variables, count and report the number of cases that fall into each category of the variable along with the total number of cases (N).

• To construct a frequency distribution for interval-ratio variables that have a wide range of values, first combine the scores into a smaller number of groups—known as class intervals—each containing a number of scores.

• Proportions and percentages are relative frequencies. To construct a proportion, divide the frequency (f) in each category by the total number of cases (N). To obtain a percentage, divide the frequency (f) in each category by the total number of cases (N) and multiply by 100.

• Percentage distributions are tables that show the percentage of observations that fall into each category of the variable. Percentage distributions are routinely added to almost any frequency table and are especially important if comparisons between groups are to be considered.

• Cumulative frequency distributions allow us to locate the relative position of a given score in a distribution. They are obtained by adding to the frequency in each category the frequencies of all the categories below it.

• Cumulative percentage distributions have wider applications than cumulative frequency distributions. A cumulative percentage distribution is constructed by adding to the percentages in each category the percentages of all the categories below it.

• One other method of expressing raw frequencies in relative terms is known as a rate. Rates are defined as the number of actual occurrences in a given time period divided by the number of possible occurrences. Rates are often multiplied by some power of 10 to eliminate decimal points and make the number easier to interpret.

KEY TERMS

cumulative frequency distribution
cumulative percentage distribution
frequency distribution
percentage

percentage distribution
proportion
rate

ON YOUR OWN

Log on to the web-based student study site at http://www.pineforge.com/frankfort-nachmiasstudy4 for additional study questions, quizzes, web resources, and links to social science journal articles reflecting the statistics used in this chapter.

SPSS DEMONSTRATIONS

Demonstration 1: Producing Frequency Distributions [ISSP00PFP]

In SPSS you can review the frequency distribution for a single variable or for several variables at once. The frequency procedure is found in the *Descriptive Statistics* menu under *Analyze.* For this chapter, we will use the ISSP00PFP data set found on your text CD.

In the Frequencies dialog box (Figure 2.5), click on the variable name(s) in the left column and transfer the name(s) to the Variable(s) box. Remember, more than one variable can be selected at one time.

Figure 2.5

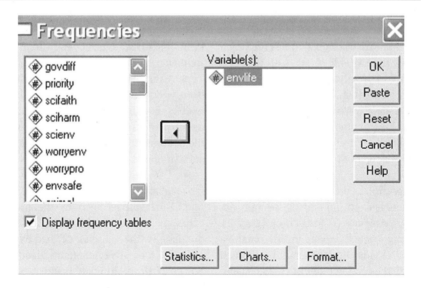

Figure 2.6

Envlife Environment: modern life harms the environment

		Frequency	Percent	Valid Percent	Cumulative Percent
Valid	1 Strongly Agree	177	11.8	12.3	12.3
	2 Agree	558	37.2	38.7	51.0
	3 Neither Agree nor Disagree	296	19.7	20.5	71.5
	4 Disagree	347	23.1	24.1	95.6
	5 Strongly Disagree	64	4.3	4.4	100.0
	Total	1442	96.1	100.0	
Missing	8 Cant Choose, DK	38	2.5		
	9 NA, refused	20	1.3		
	Total	58	3.9		
Total		1500	100.0		

For our demonstration, let's select the variable ENVLIFE ("Environment: Modern life harms the environment"). Click on *OK* to process the frequency. Respondents were asked to answer the question by indicating 1, Strongly Agree; 2, Agree; 3, Neither Agree nor Disagree; 4, Disagree; and 5, Strongly Disagree.

SPSS will produce two tables in a separate Output window, a statistics table and a frequency table. Use the Window scroll keys to move up and down the window to find the statistics and frequency tables for ENVLIFE. What level of measurement is this variable? (Refer to Chapter 1 to review definitions.)

In the first table (not shown), Statistics, SPSS identifies all the valid and missing responses to this question. Responses are coded missing if no answer was given, if the question was not applicable, or if the individual responded "don't know" to the question.

In the frequency table (see Figure 2.6), the variable is listed in the first line, along with the variable label. The first column lists the Value Label for each category of ENVLIFE. What is the value for "strongly disagree"? For "agree"? Review your output to locate all the columns and values we've just identified.

The next four columns contain important frequency information about the variable. The Frequency column shows the number of respondents who gave a particular response. Thus, we can see that 177 people strongly agree with the statement, and 64 said that they strongly disagree with the statement. Is this consistent with what you expected? How many respondents said that they agree with the statement? However, 58 of the responses are missing. The Percent column calculates what percentage of the whole sample (1,500 cases) each of the responses represents. Thus, 11.8 percent of the sample said that they strongly agree with the statement that everything we do harms the environment. Normally, data from the third column, Valid Percent, is more useful. This column removes all the cases defined as missing and recalculates percentages based only on the valid responses. Recalculated based only on valid cases (1,442), the percentage of those who strongly agree is 12.3.

The last column, Cumulative Percent, calculates cumulative percentages beginning with the first response. You can see that more than half of the sample (51%) either strongly agree or agree with the statement.

Demonstration 2: Recoding Variables

Some variables need to be collapsed or reduced into a smaller number of categories or intervals in order to better present and understand the data. We could, for example, collapse ENVLIFE into a variable with three categories: agree, neither agree nor disagree, or disagree. We would combine 1 (strongly agree) and 2 (agree) into one category, and 4 (disagree) and 5 (strongly disagree) into another. We would leave 3 (neither agree nor disagree) as its own category. To accomplish this we could use the SPSS commands *Transform–Recode into Different Variable.*

For more detailed instruction on recoding variables, please refer to the section on Recoding Variables in the SPSS Appendix on your student CD, which explains how to recode the variable EDUC (respondent's years of education).

After reviewing the SPSS Appendix on your CD, recode ENVLIFE into a new variable called RENVLIFE. Frequencies for RENVLIFE should look like Figure 2.7.

Figure 2.7

Renvlife Recoded Envlife

		Frequency	Percent	Valid Percent	Cumulative Percent
Valid	1.00 Strongly Agree or Agree	735	49.0	51.0	51.0
	2.00 Neither Agree nor Disagree	296	19.7	20.5	71.5
	3.00 Strongly Disagree or Disagree	411	27.4	28.5	100.0
	Total	1442	96.1	100.0	
Missing	System	58	3.9		
Total		1500	100.0		

SPSS PROBLEMS

[ISSP00PFP]

1. Use the SPSS *Frequencies* command to produce a frequency table for the variable MARITAL from the ISSP00PFP. How would you describe the marital status of this international sample?
   a. What percentage of the sample is married?
   b. What percentage is divorced?
   c. What percentage of the sample has ever been married?
   d. What percentage of the sample is currently unmarried? (Include all relevant categories in your total percentage.)

2. The ISSP 2000 included a series of questions on political action. The variables included PETITION (In the last 5 years, did you sign a petition about environmental issues?), DONATE (In the last 5 years, did you give money to an environmental group?), and ENVGROUP (Are you a member of a group to preserve the environment?).

a. Run frequencies for all three variables.

b. Prepare a general statement summarizing your results from the three frequency tables. Identify the level of measurement for each variable. How would you describe respondents' political activity? Are there areas where respondents were more active?

3. Produce the frequency tables for the variables WRKHRS (number of hours respondent works per week) and WORRYENV (does the respondent worry about the future state of the environment).

a. What level of measurement is each variable?

b. If you wanted to reduce WRKHRS into three to four categories, how would you redefine the variable?

c. If you wanted to reduce WORRYENV into three categories, how would you redefine the variables?

d. Recode each variable, per your answers in b and c. You may, based on Demonstration 2, recode the variables in SPSS. Report the frequencies for your revised variables. (You can also recalculate the categories with pencil/calculator.)

4. The ISSP 2000 also asked respondents their age (AGE). Run the frequency table for this variable. Notice that there are notations for values 16 and 18. For "18," the notation is that for the Great Britain sample, this corresponds to "18 years or above." Recode this interval ratio variable into an ordinal measure. How many categories do you have? Prepare a frequency and cumulative percentage table of your recoded AGE variable.

CHAPTER EXERCISES

1. Suppose you have surveyed 30 people and asked them whether they are white (W) or nonwhite (N), and how many traumas (serious accidents, rapes, or crimes) they have experienced in the past year. You also asked them to tell you whether they perceive themselves as being in the upper, middle, working, or lower class. Your survey resulted in the following raw data:

Race	*Class*	*Trauma*	*Race*	*Class*	*Trauma*
W	L	1	W	W	0
W	M	0	W	M	2
W	M	1	W	W	1
N	M	1	W	W	1
N	L	2	N	W	0
W	W	0	N	M	2
N	W	0	W	M	1
W	M	0	W	M	0
W	M	1	N	W	1
N	W	1	W	W	0
N	W	2	W	W	0
N	M	0	N	M	0
N	L	0	N	W	0
W	U	0	N	W	1
W	W	1	W	W	0

Source: Data based on General Social Survey files for 1987 to 1991.

a. What level of measurement is the variable *race*? *class*?
b. Construct raw frequency tables for race and class.
c. What proportion of the 30 individuals is nonwhite? What percentage is white?
d. What proportion of the 30 individuals identified themselves as middle class?

2. Using the data and your raw frequency tables from Exercise 1, construct a frequency distribution for class.
   a. Which is the smallest perceived class?
   b. Which two classes include the largest percentages of people?

3. Using the data from Exercise 1, construct a frequency distribution for trauma.
   a. What level of measurement is used for the trauma variable?
   b. Are people more likely to have experienced no traumas or only one trauma in the past year?
   c. What proportion has experienced one or more traumas in the past year?

4. Suppose you are using a sample from the 2002 General Social Survey (GSS) data for a research project on education in the United States. The GSS includes a question that asks for the number of years of education. Based on 997 individuals, the GSS reports the following frequency distribution for years of education:

Years of Education	Frequency
0	1
1	1
2	8
3	1
4	3
5	6
6	8
7	7
8	25
9	22
10	29
11	49
12	294
13	97
14	135
15	56
16	137
17	35
18	50
19	14
20+	19

Source: General Social Survey, 2002.

a. What is the level of measurement of years of education?
b. Construct a frequency table, with cumulative percentages, for years of education.
c. How many respondents have 8 or fewer years of education? What percentage of the sample does this value represent?

d. Assume that you are really more interested in the general level of education than in the raw number of years of education, so you would like to group the data into four categories that better reflect your interests. Assume that anyone with 12 years of education is a high school graduate, and that anyone with 16 years of education is a college graduate. Construct a cumulative frequency table for education in four categories based on these assumptions. What percentage of the sample has graduated from college? What percentage of the sample has not graduated from high school?

5. The Gallup Organization conducted a survey in May 2004, asking Americans their opinion on a constitutional amendment defining marriage as a legal union between a man and a woman. Results are provided in the table below, noting the percent of those who favor or oppose by political party affiliation. Do these data support the statement that people's views on gay marriage are related to their party affiliation? Why or why not?

	Favor (%)	*Oppose (%)*
Republicans	69	28
Independents	48	46
Democrats	38	60

6. The United States continues to receive large numbers of immigrants. Often these immigrants prefer to speak their own language, at least at home. The borough of Queens in New York City is one of the most diverse counties in the United States. The following frequency distribution shows the major immigrant languages spoken in Queens based on data from U.S. Census 2000 (Summary File 3).

Arabic	12,504
Chinese	12,690
French	46,730
German	12,054
Greek	39,418
Hungarian	4,753
Indic Languages	95,484
Italian	44,411
Japanese	5,300
Korean	57,447
Polish	20,883
Portugese	9,479
Russian	36,517
Spanish	48,504
Tagalog	25,436
Vietnamese	3,361
Yiddish	5,472

a. Compute the frequency and percentage distribution for these data.
b. Why would a cumulative frequency distribution be less useful for this table?
c. What are the four most common foreign languages in Queens?
d. Which two languages are the least common?

7. The tables below present the frequency distributions for education by gender and race based on the General Social Survey 2002. Use them to answer the following questions.

	Gender	
Education	Male	Female
Some high school	65	78
High school graduate	233	311
Some college	27	43
College graduate	116	127

	Race	
Education	White	Black
Some high school	95	36
High school graduate	433	70
Some college	52	11
College graduate	223	7

Source: General Social Survey, 2002

a. Construct frequency tables based on percentages and cumulative percentages of educational attainment for gender and race.
b. What percentage of males has continued their education beyond high school? What is the comparable percentage for females?
c. What percentage of whites has completed high school or less? What is the comparable percentage for blacks?
d. Are the cumulative percentages more similar for men and women or for blacks and whites? (In other words, where is there more inequality?) Explain.

8. In Table 2.13 you saw cumulative frequency tables for how white men and white women believe their family income compares with that for all Americans. The following two tables present relative frequency distributions for this 2002 General Social Survey question, first for males and females, then for less than high school and bachelor's degree graduates. Use the tables to answer the questions.

	Sex of Respondent			
	Male		Female	
	f	%	f	%
Far below average	7	2.9	23	8.7
Below average	53	22.1	69	26.2
Average	122	50.8	134	51.0
Above average	54	22.5	32	12.2
Far above average	4	1.7	5	1.9
Total (N)	240	100.0	263	100.0

	Educational Attainment of Respondent			
	Less than High school		Bachelor's degree	
	f	*%*	*f*	*%*
Far below average	7	10.0	1	1.2
Below average	24	34.3	14	16.9
Average	35	50.0	39	47.0
Above average	3	4.3	27	32.5
Far above average	1	1.4	2	2.4
Total (*N*)	70	100.0	83	100.0

Source: General Social Survey, 2002

a. What is the cumulative percentage of females who think their family income is "average" or "below average"? What is the percentage of males who believe their family income is "average" or "below average"?

b. Is the cumulative percentage of those with less than a high school degree who believe their family income is average or below (include those who responded "below average" and "far below average") greater or less than that for bachelor's degree respondents who hold the same belief?

c. What is the cumulative percentage of bachelor's degree graduates who believe their family income is average or above?

d. What percentage of males believes their family income is below or far below average? What percentage of males believes their family income is above or far above average? Why might these two values be so close?

9. From the ISSP 2000, we present data from three different countries on whether respondents agree with the statement, "the government should redistribute income among citizens."

	New Zealand (%)	Japan (%)	Russia (%)
Strongly agree	2.0	17.1	41.0
Agree	11.8	14.6	27.7
Neither	21.6	34.1	10.8
Disagree	33.3	9.8	12.0
Strongly disagree	31.4	24.4	8.4

Source: ISSP 2000

a. What is the level of measurement?

b. What percentage of New Zealanders, Japanese, and Russians "strongly agree" or "agree" with this statement? Calculate the percentages for each group separately.

c. What can be said about the differences in attitudes between respondents from these three countries?

10. Respondents from the United States and Great Britain were asked whether economic growth harms the environment. Individuals responded based on a five-point ordinal scale. ISSP2000 results are presented in the table below.

	United States	*Great Britain*
Strongly agree	1	1
Agree	9	7
Neither	16	15
Disagree	19	17
Strongly disagree	1	2
Total (N)	46	42

Source: ISSP 2000

a. Calculate a cumulative frequency distribution for U.S. respondents.
b. Calculate a cumulative frequency distribution for Great Britain respondents.

11. A Gallup Poll (June 9–30, 2004) compared 2,250 Americans' attitudes on minority rights and how much of a role the government should play in helping minorities improve their social and economic positions. Results from the survey are presented below. How would you characterize the differences between the attitudes of whites and minorities?

What should be the government's role in improving economic and social position of minorities?	*Non-Hispanic Whites (%)*	*Blacks (%)*	*Hispanics (%)*
Major role	32	68	67
Minor role	51	22	21
No role	16	9	8

Source: Jeffrey Jones. "Blacks more pessimistic than whites about economic opportunities."

Gallup Poll News Service, The Gallup Organization. July 9, 2004.

12. Before the Million Mom March on Washington, DC, in May 2000, the Gallup Organization asked men and women whether laws covering the sales of firearms should be made more strict, made less strict, or kept as they are now. A total of 1,031 adult Americans were surveyed during May 5–7, 2000. The following responses were collected.

	More	*Less*	*Same*	*No Opinion*
Men (*N* = 493)	256	39	193	5
Women (*N* = 538)	387	11	129	11

Transform these frequencies into percentages. How would you describe the results of this Gallup Poll? Are there any differences between women and men in their attitudes about gun laws?

13. Percentages and frequencies can also be examined in terms of their change over time. For example, have Americans grown more tolerant of gays? The Gallup Poll Archives (December 1997) reported on data from 1977 to 1996. More recently, Gallup reported data for 2001. Men and women were asked if they thought homosexuals should be hired for a select group of occupations. The percentage of respondents who said yes is reported for the set of five occupations.

	1977	*1982*	*1985*	*1987*	*1989*	*1992*	*1996*	*2001*
Salesperson	68	70	71	72	79	82	90	90
Doctor	44	50	52	49	56	53	69	75
Armed forces	51	52	55	55	60	57	65	70
Elementary school teachers	27	32	36	33	42	41	55	54
Clergy	36	38	41	42	44	43	53	54

a. Overall, is there one occupational category where Americans are more accepting of homosexuals' being employed? What do you base your answer on?

b. Overall, is there one occupational category where Americans are least tolerant of homosexual employment? What evidence do you base your answer on?

c. Since 1977, have Americans become more tolerant of homosexuals in particular work settings? Explain your answer.

14. Researchers assessed the differences between sports-related images in nonsporting men's and women's magazines. Based on their review of 59 magazines, researchers identified the number of times females or males were shown engaged in sports activities and what type of sports activity was depicted (power performance—highly organized and competitive sports such as hockey or football), pleasure participation (loosely defined athletic activities such as bowling as a family or throwing a Frisbee in the park), or mixed. Describe what differences exist between men's and women's magazines for these two variables.

Frequencies and Percentages of Sports-Related Images for All Magazines

	Men's Magazines N (%)	*Women's Magazines N (%)*
Gender Salience		
Females Shown	23 (15)	212 (65)
Males Shown	130 (85)	112 (35)
Sport Type Shown		
Power-performance	34 (42)	34 (16)
Pleasure-participation	23 (28)	149 (70)
Mixed	24 (30)	32 (15)

Source: Timothy Curry, Paula Arriagada, and Benjamin Cornwell. 2002. "Images of Sport in Popular Nonsport Magazines: Power and Performance versus Pleasure and Participation." *Sociological Perspectives* 45(4):404.

15. In the following table, we present selected items from a demographic portrait of Americans who identify themselves as either Republican or Democrat. The data were collected by the Gallup Organization (July 30, 2000) based on 10,208 surveys conducted between March and July 2000. (Either due to rounding or the omission of no response categories, not all category totals will equal 100%.)

Variables	Republicans (%)	Democrats (%)
Gender		
Male	53	43
Female	47	57
Age		
18–29	21	20
30–49	43	40
50–64	19	21
65+	15	18
Race/ethnicity		
White	93	75
Black	3	19
Other nonwhite	4	6
Education		
No college	38	47
Some college	37	29
Undergraduate degree	14	11
Postgraduate degree	12	13
Household Income		
$75,000 and over	22	15
$50,000–74,999	21	16
$30,000–49,999	24	25
$20,000–29,999	12	14
Less than $20,000	15	23
Union member		
Yes	12	18
No	87	81

Source: Lydia Saad, "Who Are the Republicans?" Poll Releases, The Gallup Organization, July 30, 2000. Copyright © 2000, The Gallup Organization, Princeton, NJ. Reprinted with permission. All rights reserved.

a. Based on these data, how would you describe differences between the profiles of Republicans and Democrats?
b. For each of the variables, identify the level of measurement.
c. Review the list of variables presented in the table. What other variables would be useful in assessing the difference between Democrats and Republicans?

Graphic Presentation

You have probably heard that "a picture is worth a thousand words." The same can be said about statistical graphs because they summarize hundreds or thousands of numbers. Many people are intimidated by statistical information presented in frequency distributions or in other tabular forms, but they find the same information to be readable and understandable when presented graphically. Graphs tell a story in "pictures" rather than in words or numbers. They are supposed to make us think about the substance rather than the technical detail of the presentation.

In this chapter you will learn about some of the most commonly used graphical techniques. We concentrate less on the technical details of how to create graphs and more on how to choose the appropriate graphs to make statistical information coherent. We also focus on how to interpret information presented graphically and how to recognize when a graph distorts what the numbers have to say. A graph is a device used to create a visual impression, and that visual impression can sometimes be misleading.

As we introduce the various graphical techniques, we also show you how to use graphs to tell a "story." The particular story we tell in this chapter is that of senior citizens in the United

States. The different types of graphs introduced in this chapter demonstrate the many facets of the aging of American society over the next four decades. People have tended to talk about seniors as if they composed a homogeneous group, but the different graphical techniques we illustrate here dramatize the wide variations in economic characteristics, living arrangements, and family status among people aged 65 and older. Most of the statistical information presented in this chapter is based on numerous reports prepared by statisticians from the U.S. Census Bureau and other government agencies that gather information about senior citizens in the United States.

Numerous graphing techniques are available to you, but here we focus on just a few of the most widely used in the social sciences. The first two, the pie and bar charts, are appropriate for nominal and ordinal variables. The next two, histograms and frequency polygons, are used with interval-ratio variables. We also discuss statistical maps and time series charts. The statistical map is most often used with interval-ratio data. Finally, time series charts are used to show how some variables change over time.

▣ THE PIE CHART: RACE AND ETHNICITY OF THE ELDERLY

The elderly population of the United States is racially heterogeneous. As the data in Table 3.1 show, of the total elderly population (defined as persons 65 years and older) in 2003, about 31.5 million were white,[1] about 3 million black, 172,247 American Indian, 993,792 Asian American, 27,399 Native Hawaiian or Pacific Islander, and 210,493 were two or more races combined.

A **pie chart** shows the differences in frequencies or percentages among categories of a nominal or an ordinal variable. The categories are displayed as segments of a circle whose pieces add up to 100 percent of the total frequencies. The pie chart shown in Figure 3.1 displays the same information that Table 3.1 presents. Although you can inspect these data in Table 3.1, you can interpret the information more easily by seeing it presented in the pie chart in Figure 3.1. It shows that the elderly population is predominantly white (87.7%), followed by black (8.3%).

✔ *Learning Check.* *Notice that the pie chart contains all of the information presented in the frequency distribution. Like the frequency distribution, charts have an identifying number, a title that describes the content of the figure, and a reference to a source. The frequency or percentage is represented both visually and in numbers.*

Pie chart A graph showing the differences in frequencies or percentages among categories of a nominal or an ordinal variable. The categories are displayed as segments of a circle whose pieces add up to 100 percent of the total frequencies.

---

[1]The U.S. Bureau of the Census identifies most Hispanic Americans as white.

Figure 3.1   Annual Estimates of the U.S. Population 65 Years and Over by Race, 2003

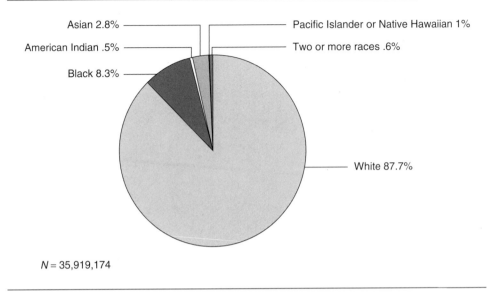

Asian 2.8%

American Indian .5%

Black 8.3%

Pacific Islander or Native Hawaiian 1%

Two or more races .6%

White 87.7%

N = 35,919,174

Source: Population Division, U.S. Census Bureau. 2004. Annual Estimates of the Population by Sex, Age and Race for the United States: April 1, 2000 to July 1, 2003 (NC-EST2003-04-01).

Table 3.1   Annual Estimates of the U.S. Population 65 Years and Over by Race, 2003

Race	Frequency (f)	Percentage (%)
White alone	31,516,040	87.7
Black alone	2,999,203	8.3
American Indian alone	172,247	.5
Asian alone	993,792	2.8
Native Hawaiian or Pacific Islander	27,399	.1
Two or more races combined	210,493	.6
Total	35,919,174	100

Source: Population Division, U.S. Census Bureau. 2004. Annual Estimates of the Population by Sex, Age and Race for the United States: April 1, 2000 to July 1, 2003 (NC-EST2003-04-01).

Notice that the percentages for several of the racial groups are under 3 percent. It might be better to combine categories—American Indian, Asian, Native Hawaiian—into an "other races" category. This will leave us with four distinct categories: White, Black, Other, and Two or More Races. The revised pie chart is presented in Figure 3.2. Confirm for yourself how the percentages are derived from Table 3.1. We can highlight the diversity of the elderly

Figure 3.2   Annual Estimates of U.S. Population 65 Years and Over by Race, 2003

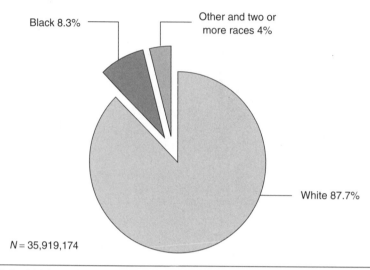

Black 8.3%

Other and two or more races 4%

White 87.7%

$N = 35,919,174$

Source: Population Division, U.S. Census Bureau. 2004. Annual Estimates of the Population by Sex, Age and Race for the United States: April 1, 2000 to July 1, 2003 (NC-EST2003-04-01).

population by "exploding" the pie chart, moving the segments representing these groups slightly outward to draw them to the viewer's attention.

> ✔ ***Learning Check.***   *Note that we could have "exploded" the segment of the pies representing the black and other population if we had wanted to highlight the proportion of whites.*

▣ THE BAR GRAPH: LIVING ARRANGEMENTS AND LABOR FORCE PARTICIPATION OF THE ELDERLY

The **bar graph** provides an alternative way to present nominal or ordinal data graphically. It shows the differences in frequencies or percentages among categories of a nominal or an ordinal variable. The categories are displayed as rectangles of equal width with their height proportional to the frequency or percentage of the category.

Let's illustrate the bar graph with an overview of the living arrangements of the elderly. Living arrangements change considerably with advancing age—an increasing number of the elderly live alone or with other relatives. Figure 3.3 is a bar graph displaying the percentage distribution of elderly males by living arrangements in 2000. This chart is interpreted similarly to a pie chart except that the categories of the variable are arrayed along the horizontal axis (sometimes referred to as the *x*-axis) and the percentages along the vertical

Figure 3.3 Living Arrangements of Males (65 and Older) in the United States, 2000

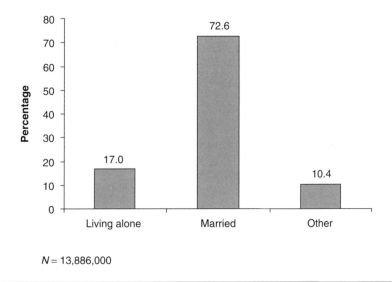

N = 13,886,000

Source: Smith, Denise. 2003. "The Older Population in the United States: March 2002." Current Population Reports, P20-546. Table 6. Washington D.C.: U.S. Government Printing Office.

axis (sometimes referred to as the *y*-axis). This bar graph is easily interpreted: It shows that in 2000, 17 percent of elderly males lived alone, 72.6 percent were married and living with their spouse, and the remaining 10.4 percent were living in other nonfamily situations.

Construct a bar graph by first labeling the categories of the variables along the horizontal axis. For these categories, construct rectangles of equal width, with the height of each proportional to the frequency or percentage of the category. Note that a space separates each of the categories to make clear that they are nominal categories.

Bar graph A graph showing the differences in frequencies or percentages among categories of a nominal or an ordinal variable. The categories are displayed as rectangles of equal width with their height proportional to the frequency or percentage of the category.

Bar graphs are often used to compare one or more categories of a variable among different groups. For example, there is an increasing likelihood that women will live alone as they age. The longevity of women is the major factor in the gender differences in living arrangements.[2] In addition, elderly widowed men are more likely to remarry than elderly widowed

---

[2]U.S. Bureau of the Census, "Marital Status and Living Arrangements: March 1996," Current Population Reports, 1998, P20-496, p. 5.

women. Also it has been noted that the current generation of elderly women has developed more protective social networks and interests.[3]

Suppose we want to show how the patterns in living arrangements differ between men and women. Figure 3.4 compares the percentage of women and men 65 years and older who lived with others or alone in 2003. It clearly shows that elderly women are more likely than elderly men to live alone.

We can also construct bar graphs horizontally, with the categories of the variable arrayed along the vertical axis and the percentages or frequencies displayed on the horizontal axis. This format is illustrated in Figure 3.5, which compares the percent of men and women 55 years and over in the civilian labor force for 2002.

From Figure 3.5, we see that for all age categories, men were more likely to be employed than women. In the age category of 65 or older, 18 percent of men were in the labor force, almost double the 9.8 percent for women.

Figure 3.4   Living Arrangement of U.S. Elderly (65 and Older) by Gender, 2003

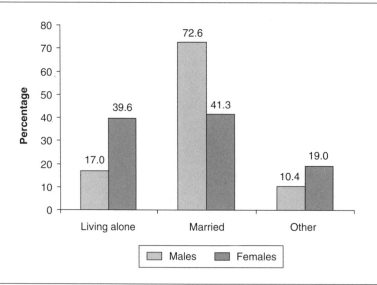

Source: Smith, Denise. 2003. "The Older Population in the United States: March 2002." Current Population Reports, P20-546. Table 6. Washington, D.C.: U.S. Government Printing Office.

◙ THE STATISTICAL MAP: THE GEOGRAPHIC DISTRIBUTION OF THE ELDERLY

Since the 1960s, the elderly have been relocating to the South and the West of the United States. It is projected that by 2020 these regions will increase their elderly population by as much as

---

[3]U.S. Bureau of the Census, "65+ in America," Current Population Reports, 1996, Special Studies P23-190, pp. 6-1, 6-8.

Figure 3.5   Percent of Men and Women 55 Years and Over in the Civilian Labor Force, 2002

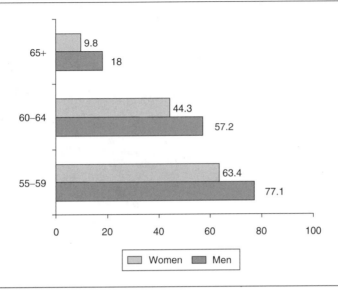

Source: Smith, Denise. 2003.

80 percent. We can display these dramatic geographical changes in American society by using a statistical map. Maps are especially useful for describing geographical variations in variables, such as population distribution, voting patterns, crime rates, or labor force composition.

Let's look at Figure 3.6. It presents a statistical map, by state, of the percent of the population 65 years and over for 2000. The variable *percent of the population 65 years and over in 2000* has four categories: under 10 percent, 10–11.9 percent, 12–13.9 percent, and 14 percent or more. Each category is represented by a different shading (or color code), and the states are shaded depending on their classification into the different categories. To make it easier to read a map that you construct and to identify its patterns, keep the number of categories relatively small—say, not more than five.

Maps can also display geographical variations on the level of cities, counties, city blocks, census tracts, and other units. Your choice of whether to display variations on the state level or for smaller units will depend on the research question you wish to explore.

✓ *Learning Check.*   *Can you think of a few other examples of data that could be described using a statistical map?*

▣ THE HISTOGRAM

The histogram is used to show the differences in frequencies or percentages among categories of an interval-ratio variable. The categories are displayed as contiguous bars, with width

Figure 3.6   Percent of the Population 65 years and Over, 2000

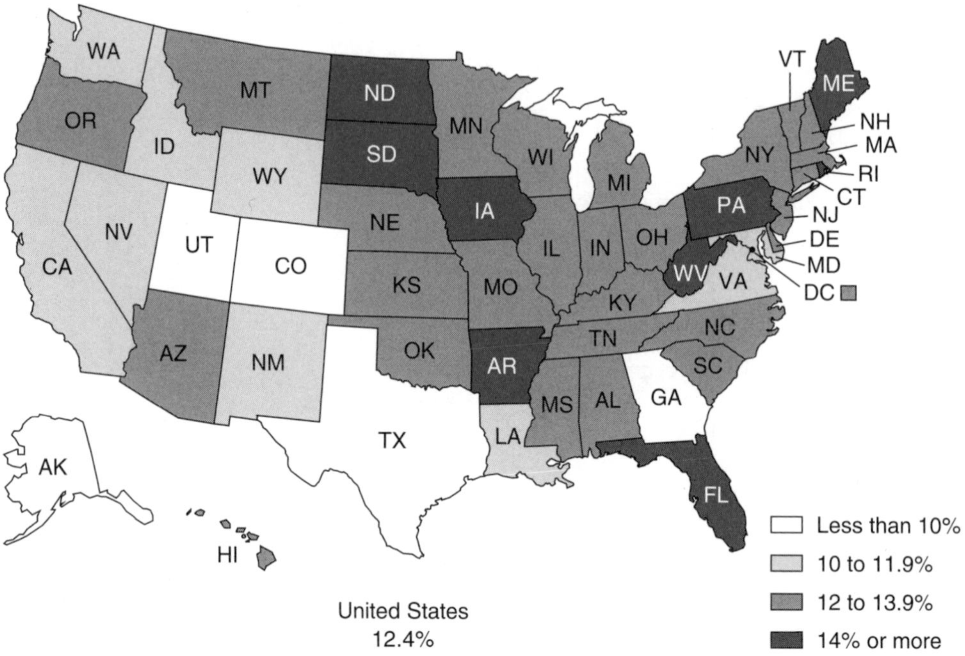

Legend:
- Less than 10%
- 10 to 11.9%
- 12 to 13.9%
- 14% or more

United States
12.4%

Source: Hetzel, Lisa and Annetta Smith. 2001. "The 65 Years and Over Population: 2000." Census 2000 Brief. C2KBR/01-10. Table 3. Washington, DC: U.S. Government Printing Office.

proportional to the width of the category and height proportional to the frequency or percentage of that category. A histogram looks very similar to a bar chart except that the bars are contiguous to each other (touching) and may not be of equal width. In a bar chart, the spaces between the bars visually indicate that the categories are separate. Examples of variables with separate categories are *marital status* (married, single); *gender* (male, female); and *employment status* (employed, unemployed). In a histogram, the touching bars indicate that the categories or intervals are ordered from low to high in a meaningful way. For example, the categories of the variables *hours spent studying, age,* and *years of school completed* are contiguous, ordered intervals.

Figure 3.7 is a histogram displaying the percentage distribution of the population 65 years and over by age. The data on which the histogram is based are presented in Table 3.2. To construct the histogram of Figure 3.7, arrange the age intervals along the horizontal axis and the percentages (or frequencies) along the vertical axis. For each age category, construct a bar with the height corresponding to the percentage of the elderly in the population in that age category. The width of each bar corresponds to the number of years that the age interval represents. The area that each bar occupies tells us the proportion of the population that falls into a given age interval. The histogram is drawn with the bars touching each other to indicate that the categories are contiguous.

Figure 3.7   Age Distribution of U.S. Population 65 Years and Over, 2000

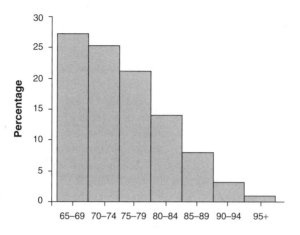

Source: Hetzel and Smith, 2001.

Table 3.2   Percentage Distribution of U.S. Population
65 Years and Over by Age, 2000

Age	Percentage
65–69	27.2
70–74	25.3
75–79	21.2
80–84	14.1
85–89	8.0
90–94	3.2
95+	1.0
Total	100.0
(*N*)	34,991,752

Histogram   A graph showing the differences in frequencies or percentages among categories of an interval-ratio variable. The categories are displayed as contiguous bars, with width proportional to the width of the category and height proportional to the frequency or percentage of that category.

✔ *Learning Check.*   *When bar charts or histograms are used to display the frequencies of the categories of a single variable, the categories are shown on the x-axis and the frequencies on the y-axis. In a horizontal bar chart or histogram this is reversed.*

◙ STATISTICS IN PRACTICE: THE "GRAYING" OF AMERICA

We can also use the histogram to depict more complex trends, as, for instance, the "graying" of America. Let's consider for a moment some of these trends: The elderly population today is ten times larger than it was in 1900, and it will more than double by the year 2030. Indeed, as a journalist has pointed out, if the automobile had existed in Colonial times, half the residents of the New Land . . . couldn't have taken a spin: One of every two people were under age 16. Most didn't live long enough to reach old age. Today, the population too young to drive has dropped to one in four while adults 65 and over account for one in eight.[4]

The histogram can give us a visual impression of these demographic trends. For an illustration, let's look at Figures 3.8 and 3.9. Both are applications of the histogram. They examine, by gender, age distribution patterns in the U.S. population for 1955 and 2010 (projected). Notice that in both figures, age groups are arranged along the vertical axis, whereas the frequencies (in millions of people) are along the horizontal axis. Each age group is classified by males on the left and females on the right. Because this type of histogram reflects age distribution by gender, it is also called an age-sex pyramid.

Figure 3.8   U.S. Population by Gender and Age, 1955 (in millions)

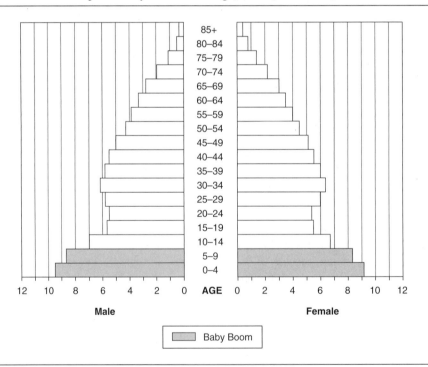

Source: U.S. Bureau of the Census, Current Population Reports, 1992, P23–178.

---

[4]*USA Today*, November 10, 1992.

Figure 3.9   U.S. Population by Gender and Age, 2010 (in millions)

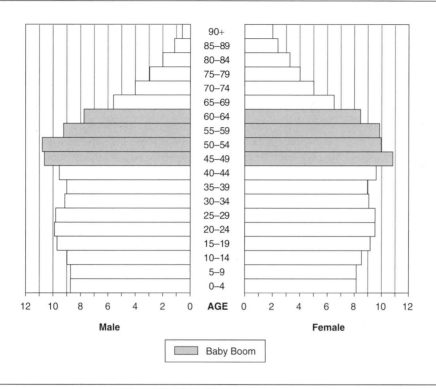

Source: U.S. Bureau of the Census, Current Population Reports, 1992, P23–178.

Visually compare the different pieces of data presented in these graphs. By observing where age groups are concentrated, you can discern major patterns in age distribution over time. Note the different shapes of Figure 3.8 and Figure 3.9. Whereas in 1955 the largest group in the population was 0 to 9 years old, in 2010 the largest group will be 45 to 54 years old. These dramatic changes reflect the "graying" of the baby boom (born 1946 to 1965) generation. Almost 84 million babies were born in the United States from 1946 to 1965, which is 60 percent more than were born during the preceding two decades. By 2010, as the baby boom generation reaches 45 to 64, the number of middle-aged and elderly Americans will increase dramatically.

Observe the differences in the number of men and women as age increases. These differences are especially noticeable in Figure 3.9. For example, between ages 70 and 74, women outnumber men 5 to 4; for those 85 years and over, women outnumber men almost 2 to 1. These differences reflect the fact that at every age male mortality exceeds female mortality.

✔ *Learning Check.*   *Notice that when we want to use the histogram to compare groups, we must show a histogram for each group (see Figures 3.8 and 3.9). When we compare groups on the bar chart, we are able to compare two or more groups on the same bar chart (see Figure 3.4).*

▣ THE FREQUENCY POLYGON

Numerical growth of the elderly population is worldwide, occurring in both developed and developing countries. In 1994 thirty nations had elderly populations of at least 2 million. Demographic projections indicate that there will be 55 such nations by 2020. Japan is one of the nations experiencing dramatic growth of its elderly population. Figure 3.10 is a frequency polygon displaying the elderly population of Japan by age.

The frequency polygon is another way to display interval-ratio distributions; it shows the differences in frequencies or percentages among categories of an interval-ratio variable. Points representing the frequencies of each category are placed above the midpoint of the category and are joined by a straight line. Notice that in Figure 3.10 the age intervals are arranged on the horizontal axis and the frequencies along the vertical axis. Instead of using bars to represent the frequencies, however, points representing the frequencies of each interval are placed above the midpoint of the intervals. Adjacent points are then joined by straight lines.

Both the histogram and the frequency polygon can be used to depict distributions and trends of interval-ratio variables. How do you choose which one to use? To some extent, the choice is a matter of individual preference, but, in general, polygons are better suited for comparing how a variable is distributed across two or more groups or across two or more time periods. For example, Figure 3.11 compares the elderly population in Japan for 2000 with the projected elderly population for the years 2010 and 2020.

Let's examine this frequency polygon. It shows that Japan's population age 65 and over is expected to grow dramatically in the coming decades. According to projections, Japan's oldest-old population, those 80 years or older, is also projected to grow rapidly, from about 4.8 million (less than 4 percent of the total population) to 10.3 million (8.4 percent) by 2020. This projected rise has already led to a reduction in retirement benefits and other adjustments to prepare for the economic and social impact of a rapidly aging society.[5]

Frequency polygon   A graph showing the differences in frequencies or percentages among categories of an interval-ratio variable. Points representing the frequencies of each category are placed above the midpoint of the category and are joined by a straight line.

✔ ***Learning Check.***   *Look closely at the frequency polygon shown in Figure 3.11. Comparing 2000 and 2020 data, how would you characterize the population increase among the Japanese elderly?*

---

[5]U.S. Bureau of the Census, "65+ in America," Current Population Reports, 1996, Special Studies, P23-190, pp. 2–3.

Figure 3.10   Population of Japan, Age 55 and Over, 2000

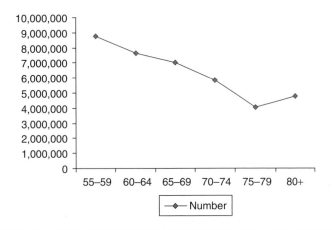

Figure 3.11   Population of Japan, Age 55 and Over, 2000, 2010, and 2020

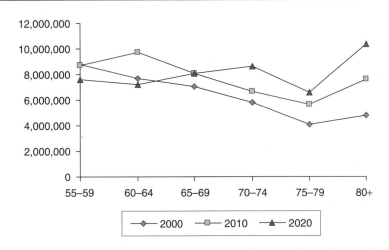

▣ TIME SERIES CHARTS

We are often interested in examining how some variables change over time. For example, we may be interested in showing changes in the labor force participation of Latina women

Figure 3.12 Percent of Total U.S. Population 65 Years and Over, 1900 to 2050

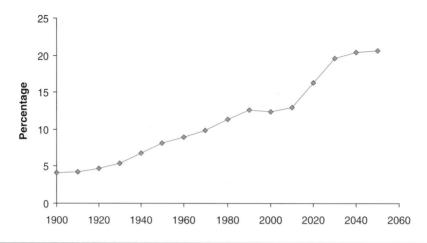

Source: Federal Interagency Forum on Aging Related Statistics, Older Americans 2004: Key Indictators of Well Being, 2004.

over the past decade, changes in the public's attitude toward abortion rights, or changes in divorce and marriage rates. A time series chart displays changes in a variable at different points in time. It involves two variables: *time,* which is labeled across the horizontal axis, and another variable of interest whose values (frequencies, percentages, or rates) are labeled along the vertical axis. To construct a time series chart, use a series of dots to mark the value of the variable at each time interval, and then join the dots by a series of straight lines.

Figure 3.12 shows a time series from 1900 to 2050 of the percentage of the total U.S. population that is 65 years or older (the figures for the years 2000 through 2050 are projections made by the Social Security Administration, as reported by the U.S. Census Bureau). This time series lets us clearly see the dramatic increase in the elderly population. The number of elderly increased from a little less than 5 percent in 1900 to about 12.4 percent in 2000. The rate is expected to increase to 20 percent of the total population. This dramatic increase in the elderly population, especially beginning in the year 2010, is associated with the "graying" of the baby boom generation. This group, which was 0 to 9 years old in 1955 (see the age pyramid in Figure 3.8), will be 55 to 64 years old in the year 2010.

The implications of these demographic changes are enormous. To cite just a few, there will be more pressure on the health-care system and on private and public pension systems. In addition, because the voting patterns of the elderly differ from those of younger people, the "graying" of America will have major political effects.

Often we are interested in comparing changes over time for two or more groups. Let's examine Figure 3.13, which charts the trends in the percentage of divorced elderly from 1960 to 2050 for men and women. This time series graph shows that the percentage of divorced elderly men and elderly women was about the same until 2000. For both groups the

Figure 3.13 Percentage Currently Divorced Among U.S. Population 65 Years and Over, by Gender, 1960 to 2040

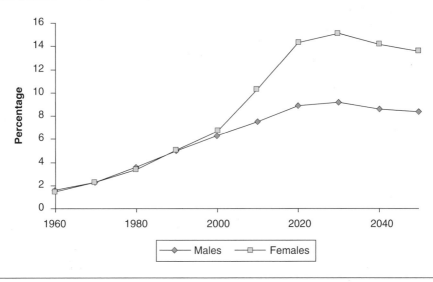

Source: U.S. Bureau of the Census, "65+ in America," Current Population Reports, 1996, Special Studies, P23-190, Table 6-1.

percentage increased from less than 2 percent in 1960 to about 5 percent in 1990.[6] According to projections, however, there will be significant increases in the percentage of men and especially women who are divorced: from 5 percent of all the elderly in 1990 to 8.4 percent of all elderly men and 13.6 percent of all elderly women by the year 2050. This sharp upturn and the gender divergence are clearly emphasized in Figure 3.13.

Time series chart   A graph displaying changes in a variable at different points in time. It shows time (measured in units such as years or months) on the horizontal axis and the frequencies (percentages or rates) of another variable on the vertical axis.

✔ ***Learning Check.***   *How does the time series chart differ from a frequency polygon? The difference is that frequency polygons display frequency distributions of a single variable, whereas time series charts display two variables. In addition, time is always one of the variables displayed in a time series chart.*

[6]Ibid., p. 6–2.

▣ DISTORTIONS IN GRAPHS

In this chapter we have seen that statistical graphs can give us a quick sense of the main patterns in the data. However, graphs not only quickly inform us; they can also quickly deceive us. Because we are often more interested in general impressions than in detailed analyses of the numbers, we are more vulnerable to being swayed by distorted graphs. But what are graphical distortions? How can we recognize them? In this section we illustrate some of the most common methods of graphical deception so you will be able to critically evaluate information that is presented graphically. To help you learn more about graphical "integrity," we highly recommend *The Visual Display of Quantitative Information* (1992), by Edward Tufte. This book not only demonstrates the many advantages of working with graphs, but also contains a detailed discussion of some of the pitfalls in the application and interpretation of graphics.

Shrinking and Stretching the Axes: Visual Confusion

Probably the most common distortions in graphical representations occur when the distance along the vertical or horizontal axis is altered in relation to the other axis.[7] Axes can be stretched or shrunk to create any desired result. Let's look at the example presented in Figure 3.14a. It is taken from a 1993 issue of *USA Today,* showing changes in cost per child enrolled in Head Start. The impression the graph gives is that from 1966 to 1993, cost per child skyrocketed! However, although the cost has indeed gone up from $271 to $3,849, these figures are not adjusted for inflation. Suppose that we want to make the increase in cost look more moderate without adjusting for inflation. We can stretch the horizontal axis to enlarge the distance between the years, as shown in Figure 3.14b. Because of this stretching, the trend appears less steep and the increase in cost appears smaller. We have changed the impression considerably without altering the data in any way. Another way to decrease the steepness of the slope is to shrink the vertical axis so that the dollar amounts are represented by smaller heights than they are in Figure 3.14a or 3.14b.

The opposite effect can be obtained by shrinking the horizontal axis and narrowing the distance between the points on the scale. That technique makes the slope look steeper.[8] Consider the graphs in Figure 3.15, which depict the increase in the number of women elected to state legislatures between 1973 and 1993. In Figure 3.15a the increase of more than 300 percent (from 424 to 1,516), although discernible, does not appear to be very great. This increase can be made to appear more dramatic by shrinking the horizontal axis so the years are moved closer together. This was done in Figure 3.15b, which represents the same data. The steeper slope, created by moving the years closer together, gives the impression of a more substantial increase.

> ✓ *Learning Check.* *When you are using a computer software program to draw a graph, the program will automatically adjust the size of the axes to avoid distortion. You can change the formats if you find that the graph is distorted.*

---

[7]R. Lyman Ott et al., *Statistics: A Tool for Social Sciences* (Boston: PWS-Kent Publishing Co., 1992), pp. 92–95.
[8]Ibid.

Figure 3.14   Cost per Child Enrolled in Head Start Program

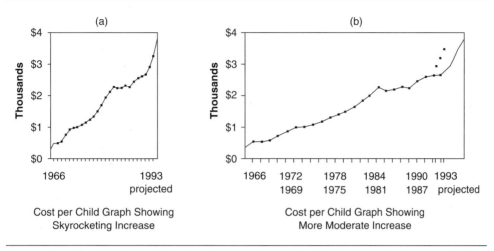

Source: Adapted from USA TODAY, March 15, 1993. Copyright 1993 USA TODAY. Reprinted with permission.

Figure 3.15   Women in U.S. Legislatures, 1973 to 1993

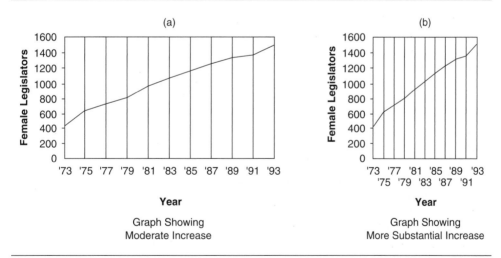

Source: Adapted from Marty Baumann, USA TODAY, February 12, 1993. Copyright 1993 USA TODAY. Reprinted with permission.

Distortions with Picture Graphs

Another way to distort data with graphs is to use pictures to represent quantitative information. The problem with picture graphs is that the visual impression received is created by

Figure 3.16   Estimated Number of HIV Infections in 1992

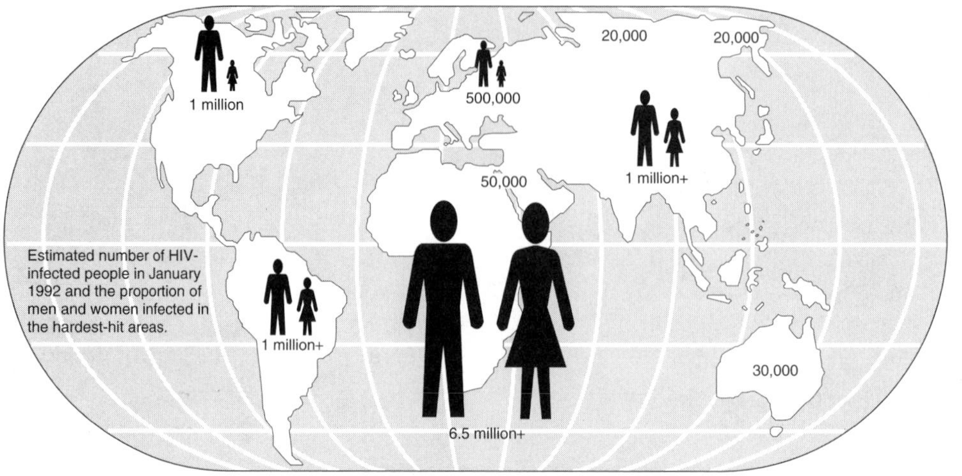

Source: Adapted from *The New York Times*, June 28, 1992. Copyright © 1992 The New York Times Co. Reprinted by permission.

the picture's total area rather than by its height (the graphs we have discussed so far rely on height).

Take a look at Figure 3.16. It shows the estimated number of HIV-infected people in some of the hardest-hit areas around the world in 1992. Note that sub-Saharan Africa, where the virus may have originated, is the hardest hit, with 6.5 million infected men and women. This number is more than six times the number of HIV infections in South and Central America, where the number of infections is about 1 million. Yet the human figures representing the number of infections for Africa are about 20 times larger in total area than the size of the human figures for South and Central America. The reason for this magnified effect is that although the data are one-dimensional (1 million compared with 6.5 million infected people in sub-Saharan Africa), the human figures representing these numbers are two-dimensional. Therefore, it is not only height that is represented but width as well, creating a false impression of the difference in the number of HIV infections.

These examples illustrate some potential pitfalls in interpreting graphs, emphasizing the point that a graph is a device used to create a visual impression, and that visual impressions can sometimes be misleading. Always interpret a graph in the context of the numerical information the graph represents.

▣ STATISTICS IN PRACTICE: DIVERSITY AT A GLANCE

We now illustrate some additional ways in which graphics can be used to visually highlight diversity. In particular, we show how graphs can help us to (1) explore the differences and similarities among the many social groups coexisting within American society and (2) emphasize the rapidly changing composition of the U.S. population. Indeed, because of the

Figure 3.17   Percentage of College Graduates among People 55 Years and Over by Age and Sex, 2002

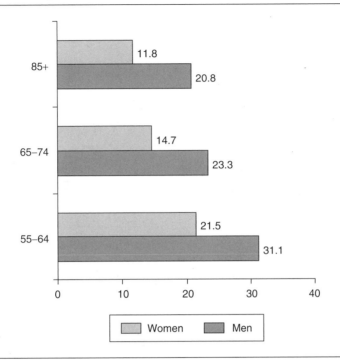

Source: Smith, 2003.

heterogeneity of American society, the most basic question to ask when you look at data is "compared to what?" This question not only is at the heart of quantitative thinking[9] but underlies inclusive thinking as well.

Three types of graphs—the bar chart, the frequency polygon, and the time series chart—are particularly suitable for making comparisons among groups. Let's begin with the bar chart displayed in Figure 3.17. It compares the college degree attainment of those 55 years and over by gender. Figure 3.17 shows that the percentage of men with a bachelor's degree or higher is higher than women in all age groups. The smallest gap is among those 65–74, of whom 23.3 percent of men had bachelor's degrees or higher compared with 14.7 percent of women.

✔ ***Learning Check.***   *Examine Figure 3.17 again. Can you think of possible explanations for the higher rates of men with bachelor's degrees than women?*

---

[9]Edward R. Tufte, *The Visual Display of Quantitative Information* (Cheshire, CT: Graphics Press, 1983), p. 53.

Figure 3.18   Years of School Completed in the United States by Race and Age, 2003

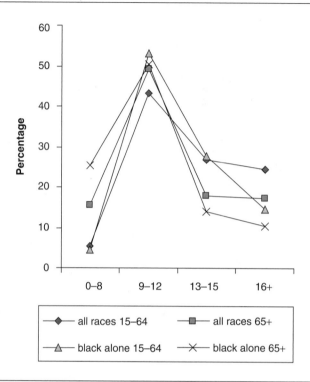

Source: Stoops, Nicole. 2004. "Educational Attainment in the United States: 2003." Current Population Reports, P20–550. Washington D.C.: U.S. Government Printing Office.

The frequency polygon provides another way of looking at differences based on gender, race/ethnicity, or other attributes such as class, age, or sexual orientation. For example, Figure 3.18 compares years of school completed by black Americans ages 15 to 64 and 65 years and older with that of all Americans in the same age groups.

The data illustrate that in the United States the percentage of Americans who have completed only 8 years of education has declined dramatically, from about 15.4 percent among Americans 65 years and older to 5.3 percent for those 15 to 64 years old. The decline for black Americans is even more dramatic, from 25.5 percent of the black elderly to about 4.4 percent for those 15 to 64 years old. The corresponding trend illustrated in Figure 3.18 is the increase in the percentage of Americans (all races as well as black Americans) who have completed 9 to 12 years of schooling, 13 to 15 years, or 16 years or more. For example, about 13.9 percent of black Americans 65 years or older completed 13 to 15 years of schooling, compared with 27.7 percent of those 15 to 64 years.

The trends shown in Figure 3.18 reflect the development of mass education in the United States during the past 50 years. The percentage of Americans who have completed 4 years of high school or more has risen from about 40 percent in 1940 to 85 percent in 2003 (record high according to the U.S. Census Bureau). Similarly, in 1940 only about 5 percent of Americans completed 4 years or more of college, compared with 27 percent in 2003.[10]

✔ *Learning Check.* *Figure 3.18 illustrates that overall, younger Americans (15 to 64 years old) are better educated than elderly Americans. However, despite these overall trends there are differences between the number of years of schooling completed by "blacks" and "all races." Examine Figure 3.18 and find these differences. What do they tell you about schooling in America?*

Finally, Figure 3.19 is a time series chart showing changes over time in the percentage of divorced white, black, and Latina women. It shows that between 1975 and 1985, the percentage of divorce among white and black women increased steadily. However, between 1985 and 1990 there was a dramatic decline in the percentage of divorce among black women, whereas the percentage of divorce among white women changed very little. In part, the apparent decline in the

Figure 3.19 Percentage of Divorced U.S. Women (after first marriage) by Race and Hispanic Origin, 1975, 1980, 1985, and 1990

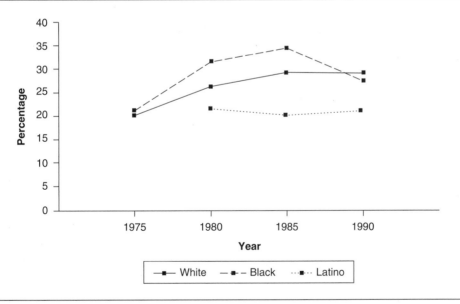

Source: U.S. Bureau of the Census, Current Population Reports, 1992, P23–178.

Note: Data not available for Hispanic women, 1975.

---

[10]Nicole Stoops, "Educational Attainment in the United States, 2003." Current Population Reports, 2004, P20-550, p. 1.

divorce rate among black women is because they are more likely to separate without divorcing. For whatever reason, however, the result is a convergence in the percentage of divorce among white and black women. In contrast, the percentage of divorce among Latina women decreased slightly between 1980 and 1985 and remained almost unchanged between 1985 and 1990.

To conclude, the three examples of graphs in this section as well as other examples throughout this chapter have illustrated how graphical techniques can portray the complexities of the social world by emphasizing the distinct characteristics of age, gender, and ethnic groups. By depicting similarities and differences, graphs help us better grasp the richness and complexities of the social world.

MAIN POINTS

- A pie chart shows the differences in frequencies or percentages among categories of nominal or ordinal variables. The categories of the variable are segments of a circle whose pieces add up to 100 percent of the total frequencies.

- A bar graph shows the differences in frequencies or percentages among categories of a nominal or an ordinal variable. The categories are displayed as rectangles of equal width with their height proportional to the frequency or percentage of the category.

- Histograms display the differences in frequencies or percentages among categories of interval-ratio variables. The categories are displayed as contiguous bars with their width proportional to the width of the category and height proportional to the frequency or percentage of that category.

- A frequency polygon shows the differences in frequencies or percentages among categories of an interval-ratio variable. Points representing the frequencies of each category are placed above the midpoint of the category (interval). Adjacent points are then joined by a straight line.

- A time series chart displays changes in a variable at different points in time. It displays two variables: time, which is labeled across the horizontal axis, and another variable of interest whose values (for example, frequencies, percentages, or rates) are labeled along the vertical axis.

KEY TERMS

bar graph
frequency polygon
histogram

pie chart
time series chart

ON YOUR OWN

Log on to the web-based student study site at http://www.pineforge.com/frankfort-nachmiasstudy4 for additional study questions, quizzes, web resources, and links to social science journal articles reflecting the statistics used in this chapter.

SPSS DEMONSTRATIONS

Demonstration 1: Producing a Bar Chart [GSS02PFP-B]

SPSS for Windows greatly simplifies and improves the production of graphics. The program offers a separate choice from the main menu bar, *Graphs,* that lists more than a dozen types of graphs that SPSS can create. We will use GSS02-B for this demonstration.

The fourth option under the *Graphs* menu is *Bar,* which will produce various types of bar charts. We will use bar charts to display the distribution of the nominal variable MARITAL (marital status of respondent). After clicking on *Graphs* and then *Bar,* you will be presented with the initial dialog box shown in Figure 3.20.

Almost all graphics procedures in SPSS begin with a dialog box that allows you to choose exactly which type of chart you want to construct. Many graph types can display more than one variable (the Clustered or Stacked choices). We will keep things simple here, so click on *Simple,* then on *Define.* When you do so, the main dialog box for simple bar charts opens (Figure 3.21).

The variable MARITAL should be placed in the box labeled "Category Axis." In the "Bars Represent" box, click on the "% of cases" radio button. This choice changes the default statistic from the number of cases to percentages, which are normally more useful for comparison purposes.

There is one more thing to do before telling SPSS to create the bar chart. Unlike the way SPSS works for the statistical procedures, SPSS automatically *includes* missing values in many graphs rather than deleting them. You can, and should, change this by clicking on *Options.* You will see the dialog box shown in Figure 3.22. Click in the box labeled "Display groups defined by missing values" to turn off this choice. Then click on *Continue,* then on *OK* to submit your request to SPSS.

The bar chart for MARITAL is presented in an output window labeled SPSS Viewer. You can see in Figure 3.23 that the bar chart for MARITAL has five bars because the only valid responses to this question are "married," "widowed," "divorced," "separated," and "never married."

Figure 3.20

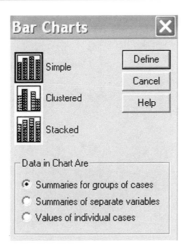

Figure 3.21

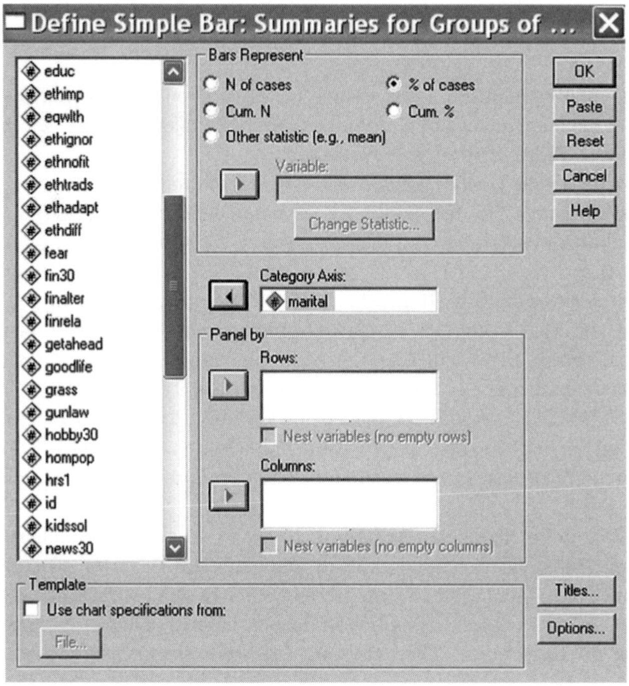

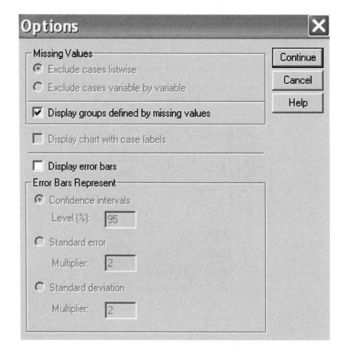

Figure 3.22

Figure 3.23

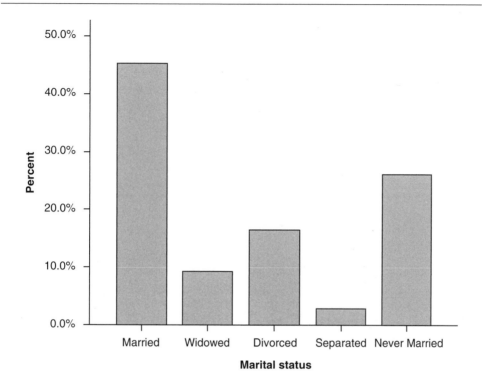

SPSS graphs can be edited by selecting *Edit,* then *SPSS Chart Object,* then *Open*, which moves the graph to its own window (*Chart Editor)* and displays various editing tools and choices.

Demonstration 2: Producing a Histogram

Histograms are used to display interval or ratio variables. We'll use the variable AGE from the 2002 GSS file. Under the *Graphs* menu in SPSS is an *Interactive* option, then select *Histogram*. Click on these choices and you will see the dialog box shown in Figure 3.24.

Histograms are created for one variable at a time (that's why there was no opening dialog box as for bar charts). You simply put (drag) the variable you want to display in the first empty box. You don't need to worry about missing values in histograms; unlike the bar chart default, SPSS automatically deletes them from the display. Notice that SPSS includes icons to indicate the level of measurement for each variable. Interval-ratio variables (or scale variables as SPSS refers to them) is matched with a ruler icon. Click on the *OK* button (on the bottom left-hand corner) to process this request. The resulting histogram is shown in Figure 3.25.

SPSS automatically decided the appropriate width for each interval, based on the range of the variable AGE and the optimal number of bars to be displayed on a screen. The number displayed under each bar is the midpoint of that interval, so, for example, the bar for 20 years of age includes everyone from 20 to 22 years of age. [The SPSS automatic formats may lead to data distortion, as we discussed in this chapter under "Distortions in Graphs." You can change the formats (interval widths) if you find that the graphs are distorted.]

Figure 3.24

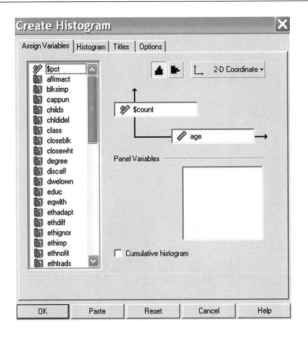

Figure 3.25

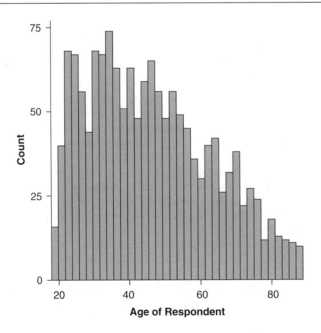

SPSS PROBLEMS

[GSS02PFP-B]

1. You have been assigned data for a presentation on Internet use. You've decided to examine the differences in use between men and women. Using data from the GSS02-B, you examine NEWS30 (number of times respondents use an Internet news site in the past 30 days) and WWWHR (number of hours spent on the Internet in a week). There are certainly other variables that you could use.
    a. Using SPSS, construct a pie chart for NEWS30. Be sure to include appropriate labels and to remove any missing responses from the chart. [*Hint:* With the SPSS menus, first choose *Data–Split File* from the menu, then choose "Organize output by groups" in the dialog box and place SEX in the Groups Based On box (and click on OK). All procedures after this will be done separately for men and women. To create your pie chart, select *Graphs–Pie*, then select "Summaries for Groups of Cases," click on "Define." In the box under "Define Slices by" click over the name of your variable, NEWS30.]
    b. Now construct a histogram for WWWHR (*Hint:* From the SPSS menus, choose *Graphs–Bar.*) Option: If you want to create one histogram with both men and women, you should return to *Data–Split File* and click on "Analyze all cases, do not create groups."
    c. Briefly describe the difference regarding Internet use between men and women, using either the pie or bar chart.

2. Is there a difference in Internet use between those with a high school diploma and those with a bachelor's degree? Based on Exercise 1, reconstruct the bar chart for WWWHR, separating the responses between high school graduates and college graduates.

    The most straightforward way to produce this output is to first split the file into groups and have SPSS analyze each separately. Choose *Data–Split File* from the menu, then choose "Organize output by groups" in the dialog box and place DEGREE in the Groups Based On box (and click on *OK*). All procedures run after this will be done separately for all degree categories.

    Your output should present separate bar charts for all degree categories, but focus only on those for high school and bachelor's degree graduates. Compare the bar charts. What difference exists in the level of internet use between these two groups?

3. Is there a difference in responses between men and women in whether they support legalization of marijuana (GRASS) and gun permits (GUNLAW)? Based on the level of measurement for each variable, determine the appropriate graphic display. Produce separate graphs for men and women (you can use the *Data–Split File* option to create separate graphs). What differences, if any, can you report?

4. Determine how best to graphically represent the following variables:
    POLVIEWS—political views of respondent
    CHILDS—number of children in household
    EQLWTH—should the government reduce income differences in the U.S.
    AFFRMACT—does respondent support affirmative action policies in the workplace

Note: Before constructing the histogram or pie chart, you may want to review the variable by first using the *Frequencies* or *Utilities–Variables* command. The levels of measurement for several variables are mislabeled in SPSS. If you are using the *Utilities–Variables* option to review each variable and its level of measurement, you should confirm the level of measurement by reviewing the variable's frequency table (*Analyze–Descriptive–Frequencies*).]

CHAPTER EXERCISES

1. The time series chart shown in Figure 3.26 displays trends in birth rates for unmarried women from 1940 to 1992.
   a. Write an 80-word report describing the variation in nonmarital birth rates for the five age categories.
   b. Can you think what causes some of the age groups to increase/decrease their birth rates across the years? Are there any important years or decades where the change in birth rates is dramatic?

Figure 3.26   Birth Rate for Unmarried Women by Age: United States, 1940–92

Source: Ventura, S. J. 1995. "Births to Unmarried Mothers: United States, 1980–1992." National Vital Health Statistics vol. 21, no. 53. Hyattsville, MD: National Center for Health Statistics.

2. We selected a sample of 30 people from the ISSP 2000. Raw data are presented for their sex (SEX), social class (CLASS), and number of household members (HOMPOP).

SEX	HOMPOP	CLASS	SEX	HOMPOP	CLASS
F	1	L	F	2	U
F	3	W	M	2	M
F	1	M	F	4	W
M	2	M	M	2	U
F	3	M	M	4	M
M	2	U	M	4	W
F	7	L	M	2	M
M	3	M	F	1	W
M	4	M	M	2	U
F	1	M	M	4	W
F	2	M	F	7	M
M	5	L	M	3	M
M	4	M	M	4	M
M	4	L	F	4	M
M	3	W	F	5	L

a. Construct a pie chart depicting the percentage distribution of sex. (*Hint:* Remember to include a title, percentages, and appropriate labels.)

b. Construct a pie chart showing the percentage distribution of social class.

c. Construct a graph with two pie charts comparing the percentage distribution of social class membership by sex.

3. We continue our analysis of birth rates for U.S. women, this time analyzing birth rates for teens 15 to 19 years of age. A bar graph chart is presented in Figure 3.27. Write a brief paragraph describing the variation in teen birth rates from 1990 to 2000.

Figure 3.27   Birth Rates per 1,000 Unmarried Women 15 to 19 Years Old, by Race: 1990 and 2000

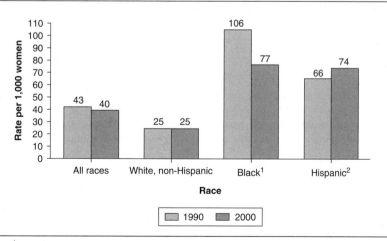

[1]Includes persons of Hispanic origin.
[2]Includes all persons of Hispanic origin of any race.

Source: Freeman, C. E. 2004. *Trends in Educational Equity for Girls and Women: 2004.* (NCES 2005–016). U.S. Department of Education, National Center for Education Statistics. Washington, D.C.: U.S. Government Printing Office.

4. Using the data from Exercise 2, construct bar graphs showing percentage distributions for sex and class. Remember to include appropriate titles, percentages, and labels.

5. Suppose you want to compare the number of household members for women and men (based on the ISSP data in Exercise 2).
   a. Construct a grouped bar graph (similar to Figure 3.4) to show the percentage distribution of the number of household members by sex.
   b. Which group reported the largest family size?
   c. Why shouldn't you construct a grouped bar chart showing the frequencies rather than the percentages?

6. During the recent presidential election, health care and health insurance were identified by voters as important issues. In fact, policy analysts have noted that the number of uninsured is increasing in the United States. Data from the National Center for Health Statistics is presented in Figure 3.28. What can be said about who did not have health insurance during 2001? How does the percent of those without health insurance vary by ethnicity/race, age, and poverty status?

7. The International Social Survey Programme 2000 asked respondents about their priorities for their country. A summary table is provided for the United States and its two neighbors, Canada and Mexico. Construct a chart or graph that best displays this information. To which country is the U.S. more similar—Canada or Mexico?

Priority	*United States*	*Canada*	*Mexico*
Order in the nation	19	17	15
Give people more say	9	19	15
Fight rising prices	7	8	19
Protect freedom of speech	11	4	7
TOTAL (*N*)	46	48	56

Source: ISSP, 2000.

8. You are writing a research paper on grandparents who had one or more of their grandchildren living with them. In 2000, 2.4 million grandparents were defined as caregivers by the U.S. Census, meaning that they had primary responsibility for raising their grandchildren under the age of 18. You discover the following information from the U.S. Census Report, "Grandparents Living with Grandchildren: 2000" (C2KBR-31, October 2003): Among grandparent caregivers, 12 percent cared for a grandchild for fewer than 6 months, 11 percent for 6 to 11 months, 23 percent for 1 to 2 years, 15 percent for 3 to 4 years, and 39 percent for 5 or more years.
   a. Construct a graph or chart that best displays this information on how long grandparents care for their grandchildren.
   b. Explain why the graph you selected is appropriate.

Figure 3.28 No Health Insurance Coverage among Persons under 65 Years of Age by Selected Characteristics: United States, 2001

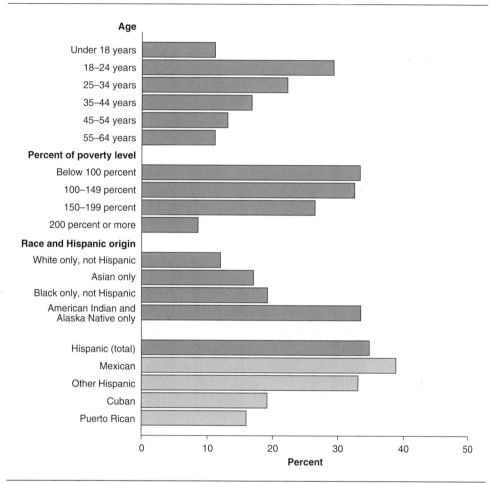

Source: Freid, V. M. Prager, K. MacKay, A. P. and Xia, H. 2003. *Chartbook on Trends in the Health of Americans. Health, United States, 2003.* Hyattsville, MD: National Center for Health Statistics.

9. Use the data on educational level in Chapter 2, Exercise 4, for this problem.
   a. What level of measurement is "years of education"? Why can you use a histogram to graph the distribution of education, in addition to a bar chart?
   b. Construct a histogram for years of education, using equal-spaced intervals of 4 years. Don't use percentages in this chart.

10. The 2002 GSS data on educational level can be further broken down by race, as follows:
    a. Construct two histograms for education, one for whites and one for blacks.
    b. Now use the two graphs to describe the differences in educational attainment by race.

Years of Education	Whites	Blacks
0	1	0
1	1	0
2	7	1
3	1	0
4	2	1
5	4	2
6	3	2
7	3	3
8	19	4
9	17	4
10	23	4
11	30	15
12	245	35
13	75	15
14	109	16
15	42	9
16	127	5
17	29	1
18	47	2
19	13	1
20	15	2

11. Use the data on minority rights attitudes from Chapter 2, Exercise 11. What would be the most appropriate graphic presentation for the data? Explain the reason for your answer.

12. Examine the time series chart concerning percent of children under 18 years of age living in poverty shown in Figure 3.29.
    a. Overall, is it accurate to say that the rate of poverty has been declining for children under 18 years of age?
    b. Compare the poverty rates for children in all families and married couple families with female-headed households. How would you characterize the difference in poverty rates for children?

Figure 3.29   Percentage of Related Children under Age 18 in Poverty by Family Structure, 1980-2001

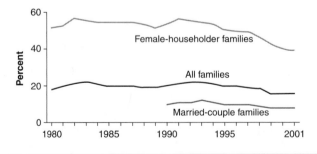

Source: Federal Interagency Forum on Child and Family Statistics. *America's Children: Key National Indicators of Well-Being, 2003.* Washington, DC: U.S. Government Printing Office.

13. Based on data on the Million Mom March (Chapter 2, Exercise 12) decide how best to graphically represent those data in an easily understood format.
    a. Would you choose to use bar charts or pie charts? Explain the reason for your answer.
    b. Construct bar or pie charts (depending on your answer) to represent all of the data. Remember to include appropriate titles, percentages, and labels.

14. As reported by Catherine Freeman (2004),[11] females have more success in post–secondary education than male students. They are more likely to enroll in college immediately after high school and have higher college graduate rates than males. In her report, Freeman provides the following time series chart (Figure 3.30), documenting the percent of women enrolled in undergraduate, graduate, and professional programs. Prepare a brief statement on the enrollment trends from 1970 to 2000.

Figure 3.30   Females as a Percent of Total Enrollment in Undergraduate, Graduate, and First-Professional Education: Various Years, Fall 1970 to Fall 2000

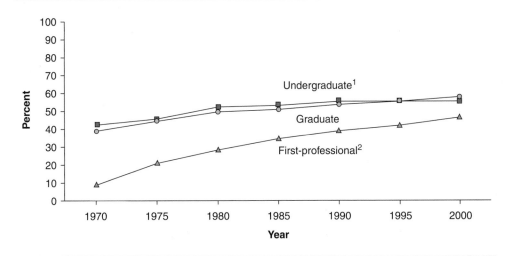

[1]Includes unclassified undergraduate students.
[2]First-professional students are enrolled in the fields of dentistry (D.D.S. or D.M.D.), medicine (M.D.), optometry (O.D.), osteopathic medicine (D.O.), pharmacy (D.Phar.), podiatric medicine (D.P.M.), veterinary medicine (D.V.M.), chiropractic medicine (D.C. or D.C.M.), law (J.D.), and the theological professions (M.Dtv. or M.H.L.).

Source: Freeman, Catherine. 2004. *Trends in Educational Equity for Girls and Women: 2004* (NCES 2005-016). U.S. Department of Education, National Center for Education Statistics. Washington, D.C.: U.S. Government Printing Office.

[11]Catherine Freeman, *Trend in Educational Equity of Girls and Women: 2004*. (Washington, D.C.: U.S. Department of Education, National Center for Education Statistics), pp. 9–10.

Measures of Central Tendency

I n Chapters 2 and 3 we learned that frequency distributions and graphical techniques are useful tools for presenting information. The main advantage of using frequency distributions or graphs is to summarize quantitative information in ways that can be easily understood even by a lay audience. Often, however, we need to describe a large set of data involving many variables for which graphs and tables may not be the most efficient tools. For instance, let's say we want to present information on the income, education, and political party affiliation of both men and women. Presenting this information might require up to six frequency distributions or graphs. The more variables we add, the more complex the presentation becomes.

Another way of describing a distribution is by selecting a single number that describes or summarizes the distribution more concisely. Such numbers describe what is typical about the distribution, for example, the average income among Latinos who are college graduates or the most common party identification among the rural poor. Numbers that describe what is average or typical of the distribution are called **measures of central tendency**.

Measures of central tendency   Numbers that describe what is average or typical of the distribution.

In this chapter we will learn about three measures of central tendency: the *mode*, the *median*, and the *mean*. You are probably somewhat familiar with these measures. The terms *median* income and *average* income, for example, are used quite a bit even in the popular media. Each describes what is most typical, central, or representative of the distribution. In this chapter we will also learn about how these measures differ from one another. We will see that the choice of an appropriate measure of central tendency for representing a distribution depends on three factors: (1) the way the variables are measured (their level of measurement), (2) the shape of the distribution, and (3) the purpose of the research.

◙ THE MODE

The mode is the category or score with the largest frequency or percentage in the distribution. Of all the averages discussed in this chapter, the mode is the easiest one to identify. Simply locate the category represented by the highest frequency in the distribution.

Mode   The category or score with the highest frequency (or percentage) in the distribution.

We can use the mode to determine, for example, the most common foreign language spoken in the United States today. English is clearly the language of choice in public communication in the United States, but you may be surprised by the U.S. Census Bureau's finding that one out of every seven people living in the United States speaks one of 329 different languages other than English at home. Record immigration from many countries since 1980 has contributed to a sharp increase in the number of people who speak a foreign language.[1]

---

[1]Douglas S. Massey. "The Social and Economic Origins of Immigration." *Annals, AAPSS*:510. July 1990.

Table 4.1   Ten Most Common Foreign Languages
Spoken in the United States, 2000

Language	*Number of Speakers*
Spanish	28,101,000
Chinese	2,022,000
French	1,644,000
German	1,383,000
Tagalog	1,224,000
Vietnamese	1,010,000
Italian	1,008,000
Korean	894,000
Russian	706,000

Source: U.S. Bureau of the Census, *Statistical Abstract of the United States*, 2003, Table 57.

What is the most common foreign language spoken in the United States today? To answer this question look at Table 4.1, which lists the 10 most commonly spoken foreign languages in the United States and the number of people who speak each language. The table shows that Spanish is the most common; more than 28 million people speak Spanish. In this example we refer to "Spanish" as the mode—the category with the largest frequency in the distribution.

The mode is always a category or score, *not* a frequency. Do not confuse the two. That is, the mode in the previous example is "Spanish," not its frequency of 28,101,000.

The mode is not necessarily the category with the majority (that is, more than 50%) of cases, as it is in Table 4.1; it is simply the category in which the largest number (or proportion) of cases fall. For example, Figure 4.1 is a pie chart showing the answers of 2002 GSS respondents to the following question: "Would you say your own health, in general, is excellent, good, fair, or poor?" Note that the highest percentage (46.24%) of respondents is associated with the answer "good." The answer "good" is therefore the mode.

The mode is used to describe nominal variables. Recall that with nominal variables—such as foreign languages spoken in the United States, race/ethnicity, or religious affiliation—we are only able to classify respondents based on a qualitative and not a quantitative property. By describing the most commonly occurring category of a nominal variable (such as Spanish in our example), the mode thus reflects the most important element of the distribution of a variable measured at the nominal level. The mode is the only measure of central tendency that can be used with nominal level variables. It can also be used to describe the most commonly occurring category in any distribution. For example, the variable *health* presented in Figure 4.1 is an ordinal variable.

In some distributions there are two scores or categories with the highest frequency. Such distributions have two modes and are said to be *bimodal*. For instance, Figure 4.2 is a bar graph showing the level of agreement of 2002 GSS respondents with the following statement:

Figure 4.1   Respondent's Health

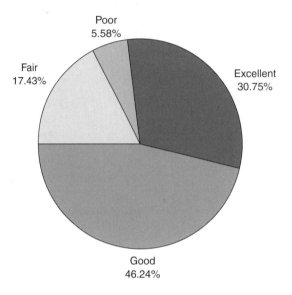

Figure 4.2   Views on Government's Responsibility to Provide Child Care

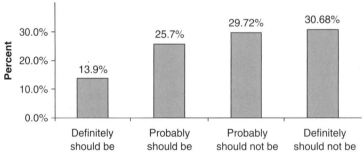

"Would you say that it is the government's responsibility to provide child care?" The same percentage of respondents (approximately 30%) expressed that "it definitely should not be" or "probably should not be" the government's responsibility to provide child care. Both response categories have the highest frequency, and therefore both are the modes. We can describe this distribution as bimodal. When two scores or categories with the highest frequencies are quite close (but not identical) in frequency, the distribution is still "essentially" bimodal. In these situations you should not rely on merely reporting the (true) mode, but instead report the two highest frequency categories.

✔ **Learning Check.** *Listed below are the political party affiliations of fifteen individuals. Find the mode.*

Democrat	Republican	Democrat	Republican	Republican
Independent	Democrat	Democrat	Democrat	Republican
Independent	Democrat	Independent	Republican	Democrat

▣ THE MEDIAN

The **median** is a measure of central tendency that can be calculated for variables that are at least at an ordinal level of measurement. The median represents the exact middle of a distribution; it is the score that divides the distribution into two equal parts so that half the cases are above it and half below it. For example, the median household income in 2002 was \$42,409.[2] This means that half the households in the United States earned more than \$42,409 and half earned less than \$42,409. Since many variables used in social research are ordinal, the median is an important measure of central tendency in social science research.

Median   The score that divides the distribution into two equal parts so that half the cases are above it and half below it.

For instance, what are the opinions of Americans about the economy in the United States? How can we describe their rating of current economic conditions? To answer these questions, the 2002 GSS asked respondents whether they felt that the national economy had improved during the past year. Respondents rated the improvement (or lack thereof) as "gotten better," "stayed the same," or "gotten worse." Rating of economic conditions is an ordered (ordinal) variable. Thus, to estimate the average rating we need to use a measure of central tendency appropriate for ordinal variables. The median is a suitable measure for those variables whose categories or scores can be arranged in order of magnitude from lowest to highest. Therefore, the median can be used with ordinal or interval-ratio variables, for which scores can be at least rank-ordered, but cannot be calculated for variables measured at the nominal level.

Finding the Median in Sorted Data

It is very easy to find the median. In most cases it can be done by a simple inspection of the sorted data. The location of the median score differs somewhat, depending on whether the

---

[2]U.S. Bureau of the Census, *Current Population Reports, Income in the United States: 2002*, September 2003.

number of observations is odd or even. Let's first consider two examples with an odd number of cases.

An Odd Number of Cases   Suppose we are looking at the responses of five people to the question "Thinking about the economy, how would you rate economic conditions in this country today?" Following are the responses of these five hypothetical persons:

Poor	Jim
Good	Sue
Only fair	Bob
Poor	Jorge
Excellent	Karen
Total (N)	5

To locate the median, first arrange the responses in order from lowest to highest (or highest to lowest):

Poor	Jim
Poor	Jorge
Only fair	Bob
Good	Sue
Excellent	Karen
Total (N)	5

The median is the response associated with the middle case. Find the middle case when N is odd by adding 1 to N and dividing by 2: $(N + 1) \div 2$. Since N is 5, you calculate $(5 + 1) \div 2 = 3$. The middle case is thus the third case (Bob), and the median is "only fair," the response associated with the third case, Bob. Notice that the median divides the distribution exactly in half so that there are two respondents who are more satisfied and two respondents who are less satisfied.

Now let's look at another example (Figure 4.3a). The following is a list of the number of hate crimes reported in the nine largest U.S. states in the year 2000.[3]

Number	*State*
1943	California
240	Florida
336	Virginia
652	New Jersey
608	New York
255	Ohio
141	Pennsylvania
287	Texas
39	North Carolina
Total (N)	9

---

[3]U.S. Bureau of the Census, *Statistical Abstract of the United States*, 2003, Table 320.

Number	*State*
39	North Carolina
141	Pennsylvania
240	Florida
255	Ohio
287	Texas
336	Virginia
608	New York
652	New Jersey
1943	California
Total (*N*)	9

To locate the median, first arrange the number of hate crimes in order from lowest to highest:

Figure 4.3a   Finding the Median Number of Hate Crimes for Nine States

1. Order the cases from lowest to highest:

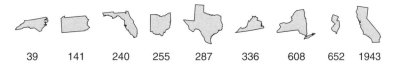

| 39 | 141 | 240 | 255 | 287 | 336 | 608 | 652 | 1943 |

2. In this situation, we need the 5th case: (9 + 1) ÷ 2 = 5.

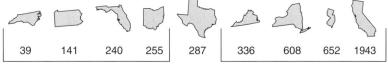

| 39 | 141 | 240 | 255 | 287 | 336 | 608 | 652 | 1943 |

4 cases on this side of the median   4 cases on this side of the median

The middle case is (9 + 1) ÷ 2 = 5, the fifth state, Texas. The median is 287, the number of hate crimes associated with Texas. It divides the distribution exactly in half so that there are four states with fewer hate crimes and four with more.

An Even Number of Cases   Now let's delete the last score to make the number of states even (Figure 4.3b). The scores have already been arranged in increasing order.

Again, to locate the median, first arrange the number of hate crimes in order from lowest to highest:

Figure 4.3b   Finding the Median Number of Hate Crimes for Eight States

1. Order the cases from lowest to highest:

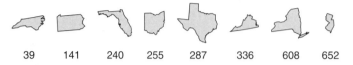

39	141	240	255	287	336	608	652

2. In this situation, we take the average of the two cases nearest the 4.5th case (8 + 1) ÷ 2 = 4.5

Median: (255 + 287)/2 = 271

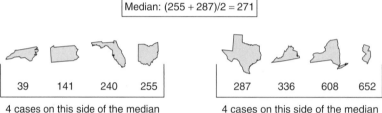

| 39 | 141 | 240 | 255 | | 287 | 336 | 608 | 652 |

4 cases on this side of the median   4 cases on this side of the median

Number	State
39	North Carolina
141	Pennsylvania
240	Florida
255	Ohio
287	Texas
336	Virginia
608	New York
652	New Jersey
Total (*N*)	8

When *N* is even (eight states), we no longer have a single middle case. The median is therefore located halfway between the two middle cases. Find the two middle cases by using the previous formula: $(N + 1) ÷ 2$, or $(8 + 1) ÷ 2 = 4.5$. In our example, this means that you average the scores for the fourth and fifth states, Ohio and Texas. The numbers of hate crimes associated with these states are 255 and 287. To find the median for an interval-ratio variable, simply average these two middle numbers:

$$\text{Median} = \frac{255 + 287}{2} = 271$$

The median is therefore 271.

As a note of caution, when data are ordinal, averaging the middle two scores is no longer appropriate. The median simply falls between two middle values.

✔ *Learning Check.*   *Find the median of the following distribution of an interval-ratio variable: 22, 15, 18, 33, 17, 5, 11, 28, 40, 19, 8, 20*

Finding the Median in Frequency Distributions

Often our data are arranged in frequency distributions. Take, for instance, the frequency distribution displayed in Table 4.2. It shows the political views of GSS respondents in 2002.

To find the median we need to identify the category associated with the observation located at the middle of the distribution. We begin by specifying N, the total number of respondents. In this particular example, $N = 1331$. We then use the formula $(N + 1) \div 2$, or $(1331 + 1) \div 2 = 666$. The median is the value of the category associated with the 666th case. The cumulative frequency (Cf) of 666 falls in the category "moderate"; thus, the median is "moderate." This may seem odd; however, the median is always the value of the response category, not the frequency.

A second approach to locating the median in a frequency distribution is to use the cumulative percentages column, as shown in the last column of Table 4.2. In this example, the percentages are cumulated from "extremely liberal" to "extremely conservative." We could also cumulate the other way, from "extremely conservative" to "extremely liberal." To find the median we identify the response category that contains a cumulative percentage value equal to 50 percent. The median is the value of the category associated with this observation.[4] Looking at Table 4.2, the percentage value equal to 50 percent falls within the category "moderate." The median for this distribution is therefore "moderate." If you are not sure why the middle of the distribution—the 50 percent point—is associated with the category "moderate," look again at the cumulative percentage column (C%). Notice that 26.1 percent of the observations are accumulated below the category "moderate" and that 65.3 percent are accumulated up to and including the category "moderate." We know, then, that the percentage value equal to 50 percent is located somewhere within the "moderate" category.

Table 4.2   Political Views of GSS Respondents, 2002

Political View	*Frequency (f)*	*Cf*	*Percentage*	*C%*
Extremely liberal	47	47	3.5	3.5
Liberal	143	190	10.7	14.2
Slightly liberal	159	349	11.9	26.1
Moderate	522	871	39.2	65.3
Slightly conservative	209	1080	15.7	81.0
Conservative	210	1290	15.8	96.8
Extremely conservative	41	1331	3.1	99.9*
Total (*N*)	1331		99.9	

*Answer varies slightly from 100.0 due to rounding.

---

[4]This rule was adapted from David Knoke and George W. Bohrnstedt, *Basic Statistics* (New York: Peacock Publishers, 1991), pp. 56–57.

✓ **Learning Check.**   *If you are confused about cumulative distributions, review Chapter 2, pp. 42–45.*

Statistics in Practice: Gendered Income Inequality

We can use the median to compare groups. Consider the significant changes that have taken place during the past few decades in the income levels of men and women in the United States. Income levels profoundly influence our lives both socially and economically. Higher income is associated with increased education and work experience for both men and women.

Figure 4.4 compares the median incomes for men and women in 1973 and in 2003. Because the median is a single number summarizing central tendency in the distribution, we can use it to note differences between subgroups of the population or changes over time. In this example, the increase in median income from 1973 to 2003 clearly shows a significant income gain for women. However, that said, in 2003 women still made nearly $10,000 less than men.

Figure 4.4   Median Income for Men and Women, 1973 and 2003 (in $)

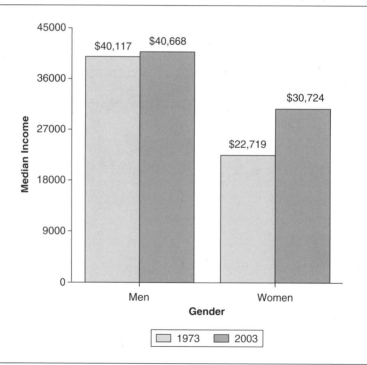

Source: U.S. Bureau of the Census. Current Population Reports P60-226, *Income, Poverty and Health Insurance Coverage in the United States:* 2003.

✔ *Learning Check.* *Examine Figure 4.4 and contrast the median incomes of women and men over the three decades. What can you learn about gender and income?*

Locating Percentiles in a Frequency Distribution

The median is a special case of a more general set of measures of location called *percentiles*. A **percentile** is a score at or below which a specific percentage of the distribution falls. The *n*th percentile is a score below which *n* percent of the distribution falls. For example, the 75th percentile is a score that divides the distribution so that 75 percent of the cases are below it. The median is the 50th percentile. It is a score that divides the distribution so that 50 percent of the cases fall below it. Like the median, percentiles require that data be ordinal or higher in level of measurement. Percentiles are easy to identify when the data are arranged in frequency distributions.

Percentile   A score below which a specific percentage of the distribution falls.

To help illustrate how to locate percentiles in a frequency distribution, we display in Table 4.3 the frequency distribution, the percentage distribution, and the cumulative percentage distribution for the number of children per family for 2002 GSS respondents.

The 50th percentile (the median) is two children, meaning that 50 percent of the respondents have two children or less (as you can see from the cumulative percentage column, 50 percent falls somewhere in the third category, associated with two children). Similarly, the 80th percentile is three children because 80 percent of the respondents have three or fewer children.

Percentiles are widely used to evaluate relative performance on standardized achievement tests, such as the SAT or ACT. Let's suppose that your ACT score was 29. To evaluate your performance for the college admissions officer, the testing service translated your score into

Table 4.3   Frequency Distribution for Number of Children: 2002 GSS Respondents

Number of Children	Frequency (f)	Percentage (%)	C%
0	799	28.9	28.9
1	469	17.0	45.9
2	657	23.8	69.7
3	481	17.4	87.1
4	185	6.7	93.8
5	73	2.6	96.4
6	40	1.4	97.8
7	22	0.8	98.6
8 or more	34	1.2	99.8*
Total (*N*)	2760	99.8*	

*Answer varies slightly from 100.0 due to rounding.

a percentile rank. Your percentile rank was determined by comparing your score with the scores of all other seniors who took the test at the same time. Suppose for a moment that 90 percent of all students received a lower ACT score than you (and 10 percent scored above you). Your percentile rank would have been 90. If, however, there were more students who scored better than you—let's say that 15 percent scored above you and 85 percent scored lower than you—your percentile rank would have been only 85.

Another widely used measure of location is the *quartile*. The lower quartile is equal to the 25th percentile and the upper quartile is equal to the 75th percentile. (Can you locate the upper quartile in Table 4.3?) A college admissions office interested in accepting the top 25 percent of its applicants based on their SAT scores could calculate the upper quartile (the 75th percentile) and accept everyone whose score is equivalent to the 75th percentile or higher. (Note that they would be calculating percentiles based on the scores of their applicants, not of all students in the nation who took the SAT.)

▣ THE MEAN

The arithmetic **mean** is by far the best-known and most widely used average. The mean is what most people call the "average." The mean is typically used to describe central tendency in interval-ratio variables such as income, age, and education. You are probably already familiar with how to calculate the mean. Simply add up all the scores and divide by the total number of scores.

Mean   The arithmetic average obtained by adding up all the scores and dividing by the total number of scores.

Crime statistics, for example, are often analyzed using the mean. Each year about 25 percent of U.S. households are victims of some form of crime. Although violent crimes are the least common types of crimes in the United States, the U.S. rate of violent crime is nonetheless the highest of any industrialized nation. For instance, murder rates in the United States are approximately five times as high as those in Europe.

Table 4.4 shows the 2001 murder rates (per 100,000 population) for the 15 largest cities in the United States. We want to summarize the information presented in this table by calculating some measure of central tendency. Because the variable "murder rate" is an interval-ratio variable, we will select the arithmetic mean as our measure of central tendency.

To find the mean murder rate for the data presented in Table 4.4, add up the murder rates for all the cities and divide the sum by the number of cities:

$$\text{Mean} = \frac{207.7}{15} = 13.85$$

The mean murder rate for the 15 largest cities in the United States is 13.85.[5]

---

[5]The rates presented in Table 4.4 are computed for aggregate units (cities) of different sizes. The mean of 13.85 is therefore called an unweighted mean. It is not the same as the murder rate for the population in the combined cities.

Table 4.4   2001 Murder Rates (per 100,000 population)
                 for the 15 Largest Cities in the United States

City	Murder Rate per 100,000
New York	8.2
Los Angeles	15.6
Chicago	22.9
Houston	13.4
Philadelphia	20.4
Phoenix	15.3
San Diego	4.0
Dallas	19.7
San Antonio	8.5
Las Vegas	11.9
Detroit	41.3
San Jose	2.4
Honolulu	2.3
Indianapolis	14.0
San Francisco	7.8

Source: U.S. Bureau of the Census, *Statistical Abstract of the United States*, 2003, Table 308.

Using a Formula to Calculate the Mean

Another way to calculate the arithmetic mean is to use a formula. Beginning with this section, we introduce a number of formulas that will help you calculate some of the statistical concepts we are going to present. A formula is a shorthand way to explain what operations we need to follow to obtain a certain result. So instead of saying "add all the scores together and then divide by the number of scores," we can define the mean by the following formula:

$$\bar{Y} = \frac{\sum Y}{N} \tag{4.1}$$

Let's take a moment to consider these new symbols because we continue to use them in later chapters. We use Y to represent the raw scores in the distribution of the variable of interest; $\bar{Y}$ is pronounced "Y-bar" and is the mean of the variable of interest. The symbol represented by the Greek letter $\sum$ is pronounced "sigma," and it is used often from now on. It is a summation sign (just like the + sign) and directs us to sum whatever comes after it. Therefore, $\sum Y$ means "add up all the raw Y scores." Finally, the letter N, as you know by now, represents the number of cases (or observations) in the distribution.

Let's summarize:

Y = the raw scores of the variable Y
$\bar{Y}$ = the mean of Y
$\sum Y$ = the sum of all the Y scores
N = the number of observations or cases

Now that we know what the symbols mean, let's work through another example. The following are the ages of the 10 students in a graduate research methods class:

$$21, 32, 23, 41, 20, 30, 36, 22, 25, 27$$

What is the mean age of the students?

For these data the ages included in this group are represented by Y; $N = 10$, the number of students in the class; and $\sum Y$ is the sum of all the ages:

$$\sum Y = 21 + 32 + 23 + 41 + 20 + 30 + 36 + 22 + 25 + 27 = 277$$

Thus, the mean age is

$$\bar{Y} = \frac{\sum Y}{N} = \frac{277}{10} = 27.7$$

The mean can also be calculated when the data are arranged in a frequency distribution. We have presented an example involving a frequency distribution in A Closer Look 4.1.

▣ A Closer Look 4.1
Finding the Mean in a Frequency Distribution

When data are arranged in a frequency distribution, we must give each score its proper weight by multiplying it by its frequency. We can use the following modified formula to calculate the mean:

$$\bar{Y} = \frac{\sum fY}{N}$$

where

Y = the raw scores of the variable Y
$\bar{Y}$ = the mean of Y
$\sum fY$ = the sum of all the fYs
N = the number of observations or cases

We now illustrate how to calculate the mean from a frequency distribution using the preceding formula. In the 2002 General Social Survey respondents were asked what they think is the ideal number of children for a family. Their responses are presented in the following table.

Ideal Number of Children: GSS 2002

Number of Children (Y)	Frequency (f)	Frequency × Y (fY)
0	13	0
1	29	29
2	447	894
3	223	669
4	81	324
5	16	80
6	5	30
7	1	7
Total	$N = 815$	$\sum fY = 2033$

Notice that to calculate the value of $\sum fY$ (column 3), each score (column 1) is multiplied by its frequency (column 2), and the products are then added together. When we apply the formula

$$\bar{Y} = \frac{\sum fY}{N} = \frac{2033}{815} = 2.5$$

we find that the mean for the ideal number of children is 2.5

✓ **Learning Check.**   If you are having difficulty understanding how to find the mean in a frequency distribution, examine this table. It explains the process without using any notation.

Finding the Mean in a Frequency Distribution

	Number of people per house	Number of houses like this	Number of people such houses contribute
	1	3	3
	2	5	10
	3	1	3
	4	1	4

Total number of people: 20
Total number of houses: 10
Mean number of people per house: 20/10 = 2

(Continued)

Here is another example. The following tables are frequency distributions of years of education for Mexican and American respondents. Try to find the mean level of education for each of the two groups by applying the formula for calculating the mean in a frequency distribution.

Years of Education for Mexican Respondents: ISSP 2000

Education (Y)	Frequency (f)	Frequency $\times$ Y (fY)
1	20	20
2	35	70
3	65	195
4	36	144
5	29	145
6	240	1440
7	14	98
8	31	248
9	229	2061
10	18	180
11	32	352
12	137	1644
13	29	377
14	19	266
15	34	510
16	51	816
17	49	833
18	16	288
19	10	190
20	8	160
21	4	84
22	3	66
25	1	25
Total	$N = 1{,}110$	$\Sigma fY = 10{,}212$
	$\overline{Y} = 9.2$	

Years of Education for American Respondents: ISSP 2000

Education (Y)	Frequency (f)	Frequency $\times$ Y (fY)
2	1	2
3	2	6
4	2	8
5	5	25
6	4	24
7	8	56

8	48	384
9	31	279
10	53	530
11	84	924
12	373	4476
13	132	1716
14	145	2030
15	60	900
16	190	3040
17	37	629
18	50	900
19	18	342
20	30	600
Total	$N = 1{,}273$	$\Sigma fY = 16{,}871$
	$\bar{Y} = 13.3$	

Examine the tables showing years of education for Mexican and American respondents. Note the similarities and differences between the two groups. Education is a major component of social class. You may want to take your analysis one step further. SPSS Problem 3 at the end of this chapter provides specific instructions that will help you explore the relationship between social class and the number of children that couples decided to have.

✔ **Learning Check.**   *The following distribution is the same as the one you used to calculate the median in an earlier Learning Check: 22, 15, 18, 33, 17, 5, 11, 28, 40, 19, 8, 20. Can you calculate the mean? Is it the same as the median, or is it different?*

Understanding Some Important Properties of the Arithmetic Mean

The following three mathematical properties make the mean the most important measure of central tendency. It is, in fact, a concept that is basic to numerous and more complex statistical operations.

Interval-Ratio Level of Measurement   Because it requires the mathematical operations of addition and division, the mean can be calculated only for variables measured at the interval-ratio level. This is the only level of measurement that provides numbers that can be added and divided.

Center of Gravity   Because the mean (unlike the mode and the median) incorporates all the scores in the distribution, we can think of it as the center of gravity of the distribution. That is, the mean is the point that perfectly balances all the scores in the distribution. If we subtract the mean from each score and add up all the differences, the sum will always be zero!

✓ **Learning Check.**   *Why is the mean considered the center of gravity of the distribution? Think of the last time you were in a park on a seesaw (it may have been a long time ago) with a friend who was much heavier than you. You were left hanging in the air until your friend moved closer to the center. In short, to balance the seesaw a light person far away from the center (the mean) can balance a heavier person who is closer to the center. Can you illustrate this principle with a simple income distribution?*

Illustrating the Seesaw Principle

a. Three people, weights 60, 120, and 180, all stand on a seesaw. The fulcrum is placed at 120. The mean is (60 + 120 + 180)/3. The seesaw balances.

060 070 080 090 100 110 120 130 140 150 160 170 180 190 200 210 220 230 240

b. The 180-pound person is replaced by a 240-pound person, but we do not move the fulcrum. The seesaw slowly falls to the right.

060 070 080 090 100 110 120 130 140 150 160 170 180 190 200 210 220 230 240

c. We move the fulcrum to 140. The new mean is (60 + 120 + 240)/3. The seesaw balances again.

060 070 080 090 100 110 120 130 140 150 160 170 180 190 200 210 220 230 240

Sensitivity to Extremes   The examples we have used to show how to compute the mean demonstrate that, unlike with the mode or the median, every score enters into the calculation of the mean. This property makes the mean sensitive to extreme scores in the distribution. The mean is pulled in the direction of either very high or very low values. A glance at Figure 4.5 should convince you of that. Figures 4.5a and 4.5b each show the incomes of 10 individuals. In Figure 4.5b, the income of one individual has shifted from $5,000 to $35,000. Notice the effect it has on the mean; it shifts from $3,000 to $6,000! The mean is disproportionately

Figure 4.5   The Value of the Mean Is Affected by Extreme Scores

a. No extreme scores: the mean is $3,000

Income (Y)		Frequency (f)	fY
1,000	👤	1	1,000
2,000	👤👤	2	4,000
3,000	👤👤👤👤	4	12,000
4,000	👤👤	2	8,000
5,000	👤	1	5,000
		$N = 10$	$\Sigma fY = 30,000$

$$\text{Mean} = \frac{\Sigma fY}{N} = \frac{30,000}{10} = \$3,000$$

Median = $3,000

b. One extreme score: the mean is $6,000

Income (Y)		Frequency (f)	fY
1,000	👤	1	1,000
2,000	👤👤	2	4,000
3,000	👤👤👤👤	4	12,000
4,000	👤👤	2	8,000
35,000	👤	1	35,000
		$N = 10$	$\Sigma fY = 60,000$

$$\text{Mean} = \frac{\Sigma fY}{N} = \frac{60,000}{10} = \$6,000$$

Median = $3,000

affected by the relatively high income of $35,000 and is misleading as a measure of central tendency for this distribution. Notice that the median's value is not affected by this extreme score; it remains at $3,000. Thus, the median gives us better information on the typical income for this group. In the next section we will see that because of the sensitivity of the mean, it is not suitable as a measure of central tendency in distributions that have a few very extreme values on one side of the distribution. (A few extreme values are no problem if they are not mostly on one side of the distribution.)

✔ *Learning Check.*   *When asked to choose the appropriate measure of central tendency for a distribution, remember that the level of measurement is not the only consideration. When variables are measured at the interval-ratio level, the mean is usually the measure of choice, but remember that extreme scores in one direction make the mean unrepresentative and the median or mode may be the better choice.*

◙ THE SHAPE OF THE DISTRIBUTION

In this chapter we have looked at the way in which the mode, median, and mean reflect central tendencies in the distribution. Distributions (this discussion is limited to distributions of interval-ratio variables) can also be described by their general shape, which can be easily represented visually. A distribution can be either symmetrical or skewed, depending on whether there are a few extreme values at one end of the distribution.

A distribution is **symmetrical** (Figure 4.6a) if the frequencies at the right and left tails of the distribution are identical, so that if it is divided into two halves, each will be the mirror image of the other. In a unimodal, symmetrical distribution the mean, median, and mode are identical.

In **skewed** distributions, there are a few extreme values on one side of the distribution. Distributions that have a few extremely low values are referred to as **negatively** skewed (Figure 4.6b). A defining feature of a negatively skewed distribution is that the value of the

Figure 4.6   Types of Frequency Distributions

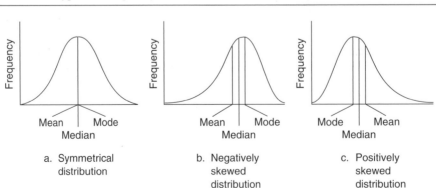

a. Symmetrical distribution

b. Negatively skewed distribution

c. Positively skewed distribution

mean is always less than the value of the median; in other words, the mean is pulled in the direction of the lower scores. Alternatively, distributions with a few extremely high values are said to be **positively** skewed (Figure 4.6c). A defining feature of a positively skewed distribution is that the value of the mean is always greater than the value of the median; the mean is pulled toward the higher scores.

We can illustrate the differences among these three types of distributions by examining three variables in the 2002 GSS.[6] The frequency distributions for these three variables are presented in Tables 4.5 through 4.7, and the corresponding graphs are depicted in Figures 4.7 through 4.9.

Symmetrical distribution   The frequencies at the right and left tails of the distribution are identical; each half of the distribution is the mirror image of the other.

Skewed distribution   A distribution with a few extreme values on one side of the distribution.

Negatively skewed distribution   A distribution with a few extremely low values.

Positively skewed distribution   A distribution with a few extremely high values.

The Symmetrical Distribution

First, let's examine Table 4.5 and Figure 4.7, displaying the distribution of the number of hours per day spent watching television. Notice that the largest number (130) watch television 2 hours per day (mode = 2.0), and about an equal number (104 and 98, respectively) reported either 1 or 3 hours watching television per day. As shown in Figure 4.7, the mode, median, and mean are almost identical, and they coincide at about the middle of the distribution.

Table 4.5   Hours Spent Watching Television

Hours Spent Watching TV	*Frequency (f)*	*fY*	*Percentage*	*C%*
1	104	104	31.3	31.3
2	130	260	39.2	70.5
3	98	294	29.5	100.0
Total	332	658	100.0	

$$\bar{Y} = \frac{\sum fY}{N} = \frac{658}{332} = 1.98$$

Median = 2.0
Mode = 2.0

---

[6]Two variables, TVHOURS and EDUC, were taken from a GSS sample. EDUC was then recoded into two distinct variables.

Figure 4.7   Hours Spent Watching Television

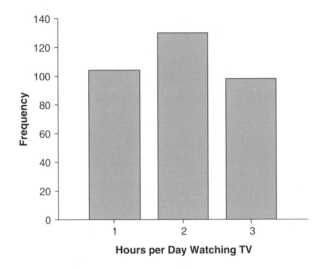

The distribution of number of hours spent per week watching television in Table 4.5 and Figure 4.7 is a nearly symmetrical distribution with the mean, median, and mode being almost identical.

The Positively Skewed Distribution

Now let's examine Table 4.6 and Figure 4.8, displaying the distribution of the number of years respondents spent in school once they earned a high school diploma. Note that the largest number of respondents (207) is concentrated at the low end of the scale (2 years), with few people reporting a high number of years (8 years). Notice also that in this distribution the mean, median, and mode have different values, with the mean having the highest value (mean = 3.31), the median having the second highest value (median = 3.0), and the mode the lowest value (mode = 2.0).

The distribution as depicted in Table 4.6 and Figure 4.8 is positively skewed. As a general rule, for skewed distributions the mean, median, and mode do not coincide. The mean, which is always pulled in the direction of extreme scores, falls closest to the tail of the distribution where a small number of extreme scores are located.

The Negatively Skewed Distribution

Now examine Table 4.7 and Figure 4.9 for the number of years spent in school among those respondents who did not finish high school. Here you can see the opposite pattern. The distribution of the number of years spent in school for those without a high school diploma is a negatively skewed distribution. First, note that the largest number of years spent in school are concentrated at the high end of the scale (11 years) and that there are fewer respondents at the low end. The mean, median, and mode also differ in values as they

Table 4.6   Years in School Post High School

Years in School Post High School	Frequency (f)	fY	Percentage	C%
1	148	148	18.4	18.4
2	207	414	25.7	44.1
3	83	249	10.3	54.4
4	196	784	24.3	78.7
5	43	215	5.3	84.0
6	74	444	9.2	93.2
7	23	161	2.9	96
8	32	256	4.0	100.1
Total	806	2671	100.1*	

$$\bar{Y} = \frac{\sum fY}{N} = \frac{2671}{806} = 3.31$$

Median = 3.0
Mode = 2.0

*Answer varies slightly from 100.0 due to rounding.

Figure 4.8   Years in School Post High School

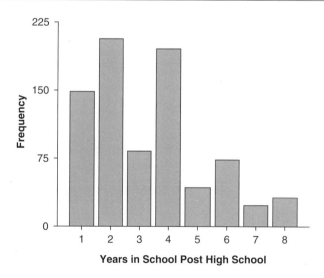

Years in School Post High School

did in the previous example. However, here the mode has the highest value (mode = 11.0), the median has the second highest (median = 10.0), and the mean has the lowest value (mean = 8.91).

Table 4.7   Years of School Among Respondents Without a High School Diploma

Years of School	Frequency (f)	fY	Percentage	C%
3 or fewer	15	45	6.2	6.2
4	5	20	2.1	8.3
5	7	35	2.9	11.2
6	14	84	5.8	17.0
7	9	63	3.7	20.7
8	32	256	13.2	33.9
9	29	261	11.9	45.8
10	51	510	21.0	66.8
11	81	891	33.3	100.1*
Total	243	2165	100.1*	

$$\bar{Y} = \frac{\sum fY}{N} = \frac{2165}{243} = 8.91$$

Median = 10.0
Mode = 11.0

*Answer varies slightly from 100.0 due to rounding.

Figure 4.9   Years of School Among Respondents Without a High School Diploma

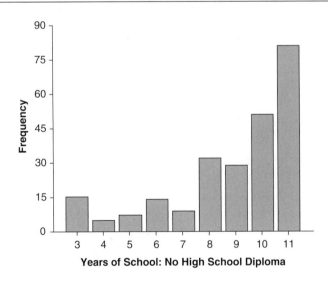

Guidelines for Identifying the Shape of a Distribution

Following are some useful guidelines for identifying the shape of a distribution.

1. In unimodal distributions, when the mode, median, and mean coincide or are almost identical, the distribution is symmetrical.

2. When the mean is higher than the median (or is positioned to the right of the median), the distribution is positively skewed.

3. When the mean is lower than the median (or is positioned to the left of the median), the distribution is negatively skewed.

✔ *Learning Check.*   *To identify positively and negatively skewed distributions, look at the tail on the chart. If the tail points to the right (the positive end of the x-axis), the distribution is positively skewed. If the tail points to the left (the negative, or potentially negative, end of the x-axis), the distribution is negatively skewed.*

▣ CONSIDERATIONS FOR CHOOSING A MEASURE OF CENTRAL TENDENCY

So far we have considered three basic kinds of averages: the mode, the median, and the mean. Each can represent the central tendency of a distribution. But which one should we use? The mode? The median? The mean? Or, perhaps, all of them? There is no simple answer to this question. However, in general, we tend to use only one of the three measures of central tendency, and the choice of the appropriate one involves a number of considerations. These considerations and how they affect our choice of the appropriate measure are presented in the form of a decision tree in Figure 4.10.

Level of Measurement

One of the most basic considerations in choosing a measure of central tendency is the variable's level of measurement. The valid use of any of the three measures requires that the data be measured at the level appropriate for that measure or higher. Thus, as shown in Figure 4.10, with nominal variables our choice is restricted to the mode as a measure of central tendency.

However, with ordinal data we have two choices: the mode or the median (or sometimes both). Our choice depends on what we want to know about the distribution. If we are interested in showing what is the most common or typical value in the distribution, then our choice

Figure 4.10 How to Choose a Measure of Central Tendency

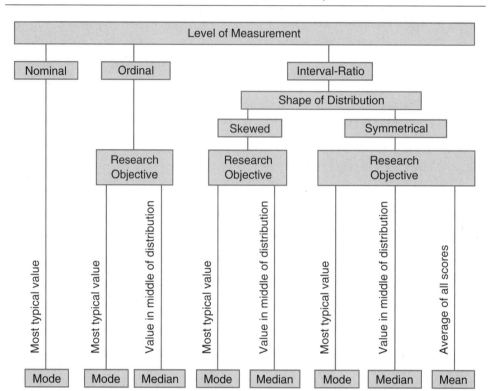

is the mode. If, however, we want to show which value is located exactly in the middle of the distribution, then the median is our measure of choice.

When the data are measured on an interval-ratio level, the choice between the appropriate measures is a bit more complex and is restricted by the shape of the distribution.

Skewed Distribution

When the distribution is skewed, the mean may give misleading information on the central tendency because its value is affected by extreme scores in the distribution. The median (see, for example, A Closer Look 4.2) or the mode can be chosen as the preferred measure of central tendency because neither is influenced by extreme scores. For instance, when examining the number of years that GSS respondents have spent in school after graduating from high school (Table 4.6, Figure 4.8), the mean does not provide as accurate a representation of the "typical" number of years that an individual has spent in school post high school as the median and the mode. Thus, either one could be used as an "average," depending on the research objective.

Symmetrical Distribution

When the distribution we want to analyze is symmetrical, we can use any of the three averages. Again, our choice depends on the research objective and what we want to know about the distribution. In general, however, the mean is our best choice because it contains the greatest amount of information and is easier to use in more advanced statistical analyses.

▣ **A Closer Look 4.2**
Statistics in Practice: Median Annual Earnings Among Ethnic/Racial Subgroups

Personal income is frequently positively skewed because there are fewer people with high income; therefore, studies on earnings often report median income. The mean tends to overestimate both the earnings of the most typical earner (the mode) and the earnings represented by the 50th percentile (the median). In the following example the median is used to compare annual household earnings for whites, blacks, and Latinos.

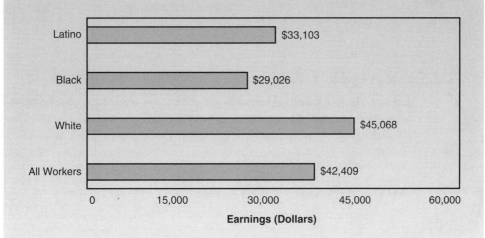

The graph compares the 2002 median annual earnings of white, black and Latino households in the United States. Since the earnings of whites are the highest in comparison with all other groups, it is useful to look at each group's median earnings relative to the earnings of whites. For example, blacks were paid just 64 cents ($29,026/$45,068) and Latinos ($33,103/$45,068) were paid 73 cents for every $1 paid to whites.

Source: U.S. Bureau of the Census, Current Population Reports, *Income in the United States, 2003,* Table A-1

MAIN POINTS

- The mode, the median, and the mean are measures of central tendency—numbers that describe what is average or typical about the distribution.

- The mode is the category or score with the largest frequency (or percentage) in the distribution. It is often used to describe the most commonly occurring category of a nominal level variable.

- The median is a measure of central tendency that represents the exact middle of the distribution. It is calculated for variables measured on at least an ordinal level of measurement.

- The mean is typically used to describe central tendency in interval-ratio variables, such as income, age, or education. We obtain the mean by summing all the scores and dividing by the total (N) number of scores.

- In a symmetrical distribution the frequencies at the right and left tails of the distribution are identical. In skewed distributions there are either a few extremely high (positive skew) or a few extremely low (negative skew) values.

KEY TERMS

measures of central tendency
mean
median
mode
negatively skewed distribution

percentile
positively skewed distribution
skewed distribution
symmetrical distribution

ON YOUR OWN

Log on to the web-based student study site at http://www.pineforge.com/frankfort-nachmiasstudy4 for additional study questions, quizzes, web resources, and links to social science journal articles reflecting the statistics used in this chapter.

SPSS DEMONSTRATIONS

[GSS02PFP-A]

Demonstration 1:
Producing Measures of Central Tendency with Frequencies

The Frequencies command, which we demonstrated in Chapter 2, also has the ability to produce the three measures of central tendency discussed in this chapter. We will use Frequencies to calculate measures of central tendency for AGE and FEFAM ("It is better for the men to work, and women to tend the home").

Click on *Analyze, Descriptive Statistics*, then *Frequencies*. Place AGE and FEFAM in the Variable(s) box. Then click on the *Statistics* button. You will see that the Central Tendency box (Figure 4.11) lists four choices, but we will click on only the first three. Then click on *Continue*, then on *OK* to process this request.

The frequency table for AGE is quite lengthy, so only the Statistics box is displayed here (Figure 4.12). AGE is an interval-ratio variable, which means that the mode, median, and mean are all

Figure 4.11

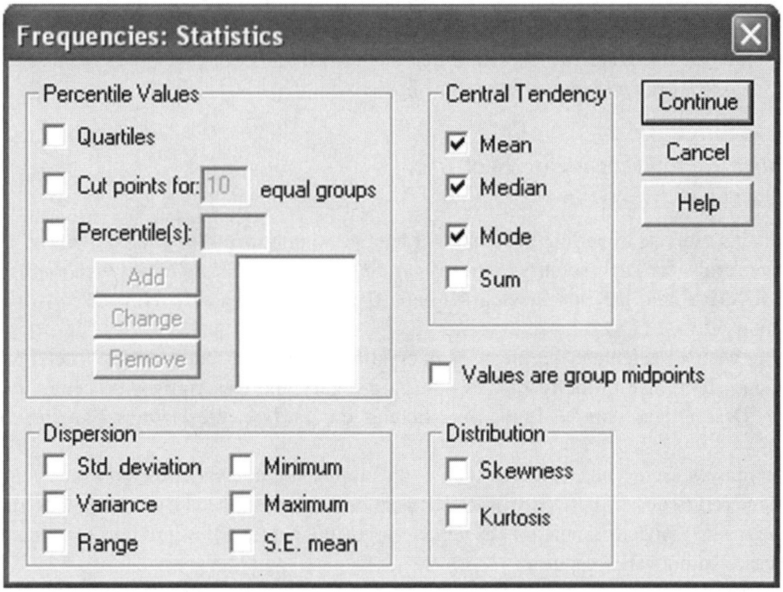

Figure 4.12

		Age of Respondent	FEFAM Better for Man to Work, Woman Tend Home
N	Valid	1496	498
	Missing	4	1002
Mean		46.21	270
Median		44.00	3.00
Mode		34	3

appropriate measures of central tendency. The mean of AGE is 46.21, but remember that the GSS file includes only adults, so the mean value (and the other measures of central tendency) should not be taken as representative of the American population as a whole.

The median for the variable AGE is 44, which is close to the mean value. Since we know the value of the median, we can say that roughly half of the respondents are older than 44 and half are younger than 44. In addition, since the median is less than the mean (44.00 < 46.21), the distribution of AGE is positively skewed. Finally, the mode is 34, so more people in this distribution are 34 years old than any other age.

In the same Statistics box, measures of central tendency are also displayed for FEFAM. SPSS has no idea that this variable is measured on an ordinal scale. In other words, SPSS produces exactly the output we asked for, without regard for whether the output is correct for this type of variable. It's up to you to select the proper measure of central tendency.

Since FEFAM is an ordinal variable, we can use the median and mode to summarize its distribution. The median is 3.0, which corresponds to the response "disagree." We can confirm from the cumulative percentages (not shown) that about 61 percent either "strongly disagree" or "disagree" with this statement; the remaining 39 percent (which is not shown here) are either neutral or "agree" or "strongly agree." The mode is "disagree," which means that the most frequent response was to "disagree" that it is better for men to work and women to tend the home.

Demonstration 2: Producing Measures of
Central Tendency with Descriptives [ISSP00PFP]

We begin this exercise by telling the computer that we want our results split by country. That is, we want separate results for each country. Select *Data, Split File, Organize Output by Groups*. Insert the variable COUNTRY into the box labeled "Groups Based on." Click *OK*. Now SPSS will filter our results by country.

When you want to calculate the mean of interval-ratio variables but you don't need to view the actual frequency table listing the responses in each category, the Descriptives procedure is often the best choice. Descriptives can be found by clicking on *Analyze, Descriptive Statistics,* and then *Descriptives*.

The Descriptives dialog box (Figure 4.13) is uncomplicated and requires only that you place the variables of interest (AGE, EDUCYRS, WRKHRS) in the Variable(s) box. By default, Descriptives will calculate the mean, standard deviation (to be discussed in Chapter 5), minimum, maximum, and the number of cases with a valid response.

You will need to scroll through your SPSS output to located the descriptive statistics for U.S. respondents. Figure 4.14 displays these descriptive statistics for AGE, EDUCYRS (years of education), and WRKHRS (hours worked per week) for U.S. ISSP respondents in 2000.

The output from Descriptives automatically lists the variables in the order we specified in the dialog box. Based on the output, we can determine that on average, U.S. respondents were 47 years old, had approximately 13 years of education, and worked about 40 hours per week.

Figure 4.13

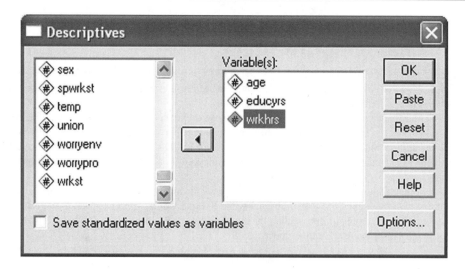

Figure 4.14

Descriptive Statistics

	N	Minimum	Maximum	Mean	Std.Deviation
V201 AGE R: Age	49	19	85	46.67	16.307
V204 EDUCYRS R: Education: years in school	49	8	20	13.29	2.264
V238 WRKHRS R: Hours worked weekly	32	10	70	40.25	11.438
Valid N (listwise)	32				

SPSS PROBLEMS

[GSS02PFP-A].

1. Create a frequency distribution, including any appropriate measures of central tendency, for HOMOSEX.
   a. Which measure of central tendency, mean or median, is most appropriate to summarize the distribution of HOMOSEX? Explain why.
   b. Suppose we are interested in whether males and females have the same attitudes about homosexual relations? Create a frequency distribution, including any appropriate measures of central tendency, for HOMOSEX, this time separating results for men and women. (Use the *Data–Split File* command by clicking on *Data*, *Split File*, *Organize Output by Groups*, insert the variable SEX into the box labeled "Groups Based on," and click *OK*. Are there any differences in their measures of central tendency? Explain. (Remember, once you have completed this exercise, reset the Split File command to include all cases by clicking on *Data*, *Split File*, *Analyze All Cases, Do Not Create Groups*.)
   c. Repeat b, this time using PREMARSX (acceptance of premarital sex) as your variable.

2. We are interested in investigating whether males and females have equal levels of education. Use the variable EDUC with the Frequencies procedure to produce frequency tables and the mean, median, and mode separately for males and females (as described in 1b).
   a. On average, do men and women have equal levels of education? Use all the available information to answer this question.
   b. When we use statistics to describe the social world, we should always go beyond merely using statistics to describe the condition of various social groups. Just as important is our interpretation of the statistics and some judgment as to whether any differences we find between groups seem of practical importance; that is, do they make a practical difference in the world? Do you think any differences you discovered between male and female educational levels are important enough to have an effect on such things as the ability to get a job or the salary that someone makes? Defend your answer.

3. Some people believe that social class influences the number of children the couples decided to have. Use SPSS to investigate this question with either GSS data file. (The variable CHILDS records the respondent's number of children.) To get the necessary information, have SPSS split the file by CLASS, and then run Frequencies for CHILDS.
   a. What is the best measure of central tendency to represent the number of children in a household? Why?

    b. Which social class has more children per respondent?

    c. Rerun your analysis, this time with the variable CHLDIDEL (ideal number of children). Is there a difference among the social class categories? Explain.

4. Picking an appropriate statistic to describe the central tendency of a distribution is a critical skill. Based on the GSS02PFP-B, determine the appropriate measure(s) of central tendency for the following variables:

    a. How often do respondents watch television? [TVHOURS]

    b. Respondents' political views. [POLVIEWS]

    c. The marital status of those sampled. [MARITAL]

    d. Whether respondents have a gun in their home. [OWNGUN]

    e. Number of brothers and sisters of those sampled [SIBS]?

    f. Does respondent support legalization of marijuana? [GRASS]

5. The educational attainment of Mexican and American respondents was compared in A Closer Look 4.1. You can use SPSS to do similar comparisons with other variables. For example, it may be interesting to look at levels of marital satisfaction and compare the number of hours that men and women spend on the Internet each week. There is more than one method to get the frequency distributions of Internet hours per week for these groups, but the easiest might be to use the *Split File* command (as described in 1b). Use the GSS02PFP-A for this exercise.

    Click on *Data, Split File, Organize Output by Groups,* place the variables HAPMAR and SEX (in that order) in the "Groups Based on" box, and then click *OK.* HAPMAR has three categories of interest: 1 = very happy, 2 = pretty happy, 3 = not too happy. SEX has two valid values, with male = 1 and female = 2.

    Essentially, you have told SPSS to create a separate set of output for each group defined by the combination of the values of HAPMAR and SEX.

    a. Create a frequency distribution by clicking on *Analyze, Descriptive Statistics, Frequencies.* Select the variable WWWHR, as well as all appropriate measures of central tendency under the *Statistics* option. Click *OK.* SPSS will create a great deal of output; all you need to do is find the appropriate frequency tables and measures of central tendency. For example, to find and report the frequency table for males who are pretty happy, look for the section with values of HAPMAR = 2 and SEX = 1.

    b. Do you notice any gaps in Internet hours between men and women at different levels of marital satisfaction? How about just for women at various levels of marital satisfaction?

CHAPTER EXERCISES

1. The following frequency distribution contains information about attitudes on the genetic manipulation of certain food crops.

Genetic Manipulation of Crops	Frequency	Percent	Cumulative Percent
1 Extremely dangerous	4851	19.4	19.4
2 Very dangerous	7103	28.5	47.9
3 Somewhat dangerous	7570	30.3	78.2
4 Not very dangerous	4404	17.6	95.8
5 Not dangerous at all	1029	4.1	99.9*
Total	24957	99.9*	

Source: International Social Survey Programme 2000.

*Answer varies slightly from 100.0 due to rounding.

a. Find the mode.
b. Find the median.
c. Interpret the mode and the median.
d. Why would you not want to report the mean for this variable?

2. Same-sex unions have increasingly become a heated political issue. The 2002 General Social Survey asked respondents' opinions on homosexual relations. Five response categories ranged from "Always Wrong" to "Not Wrong at All." See the following frequency distribution:

Homosexual Relations	Frequency	Percent	Cumulative Percent
1 Always Wrong	486	55.0	55.0
2 Almost Always Wrong	43	4.9	59.9
3 Sometimes Wrong	63	7.1	67.0
4 Not Wrong at All	292	33.0	100.0
Total	884	100.0	

a. At what level is this variable measured? What is the mode for this variable?
b. Calculate the median for this variable. In general, how would you characterize the public's attitude about homosexual relations?

3. The following frequency distribution contains information on the number of hours worked per week for a sample of 75 Japanese adults.

Hours Worked per Week	Frequency	Percent	Cumulative Percent
7	1	1.3	1.3
12	1	1.3	2.6
15	2	2.7	5.3
20	3	4.0	9.3
24	1	1.3	10.6
30	7	9.3	19.9
32	1	1.3	21.2
35	2	2.7	23.9
36	2	2.7	26.6
38	2	2.7	29.3
40	13	17.3	46.6
41	1	1.3	47.9
42	2	2.7	50.6
43	1	1.3	51.9
45	7	9.3	61.2
48	2	2.7	63.9
50	14	18.7	82.6
54	2	2.7	85.3
55	1	1.3	86.6
60	6	8.0	94.6
65	1	1.3	95.9
66	1	1.3	97.2
68	1	1.3	98.5
72	1	1.5	100.0
Total	75	100.0	

Source: International Social Survey Programme 2000.

a. What is the level of measurement for "hours worked weekly"? What is the mode for hours worked weekly? What is the median for hours worked weekly?

b. Construct quartiles for hours worked weekly. What is the 25th percentile? The 50th percentile? The 75th percentile? Why don't you need to calculate the 50th percentile to answer this question?

4. Using a random sample from the 2002 General Social Survey, you find the following grouped distribution for respondent's age:

Age Category	Frequency
18–29	54
30–39	54
40–49	60
50–59	36
60–69	31
70–89	27

a. Calculate the median age category.
b. Also, report which age category contains the 20th and 80th percentiles.

5. Religion has been and continues to be important to many Americans. There are demographic differences in religious behavior, including age. The following table, taken from the 2002 General Social Survey, depicts how often people pray within various age groups (not all ages are displayed).

| Prayer Frequency | Age Group | | | |
	18–29	30–39	40–49	50–59
Several times a day	42	61	67	66
Once a day	69	74	93	72
Several times a week	30	47	36	18
Once a week	28	27	19	19
Less than once a week	89	66	50	41
Never	2	2	1	0

a. Calculate the median and mode for each age group.
b. Use this information to characterize how prayer behavior varies by age. Does the median or mode provide a better description of the data? Do the statistics support the idea that there is a prayer "generation gap," such that some age groups engage in more prayer than others?

6. AIDS is a serious health problem for this country (and many others). Data from the Centers for Disease Control and Prevention for 2002 and 2003 show the number of AIDS cases in various metropolitan statistical areas (MSAs) (the top 10 MSAs are listed).

Metropolitan Statistical Area	2002	2003
New York	5,649	5,580
Miami	1,139	1,072
Newark	569	534

Washington, DC	1,832	1,743
Baltimore	1,257	1,028
Chicago	1,849	1,527
Los Angeles	1,549	2,558
Boston	721	664
Fort Lauderdale	750	690
Philadelphia	1,411	1,288

Source: Data adapted from Centers for Disease Control and Prevention, *HIV/AIDS Surveillance Report*, Vol. 15 (2004), Table 15.

   a. Calculate the mean number of AIDS cases in these urban areas for both 2002 and 2003. How would you characterize the difference in the number of AIDS cases between 2002 and 2003? Does the mean adequately represent the central tendency of the distribution of AIDS cases in each year? Why or why not?
   b. Recalculate the mean for each year after removing the New York MSA. Is the mean now a better representation of central tendency for the remaining nine MSAs? Explain.

7. U.S. households have become smaller over the years. The following table from the 2002 General Social Survey contains information on the number of people currently aged 18 years or older living in a respondent's household.

Calculate the mean number of people living in a U.S. household in 2002.

Household Size	*Frequency*
1	870
2	941
3	454
4	291
5	144
6	41
7	16
8	3
9	3
10	2
Total	2765

8. In Exercise 6 you calculated the mean number of AIDS cases. We now want to test whether the distribution of AIDS cases is symmetrical or skewed.
   a. Calculate the median and mode for each year, using all MSAs. Based on these results and the means, how would you characterize the distribution of AIDS cases for each year?
   b. What value best represents the central tendency of each distribution?
   c. If you found the distributions to be skewed, what is the statistical cause?

9. In Exercise 7 you examined U.S. household size in 2002. Using these data, construct a histogram to represent the distribution of household size.
   a. From the appearance of the bar chart, would you say the distribution is positively or negatively skewed? Why?
   b. Now calculate the median for the distribution and compare this value with the value of the mean from question 7. Do these numbers provide further evidence to support your decision about how the distribution is skewed? Why do you think the distribution of household size is asymmetrical?

10. Exercise 3 used ISSP data on the number of hours worked per week for a sample of 75 Japanese adults.
    a. Calculate the mean for hours worked per week.
    b. Compare the value of the mean with those you have already calculated for the median. Without constructing a histogram, describe whether and how the distribution of hours worked per week is skewed.

11. You listen to a debate between two politicians discussing the economic health of the United States. One politician says that the average income of U.S. adults is $37,000; the other says that the average American makes only $30,000, so Americans are not as well off as the first politician claims. Is it possible for both these politicians to be correct? If so, explain how.

12. Discuss the advantages and disadvantages of all three measures of central tendency. Are you confident that one of these three is the best measure of central tendency? If so, why?

13. Do murder rates in cities vary with city size? Investigate this question using the following data for selected U.S. cities grouped by population size, the top 10 cities and the bottom 10 (all among the largest U.S. cities). Calculate the mean and median for each group of cities. Where is the murder rate highest? Do the mean and median have the same pattern for the two groups?

Murder Rate per 100,000 in 2001			
Top 10 by Population	*Murder Rate*	*Bottom 10 by Population*	*Murder Rate*
New York	8.2	Aurora, CO	6.0
Los Angeles	15.6	Corpus Christi, TX	6.7
Chicago	22.9	Raleigh, NC	3.6
Houston	13.4	Newark, NJ	32.6
Philadelphia	20.4	Anchorage, AK	3.8
Phoenix	15.3	Lexington, KY	9.2
San Diego	4.0	Riverside, CA	7.7
Dallas	19.7	Louisville, KY	9.7
San Antonio	8.5	Mobile, AL	16.4
Las Vegas	11.9	St. Petersburg, FL	8.2

Source: U.S. Bureau of the Census, *Statistical Abstract of the United States*, 2003, Table 308.

14. When the United States declared war on Iraq in 2003 the German government was quick to denounce the actions of the United States and insist upon an international agreement in the form of a resolution backed by the United Nations. Imagine that you are interested in understanding American and German attitudes on forming international agreements. The following table contains information on the attitudes of American and German citizens.

International Agreements Should Be Made	*United States*	*Germany*
Strongly agree	278	950
Agree	619	447
Neither agree nor disagree	166	32
Disagree	60	5
Strongly disagree	9	0
Total	1,132	1,434

Source: International Social Survey Programme 2000

Calculate the appropriate measures of central tendency for Americans and Germans. How would you describe their positions on the issue of whether international agreements should be made?

15. The following table shows the number of marriages and divorces for a sample of respondents from seven countries.

Country	Number of Marriages	Number of Divorces
Canada	813	58
England	548	78
Japan	822	26
Mexico	816	41
Netherlands	937	108
Russia	964	144
United States	570	200

Source: Adapted from the International Social Survey Programme 2000

a. Calculate the mean number of marriages and divorces for the seven countries.
b. Now let's drop one of the outliers and see how the results change. Recalculate the mean number of marriages without England. Recalculate the number of divorces without the United States. Explain how your mean numbers have changed.

Measures of Variability

I n the previous chapter we looked at measures of central tendency: the mean, the median, and the mode. With these measures, we can use a single number to describe what is average for or typical of a distribution. Although measures of central tendency can be very helpful, they tell only part of the story. In fact, when used alone they may mislead rather than inform. Another way of summarizing a distribution of data is by selecting a single number that describes how much variation and diversity there is in the distribution. Numbers that describe diversity or variation are called measures of variability. Researchers often use measures of central tendency along with measures of variability to describe their data.

Measures of variability   Numbers that describe diversity or variability in the distribution.

In this chapter we discuss five measures of variability: the index of qualitative variation, the range, the interquartile range, the standard deviation, and the variance. Before we discuss these measures, let's explore why they are important.

▣ THE IMPORTANCE OF MEASURING VARIABILITY

The importance of looking at variation and diversity can be illustrated by thinking about the differences in the experiences of U.S. women. Are women united by their similarities or divided by their differences? The answer is *both*. To address the similarities without dealing with differences is "to misunderstand and distort that which separates as well as that which binds women together."[1] Even when we focus on one particular group of women, it is important to look at the differences as well as the commonalities. Take, for example, Asian-American women. As a group they share a number of characteristics.

> Their participation in the workforce is higher than that of women in any other ethnic group. Many . . . live life supporting others, often allowing their lives to be subsumed by the needs of the extended family. . . . However, there are many circumstances when these shared experiences are not sufficient to accurately describe the condition of a particular Asian-American woman. Among Asian-American women there are those who were born in the United States . . . and . . . those who recently arrived in the United States. Asian-American women are diverse in their heritage or country of origin: China, Japan, the Philippines, Korea . . . and . . . India. . . . Although the majority of Asian-American women are working class—contrary to the stereotype of the "ever successful" Asians—there are poor, "middle-class," and even affluent Asian-American women.[2]

As this example illustrates, one basis of stereotyping is treating a group as if it were totally represented by its central value, ignoring the diversity within the group. Sociologists often contribute to this type of stereotyping when their empirical generalizations, based on a statistical difference between averages, are interpreted in an overly simplistic way. All this argues for the importance of using measures of variability as well as central tendency whenever we want to characterize or compare groups. Whereas the similarities and commonalities in the experiences of Asian-American women are depicted by a measure of central tendency, the diversity of their experiences can be described only by using measures of variation.

The concept of variability has implications not only for describing the diversity of social groups such as Asian-American women, but also for issues that are important in your everyday life. One of the most important issues facing the academic community is how to reconstruct the curriculum to make it more responsive to students' needs. Let's consider the issue of statistics instruction on the college level.

---

[1] Johnneta B. Cole, "Commonalities and Differences," in *Race, Class, and Gender*, ed. Margaret L. Andersen and Patricia Hill Collins (Belmont, CA: Wadsworth, 1998), pp. 128–129.
[2] Ibid., pp. 129–130.

Figure 5.1   Distribution of Grades for Professors Brown and Yamato's Statistics Classes

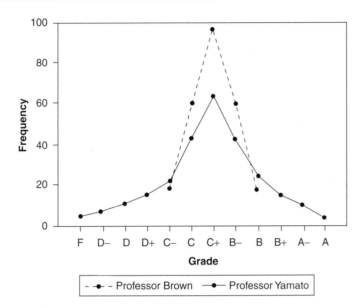

Statistics is perhaps the most anxiety-provoking course in any social science curriculum. Statistics courses are often the last "roadblock" preventing students from completing their major requirements. One factor, identified in numerous studies as a handicap for many students, is the "math anxiety syndrome." This anxiety often leads to a less than optimum learning environment, with students often trying to memorize every detail of a statistical procedure rather than understand the general concept involved.

Let's suppose that a university committee is examining the issue of how to better respond to the needs of students. In its attempt to evaluate statistics courses offered in different departments, the committee compares the grading policy in two courses. The first, offered in the sociology department, is taught by Professor Brown; the second, offered through the school of social work, is taught by Professor Yamato. The committee finds that over the years the average grade for Professor Brown's class has been C+. The average grade in Professor Yamato's class is also C+. We could easily be misled by these statistics into thinking that the grading policy of both instructors is about the same. However, we need to look more closely into how the grades are distributed in each of the classes. The differences in the distribution of grades are illustrated in Figure 5.1, which displays the frequency polygon for the two classes.

Compare the shapes of these two distributions. Notice that while both distributions have the same mean, they are shaped very differently. The grades in Professor Yamato's class are more spread out, ranging from A to F, while the grades for Professor Brown's class are clustered around the mean and range only from B to C–. Although the means for both distributions are identical, the grades in Professor Yamato's class vary considerably more than the grades given by Professor Brown. The comparison between the two classes is more complex than we first thought it would be.

As this example demonstrates, information on how scores are spread from the center of a distribution is as important as information about the central tendency in a distribution. This type of information is obtained by measures of variability.

✓ **Learning Check.** *Look closely at Figure 5.1. Whose class would you choose to take? If you were worried that you might fail statistics, your best bet would be Professor Brown's class where no one fails. However, if you want to keep up your GPA and are willing to work, Professor Yamato's class is the better choice. If you had to choose one of these classes based solely on the average grades, your choice would not be well informed.*

▣ THE INDEX OF QUALITATIVE VARIATION (IQV): A BRIEF INTRODUCTION

The United States is undergoing a demographic shift from a predominantly European population to one characterized by increased racial, ethnic, and cultural diversity. These changes challenge us to rethink every conceptualization of society based solely on the experiences of European populations and force us to ask questions that focus on the experiences of different racial/ethnic groups. For instance, we may want to compare the racial/ethnic diversity in different cities, regions, or states or to find out if a group has become more racially and ethnically diverse over time.

The **index of qualitative variation** (IQV) is a measure of variability for nominal variables like race and ethnicity. The index can vary from 0.00 to 1.00. When all the cases in the distribution are in one category, there is no variation (or diversity) and the IQV is 0.00. In contrast, when the cases in the distribution are distributed evenly across the categories, there is maximum variation (or diversity) and the IQV is 1.00.

Index of qualitative variation (IQV) A measure of variability for nominal variables. It is based on the ratio of the total number of differences in the distribution to the maximum number of possible differences within the same distribution.

Suppose you live in Maine, where the majority of residents are white and a small minority are Latino or Asian. Also suppose that your best friend lives in Hawaii, where the majority of the population are either Asians or Native Hawaiians. The distributions for these two states are presented in Table 5.1. Which is more diverse? Clearly, Hawaii, where the majority of the population are either Asians or Native Hawaiians, is more diverse than Maine, where Asians and Latinos are but a small minority. You can also get a visual feel for the relative diversity in the two states by examining the two bar charts presented in Figure 5.2.

Table 5.1   Top Three Racial/Ethnic Groups for Two States by Percentage

Race/Ethnic Group	Maine	Hawaii
White	98.4	33.5
Latino	0.8	n.d.
Asian	0.8	55.1
Native Hawaiian or Pacific Islander	n.d.	11.4
Total	100.0	100.0

Note: Tables include only the three largest racial/ethnic groups per state.

Source: U.S. Bureau of the Census, *Statistical Abstract of the United States*, 2003, Tables 21 & 22.

Figure 5.2   Top Three Racial/Ethnic Groups in Maine and Hawaii

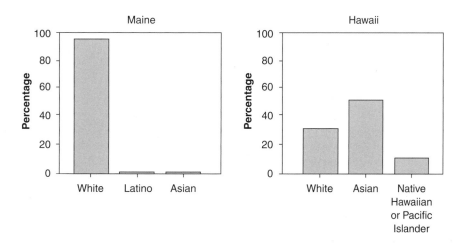

Steps for Calculating the IQV

To calculate the IQV, we use this formula:

$$\text{IQV} = \frac{K(100^2 - \sum Pct^2)}{100^2(K-1)} \tag{5.1}$$

where

K = the number of categories

$\sum Pct^2$ = the sum of all squared percentages in the distribution

Table 5.2   Squared Percentages for Three Racial/Ethnic Groups for Two States

| | Maine | | Hawaii | |
Race/Ethnic Group	%	(%)²	%	(%)²
White	98.4	9,682.56	33.5	1122.25
Latino	0.8	0.64	n.d.	n.d.
Asian	0.8	0.64	55.1	3036.01
Native Hawaiian or Pacific Islander	n.d.	n.d.	11.4	129.96
Total	100.0	9683.84	100.0	4288.22

In Table 5.2 we present the squared percentages for each racial/ethnic group for Maine and Hawaii.

The IQV for Maine is

$$\text{IQV} = \frac{K(100^2 - \sum Pct^2)}{100^2(K - 1)} = \frac{3(100^2 - 9,683.84)}{100^2(3 - 1)} = \frac{948.48}{20,000} = 0.05$$

The IQV for Hawaii is

$$\text{IQV} = \frac{K(100^2 - \sum Pct^2)}{100^2(K - 1)} = \frac{3(100^2 - 4,288.22)}{100^2(3 - 1)} = \frac{17,135.34}{20,000} = 0.86$$

Notice that the values of the IQV for the two states support our earlier observation: in Hawaii, where the IQV = 0.86, there is considerably more racial/ethnic variation than in Maine, where the IQV = 0.05.

It is important to remember that the IQV is partially a function of the number of categories. In this example, we used three racial/ethnic categories. Had we used, six categories, the IQV for both states would have been considerably less.

To summarize, these are the steps we follow to calculate the IQV:

1. Construct a percentage distribution.

2. Square the percentages for each category.

3. Sum the squared percentages.

4. Calculate the IQV using the formula

$$\text{IQV} = \frac{K(100^2 - \sum Pct^2)}{100^2(K - 1)}$$

◙ A Closer Look 5.1
Statistics in Practice: Diversity at Berkeley Through the Years*

"Berkeley, Calif.—The photograph in Sproul Hall of the 10 Cal 'yell leaders' from the early 1960s, in their Bermuda shorts and letter sweaters, leaps out like an artifact from an ancient civilization. They are all fresh-faced, and in a way that is unimaginable now, they are all white."[†]

On the flagship campus of the University of California system, the center of the affirmative action debate in higher education today, the ducktails and bouffant hairdos of those 1960s cheerleaders seem indeed out of date. The University of California's Berkeley campus was among the first of the nation's leading universities to embrace elements of affirmative action in its admission policies and now boasts that it has one of the most diverse campuses in the United States.

The following pie charts show the racial and ethnic breakdown of undergraduates at U.C. Berkeley for 1984 and 2003. The IQVs were calculated using the percentage distribution (as shown in the pie charts) for race and ethnicity for each year. Not only has the modal category of Berkeley's student body changed from white in 1984 to Asian in 2003, but the campus has become one of the most diverse in the United States.

Racial/Ethnic Composition of Student Body at U.C. Berkeley, 1984 and 2003

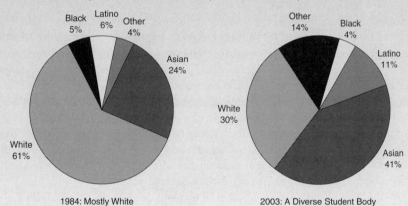

*Adapted from *The New York Times,* June 4, 1995. Copyright © 1995 by The New York Times Co. Reprinted by permission.

*2003 data are from the Office of Planning and Analysis, University of California, Berkeley. Cal Stats 2004.

[†]Ibid.

$$IQV_{1984} = \frac{K(100^2 - \sum Pct^2)}{100^2(K-1)} = \frac{5(100^2 - 4{,}374)}{100^2(5-1)} = \frac{28{,}130}{40{,}000} = 0.70$$

$$IQV_{2003} = \frac{K(100^2 - \sum Pct^2)}{100^2(K-1)} = \frac{5(100^2 - 2{,}914)}{100^2(5-1)} = \frac{35{,}430}{40{,}000} = 0.88$$

Expressing the IQV as a Percentage   The IQV can also be expressed as a percentage rather than a proportion: simply multiply the IQV by 100. Expressed as a percentage, the IQV would reflect the percentage of racial/ethnic differences relative to the maximum possible differences in each distribution. Thus, an IQV of 0.05 indicates that the number of racial/ethnic differences in Maine is 5.0 percent (0.05 × 100) of the maximum possible differences. Similarly, for Hawaii, an IQV of 0.86 means that the number of racial/ethnic differences is 86.0 percent (0.86 × 100) of the maximum possible differences.

✔ *Learning Check.*   *Examine A Closer Look 5.1 and consider the impact that the number of categories of a variable has on the IQV. What would happen to the Berkeley case if Asians were broken down into two categories with 20 percent in one and 21 percent in the other? (To answer this question you will need to recalculate the IQV with these new data.)*

Statistics in Practice: Diversity in U.S. Society

According to demographers' projections, by the middle of this century the United States will no longer be a predominantly white society. The combined population of the four largest minority groups—African Americans, Asian Americans, Latino Americans, and Native Americans—reached an estimated 89 million in 2002.[3] Population shifts during the 1990s indicate geographic concentration of minority groups in specific regions and metropolitan areas of the United States.[4] Demographers call it chain migration: Essentially, migrants utilize social capital specific knowledge of the migration process (i.e.) to move from one area and settle in another.[5] For example, Los Angeles is home to one-fifth of the Latino population, placing first in total growth.

How do you compare the amount of diversity in different cities or states? Diversity is a characteristic of a population many of us can sense intuitively. For example, the ethnic diversity of a large city is seen in the many members of various groups encountered when walking down its streets or traveling through its neighborhoods.[6]

We can use the IQV to measure the amount of diversity in different states. Table 5.3 displays the 2002 percentage breakdown of the population by race for all 50 states. Based on the data in Table 5.3, and using Formula 5.1 as in our earlier example, we have calculated the IQV for each state in Table 5.4.[7]

---

[3]U.S. Bureau of the Census, *Statistical Abstract of the United States*, 2003, Tables 21 & 22.

[4]William H. Frey, "The Diversity Myth," *American Demographics* 20, no. 6 (1998): 39–43.

[5]Douglas S. Massey and Kristin Espinosa, "What's Driving Mexico-U.S. Migration? A Theoretical, Empirical, and Policy Analysis." *American Journal of Sociology* 102 (1997):939–999.

[6]Michael White, "Segregation and Diversity Measures in Population Distribution," *Population Index* 52, no. 2 (1986): 198–221.

[7]Because IQV is partially a function of the number of categories, the IQVs for Maine and Hawaii in Table 5.4 differ slightly from our earlier calculations. Table 5.4 utilizes five racial/ethnic categories. Table 5.1 utilized only the top three racial/ethnic categories.

Table 5.3   Percentage Makeup of Population for States by Race, 2002

State	White	Black	Latino	Asian	Other
Alabama	70.7	26.0	1.9	0.8	0.6
Alaska	71.4	3.8	4.4	4.2	16.3
Arizona	69.9	2.6	21.5	1.6	4.3
Arkansas	79.3	15.6	3.5	0.9	0.7
California	58.9	5.3	25.8	8.8	1.2
Colorado	77.6	3.6	15.6	2.1	1.1
Connecticut	78.7	9.2	9.2	2.6	0.4
Delaware	73.6	18.7	4.9	2.4	0.4
Florida	69.1	13.5	15.4	1.6	0.4
Georgia	64.3	27.4	5.7	2.2	0.4
Hawaii	29.8	2.6	8.3	48.9	10.4
Idaho	89.1	0.5	7.9	1.0	1.5
Illinois	70.8	13.6	11.9	3.4	0.3
Indiana	86.6	8.3	3.7	1.1	0.3
Iowa	93.0	2.2	3.0	1.5	0.4
Kansas	84.3	5.6	7.2	1.9	1.0
Kentucky	89.8	7.4	1.7	0.8	0.3
Louisiana	63.2	32.3	2.5	1.4	0.6
Maine	97.2	0.6	0.8	0.9	0.6
Maryland	63.6	27.2	4.6	4.3	0.4
Massachusetts	82.5	6.3	6.8	4.0	0.3
Michigan	79.9	14.1	3.4	2.0	0.6
Minnesota	88.9	3.7	3.1	3.1	1.2
Mississippi	60.8	36.5	1.5	0.8	0.4
Missouri	84.6	11.4	2.2	1.3	0.5
Montana	90.6	0.3	2.1	0.5	6.5
Nebraska	87.9	4.0	5.7	1.4	0.9
Nevada	70.7	5.8	17.9	4.0	1.6
New Hampshire	95.6	0.9	1.8	1.5	0.3
New Jersey	68.6	12.9	12.6	5.6	0.3
New Mexico	60.1	1.6	30.3	0.9	7.1
New York	64.5	15.5	14.0	5.5	0.6
North Carolina	71.1	20.9	5.1	1.6	1.3
North Dakota	92.4	0.8	1.3	0.6	4.9
Ohio	84.8	11.6	2.0	1.4	0.3
Oklahoma	77.5	7.6	5.4	1.6	7.9
Oregon	85.4	1.7	8.3	3.1	1.6
Pennsylvania	84.4	10.1	3.3	2.0	0.2
Rhode Island	83.0	5.4	8.6	2.4	0.6
South Carolina	66.6	29.3	2.6	1.0	0.4
South Dakota	88.7	0.8	1.6	0.7	8.3
Tennessee	79.8	16.4	2.4	1.1	0.3
Texas	63.1	8.8	25.3	2.2	0.6
Utah	86.5	0.9	8.9	1.7	2.0

(Continued)

Table 5.3   (Continued)

State	White	Black	Latino	Asian	Other
Vermont	97.0	0.6	1.0	1.0	0.4
Virginia	71.4	19.3	5.0	3.9	0.4
Washington	81.5	3.3	7.7	5.6	2.0
West Virginia	95.1	3.3	0.7	0.6	0.2
Wisconsin	87.8	5.7	3.7	1.8	0.9
Wyoming	90.0	0.8	6.3	0.6	2.4

Source: U.S. Bureau of the Census, *Statistical Abstract of the United States*, 2003, Tables 21 & 22.

Table 5.4   Racial Diversity as Measured by the IQVs for the 50 States, 2002

State	IQV	State	IQV
Hawaii	0.82	Tennessee	0.42
California	0.72	Washington	0.41
New Mexico	0.68	Massachusetts	0.38
New York	0.67	Rhode Island	0.38
Texas	0.66	Kansas	0.35
Maryland	0.65	Pennsylvania	0.35
Georgia	0.63	Missouri	0.34
Mississippi	0.62	Ohio	0.34
Louisiana	0.62	Oregon	0.33
New Jersey	0.62	Utah	0.30
Florida	0.60	Indiana	0.30
South Carolina	0.59	Wisconsin	0.28
Illinois	0.58	Nebraska	0.28
Nevada	0.58	South Dakota	0.26
Arizona	0.58	Minnesota	0.26
Alaska	0.58	Idaho	0.25
Virginia	0.56	Kentucky	0.23
North Carolina	0.56	Wyoming	0.23
Alabama	0.54	Montana	0.22
Delaware	0.53	North Dakota	0.18
Oklahoma	0.48	Iowa	0.17
Colorado	0.46	West Virginia	0.12
Connecticut	0.46	New Hampshire	0.11
Arkansas	0.43	Vermont	0.07
Michigan	0.43	Maine	0.07

The advantage of using a single number to express diversity is demonstrated in Figure 5.3, which depicts the regional variations in diversity as expressed by the IQVs from Table 5.4. Figure 5.3 shows the wide variation in racial diversity that exists in the United States. Notice that Hawaii, with an IQV of 0.82, is the most diverse state. At the other extreme Vermont and Maine, whose populations are overwhelmingly white, have IQVs of less than 0.10 and are the most homogeneous states.

Figure 5.3 Racial Diversity in the United States in 2000 (IQV)

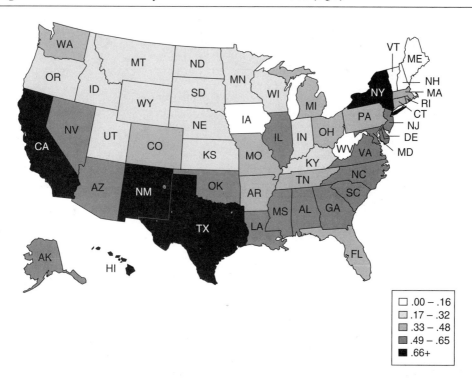

☐ .00 − .16
.17 − .32
.33 − .48
.49 − .65
■ .66+

✔ ***Learning Check.*** *What regional variations in racial diversity are depicted in Figure 5.3? Can you think of at least two explanations for these patterns?*

▣ THE RANGE

The simplest and most straightforward measure of variation is the **range**, which measures variation in interval-ratio variables. It is the difference between the highest (maximum) and lowest (minimum) scores in the distribution:

Range = highest score − lowest score

In the 2002 GSS, the oldest person included in the study was 89 years old and the youngest was 18. Thus, the range was 89 − 18, or 71 years.

Range A measure of variation in interval-ratio variables. It is the difference between the highest (maximum) and lowest (minimum) scores in the distribution.

The range can also be calculated on percentages. For example, since the 1980s relatively large communities of the elderly have become noticeable not just in the traditional retirement

Table 5.5   Percentage Change in the Population 65 Years and Over by Region and State, 1990–2000

Region, Division, and State	Percentage Change	Region, Division, and State	Percentage Change
United States	12.0		
		South (cont.)	
Northeast	5.4	Georgia	20.0
Connecticut	5.4	Kentucky	8.1
Delaware	26.0	Louisiana	10.2
Maine	12.3	Maryland	15.8
Massachusetts	5.0	Mississippi	6.9
New Hampshire	18.3	North Carolina	20.5
New Jersey	7.9	Oklahoma	7.5
New York	3.6	South Carolina	22.3
Pennsylvania	4.9	Tennessee	13.7
Rhode Island	1.2	Texas	20.7
Vermont	17.2	Virginia	19.2
		Washington, DC	−10.2
Midwest	6.6	West Virginia	3.0
Indiana	8.1		
Illinois	4.4	**West**	19.9
Iowa	2.4	Alaska	59.8
Kansas	4.0	Arizona	39.5
Michigan	10.0	California	14.7
Minnesota	8.7	Colorado	26.3
Missouri	5.3	Hawaii	28.5
Nebraska	4.1	Idaho	20.3
North Dakota	3.8	Montana	13.6
Ohio	7.2	Nevada	71.5
South Dakota	5.7	New Mexico	30.1
Wisconsin	7.9	Oregon	12.0
		Utah	26.9
South	16.0	Washington	15.1
Alabama	10.9	Wyoming	22.2
Arkansas	6.8		
Florida	18.5		

Source: U.S. Bureau of the Census, "65+ in the United States," *Current Population Reports*, 2001, P1-8, Table 3.

meccas of the Sun Belt, but also in the Ozarks of Arkansas and the mountains of Colorado and Montana. The number of elderly persons increased in every state during the 1990s (Washington, DC, is the exception), but by different amounts. Table 5.5 displays the percentage increase in the elderly population from 1990 to 2000 by region and by state.[8]

---

[8]The percentage increase in the population 65 years and over for each state and region was obtained by the following formula:

Percentage Increase = [(2000 population − 1990 population) / 1990 population] × 100

What is the range in the percentage increase in state elderly population for the United States? To find the ranges in a distribution, simply pick out the highest and lowest scores in the distribution and subtract. Nevada has the highest percentage increase, with 71.5 percent, and the Washington, DC, has the lowest increase, with –10.2 percent. The range is 81.7 percentage points, or 71.5 percent – (–10.2 percent).

Although the range is simple and quick to calculate, it is a rather crude measure because it is based on only the lowest and highest scores. These two scores might be extreme and rather atypical, which might make the range a misleading indicator of the variation in the distribution. For instance, notice that among the 50 states and Washington, DC, listed in Table 5.5, no state has a percentage decrease as does the Washington, DC's, and only Alaska has a percentage increase nearly as high as Nevada's. The range of 81.7 percentage points does not give us information about the variation in states between the Washington, DC, and Alaska.

✓ *Learning Check.* *Why can't we use the range to describe diversity in nominal variables? The range can be used to describe diversity in ordinal variables (for example, we can say that responses to a question ranged from "somewhat satisfied" to "very dissatisfied"), but it has no quantitative meaning. Why not?*

▣ THE INTERQUARTILE RANGE: INCREASES IN ELDERLY POPULATIONS

To remedy this limitation we can employ an alternative to the range—the *interquartile range*. The **interquartile range (IQR)**, a measure of variation for interval-ratio variables, is the width of the middle 50 percent of the distribution. It is defined as the difference between the lower and upper quartiles ($Q1$ and $Q3$).

$$IQR = Q3 - Q1$$

Recall that the first quartile ($Q1$) is the 25th percentile, the point at which 25 percent of the cases fall below it and 75 percent above it. The third quartile ($Q3$) is the 75th percentile, the point at which 75 percent of the cases fall below it and 25 percent above it. The interquartile range, therefore, defines variation for the middle 50 percent of the cases.

Like the range, the interquartile range is based on only two scores. However, because it is based on intermediate scores, rather than on the extreme scores in the distribution, it avoids some of the instability associated with the range.

These are the steps for calculating the IQR:

1. To find $Q1$ and $Q3$, order the scores in the distribution from the highest to the lowest score, or vice versa. Table 5.6 presents the data of Table 5.5 arranged in order from Nevada (71.5%) to Washington, DC (–10.2%).

2. Next, we need to identify the first quartile, $Q1$ or the 25th percentile. We have to identify the percentage increase in the elderly population associated with the state that

Table 5.6   Percentage Increase in the Population 65 Years and Over, 1990–2000, by State, Ordered from Highest to Lowest

State	Percentage Change	State	Percentage Change	State	Percentage Change
Nevada	71.5	Vermont	17.2	Ohio	7.2
Alaska	59.8	Maryland	15.8	Mississippi	6.9
Arizona	39.5	Washington	15.1	Arkansas	6.8
New Mexico	30.1	California	14.7	South Dakota	5.7
Hawaii	28.5	Tennessee	13.7	Connecticut	5.4
Utah	26.9	Montana	13.6	Missouri	5.3
Colorado	26.3	Maine	12.3	Massachusetts	5.0
Delaware	26.0	Oregon	12.0	Pennsylvania	4.9
South Carolina	22.3	Alabama	10.9	Illinois	4.4
Wyoming	22.2	Louisiana	10.2	Nebraska	4.1
Texas	20.7	Michigan	10.0	Kansas	4.0
North Carolina	20.5	Minnesota	8.7	North Dakota	3.8
Idaho	20.3	Indiana	8.1	New York	3.6
Georgia	20.0	Kentucky	8.1	West Virginia	3.0
Virginia	19.2	New Jersey	7.9	Iowa	2.4
Florida	18.5	Wisconsin	7.9	Rhode Island	1.2
New Hampshire	18.3	Oklahoma	7.5	Washington, DC	−10.2

divides the distribution so that 25 percent of the states are below it and 75 percent of the states are above it. To find $Q1$ we multiply N by .25:

$$(N)\,(.25) = (51)\,(.25) = 12.75$$

The first quartile falls between the 12th and 13th states. Counting from the bottom, the 12th state is Missouri, and the percentage increase associated with it is 5.3. The 13th state is Connecticut, with a percentage increase of 5.4. To find the first quartile we take the average of 5.3 and 5.4. Therefore, $(5.3 + 5.4) \div 2 = 5.35$ is the first quartile ($Q1$).

3. To find $Q3$, we have to identify the state that divides the distribution in such a way that 75 percent of the states are below it and 25 percent of the states are above it. We multiply N this time by .75:

$$(N)(.75) = (51)(.75) = 38.25$$

The third quartile falls between the 38th and 39th states. Counting from the bottom, the 38th state is Georgia, and the percentage increase associated with it is 20.0. The 39th state is Idaho, with a percentage increase of 20.3. To find the third quartile we take the average of 20.0 and 20.3. Therefore, $(20.0 + 20.3) \div 2 = 20.15$ is the third quartile ($Q3$).

4. We are now ready to find the interquartile range:

$$IQR = Q3 - Q1 = 20.15 - 5.35 = 14.80$$

The interquartile range of percentage increase in the elderly population is 14.8 percentage points.

Notice that the IQR gives us better information than the range. The range gave us an 81.7-point spread, from 71.5 percent to –10.2 percent, but the IQR tells us that half the states are clustered between 20.15 and 5.35—a much narrower spread. The extreme scores represented by Nevada and Washington, DC, have no effect on the IQR because they fall at the extreme ends of the distribution (see also Figure 5.4).

Interquartile range (IQR)   The width of the middle 50 percent of the distribution. It is defined as the difference between the lower and upper quartiles ($Q1$ and $Q3$).

✓ **Learning Check.**   *Why is the IQR better than the range as a measure of variability, especially when there are extreme scores in the distribution? To answer this question you may want to examine Figure 5.4.*

Figure 5.4   The Range Versus the Interquartile Range: Number of Children Among Two Groups of Women

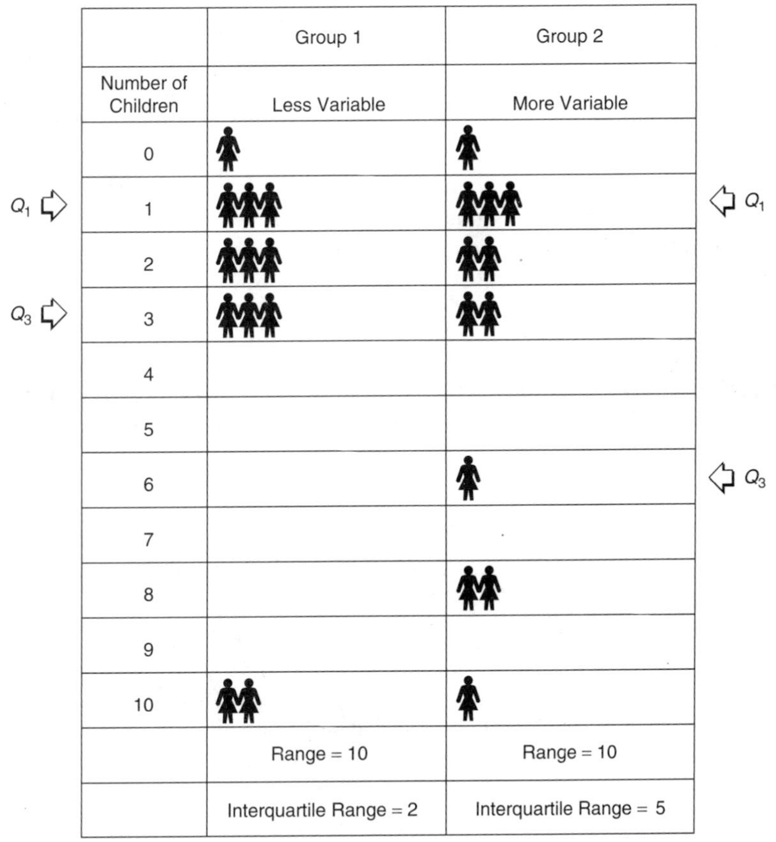

Number of Children	Group 1 Less Variable	Group 2 More Variable
0	👤	👤
1 (Q_1 Group 1)	👤👤👤	👤👤👤 (Q_1 Group 2)
2	👤👤👤	👤👤
3 (Q_3 Group 1)	👤👤👤	👤👤
4		
5		
6		👤 (Q_3 Group 2)
7		
8		👤👤
9		
10	👤👤	👤
	Range = 10	Range = 10
	Interquartile Range = 2	Interquartile Range = 5

▣ THE BOX PLOT

A graphic device called the *box plot* can visually present the range, the interquartile range, the median, the lowest (minimum) score, and the highest (maximum) score. The box plot provides us with a way to visually examine the center, the variation, and the shape of distributions of interval-ratio variables.

Figure 5.5 is a box plot of the distribution of the 1990–2000 percentage increase in elderly population displayed in Table 5.6. To construct the box plot in Figure 5.5 we used the lowest and highest values in the distribution, the upper and lower quartiles, and the median. We can easily draw a box plot by hand following these instructions:

1. Draw a box between the lower and upper quartiles.

2. Draw a solid line within the box to mark the median.

3. Draw vertical lines (called whiskers) outside the box, extending to the lowest and highest values.

What can we learn from creating a box plot? We can obtain a visual impression of the following properties: First, the center of the distribution is easily identified by the solid line inside the box. Second, since the box is drawn between the lower and upper quartiles, the interquartile range is reflected in the height of the box. Similarly, the length of the vertical lines drawn outside the box (on both ends) represents the range of the distribution.[9] Both the interquartile range and the range give us a visual impression of the spread in the distribution. Finally, the relative position of the box and the position of the median within the box tell us whether the distribution is symmetrical or skewed. A perfectly symmetrical distribution would have the box at the center of the range as well as the median in the center of the box. When the distribution departs from symmetry, the box and/or the median will not be centered; it will be closer to the lower quartile when there are more cases with lower scores or to the upper quartile when there are more cases with higher scores.

✔ *Learning Check.*   *Is the distribution shown in the box plot in Figure 5.5 symmetrical or skewed?*

Box plots are particularly useful for comparing distributions. To demonstrate box plots that are shaped quite differently, in Figure 5.6 we have used the data on the percentage increase

---

[9]The extreme values at either end are referred to as outliers. SPSS will include outliers in box plots and in the calculation of the interquartile range; however, SPSS extends whiskers from the box edges to 1.5 times the box width (the IQR). If there are additional values beyond 1.5 times the IQR, SPSS displays the individual cases. It is important to keep this in mind when examining the shape of a distribution from a box plot.

Figure 5.5   Box Plot of the Distribution of the Percentage Increase in Elderly
Population, 1990–2000

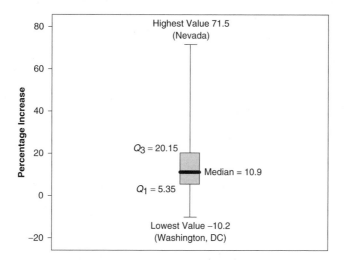

Figure 5.6   Box Plots of the Percentage Increase in the Elderly Population, 1990–2000,
for Northeast and West Regions

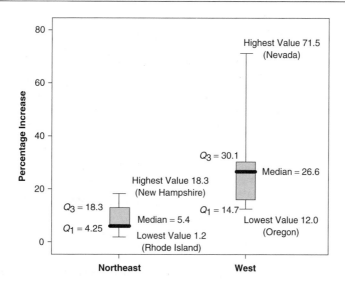

in the elderly population (Table 5.5) to compare the pattern of change occurring between 1990
and 2000 in the Northeastern and Western regions of the United States.

As you can see, the box plots differ from each other considerably. What can you learn from
comparing the box plots for the two regions? First, the positions of the medians highlight the

dramatic increase in the elderly population in the western United States. While the Northeast (median = 5.4%) has experienced a steady rise in its elderly population, the West shows a much higher percentage increase (median = 26.6%). Second, both the range (illustrated by the position of the whiskers in each box plot) and the interquartile range (illustrated by the height of the box) are much wider in the West (range = 59.5%; IQR = 15.4 %) than in the Northeast (range = 17.1%; IQR = 14.05%), indicating that there is more variability among states in the West than among those in the Northeast. Finally, the relative positions of the boxes tell us something about the different shapes of these distributions. Because its box is at about the center of its range, the Northeast distribution is almost symmetrical. In contrast, with its box off center and closer to the lower end of the distribution, the distribution of percentage change in the elderly population for the Western states is positively skewed.

◙ THE VARIANCE AND THE STANDARD DEVIATION: CHANGES IN THE ELDERLY POPULATION

The elderly population in the United States today is 10 times as large as in 1900 and is projected to more than double from 1990 to 2030. The pace and direction of these demographic changes will create compelling social, economic, and ethical choices for individuals, families, and governments.

Table 5.7 presents the percentage change in the elderly population for all regions of the United States.

Table 5.7   Percentage Change in Elderly Population by Region, 1990–2000

Region	Percentage
Northeast	5.4
South	16.0
Midwest	6.6
West	19.9
Mean ($\bar{Y}$)	11.975

Source: U.S. Bureau of the Census, "65+ in the United States," *Current Population Reports*, 2001, P1-8, Table 3.

These percentage changes were calculated by the U.S. Census Bureau using the following formula:

$$\text{Percentage Change} = \left[\frac{(2000\ \text{population} - 1990\ \text{population})}{1990\ \text{population}} \right] \times 100$$

For example, the elderly population in the West region was 5,773,363 in 1990. In 2000 the elderly population increased to 6,922,129. Therefore, the percentage change from 1990 to 2000 is

$$\text{Percentage Change} = \left[\frac{(6,922,129 - 5,773,363)}{5,773,363} \right] \times 100 = 19.9$$

Table 5.7 shows that between 1990 and 2000 the size of the elderly population in the United States increased by an average of 11.975 percent. But this average increase does not inform us about the regional variation in the elderly population. For example, do the Northeastern states show a smaller-than-average increase because of the outmigration of the elderly population to the warmer climate of the Sun Belt states? Is the increase higher in the South because of the immigration of the elderly?

Although it is important to know the average percentage increase for the nation as a whole, you may also want to know whether regional increases differ from the national average. If the regional increases are close to the national average, the figures will cluster around the mean, but if the regional increases deviate much from the national average, they will be widely dispersed around the mean.

Table 5.7 suggests that there is considerable regional variation. The percentage change ranges from 19.9 percent in the West to 5.4 percent in the Northeast, so the range is 14.5 percent (19.9 percent − 5.4 percent = 14.5 percent). Moreover, most of the regions deviate considerably from the national average of 11.975 percent. How large are these deviations on the average? We want a measure that will give us information about the overall variations among all regions in the United States and, unlike the range or the interquartile range, will not be based on only two scores.

Such a measure will reflect how much, on the average, each score in the distribution deviates from some central point, such as the mean. We use the mean as the reference point rather than other kinds of averages (the mode or the median) because the mean is based on all the scores in the distribution. Therefore, it is more useful as a basis from which to calculate average deviation. The sensitivity of the mean to extreme values carries over to the calculation of the average deviation, which is based on the mean. Another reason for using the mean as a reference point is that more advanced measures of variation require the use of algebraic properties that can be assumed only by using the arithmetic mean.

The *variance* and the *standard deviation* are two closely related measures of variation that increase or decrease based on how closely the scores cluster around the mean. The **variance** is the average of the squared deviations from the center (mean) of the distribution, and the **standard deviation** is the square root of the variance. Both measure variability in interval-ratio variables.

Variance   A measure of variation for interval-ratio variables; it is the average of the squared deviations from the mean.

Standard deviation   A measure of variation for interval-ratio variables; it is equal to the square root of the variance.

Calculating the Deviation from the Mean

Consider again the distribution of the percentage change in the elderly population for the four regions of the United States. Because we want to calculate the average difference of all the regions from the national average (the mean), it makes sense to first look at the difference between each region and the mean. This difference, called a deviation from the mean, is symbolized as $(Y - \bar{Y})$. The sum of these deviations can be symbolized as $\sum(Y - \bar{Y})$.

The calculations of these deviations for each region are displayed in Table 5.8 and Figure 5.7. We have also summed these deviations. Note that each region has either a positive or a negative deviation score. The deviation is positive when the percentage change in the elderly home population is above the mean. It is negative when the percentage change is below the mean. Thus, for example, the Northeast's deviation score of -6.575 means that its percentage change in the elderly population was 6.575 percentage points below the mean.

Table 5.8   Percentage Change in the Elderly Population, 1990-2000, by Region and Deviation from the Mean

Region	Percentage	$Y - \bar{Y}$
Northeast	5.4	$5.4 - 11.975 = -6.575$
South	16.0	$16.0 - 11.975 = 4.025$
Midwest	6.6	$6.6 - 11.975 = -5.375$
West	19.9	$19.9 - 11.975 = 7.925$
	$\sum Y = 47.9$	$\sum(Y - \bar{Y}) = 0$
	$\bar{Y} = \dfrac{\sum Y}{N} = \dfrac{47.9}{4} = 11.975$	

Figure 5.7   Illustrating Deviations from the Mean

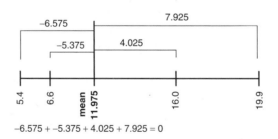

$$-6.575 + -5.375 + 4.025 + 7.925 = 0$$

You may wonder if we could calculate the average of these deviations by simply adding up the deviations and dividing them. Unfortunately we cannot, because the sum of the deviations of scores from the mean is always zero, or algebraically $\sum(Y - \bar{Y}) = 0$. In other words, if we were to subtract the mean from each score and then add up all the deviations as we did in Table 5.8, the sum would be zero, which in turn would cause the average deviation (that is, average difference) to compute to zero. This is always true because the mean is the center of gravity of the distribution.

Table 5.9 Percentage Change in the Elderly Population, 1990–2000, by Region and Deviation from the Mean

Region	Percentage	$Y - \bar{Y}$	$(Y - \bar{Y})^2$
Northeast	5.4	$5.4 - 11.975 = -6.575$	43.231
South	16.0	$16.0 - 11.975 = 4.025$	16.201
Midwest	6.6	$6.6 - 11.975 = -5.375$	28.891
West	19.9	$19.9 - 11.975 = 7.925$	62.806
	$\sum Y = 47.9$	$\sum(Y - \bar{Y}) = 0$	$\sum(Y - \bar{Y})^2 = 151.129$

$$\text{Mean} = \bar{Y} = \frac{\sum Y}{N} = \frac{47.9}{4} = 11.975$$

Mathematically, we can overcome this problem either by ignoring the plus and minus signs, using instead the absolute values of the deviations, or by squaring the deviations—that is, multiplying each deviation by itself to get rid of the negative sign. Since absolute values are difficult to work with mathematically, the latter method is used to compensate for the problem.

Table 5.9 presents the same information as Table 5.8, but here we have squared the actual deviations from the mean and added together the squares. The sum of the squared deviations is symbolized as $\sum(Y - \bar{Y})^2$. Note that by squaring the deviations we end up with a sum representing the deviation from the mean, which is positive. (Note that this sum will equal zero if all the cases have the same value as the mean case.) In our example, this sum is $\sum(Y - \bar{Y})^2 = 151.129$.

✔ *Learning Check.* *Examine Table 5.9 again and note the disproportionate contribution of the Western region to the sum of the squared deviations from the mean (it actually accounts for almost 42% of the sum of squares). Can you explain why? Hint: It has something to do with the sensitivity of the mean to extreme values.*

Calculating the Variance and the Standard Deviation

The average of the squared deviations from the mean is known as the *variance*. The variance is symbolized as S_Y^2. Remember that we are interested in the *average* of the squared deviations from the mean. Therefore, we need to divide the sum of the squared deviations by the number of scores (N) in the distribution. However, unlike the calculation of the mean, we will use $N - 1$ rather than N in the denominator.[10] The formula for the variance can be stated as

$$S_Y^2 = \frac{\sum(Y - \bar{Y})^2}{N - 1} \tag{5.2}$$

---

[10]$N - 1$ is used in the formula for computing variance because usually we are computing from a sample with the intention of generalizing to a larger population. $N - 1$ in the formula gives a better estimate and is also the formula used in SPSS.

where

$$S_Y^2 = \text{the variance}$$
$$Y - \overline{Y} = \text{the deviation from the mean}$$
$$\Sigma(Y - \overline{Y})^2 = \text{the sum of the squared deviations from the mean}$$
$$N = \text{the number of scores}$$

Notice that the formula incorporates all the symbols we defined earlier. This formula means: The variance is equal to the average of the squared deviations from the mean.

Follow these steps to calculate the variance:

1. Calculate the mean, $\overline{Y} = \dfrac{\Sigma Y}{N}$.

2. Subtract the mean from each score to find the deviation, $(Y - \overline{Y})$.

3. Square each deviation, $(Y - \overline{Y})^2$.

4. Sum the squared deviations, $\Sigma(Y - \overline{Y})^2$.

5. Divide the sum by $N - 1$, $\dfrac{\Sigma(Y - \overline{Y})^2}{N - 1}$.

6. The answer is the variance.

To assure yourself that you understand how to calculate the variance, go back to Table 5.9 and follow this step-by-step procedure for calculating the variance. Now plug the required quantities into Formula 5.2. Your result should look like this:

$$S_Y^2 = \frac{\Sigma(Y - \overline{Y})^2}{N - 1} = \frac{151.129}{3} = 50.38$$

One problem with the variance is that it is based on squared deviations and therefore is no longer expressed in the original units of measurement. For instance, it is difficult to interpret the variance of 50.38, which represents the distribution of the percentage change in the elderly population, because this figure is expressed in squared percentages. Thus, we often take the square root of the variance and interpret it instead. This gives us the *standard deviation*, S_Y.

The standard deviation, symbolized as S_Y, is the square root of the variance, or

$$S_Y = \sqrt{S_Y^2}$$

The standard deviation for our example is

$$S_Y = \sqrt{S_Y^2} = \sqrt{50.38} = 7.0$$

The formula for the standard deviation uses the same symbols as the formula for the variance:

$$S_Y = \sqrt{\frac{\sum(Y - \overline{Y})^2}{N - 1}} \qquad (5.3)$$

As we interpret the formula, we can say that the standard deviation is equal to the square root of the average of the squared deviations from the mean.

The advantage of the standard deviation is that unlike the variance, it is measured in the same units as in the original data. For instance, the standard deviation for our example is 7.10. Because the original data were expressed in percentages, this number is expressed as a percentage as well. In other words, you could say, "The standard deviation is 7.10 percent." But what does this mean? The actual number tells us very little by itself, but it allows us to evaluate the dispersion of the scores around the mean.

In a distribution where all the scores are identical, the standard deviation is zero (0). Zero is the lowest possible value for the standard deviation. In an identical distribution, all of the points would be the same, with the same mean, mode, and median. There is no variation or dispersion in the scores.

The more the standard deviation departs from zero, the more variation there is in the distribution. There is no upper limit to the value of the standard deviation. In our example, we can conclude that a standard deviation of 7.10 percent means that the percentage change in the elderly population for the four regions of the United States is widely dispersed around the mean of 11.975 percent.

The standard deviation can be considered a standard against which we can evaluate the positioning of scores relative to the mean and to other scores in the distribution. As we will see in more detail in Chapter 9, in most distributions, unless they are highly skewed, about 34 percent of all scores fall between the mean and one standard deviation above the mean. Another 34 percent of scores fall between the mean and one standard deviation below it. Thus we would expect the majority of scores (68 percent) to fall within one standard deviation of the mean. For example, let's consider the distribution of education for GSS respondents in 2002. This variable has a mean of 13.36 years and a standard deviation of 2.97 years. We can expect about 68 percent of all respondents to fall within the range of 10.39 (13.36 − 2.97) to 16.33 (13.36 + 2.97) years of education. Hence, based on the mean and the standard deviation, we have a pretty good indication of what would be considered a typical level of education for the majority of cases. For example, we would consider a person with some graduate training (17 years of education) to have a relatively high level of education in comparison with other respondents. More than two thirds of all respondents fall closer to the mean than this person.

Another way to interpret the standard deviation is to compare it with another distribution. For instance, Table 5.10 displays the means and standard deviations of employee age for two samples drawn from a Fortune 100 corporation. Samples are divided into female clerical and female technical. Note that the mean ages for both samples are about the same—approximately 39 years of age. However, the standard deviations suggest that the distribution of age is dissimilar between the two groups. Figure 5.8 loosely illustrates this dissimilarity in the two distributions.

The relatively low standard deviation for female technical indicates that this group is relatively homogeneous in age. That is to say, most of the women's ages, while not identical, are fairly

Table 5.10   Age Characteristics of Female Clerical and Technical Employees

Characteristics	Female Clerical N = 22	Female Technical N = 39
Mean age	39.46	39.87
Standard deviation	7.80	3.75

Source: Adapted from Marjorie Armstrong-Stassen, "The Effect of Gender and Organizational Level on How Survivors Appraise and Cope with Organizational Downsizing," *Journal of Applied Behavioral Science* 34, no. 2 (June 1998): 125–142. Reprinted by permission of Sage Publications, Inc.

similar. The average deviation from the mean age of 39.87 is 3.75 years. In contrast, the standard deviation for female clerical employees is about twice the standard deviation for female technical. This suggests a wider dispersion or greater heterogeneity in the ages of clerical workers. We can say that the average deviation from the mean age of 39.46 is 7.80 years for clerical workers. The larger standard deviation indicates a wider dispersion of points below or above the mean. On average, clerical employees are farther in age from their mean of 39.46.

✔ *Learning Check.*   *Take time to understand the section on standard deviation and variance. You will see these statistics in more advanced procedures. Although your instructor may require you to memorize the formulas, it is more important for you to understand how to interpret standard deviation and variance and when they can be appropriately used. Many hand calculators and all statistical software programs will calculate these measures of diversity for you, but they won't tell you what they mean. Once you understand the meaning behind these measures, the formulas will be easier to remember.*

Figure 5.8   Illustrating the Means and Standard Deviations for Age Characteristics

Female clerical: mean = 39.46, standard deviation = 7.80

Female technical: mean = 39.87, standard deviation = 3.75

◙ **A Closer Look 5.2**
Computational Formulas for the
Variance and the Standard Deviation

We have learned how to use the definitional formulas for the standard deviation and the variance. These formulas are easy to follow conceptually, but they are tedious to compute, especially when working with a large number of scores. The following computational formulas are easier and faster to use and give exactly the same result:

$$S_Y^2 = \left(\frac{\sum Y^2}{N-1}\right) - \left(\frac{N}{N-1}\right)\left(\frac{\sum Y}{N-1}\right)^2$$

$$S_Y^2 = \sqrt{\left(\frac{\sum Y^2}{N-1}\right) - \left(\frac{N}{N-1}\right)\left(\frac{\sum Y}{N}\right)^2}$$

where

$\sum Y^2$ = the sum of the squared scores (find this by first squaring each score and adding up the squared scores)

N = the number of scores in the distribution

$\left(\frac{\sum Y}{N}\right)^2$ = the sum of the scores divided by N and then squared (find this quantity by first dividing the sum of the scores by N and then squaring the answer, which is equivalent to squaring the mean)

To illustrate how to calculate the variance and standard deviation using the computational formulas, we will use data on the burglary rates per 100,000 population in the nine largest states in the United States. In the following table, we have added a column to the original data to help us generate the following quantities required by the formula: Y^2 and $\sum Y^2$.

Burglary Rates per 100,000 Population in the Nine Largest U.S. States for 2001

State	*Burglary Rate (Y)*	Y^2
New Jersey	552	304,704
New York	423	178,929
Ohio	852	725,904
Michigan	721	519,841
Pennsylvania	442	195,364
California	672	451,584

Georgia	856	732,736
Texas	958	917,764
Florida	1,074	1,153,476

$$\sum Y = 6550 \qquad \sum Y^2 = 5{,}180{,}302$$

$$\text{Mean} = \bar{Y} = \frac{\sum Y}{N} = \frac{6550}{9} = 727.8$$

Source: U.S. Bureau of the Census, *Statistical Abstract of the United States*, 2003, Table 307.

Now plug the results into the formula for the variance:

$$S_Y^2 = \left(\frac{\sum Y^2}{N-1}\right) - \left(\frac{N}{N-1}\right)\left(\frac{\sum Y}{N}\right)^2 = \frac{5{,}180{,}302}{8} - \left(\frac{9}{8}\right)\left(\frac{6550}{9}\right)^2$$

$$= 647{,}537.75 - (1.13)(727.78)^2$$

$$= 647{,}537.75 - 598{,}520.01$$

$$= 49{,}017.74$$

The standard deviation can be found by taking the square root of the variance. For our example, the standard deviation is

$$S_Y = \sqrt{S_Y^2} = \sqrt{49{,}017.74} = 221.4$$

Hence, for the nine largest states, the burglary rate has a mean* of 727.8 burglaries per 100,000 population and a standard deviation of 221.4 burglaries per 100,000 population.

*Unweighted mean

◉ CONSIDERATIONS FOR CHOOSING A MEASURE OF VARIATION

So far we have considered five measures of variation: the IQV, the range, the interquartile range, the variance, and the standard deviation. Each measure can represent the degree of variability in a distribution. But which one should we use? There is no simple answer to this question. However, in general, we tend to use only one measure of variation, and the choice of the appropriate one involves a number of considerations. These considerations and how they affect our choice of the appropriate measure are presented in the form of a decision tree in Figure 5.9.

As in choosing a measure of central tendency, one of the most basic considerations in choosing a measure of variability is the variable's level of measurement. Valid use of any of the measures requires that the data are measured at the level appropriate for that measure or higher, as shown in Figure 5.9.

Figure 5.9 How to Choose a Measure of Variation

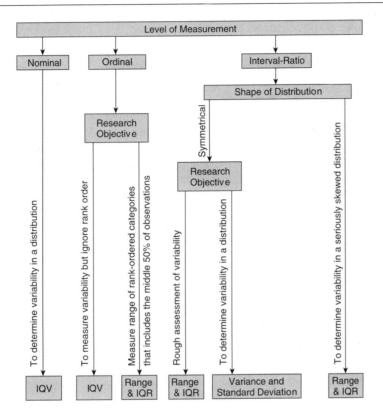

Nominal level. With nominal variables, your choice is restricted to the IQV as a measure of variability.

Ordinal level. The choice of measure of variation for ordinal variables is more problematic. The IQV can be used to reflect variability in distributions of ordinal variables, but because it is not sensitive to the rank ordering of values implied in ordinal variables, it loses some information. Another possibility is to use the interquartile range. However, the interquartile range relies on distance between two scores to express variation, information that cannot be obtained from ordinal measured scores. The compromise is to use the interquartile range (reporting $Q1$ and $Q3$) alongside the median, interpreting the interquartile range as the range of rank-ordered values that includes the middle 50 percent of the observations.[11]

Interval-ratio level. For interval-ratio variables, you can choose the variance (or standard deviation), range, or interquartile range. Because the range, and to a lesser extent the interquartile range, is based on only two scores in the distribution (and therefore tends to be sensitive if either of the two points is extreme), the variance and/or standard deviation is usually preferred. However, if a distribution is extremely skewed so that the mean is no longer representative of the central tendency in the distribution, the range and the interquartile range can be used. The

---

[11]Herman J. Loether and Donald G. McTavish, *Descriptive and Inferential Statistics: An Introduction* (Boston: Allyn and Bacon, 1980), pp. 160–161.

range and the interquartile range will also be useful when you are reading tables or quickly scanning data to get a rough idea of the extent of dispersion in the distribution.

▣ READING THE RESEARCH LITERATURE: GENDER DIFFERENCES IN CAREGIVING

In Chapter 2 we discussed how frequency distributions are presented in the professional literature. We noted that most statistical tables presented in the social science literature are considerably more complex than those we describe in this book. The same can be said about measures of central tendency and variation. Most research articles use measures of central tendency and variation in ways that go beyond describing the central tendency and variation of a single variable. In this section, we refer to both the mean and standard deviation because in most research reports the standard deviation is given along with the mean.

Table 5.11   Gender and Caregiving: Number and Hours Helped per Month

	Wives		Husbands	
	Mean	*Standard Deviation*	*Mean*	*Standard Deviation*
I. INFORMAL CAREGIVING				
Number helped				
Kin†	5.25	2.86	4.02	2.59
Friends	3.44	2.54	2.26	2.17
Total people	8.71	5.34	6.29	3.58
Hours helped				
All kin†	42.77	30.82	15.06	14.02
Parents	11.52	19.10	3.76	3.74
Parents-in-law	6.20	12.23	3.95	7.47
Adult children	20.79	37.27	4.60	10.45
Friends	10.93	14.48	6.35	10.30
II. FORMAL CAREGIVING				
Number of groups	2.08	2.26	3.14	3.54
Volunteer hours	8.09	13.68	9.41	13.67
III. TOTAL CAREGIVING				
Total hours‡	61.79	60.49	30.82	27.76

NOTE: Measures computed for all respondents ($N = 273$), except hours helped parents (includes only those with at least one living parent, $N = 165$); parents-in-law (includes only those with at least one living parent-in-law, $N = 162$); and adult children (includes only those with at least one adult child, $N = 126$).

†Kin includes parents, parents-in-law, adult children, siblings, grandparents, aunts, uncles, and any other kin mentioned.

‡Total hours = Informal (hours for all kin and friends) + Formal (volunteer hours).

Source: Adapted from Naomi Gerstel and Sally Gallagher, "Caring for Kith and Kin: Gender, Employment, and the Privatization of Care," *Social Problems* 41, no. 4 (November 1994): 525.

Table 5.11, taken from an article by Professors Naomi Gerstel and Sally Gallagher,[12] illustrates a common research application of the mean and standard deviation. Gerstel and Gallagher examined gender differences in caregiving to relatives and friends, as well as in volunteering to groups. Despite the growing acceptance among Americans of governmental aid for the disabled, the majority of Americans continues to believe it is the responsibility of women to provide personal and household assistance to elderly parents and in-laws, as well as to aging siblings and adult children.

Gerstel and Gallagher's major hypothesis is "that wives will give care in far greater breadth . . . and depth than husbands. That is, wives are far more likely to give help to a larger number of people than husbands, including to more relatives, more friends, as well as more volunteer groups."[13] The researchers (1) assess the amount and types of care provided by wives compared with husbands and (2) look at the relevance of employment status to the amount and type of care provided by wives and husbands.

Data for this study come from household interviews conducted in 1990 with 273 married respondents—179 married women and 94 of their husbands. The sample was limited to whites (86 percent) and blacks (14 percent) over the age of 21 in Springfield, Massachusetts. Table 5.11 lists the most important variables used in the study. It presents the means and standard deviations for the breadth and depth of informal and formal caregiving.

To measure the breadth of informal caregiving, interviewers named a number of different categories of people—including mother and father, adult children, other relatives, and friends. After naming a category (for example, mother), the interviewer gave the respondent a list of tasks and asked if she or he had done each task for the person named within the past month. The total number of people given care and the number of people in each category provide a measure of the breadth of informal caregiving. In addition, respondents were asked how many hours in the past month they provided care to each category of person. The total number of hours of care given to all kin and hours given to people in each category (parents, parents-in-law, adult children) is a measure of the depth of informal caregiving. To measure the breadth and depth of formal caregiving, respondents were asked to list the number of voluntary organizations to which they belonged and in which they did charity and volunteer work, and how many hours they spent on that work. Finally, because gender is a central focus of this study, the means and standard deviations are reported separately for men (husbands) and women (wives).

What can you conclude from examining the standard deviations for these variables? The first thing to look for is variables with a great deal of variation, based on a large standard deviation score. Based on the summary in Table 5.11, you can see that this is the case with the variable "hours helped" in all categories, as well as with both aspects of formal caregiving. Notice that for both men and women (except for parent hours helped for men), the standard deviations are larger than the mean. Under the category of adult children for "hours helped," wives have a mean of 20.79 hours with 37.27 hours as the standard deviation. On the other hand, husbands reported an average of 4.60 helping hours (about a fifth of the time wives reported), with a standard deviation of 10.45 hours. This indicates that there is a great

---

[12]Naomi Gerstel and Sally Gallagher, "Caring for Kith and Kin: Gender, Employment, and Privatization of Care," *Social Problems* 41, no. 4 (November 1994): 519–537.

[13]Ibid., p. 522.

deal of variation in the hours of care among both men and women in the study. Since hours cannot be negative, a standard deviation larger than the mean indicates a positively skewed deviation.

In describing the data displayed in Table 5.11, the researchers focused on the differences between men and women for each of the variables:

> The table shows striking differences between wives and husbands in the breadth, depth, and distribution of caregiving. Compared to husbands, wives help . . . a larger number of people, both kin and friends. Moreover, wives give . . . more hours of care to friends. The differences in the amount of time wives, compared to husbands, spend providing for their own parents are even larger. . . . Mothers spend more than four times more hours than fathers helping their adult children. Overall, wives give help to more relatives and spend almost three times as much time doing so. Clearly, wives are the major caregivers.[14]

MAIN POINTS

- Measures of variability are numbers that describe how much variation or diversity there is in a distribution.

- The index of qualitative variation (IQV) is used to measure variation in nominal variables. It is based on the ratio of the total number of differences in the distribution to the maximum number of possible differences within the same distribution. IQV can vary from 0.00 to 1.00.

- The range measures variation in interval-ratio variables and is the difference between the highest (maximum) and lowest (minimum) scores in the distribution. To find the range, subtract the lowest from the highest score in a distribution. For an ordinal variable, just report the lowest and highest values without subtracting.

- The interquartile range (IQR) measures the width of the middle 50 percent of the distribution. It is defined as the difference between the lower and upper quartiles ($Q1$ and $Q3$). For an ordinal variable, just report $Q1$ and $Q3$ without subtracting.

- The box plot is a graphical device that visually presents the range, the interquartile range, the median, the lowest (minimum) score, and the highest (maximum) score. The box plot provides us with a way to visually examine the center, the variation, and the shape of a distribution.

- The variance and the standard deviation are two closely related measures of variation for interval-ratio variables that increase or decrease based on how closely the scores cluster around the mean. The variance is the average of the squared deviations from the center (mean) of the distribution; the standard deviation is the square root of the variance.

KEY TERMS

index of qualitative variation (IQV)
interquartile range (IQR)
measures of variability

range
standard deviation
variance

---

[14]Ibid., p. 525.

SPSS DEMONSTRATIONS

Demonstration 1: Producing Measures of Variability with Frequencies [GSS02PFP-A]

Figure 5.10

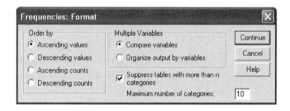

Figure 5.11

Except for the IQV, the SPSS Frequencies procedure can produce all the measures of variability we've reviewed in this chapter. (SPSS can be programmed to calculate the IQV, but the programming procedures are beyond the scope of our book.)

We'll begin with Frequencies and calculate various statistics for AGE. If we click on *Analyze, Descriptive Statistics, Frequencies*, then on the *Statistics* button, we can select the appropriate measures of variability.

The measures of variability available are listed in the Dispersion box on the bottom of the dialog box (see Figure 5.10). We've selected the standard deviation, variance, and range, plus the mean (in the Central Tendency box) for reference. In the Percentile Values box, we've selected Quartiles to tell SPSS to calculate the values for the 25th, 50th, and 75th percentiles. SPSS also allows us to specify exact percentiles in this section (such as the 34th percentile) by typing a number in the box after "Percentile(s)" and then clicking on the *Add* button.

We have already seen the frequency table for the variable AGE, so after clicking on *Continue*, we click on *Format* to turn off the display table. This is done by clicking on the button for "Suppress tables

Figure 5.12

Statistics

AGE OF RESPONDENT		
N	Valid	1496
	Missing	4
Std. Deviation		17.454
Variance		304.64
Range		71
Percentiles	25	32.00
	50	44.00
	75	58.75

with more than 10 categories" (see Figure 5.11). There are other formatting options here that you may explore later when using SPSS.

Click on *Continue*, then *OK* to run the procedure. SPSS produces the mean and the other statistics we requested (Figure 5.12). The range of age is 71 years (from 18 to 89). The standard deviation is 17.454, which indicates that there is a moderate amount of dispersion in the ages (this can also be seen from the histogram of AGE in Chapter 3). The variance, 304.64, is the square of the standard deviation.

The value of the 25th percentile is 32, the value of the 50th percentile (which is also the median) is 44, and the value of the 75th percentile is 58.75. Although Frequencies does not calculate the interquartile range, it can easily be calculated by subtracting the value of the 25th percentile from the 75th percentile, which yields a value of 26.75 years. Compare this value with the standard deviation.

Demonstration 2: Producing Variability Measures and Box Plots with Explore [ISSP00PFP]

Another SPSS procedure that can produce the usual measures of variability is Explore, which also produces box plots. The Explore procedure is located in the *Descriptive Statistics* section of the

Figure 5.13

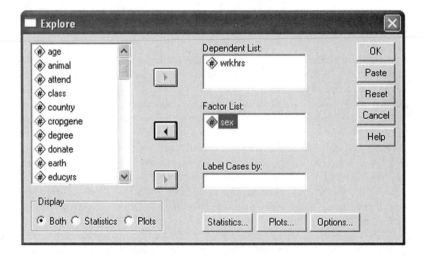

Figure 5.14

SEX			Statistic	Std. Error
1 MALE	Mean		42.92	.586
	95% Confidence	Lower Bound	41.77	
	Interval for Mean	Upper Bound	44.07	
	5% Trimmed Mean		43.12	
	Median		40.00	
	Variance		139.125	
	Std. Deviation		11.795	
	Minimum		7	
	Maximum		80	
	Range		73	
	Interquartile Range		8.00	
	Skewness		–.220	.121
	Kurtosis		1.955	.242

Analyze menu. In its main dialog box (Figure 5.13), the variables for which you want statistics are placed in the Dependent List box. You have the option of putting one or more nominal variables in the Factor List box. Explore will display separate statistics for each category of the nominal variable(s) you've selected.

Place the variable WRKHRS (hours worked per week) in the Dependent box and SEX in the Factor box, to provide separate output for males and females. Click *OK*. By default, Explore will produce statistics and plots, so we don't need to make any other choices. Although our request will not produce percentiles or create a histogram, Explore has options to do both plus several other tasks.

Selected output for males is shown in Figure 5.14. Though not replicated here, you'll notice that the first table is the Case Processing Summary Table. It indicates that 405 males answered this question. The valid sample of females is also reported, 372. Based on the second table, Descriptives, we know that for males, the mean number of hours worked per week is 42.92; the median is 40. The standard deviation is 11.795, the range is 73, and the IQR is 8, which is quite narrow compared with the range or standard deviation. (A stem and leaf plot—another way to visually present and review data—is also displayed by default. However, we will not cover stem and leaf plots in this text. The option for the stem and leaf plot can be changed so that it will not be displayed.)

Although not displayed here, the mean number of hours worked per week for females is 35.00; the median is 38. The standard deviation is 14.165, the IQR is 12, and the range, 82—values somewhat close to those for males. The variation in the number of hours worked per week is slightly larger for females.

Explore displays separate box plots for males and females in the same window for easy comparison (Figure 5.15). Although the SPSS box plot has some differences from those discussed in this chapter, some things are the same. The solid dark line is the value of the median. The width of the shaded box (in color on the screen) is the IQR (8 for males, 12 for females).

Unlike the box plots in this chapter—in which the "whiskers" extend out to the minimum and max-imum values—SPSS extends whiskers from the box edges to 1½ times the box width (the IQR). If there are additional values beyond 1½ times the IQR, SPSS displays the individual cases. Those that are somewhat extreme (1½ to 3 box widths from the edge of the box) are marked with an open circle; those considered very extreme (more than 3 box widths from the box edge) are marked with an asterisk.

The box plot shows us that variability in hours worked per week for males and females is similar. The IQR for males runs from 40 to 48, while the IQR for females runs from 28 to 40. Both genders have outlying cases beyond the edge of the whiskers.

Figure 5.15

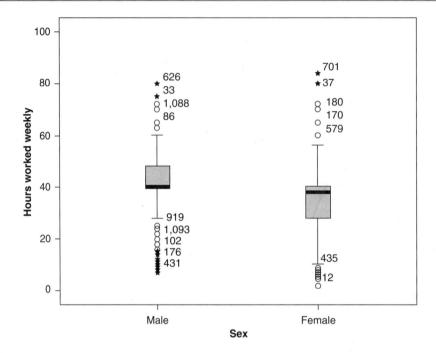

SPSS PROBLEMS

[GSS02PFP-A]

1. Use the Frequencies procedure to investigate the variability of the respondent's current age (AGE) and age when the respondent's first child was born (AGKDBRN). Click on *Analyze, Descriptive Statistics, Frequencies,* and then *Statistics.* Select the appropriate measures of variability.
   a. Which variable has more variability? Use more than one statistic to answer this question.
   b. Why should one variable have more variability than the other, from a societal perspective?

2. Using the Explore procedure, separate the statistics for AGKDBRN for men and women, selecting SEX as a factor variable in the Explore window. Click on *Analyze, Descriptive Statistics, Explore,* and then insert AGKDBRN into the Dependent List and SEX in the Factor List. What differences exist in the age of men and women at the birth of their first child? Assess the differences between men and women based on measures of central tendency and variability.

3. Repeat the procedure in Exercise 2, investigating the dispersion in the variables EDUC (education) and PRESTG80 (occupational prestige score). Select your own factor (nominal) variable to make the comparison (such as CLASS, RACECEN1, or some other factor). Click on *Analyze, Descriptive Statistics, Explore,* insert EDUC and PRESTG80 into the Dependent List and your factor variable of choice in the Factor List. Write a brief statement summarizing your results.

ordinal *ordinal* *nominal*

4. Investigate respondents' religious beliefs and practices, based on three GSS variables: BIBLE (Is the Bible the inspired word of God?), ATTEND (How often does the respondent attend religious services?), and POSTLIFE (Does the respondent believe in life after death?). [GSS02PFP-A]

   a. First, identify the level of measurement for each variable.
   b. Based on the level of measurement, what would be the appropriate set of measures of central tendency and variability?
   c. Use SPSS (and your calculator, if you need to calculate the IQV) to calculate the different sets of measures of central tendency and variability for blacks/whites *or* women/men. Is there a difference in attitudes between the groups on these three variables?

5. Use GSS02PFP-B to study the number of hours that blacks and whites work each week. The variable HRS1 measures the number of hours a respondent worked the week before the interview. Use the Explore procedure to study the variability of hours worked, comparing blacks and whites (RACECEN1) in the GSS sample.

   a. Is there much difference between the two groups in the variability of work hours?
   b. Write a short paragraph describing the box plot that SPSS created as if you were writing a report and had included the box plot as a chart to support your conclusions about the difference between blacks and whites in the variability (and central tendency) of hours worked.

CHAPTER EXERCISES

1. Americans often think of themselves as quite diverse in their political opinions, within the continuum of liberal to conservative. Let's use data from the 2002 General Social Survey to investigate the diversity of political views. The frequency table shown displays respondents' self-rating of their political position. (The statistics box is not displayed; cases with no response were removed for this example.)

Political Views	*Percent*
Extremely liberal	3.5
Liberal	10.7
Slightly liberal	11.9
Moderate	39.2
Slightly conservative	15.7
Conservative	15.8
Extremely conservative	3.2
Total	100.0

   a. How many categories (K) are we working with?
   b. Calculate the sum of the squared percentages, or $\sum Pct^2$.
   c. What is the IQV for this variable? Do you find it to be higher (closer to 1) or lower (closer to 0) than you might have expected for political views? Or to put it another way, did you expect that Americans would be diverse in their political views, or more narrowly concentrated in certain categories? Does this IQV support your expectation and what you observe from the table?

2. Up until now, we have been relying on data from the General Social Survey to illustrate some of the topics that have been discussed. As will be discussed in Chapter 10, we assume (through the use of sampling techniques) that our data accurately represent the population that we are trying to study. Using the information listed below, answer the following questions to get an idea about the educational attainment, by percentage, of GSS respondents in 2002.

Highest Educational Degree	Male	Female
Less than high school	17.0	13.6
High school graduate	55.8	55.8
Junior college	7.8	7.4
Bachelor's degree	9.1	15.7
Graduate degree	10.3	7.5
Total	100.0	100.0

a. What is the value of K?
b. Calculate the sum of the squared percentages, or $\sum Pct^2$, for both males and females.
c. Use the values you calculated in (a) and (b) to calculate the IQV for males and females. Is there more diversity by degree for males or females?

3. The U.S. Census Bureau annually estimates the percentage of Americans below the poverty level for various geographic areas. Use the information in the following table to characterize poverty in the southern versus the western portions of the United States.

Percentage of Americans Below the Poverty Level: Southern and Western Regions, 1999

South	Percent Below Poverty Level	West	Percent Below Poverty Level
Alabama	16.1	Alaska	16.1
Florida	12.5	Arizona	13.9
Georgia	13.0	California	14.2
Kentucky	15.8	Idaho	11.8
Louisiana	19.6	Montana	14.6
Mississippi	10.5	Nevada	10.5
North Carolina	12.3	New Mexico	18.4
South Carolina	14.1	Oregon	11.6
Tennessee	13.5	Utah	9.4
		Washington	10.6

Source: U.S. Bureau of the Census, *Statistical Abstract of the United States*, 2003, Table 698.

a. What is the range of poverty rates in the South? The West? Which is greater?
b. What is the interquartile range (IQR) for the South? For the West? Which is greater?
c. Using these calculations, compare the variability of the poverty rate of the states in the West with those in the South.

d. Calculate the standard deviation for each region.
e. Which region appears to have more variability as measured by the standard deviation? Are these results consistent with what you found using the range and the IQR?

4. Use the data from Exercise 3 again. This time your task is to create box plots to display the variation in poverty level by region.
   a. First combine both regions and create a box plot for all the states. What is the 75th percentile for the percentage living below the poverty level?
   b. Now create a separate box plot for the West and one for the South. Do these box plots add to your discussion from Exercise 3? If so, how?

5. Use Table 5.5 from the chapter for this exercise to continue comparisons by region. Use only the information for states in the West and Midwest.
   a. Compare the Western states with those in the Midwest on the percentage increase in the elderly population by calculating the range. Which region had a greater range?
   b. Calculate the IQR for each region. Which is greater?
   c. Use the statistics to characterize the variability in population increase of the elderly in the two regions. Why do you think one region is more variable than another?

6. Occupational prestige is a statistic developed by sociologists to measure the status of one's occupation. Occupational prestige is also a component of what sociologists call socioeconomic status, a composite measure of one's status in society. On average, people with more education tend to have higher occupational prestige than people with less education. We investigate this using the 2002 GSS variable PRESTG80 and the Explore procedure to generate the selected SPSS output shown in Figure 5.16.
   a. Notice that SPSS supplies the interquartile range (IQR), the median, and the minimum and maximum values for each group. Looking at the values of the mean and median, do you think the distribution of prestige is skewed for respondents with a high school diploma? For respondents with a bachelor's degree? Why or why not?
   b. Explain why you think there is more variability of prestige for either group, or why the variability of prestige is similar for the two groups.

Figure 5.16

PRESTG80			Statistic
OCCUPATIONAL PRESTIGE SCORE	High School Diploma	Mean	40.39
		Median	40.00
		Std. Deviation	11.44
		Minimum	17
		Maximum	86
		Range	69
		Interquartile Range	16
	Bachelor's Degree	Mean	53.19
		Median	52.00
		Std. Deviation	12.48
		Minimum	19
		Maximum	75
		Range	56
		Interquartile Range	17

7. There is growing concern these days that Americans, as well as other world citizens, consume too much. Because our way of life is not sustainable, eventually the world's resources will be exhausted. The 2000 International Social Survey Programme asked respondents from several countries if they agreed with the statement that modern life harms the environment. Their responses are listed below by percentage.

Country	Percentage That Agree That Modern Life Harms Environment
Austria	40.7
Ireland	49.1
Sweden	47.9
Spain	62.9
Portugal	78.7
Denmark	44.4
Bulgaria	55.0
Russia	64.4
Latvia	39.5
Slovenia	42.9

a. Calculate and interpret the mean and standard deviation.
b. Calculate the 25th and the 75th percentiles. Interpret each value.
c. Compare the response of western European and eastern European countries. Calculate the mean and standard deviation for western European countries: Austria, Ireland, Sweden, Spain, Portugal, and Denmark. Calculate the mean and standard deviation for eastern European countries: Bulgaria, Russia, Latvia, and Slovenia. Use these values to discuss differences in respondents' attitudes from a statistical standpoint.

8. A group of investigators has just finished a study that measured the amount of time each partner in a marriage spends doing housework. The investigators classified each couple as traditional or nontraditional, depending on the attitudes of both partners. (Traditional couples commonly grant more authority to the male; nontraditional couples share more in decision making.) The investigators provide you with the following data for males only.

Hours of Housework per Week	
Traditional Family	Nontraditional Family
$\bar{Y} = 6.3$	$\bar{Y} = 12.4$
$\sum Y^2 = 1,104$	$\sum Y^2 = 2,889$
$\sum Y = 63$	$\sum Y = 186$
$N = 10$	$N = 15$

a. Calculate the variance and standard deviation from these statistics for each family type.
b. What can you say about the variability in the amount of time men spend doing housework in traditional versus nontraditional marriages? Why might there be a difference? Why might there be more variability for one type of family than another?
c. Was it necessary in this problem to provide you with the value to calculate the variance and standard deviation?

9. You are interested in studying the variability of crimes committed and police expenditures in the eastern and midwestern United States. The U.S. Census Bureau collected the following statistics on these two variables for 21 states in the East and Midwest in 1999.

State	Number of Crimes	Police Protection Expenditures per Capita (dollars), per 10,000 People
Maine	2,875	122.5
New Hampshire	2,282	141.8
Vermont	2,819	102.8
Massachusetts	3,262	218.7
Rhode Island	3,583	179.2
Connecticut	3,389	193.6
New York	3,279	292.4
New Jersey	3,400	236.6
Pennsylvania	3,114	171.2
Ohio	3,997	179.4
Indiana	3,766	124.5
Illinois	4,515	224.4
Michigan	4,325	172.3
Wisconsin	3,296	196.6
Minnesota	3,598	166.8
Iowa	3,224	135.8
Missouri	4,578	153.9
North Dakota	2,394	102.9
South Dakota	2,644	115.3
Nebraska	4,108	128.8
Kansas	4,439	161.6

Source: U.S. Bureau of the Census, *Statistical Abstract of the United States*, 2003, Tables 307 and 341.

a. Calculate the mean for each variable.
b. Calculate the standard deviation for each variable.
c. Calculate the interquartile range for each variable.
d. Compare the mean with the standard deviation and IQR for each variable. Does there appear to be more variability in the number of crimes or in police expenditures per capita in these states? Which states contribute more to this greater variability?
e. Suggest why one variable has more variability than the other. In other words, what social forces would cause one variable to have a relatively large standard deviation?

10. Construct a box plot for both variables in Exercise 9. Discuss how the box plot reinforces the conclusions you drew about the variability of crimes committed and police expenditures per capita.

11. Use the data in Table 5.6 from the chapter for this exercise.
a. Calculate the standard deviation for the percentage increase in the elderly population from 1990 to 2000.
b. Compare this statistic with the IQR and the box plot shown in Figure 5.6. Which is larger, the IQR or the standard deviation?
c. Would the standard deviation lead you to the same conclusion about the variability of the increase in the elderly population as the IQR and the box plot?

12. You decide to use GSS 2002 data to investigate how Americans feel about spending federal government money to halt the rising crime rate (NATCRIME) and on welfare programs (NAT-FARE). You obtain the following (Figures 5.17 (a) and (b)) selected output shown below, where "Too little" means that the federal government is spending too little, "About right" means that the level of government spending on this issue is about right, and "Too much" means the government is spending too much. (NAP stands for "Not Applicable," DK stands for "Don't Know," and NA for "No Answer.")

a. What would an appropriate measure of variability be for these variables? Why?
b. Calculate the appropriate measure of variability for each variable.
c. In 2002, was there more variability in attitudes toward spending on reducing crime or welfare?

Figure 5.17 (a)

NATCRIME HALTING RISING CRIME RATE

		Frequency	Percent	Valid Percent	Cumulative Percent
Valid	1 Too Little	758	27.4	57.4	57.4
	2 About Right	473	17.1	35.8	93.2
	3 Too Much	90	3.3	6.8	100.0
	Total	1321	47.8	100.0	
Missing	0 NAP	1407	50.9		
	8 DK	35	1.3		
	9 NA	2	0.1		
	Total	1444	52.2		
Total		2765	100.0		

Figure 5.17 (b)

NATFARE WELFARE

		Frequency	Percent	Valid Percent	Cumulative Percent
Valid	1 Too Little	279	10.1	21.2	21.2
	2 About Right	502	18.2	38.2	59.4
	3 Too Much	533	19.3	40.6	100.0
	Total	1314	47.5	100.0	
Missing	0 NAP	1407	50.9		
	8 DK	41	1.5		
	9 NA	3	0.1		
	Total	1451	52.5		
Total		2765	100.0		

13. Average hours worked per week by women is reported for 10 countries. What difference exists between average hours worked per week between European and non-European countries? Calculate the appropriate measures of central tendency and variability.

Country	Average Hours Worked per Week
European Countries	
England	33.25
Ireland	31.89
Portugal	42.11
Spain	35.38
Netherlands	24.68
Non-European Countries	
Japan	35.24
New Zealand	32.80
Canada	34.39
Philippines	42.18
United States	38.75

Source: International Social Survey Programme 2000.

14. You have been asked by your employer to prepare a brief statement on Internet usage at your company. Using the statistics from the following table, comment on your company's Internet usage characteristics.

Hours per Week	N	Mean	Std. Deviation
Hours spent surfing the Internet per week	1904	1.14	0.344
Hours spent on email per week	1881	4.15	7.235
Hours spent in chat rooms per week	313	2.12	5.492

Source: General Social Survey, 2002.

15. In 2003, a U.S. Census report examined several percentile ranks for income in the United States over the past 10 years. Selected findings are listed below.

Percentile Rank	2003	1998	1993
10th	10,536	10,979	9,635
20th	17,984	18,164	16,256
50th	43,318	43,825	39,165
80th	86,867	84,529	75,594
90th	118,200	114,396	103,010
95th	154,120	148,995	131,178

Source: U.S. Bureau of the Census. Current Population Reports P60-226, *Income, Poverty and Health Insurance Coverage in the United States: 2003*.

a. What is the difference, in dollars, between those at the 80th percentile and those at the 20th percentile in 2003? In 1993?

b. Your answer from 15a gives us a pretty good idea of the value of one of the measures of variability that has been discussed in this chapter. To which measure are we referring? Why?

16. The following table summarizes gender differences in education and pay in Canada. Based on the means and standard deviations (in parentheses), what conclusions can be drawn about the gender differences in pay?

	Women	Men
Education (years)	13.20	13.39
	(3.253)	(3.697)
Pay	$47,420	$52,557
	($24,300)	($22,572)

Source: International Social Survey Programme 2000.

Chapter 6

Relationships Between Two Variables: Cross-Tabulation

Bivariate analysis is a method designed to detect and describe the relationship between two variables. Before now, you may have had an intuitive sense of the terms *relationship* and *association*. We have seen in earlier chapters that by comparing the properties of different groups one can often think in terms of "relationships." In this and the following two chapters we look at the concept of relationships between variables in more depth.

> **Bivariate analysis**   A statistical method designed to detect and describe the relationship between two variables.

You should be familiar with the idea of "relationship" simply because you are aware that in the world around you things (and people) "go together." For example, as children age, their weight increases; larger cities have more crime than smaller cities. In fact, many of the reports in our daily newspapers are statements about relationships. For example, the 2000 International Social Survey Programme asked a sample of international respondents whether they were concerned about the current state of the environment. We adapted and compared these answers to a second question: whether respondents were willing to take a cut in their standard of living to protect the environment. Notice the pattern in Figure 6.1. In the year 2000, respondents who were worried about the state of the environment were more willing to take a cut in their standard of living (57.3%) relative to those who were not worried about the state of the environment (31.4%). This simple example illustrates a relationship or association between the variables *environmental concern* (worried vs. not worried) and *willingness to take a cut in standard of living* (willing vs. unwilling).[1]

One of the main objectives of social science is to make sense out of human and social experience by uncovering regular patterns among events. Therefore, the language of relationships is at the heart of social science inquiry. Consider the following examples from articles and research reports:

Figure 6.1   Environmental Concern and Willingness to Take Cut in Standard of Living: ISSP 2000

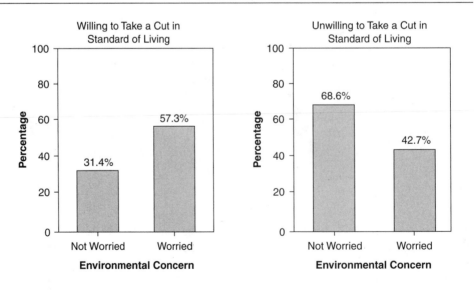

[1]For the purposes of this example, the original categories for each variable have been recoded.

Example 1 Students who had a history of earning good grades were less likely to miss class than students who did not.[2] (This example indicates a relationship between grade point average and absenteeism among college students.)

Example 2 Contrary to the stereotype, whites use government safety net programs more than blacks or Latinos, and they are more likely than minorities to be lifted out of poverty by the taxpayer money they get.[3] (This example indicates a relationship between race and receipt of government aid.)

Example 3 On average, blacks and Latinos have a lower likelihood of access to health care than whites.[4] (This example indicates a relationship between race and access to health care.)

In each of these examples, a relationship means that certain values of one variable tend to "go together" with certain values of the other variable. In Example 1, lower grades "go together" with irregular class attendance; higher grades go with regular class attendance. In Example 2, being white "goes together" with frequent use of government aid; being black or Latino goes with less frequent use of government aid. Finally, in Example 3, white "goes together" with greater likelihood of access to health care; being black or Latino "goes together" with a lower likelihood of access.

In this chapter we introduce one of the most common techniques used in the analysis of relationships between two variables: *cross-tabulation*. **Cross-tabulation** is a technique for analyzing the relationship between two variables that have been organized in a table. We demonstrate not only how to detect whether two variables are associated, but also how to determine the strength of the association and, when appropriate, its direction. We will also see how these methods are applied in "real" research situations.[5]

Cross-tabulation   A technique for analyzing the relationship between two variables that have been organized in a table.

▣ INDEPENDENT AND DEPENDENT VARIABLES

In the social sciences, an important aspect in research design and statistics is the distinction between the *independent variable* and the *dependent variable*. These terms, first introduced in Chapter 1, are used throughout this chapter as well as in the following chapters, and therefore it is important that you understand the distinction between them. Let's take our example about environmental concern and willingness to take a cut in one's standard of living, in which environmental concern—whether a person is concerned with the current state of the

---

[2]Gary Wyatt, "Skipping Class: An Analysis of Absenteeism Among First-Class College Students," *Teaching Sociology* 20 (July 1992): 201–207.

[3]*USA Today*, October 9, 1992.

[4]National Center for Health Statistics, *Summary Health Statistics for U.S. Population: National Health Interview Survey*, United States, 2002.

[5]Full consideration of the question of detecting the presence of a bivariate relationship requires the use of inferential statistics. Inferential statistics is discussed in Chapters 11 through 14.

environment—has some influence on the level of willingness to take a cut in one's standard of living. Even though environmental concern is not necessarily a direct cause of willingness to take a cut in one's standard of living, environmental concern is assumed to be connected to willingness to take a cut in one's standard of living through a complex set of experiences—such as education, employment, and other socialization experiences—all of which do have an influence on the acquisition of attitudes and opinions. If we hypothesize that willingness to take a cut in one's standard of living (which is the variable to be explained by the researcher) varies by whether a person is concerned about the state of the environment (which is the variable assumed to influence willingness to take a cut in one's standard of living), then environmental concern is the independent variable and willingness to take a cut in one's standard of living is the dependent variable.

In each of the illustrations given, there are two variables, an independent and a dependent variable. In Example 1, the purpose of the research is to explain absenteeism. One of the variables hypothesized as being connected to absenteeism is grades. Therefore, absenteeism is the dependent variable and grades the independent variable. In Example 2, the object of the investigation is to examine the common stereotype that people of color use government aid more than white Americans. The investigator is trying to explain differences in utilization of government aid using race as an explanatory variable. Therefore, utilization of government aid is the dependent variable and race is the independent variable. Similarly, in Example 3, access to health care is the dependent variable because it is the variable to be explained, whereas race, the explanatory variable, is the independent variable.

The statistical techniques discussed in this and the following two chapters help the researcher decide the strength of the relationship between the independent and dependent variables.

✓ *Learning Check.*   *For some variables, whether it is the independent or dependent variable depends on the research question. If you are still having trouble distinguishing between an independent and a dependent variable, go back to Chapter 1 (pp. 9–10) for a detailed discussion.*

▣ HOW TO CONSTRUCT A BIVARIATE TABLE: RACE AND HOME OWNERSHIP

A **bivariate table** displays the distribution of one variable across the categories of another variable. It is obtained by classifying cases based on their joint scores for two variables. It can be thought of as a series of frequency distributions joined to make one table. The data in Table 6.1 represent a sample of General Social Survey (GSS) respondents by race and whether they own or rent their home.

To make sense out of these data we must first construct the table in which these individual scores will be classified. In Table 6.2, the 15 respondents have been classified according to joint scores on race and home ownership.

Table 6.1   Race and Home Ownership for 15 GSS Respondents

Respondent	Race	Home Ownership
1	Black	Own
2	Black	Own
3	White	Rent
4	White	Rent
5	White	Own
6	White	Own
7	White	Own
8	Black	Rent
9	Black	Rent
10	Black	Rent
11	White	Own
12	White	Own
13	White	Rent
14	White	Rent
15	Black	Rent

Table 6.2   Home Ownership by Race (Absolute Frequencies)

HOME OWNERSHIP	RACE			
	Black	White		
Own	2	5	7	Row Marginals
Rent	4	4	8	(Row Total)
	6	9	15	Total Cases (*N*)
	Column Marginals (Column Total)			

The table has the following features typical of most bivariate tables:

1. The table's title is descriptive, identifying its content in terms of the two variables.

2. It has two dimensions, one for race and one for home ownership. The variable *home ownership* is represented in the rows of the table, with one row for owners and another for renters. The variable *race* makes up the columns of the table, with one column for each racial group. A table may have more columns and more rows, depending on how many categories the variables represent. For example, had we included a group of

Latinos, there would have been three columns (not including the Row Total column). Usually, the independent variable is the **column variable** and the dependent variable is the **row variable.**

3. The intersection of a row and a column is called a **cell**. For example, the two individuals represented in the upper left cell are blacks who are also homeowners.

4. The column and row totals are the frequency distribution for each variable, respectively. The column total is the frequency distribution for *race*, the row total for *home ownership*. Row and column totals are sometimes called **marginals**. The total number of cases (N) is the number reported at the intersection of the row and column totals. (These elements are all labeled in the table.)

5. The table is a 2×2 table because it has two rows and two columns (not counting the marginals). We usually refer to this as an $r \times c$ table, in which r represents the number of rows and c the number of columns. Thus, a table in which the row variable has three categories and the column variable two categories would be designated as a 3×2 table.

6. The source of the data should also be clearly noted in a source note to the table. This is consistent with what we reviewed in Chapter 2, Organization of Information.

Bivariate table   A table that displays the distribution of one variable across the categories of another variable.

Column variable   A variable whose categories are the columns of a bivariate table.

Row variable   A variable whose categories are the rows of a bivariate table.

Cell   The intersection of a row and a column in a bivariate table.

Marginals   The row and column totals in a bivariate table.

✔ *Learning Check.*   *Examine Table 6.2. Make sure you can identify all of the parts just described and that you understand how the numbers were obtained. Can you identify the independent and dependent variables in the table? You will need to know this to convert the frequencies to percentages.*

回 HOW TO COMPUTE PERCENTAGES IN A BIVARIATE TABLE

To compare home ownership status for blacks and whites, we need to convert the raw frequencies to percentages because the column totals are not equal. Recall from Chapter 2 that percentages are especially useful for comparing two or more groups that differ in size. There are two basic rules for computing and analyzing percentages in a bivariate table:

1. Calculate percentages within each category of the independent variable.

2. Interpret the table by comparing the percentage point difference for different categories of the independent variable.

Calculating Percentages Within Each Category of the Independent Variable

The first rule means that we have to calculate percentages within each category of the variable that the investigator defines as the independent variable. When the independent variable is arrayed in the *columns*, we compute percentages within each column separately. The frequencies within each cell and the row marginals are divided by the total of the column in which they are located, and the column totals should sum to 100 percent. When the independent variable is arrayed in the *rows*, we compute percentages within each row separately. The frequencies within each cell and the column marginals are divided by the total of the row in which they are located, and the row totals should sum to 100 percent.

In our example, we are interested in *race* as the independent variable and in its relationship with *home ownership*. Therefore, we are going to calculate percentages by using the column total of each racial group as the base of the percentage. For example, the percentage of black respondents who own their homes is obtained by dividing the number of black homeowners by the total number of blacks in the sample:

$$(100)\frac{2}{6} = 33.3\%$$

Table 6.3 presents percentages based on the data in Table 6.2. Notice that the percentages in each column add up to 100 percent, including the total column percentages. Always show the *N*s that are used to compute the percentages—in this case, the column totals.

Table 6.3   Home Ownership by Race (in Percentages)

HOME OWNERSHIP	RACE		
	Black	*White*	*Total*
Own	33.3%	55.6%	46.7%
Rent	66.7%	44.4%	53.3%
Total	100% (6)	100% (9)	100% (15)

Comparing the Percentages Across Different Categories of the Independent Variable

The second rule tells us to compare how home ownership varies between blacks and whites. Comparisons are made by examining differences between percentage points across different categories of the independent variable. Some researchers limit their comparisons to categories with at least a 10 percentage point difference. In our comparison, we can see that there is a 22.3 percentage point difference between the percentage of white homeowners (55.6%) versus black homeowners (33.3%). In other words, in this group whites are more likely to be homeowners than blacks[6]. Therefore, we can conclude that one's race appears to be associated with the likelihood of being a homeowner.

Notice that the same conclusion would be drawn had we compared the percentage of black and white renters. However, since the percentages of homeowners and renters within each racial group sum to 100 percent, we need to make only one comparison. In fact, for any 2 × 2 table only one comparison needs to be made to interpret the table. For a larger table, more than one comparison can be made and used in interpretation.

✔ *Learning Check.* *Practice constructing a bivariate table. Use Table 6.1 to create a percentage bivariate table. Compare your table with Table 6.3. Did you remember all of the parts? Are your calculations correct? If not, go back and review this section. It might be helpful to examine A Closer Look 6.1. It illustrates the process of constructing and percentaging bivariate tables. Remember, you must correctly identify the independent variable so you know whether to percentage across the rows or down the columns.*

▣ HOW TO DEAL WITH AMBIGUOUS RELATIONSHIPS BETWEEN VARIABLES

Sometimes it isn't apparent which variable is independent or dependent; sometimes the data can be viewed either way. In this case, you might compute both row and column percentages. For example, Table 6.4 presents three sets of figures for the variables *attitude toward abortion* and *job security* for a sample of 89 GSS 2002 respondents: (a) the absolute frequencies, (b) the column percentages, and (c) the row percentages. *Job security* (labeled "Job Find") is measured with the survey question "About how easy would it be for you to find a job with another employer with approximately the same income and fringe benefits you now have?" The variable *attitude toward abortion* is measured in terms of the respondent's approval or disapproval of three reasons for obtaining an abortion: (1) the woman does not want the baby because the family has a very low income and cannot afford more children; (2) the woman is

---

[6]Note that this group is but a small sample taken from the GSS national sample. The relationship between home ownership and race noted here may not necessarily hold true in other (larger) samples.

▣ A Closer Look 6.1
Percentaging a Bivariate Table

1. Black and white homeowners and renters:

Owners

Renters

2. Divide respondents into two groups by race (the independent variable); count the number in each group to get the column totals.

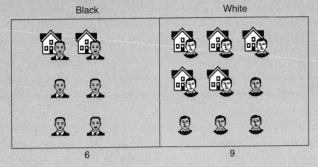

Black   White

6   9

3. Divide each group into homeowners and renters (the dependent variable); count the number in each group to get the row totals.

Black   White

Owners   7

Renters   8

6   9

4. Count each cell:

	Black	White	
Owners	2	5	7
Renters	4	4	8
	6	9	

5. % of blacks who are owners: (100)2/6 = 33%
   % of whites who are owners: (100)5/9 = 56%
   % of blacks who are renters: (100)4/6 = 67%
   % of whites who are renters: (100)4/9 = 44%

6. Compare percentages: 33% vs. 56%
                         67% vs. 44%

Table 6.4 The Different Ways Percentages Can Be Computed: Support for Abortion by Job Security

a. Absolute Frequencies

Abortion	Job Find Easy	Job Find Not Easy	Row Total
Yes	7	33	40
No	26	23	49
Column total	33	56	89

b. Column Percentages (column totals as base)

Abortion	Job Find Easy	Job Find Not Easy	Row Total
Yes	21%	59%	45%
No	79%	41%	55%
Column total	100%	100%	100%
	(33)	(56)	(89)

c. Row Percentages (row totals as base)

Abortion	Job Find Easy	Job Find Not Easy	Row Total
Yes	18%	82%	100% (40)
No	53%	47%	100% (49)
Column total	37%	63%	100% (89)

Source: General Social Survey, 2002.

not married and does not want to marry the father; and (3) the woman does not want to have more children. Table 6.4b shows that respondents who feel less secure economically are more likely to support the right to abortion than those who feel more secure economically (59% compared with 21%). Table 6.4c shows that individuals who support abortion are less likely to feel economically secure than those who are against abortion (18% compared with 53%).

Thus, percentaging within each *column* (Table 6.4b) allows us to examine the hypothesis that job security (the independent variable) is associated with support for abortion (the dependent variable). When we percentage within each *row* (Table 6.4c), the hypothesis is that attitudes toward abortion (the independent variable) may be related to one's sense of job security (the dependent variable).[7] Figures 6.2a and 6.2b are simple bar charts illustrating the two methods of calculating and comparing percentages depicted in Tables 6.4b and 6.4c.

---

[7]One other way in which percentages are sometimes expressed is with the total number of cases (N) used as the base. These overall percentages express the proportion of the sample who share two properties. For example, 7 of 89 respondents (7.9%) support abortion and have job security. Overall percentages do not have as much research utility as row and column percentages and are used less frequently.

Figure 6.2   Percentage Who Support Abortion by Job Find (column percent)

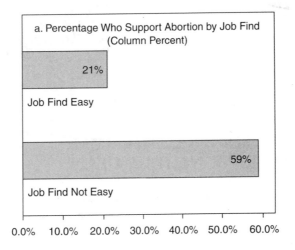

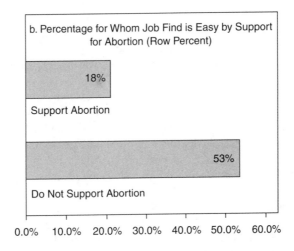

Source: General Social Survey, 2002.

Finally, it is important to understand that ultimately what guides the construction and interpretation of bivariate tables is the theoretical question posed by the researcher. Although the particular example in Table 6.4 makes sense if interpreted using row or column percentages, not all data can be interpreted this way. For example, a table comparing women's and men's attitudes toward sexual harassment in the workplace could provide a sensible explanation in only one direction. Gender might influence a person's attitude toward sexual harassment, but a person's attitude toward sexual harassment certainly couldn't influence her or his gender. Therefore, either row or column percentages are appropriate, depending on the way the variables are arrayed, but not both.

✔**Learning Check.**   *Figures 6.2a and 6.2b each show only one set of bars. Figure 6.2a shows only the percentage of each job group who support abortion. Figure 6.2b shows only the percentage for whom job find is easy in each abortion attitude group. For what percentage of each group is job find not easy? What percentage of each group does not have job security (job find not easy)? Hint: The percentages add to 100.0 percent within each category of the variable treated as independent.*

▣ READING THE RESEARCH LITERATURE: MEDICAID USE AMONG THE ELDERLY

The guidelines for constructing and interpreting bivariate tables discussed in this chapter are not always strictly followed. Most bivariate tables presented in the professional literature are a good deal more complex than those we have just been describing. Let's conclude this section with a typical example of how bivariate tables are presented in social science literature. The following example is drawn from a study by Madonna Harrington Meyer on Medicaid use among the elderly.[8]

Access to health care for all Americans is at the top of the U.S. domestic agenda today. The rise of long-term care as a politically salient topic is fueled by the increase in the elderly population, which is 10 times larger than it was in 1900 and will more than double by the year 2030. Financing of long-term care is a problem for older persons because Medicare, the universal health-care program for the aged, excludes most long-term care. Only Medicaid, the poverty-based health-care program, includes long-term care coverage. Therefore, only the poor elderly receive assistance from the state for long-term care.[9]

In this study, Meyer explores the distribution of Medicaid benefits to the frail elderly. She examines the hypothesis that gender and race are important determinants of Medicaid use "because the U.S. long-term care system stratifies by gender and race by perpetuating, rather than alleviating, inequality created by social and market forces."[10]

The study examines differences in Medicaid use in 1984 by age, education, marital status, gender, and race. Meyer analyzed data from the National Long Term Care survey conducted by the Department of Health and Human Services in 1982 and 1984. The data set is based on a national random sample of 6,000 functionally impaired older persons who resided in the community in 1982. By 1984, respondents were either still living in the community, had entered a nursing home, or had died.

---

[8]Madonna Harrington Meyer, "Gender, Race, and the Distribution of Social Assistance: Medicaid Use Among the Frail Elderly," *Gender & Society* 8, no. 1 (1994): 8–28.

[9]Ibid., p. 9.

[10]Ibid., p. 12.

Table 6.5 Percentage Medicaid Use by Functionally Impaired Older Persons in 1984, by Age, Education, Marital Status, Gender, and Race

| | Received Medicaid in 1984 | | |
	Yes	*No*	*N*
Age			
65–74	19.5	80.5	1,561
75–84	23.8	76.2	1,943
85+	25.4	74.6	1,007
Education			
8th grade or less	29.5	70.5	2,326
9th–12th grade	16.5	83.5	1,523
Some college	8.1	91.5	530
Marital Status			
Married	13.7	86.3	1,947
Widowed	28.6	71.4	2,079
Divorced, separated, never married	34.6	65.4	437
Gender			
Men	17.1	82.9	1,488
Women	25.4	74.6	3,024
Race			
White	19.1	80.9	3,942
Black and Latino	47.7	52.3	570

Source: Adapted from Madonna Harrington Meyer, "Gender, Race, and the Distribution of Social Assistance: Medicaid Use Among the Frail Elderly," *Gender & Society* 8, no. 1 (1994): 8–28. Used by permission of Sage Publications, Inc.

Table 6.5 shows the results of the survey. Follow these steps in examining it:

1. Identify the dependent variable and the type of unit of analysis it describes (such as individual, city, or child). Here the dependent variable is *received Medicaid in 1984*. The categories for this variable are "yes" and "no." The type of unit used in this table is individual.

2. Identify the independent variables included in the table and the categories of each. There are five independent variables: age, education, marital status, gender, and race. Age consists of three categories: 65–74, 75–84, and 85+. Education consists of the categories "8th grade or less," "9th to 12th grade," and "some college." "Married," "widowed," and "divorced, separated, never married" are the categories for marital status. Gender consists of "men" and "women"; and finally, "white" and "black and Latino" are the categories for race.

3. Clarify the structure of the table. Note that the independent variables are arrayed in the rows of the table and the dependent variable, *received Medicaid in 1984*, is arrayed in the

columns. The table is divided into five panels, one for each independent variable. There are actually five bivariate tables here—one for each independent variable.

Since the independent variables are arrayed in the rows, percentages are calculated within each row separately, with the row totals serving as the base for the percentages. For example, there were 1,561 respondents who were 65 to 70 years old. Of these, 19.5 percent received Medicaid in 1984 and 80.5 percent did not. Similarly, of the 1,943 respondents who were 75 to 84, 23.8 percent received Medicaid in 1984 and 76.2 percent did not. Although not shown in the table, the percentages within each row add to 100 percent.

4. Using Table 6.5, we can make a number of comparisons, depending on which independent variable we are examining. For example, to determine the relationship between age and the propensity to use Medicaid, compare the percentages of respondents in the different age groups who received Medicaid in 1984 (19.5%, 23.8%, and 25.4%). Alternatively, you can compare the percentages of respondents in the three age groups who did not receive Medicaid (80.5%, 76.2%, and 74.6%). Based on these percentage comparisons, we can conclude that among the frail elderly age is associated with Medicaid use: the oldest-old are more likely than the youngest-old to receive Medicaid.

Next look at the relationship between education and the propensity to use Medicaid. You can compare the percentage of respondents with an 8th-grade education or less who received Medicaid (29.5%) with the percentage of respondents with a 9th- to 12th-grade education (16.5%) or with some college (8.1%) who received Medicaid. You can make similar comparisons to determine the association between marital status, gender, or race and the receipt of Medicaid.

The bivariate relations between all five independent variables—age, education, marital status, gender, and race—and receipt of Medicaid presented in Table 6.5 are also illustrated in Figure 6.3. Notice that only the percentages of elderly who received Medicaid in 1984 ("yes") are shown. Since the percentage of elderly who responded "yes" and the percentage that responded "no" sum to 100, there is no need to show both sets of figures.

5. Finally, what conclusions can you draw about variations in the propensity to use Medicaid? The author offers this interpretation of the findings presented in the table:

> Advancing age is a . . . determinant of Medicaid use, because with age, the need for chronic care increases while available resources decrease. The magnitude of the relationship is small, however; Medicaid use is only slightly higher for the oldest old than for the youngest old. Level of education is also . . . related to Medicaid use, in part because of the link between education and income. Those with an 8th grade level of education or less are nearly 4 times as likely as those with some college education to receive Medicaid. Marital status is . . . [also] related to Medicaid use. Widowed persons, for example, are more than twice as likely as married persons to rely on Medicaid. Older persons who are divorced, separated, or never married are most likely to receive Medicaid. . . . Finally, gender and race are also . . . predictors of Medicaid use. Women are somewhat more likely than men to rely on Medicaid, and other races are considerably more likely than whites to rely on Medicaid.[11]

---

[11]Ibid., pp. 14–15.

Figure 6.3   Percentage Who Received Medicaid in 1984 by Age, Education, Marital
Status, Gender, and Race

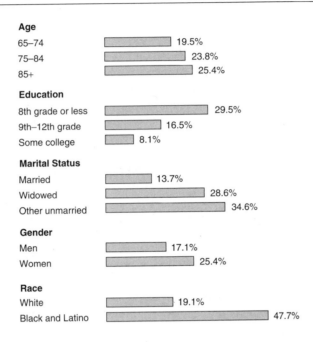

✔ *Learning Check.* *Use Table 6.5 to verify each of the following conclusions drawn by the researcher about Medicaid use among the elderly: (1) Advancing age is a determinant of Medicaid use. (2) Level of education is related to Medicaid use. (3) Marital status is related to Medicaid use. (4) Gender and race are also related to Medicaid use. Can you explain these patterns? What other questions do these patterns raise about Medicaid use among the elderly?*

▣ THE PROPERTIES OF A BIVARIATE RELATIONSHIP

So far we have looked at the general principles of a bivariate relationship as well as the more specific "mechanics" involved in examining bivariate tables. In this section we present some detailed observations we may want to make about the "properties" of a bivariate association. These properties can be expressed as three questions to ask when examining a bivariate relationship[12]:

---

[12]The same three properties are also discussed by Joseph F. Healey in *Statistics: A Tool for Social Research*, 5th ed. (Belmont, CA: Wadsworth, 1999), pp. 314–320.

1. Does there appear to be a relationship?
2. How strong is it?
3. What is the direction of the relationship?

The Existence of the Relationship

We have seen earlier in this chapter that calculating percentages and comparing them are the two operations necessary to analyze a bivariate table. Based on Table 6.6, we want to examine whether the frequency of church attendance by respondents had an effect on their support for abortion. Support for abortion was measured with the following question: "Please tell me whether or not you think it should be possible for a pregnant woman to obtain a legal abortion if the woman wants it for any reason." Frequency of church attendance was determined by asking respondents to indicate how often they attend religious services.[13]

Let's hypothesize that those who attend church frequently are more likely to be pro-life. We are not suggesting that church attendance necessarily "causes" pro-life attitudes, but that perhaps there is an indirect connection between the two. For example, perhaps those who attend church less than frequently are more likely to want decisions about the body to be made on an individual basis through the right to choose an abortion. (Indirect associations often can be elaborated further by looking at other variables. We discuss elaboration in more detail later in this chapter.)

In this formulation church attendance is said to "influence" attitudes toward abortion, so it is the independent variable; therefore, percentages are calculated within each category of church attendance (church attendance is the column variable). We now want to establish whether a relationship exists between the two variables.

A relationship is said to exist between two variables in a bivariate table if the percentage distributions vary across the different categories of the independent variable, church attendance. We can easily see that the percentage that supports abortion changes across the different levels of church attendance. Of those who never attend church, 55 percent are pro-choice; of those who infrequently attend church, 50.1 percent are pro-choice; and of those who frequently attend church, 26 percent are pro-choice.

Table 6.6 indicates that church attendance and support for abortion are associated, as hypothesized.

Table 6.6   Support for Abortion by Church Attendance

ABORTION	CHURCH ATTENDANCE			
	Never	Infrequently	Frequently	Total
Yes	55.0%	50.1%	26.0%	42.5%
No	45.0%	49.9%	74.0%	57.5%
	100% (100)	100% (208)	100% (181)	100% (489)

Source: General Social Survey, 2002.

---

[13]Church attendance has been recoded into three categories.

Table 6.7   Support for Abortion by Church Attendance (a Hypothetical Illustration of No Relationship)

ABORTION	*CHURCH ATTENDANCE*			
	Never	*Infrequently*	*Frequently*	*Total*
Yes	42.5%	42.5%	42.5%	42.5%
No	57.5%	57.5%	57.5%	57.5%
	100%	100%	100%	100%
	(100)	(208)	(181)	(489)

If church attendance were unrelated to attitudes toward abortion among GSS respondents, then we would expect to find equal percentages of respondents who are pro-choice (or anti-choice), regardless of the level of church attendance. Table 6.7 is a fictional representation of a strictly hypothetical pattern of no association between abortion attitudes and church attendance. The percentage of respondents who are pro-choice in each category of church attendance is equal to the overall percentage of respondents in the sample who are pro-choice (42.5%).

The Strength of the Relationship

In the preceding section we saw how to establish whether an association exists in a bivariate table. If it does, how do we determine the strength of the association between the two variables? A quick method is to examine the percentage difference across the different categories of the independent variable. The larger the percentage difference across the categories, the stronger the association.

In the hypothetical example of no relationship between church attendance and attitude toward abortion (Table 6.7), there is a 0 percent difference between the columns. At the other extreme, if all respondents who never attended church were pro-choice, and none of the respondents who frequently attended church were pro-choice, a perfect relationship would be manifested in a 100 percent difference. Most relationships, however, will be somewhere in between these two extremes. In fact, we rarely see a situation with either a 0 percent or a 100 percent difference. Going back to the observed percentages in Table 6.6, we find the largest percentage difference between respondents who never attend church and respondents who frequently attend church (55.0% − 26.0% = 29.0%). The difference between respondents who infrequently attend church and respondents who frequently attend church (50.1% − 26.0% = 24.1%), though not as large, is nonetheless substantial, indicating a moderate relationship between church attendance and attitudes toward abortion.

Percentage differences are a rough indicator of the strength of a relationship between two variables. In later chapters we discuss measures of association that provide a more standardized indicator of the strength of an association.

The Direction of the Relationship

When both the independent and dependent variables in a bivariate table are measured at the ordinal level or higher, we can talk about the relationship between the variables as being either positive or negative. A **positive** bivariate relationship exists when the variables vary in the same direction. Higher values of one variable "go together" with higher values of the other variable. In a **negative** bivariate relationship the variables vary in opposite directions: higher values of one variable "go together" with lower values of the other variable (and the lower values of one go together with the high values of the other).

Positive relationship   A bivariate relationship between two variables measured at the ordinal level or higher in which the variables vary in the same direction.

Negative relationship   A bivariate relationship between two variables measured at the ordinal level or higher in which the variables vary in opposite directions.

Table 6.8, from the 2000 International Social Survey Programme, displays a positive relationship between willingness to pay higher taxes and willingness to pay higher prices. Examine each category separately. For respondents who are unwilling to pay higher prices, an unwillingness to pay higher taxes is most typical (91.5%). For respondents who are indifferent to paying higher prices, the most common response is to be indifferent to paying higher taxes (55.1%); and finally, for respondents who are willing to pay higher prices, a willingness to pay higher taxes is most typical (57.8%). This is a positive relationship, with a willingness to pay higher prices associated with a willingness to pay higher taxes and an unwillingness to pay higher prices associated with an unwillingness to pay higher taxes.

Table 6.8   Willingness to Pay Higher Taxes by Willingness to Pay Higher Prices: A Positive Relationship

WILLINGNESS TO PAY HIGHER TAXES	WILLINGNESS TO PAY HIGHER PRICES		
	Unwilling	*Indifferent*	*Willing*
Unwilling	91.5%	36.4%	23.6%
Indifferent	5.1%	55.1%	18.6%
Willing	3.4%	8.5%	57.8%
	100% (529)	100% (352)	100% (532)

Source: ISSP 2000.

Table 6.9   Support for Attendance of Religious Services by Educational Level: A
Negative Relationship

ATTENDANCE OF RELIGIOUS SERVICES	*EDUCATIONAL LEVEL*		
	None	*Secondary Degree*	*University Degree*
Never	5.2%	32.5%	37.3%
Infrequently	28.6%	35.0%	34.9%
2-3 Times per Month or More	66.2%	32.5%	27.8%
	100% (77)	100% (237)	100% (126)

Source: ISSP 2000.

Table 6.9, also from the International Social Survey Programme, shows a negative association between educational level and attendance of religious services for a sample of about 400 international respondents.[14] Individuals with no education typically attended religious services two to three times per month or more (66.2%). Individuals with a secondary degree (i.e., roughly, the U.S. equivalent to high school) typically attended religious services infrequently, ranging from monthly to several times a year (35.0%); and for individuals who had completed work at a university the most common category was "never," meaning they never attend religious services (37.3%). The relationship is a negative one because as educational level increases the frequency of attendance of religious services decreases.

In the next chapter we will see that measures of relationship for ordinal or interval-ratio variables take on a positive or a negative value, depending on the direction of the relationship.

> ✔ *Learning Check.*   *Based on Table 6.9, collapse the attendance of religious services categories into two: never vs. sometimes. Recalculate the bivariate table, estimating the percentages. Compare your results to Table 6.9. What can you say about the changes in the relationship between educational level and attendance of religious services?*

▣ ELABORATION

In the preceding sections we have looked at relationships between two variables—an independent and a dependent variable. The examination of a possible relationship between two

---

[14]For purposes of illustration, only selected categories of educational level and attendance of religious services are shown.

variables, however, is only a first step in data analysis. Having established through bivariate analysis that the independent and dependent variables are associated, we seek to further interpret and understand the nature of this relationship. In this section we discuss a procedure called *elaboration*. **Elaboration** is a process designed to further explore a bivariate relationship, involving the introduction of additional variables, called **control variables**. By adding a control variable to our analysis, we are considering or "controlling" for the variable's effect on the bivariate relationship. Each potential control variable represents an alternative explanation for the bivariate relationship under consideration.

Elaboration   A process designed to further explore a bivariate relationship; it involves the introduction of control variables.

Control variable   An additional variable considered in a bivariate relationship. The variable is controlled for when we take into account its effect on the variables in the bivariate relationship.

The introduction of additional control variables into a bivariate relationship serves three primary goals in data analysis.

- Elaboration allows us to test for nonspuriousness. Establishing cause-and-effect relations requires not only showing that an independent and a dependent variable are associated, but also establishing the time order between them and providing theoretical and empirical evidence that the association is nonspurious—that is, that it cannot be "explained away" by other variables.
- Elaboration clarifies the causal sequence of bivariate relationships by introducing variables hypothesized to intervene between the independent and dependent variables.
- Elaboration specifies the different conditions under which the original bivariate relationship might hold.

In the preceding sections we learned how to establish that two variables are associated; in this section we explore the theoretical and statistical considerations involved in elaborating bivariate relationships. We illustrate the process of elaboration using three examples. The first is an example of testing for nonspuriousness; the second is a research example illustrating a causal sequence in which a third variable intervenes between the independent and dependent variables; and finally, the third research example illustrates how elaboration can uncover conditional relationships.

Testing for Nonspuriousness: Firefighters and Property Damage

Let's begin with a favorite example of a spurious relationship. Researchers have confirmed a strong bivariate relationship between *number of firefighters* (the independent variable) at a fire site and *amount of property damage* (the dependent variable). The more firefighters at the site, the greater the amount of damage. This association might lead you to the embarrassing conclusion (depicted in Figure 6.4) that firefighters cause property damage at fire sites.

Figure 6.4   The Bivariate Relation Between Number of Firefighters and Property
Damage

Figure 6.5   Spurious Relationship

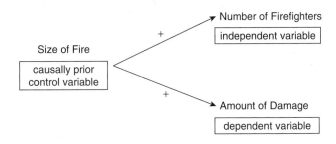

Figure 6.4 depicts what might be a *direct causal relationship* between firefighters and the amount of damage. The relationship between two variables is said to be a **direct causal relationship** when it cannot be accounted for by other theoretically relevant variables. Clearly, in this case the relationship between the number of firefighters and amount of damage can be accounted for by a third, causally prior variable—the size of the fire. When the fire is large, more firefighters are sent to the site and there is a great deal of property damage. Similarly, when the fire is small, fewer firefighters are at the site and there is probably very little damage.

This alternative explanation is shown in Figure 6.5. Note that according to the hypothesized causal order suggested in Figure 6.5, the number of firefighters and the extent of property damage are both related to the variable *size of fire* but are not related to each other. The size of the fire is called a *control variable*, and the relation between the number of firefighters and property damage as depicted in Figure 6.5 is *spurious*. A **spurious relationship** is a relationship between two variables in which both the independent and dependent variables are influenced by a causally prior control variable and there is no causal link between them. The bivariate relationship between the independent and dependent variables can thus be "explained away" through the introduction of the control variable.

Direct causal relationship   A bivariate relationship that cannot be accounted for by other theoretically relevant variables.

Spurious relationship   A relationship in which both the independent and dependent variables are influenced by a causally prior control variable and there is no causal link between them. The relationship between the independent and dependent variables is said to be "explained away" by the control variable.

Researchers have adopted the following rule of thumb for determining whether a relationship between two variables is either direct (causal) or spurious: If the bivariate relationship between the two variables remains about the same after controlling for the effect of one or more causally prior and theoretically relevant variables, then the original bivariate relationship is said to be a direct (causal relationship) association. On the other hand, if the original bivariate relationship decreases considerably (or vanishes), then the bivariate relationship is said to be spurious.

Let's see how we can apply this rule of thumb to the firefighter example. One way to control for the effect of the size of the fire on the relationship between the number of firefighters and the extent of damage is to divide the fire sites into large and small fires and then reexamine the bivariate association between the other two variables within each group of fire sites. If the original bivariate relationship vanishes (or diminishes considerably), then the explanation suggested by Figure 6.5 would seem more likely. If, however, the original relationship is maintained, then we may need to hold on to the original explanation suggested by Figure 6.4 or go back to the drawing board and think of other alternative explanations for the puzzling relationship between the number of firefighters and the extent of property damage.

Figure 6.6 illustrates the bivariate association between the number of firefighters and the extent of property damage (6.6a), and the process of controlling for the variable *size of fire* (6.6b). Note that the control for *size of fire* resulted in a substantial decrease (from 40% to 12% difference) in the size of the relationship between the number of firefighters and property damage. This result supports the notion, as depicted in Figure 6.5, that the size of the fire explains both the number of firefighters and the extent of property damage, and that the relationship between the number of firefighters and property damage is therefore spurious.

The introduction of the control variable *size of fire* into the original bivariate relationship between *number of firefighters* and *amount of damage* illustrates the process of elaboration. These are the steps:

1. Divide the observations into subgroups on the basis of the control variable. We have as many subgroups as there are categories in the control variable. (In our case there were two subgroups: small and large fires.)

2. Reexamine the relationship between the original two variables separately for the control variable subgroups. The separate tables are called **partial tables**; they display the **partial relationship** between the independent (number of firefighters) and dependent (amount of damage) variables within each specific category of the control variable (small versus large fire size).

3. Compare the partial relationships with the original bivariate relationship for the total group. In a direct causal pattern, the partial relationships will be very close to the original bivariate relationship. In a spurious pattern, the partial relationship will be much weaker than the original bivariate relationship.

Partial tables   Bivariate tables that display the relationship between the independent and dependent variables while controlling for a third variable.

Partial relationship   The relationship between the independent and dependent variables shown in a partial table.

Figure 6.6   Elaborating a Bivariate Relationship

1. A bivariate relationship between the number of firefighters and the extent of the property damage at 20 fire sites.

(a)

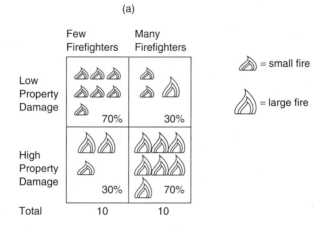

% difference = 70% − 30% = 40% (column percentage)

2. Control for size of fire: divide fire sites into small and large fires. In each group, recalculate the bivariate relationship between the number of firefighters and the extent of the property damage.

(b)

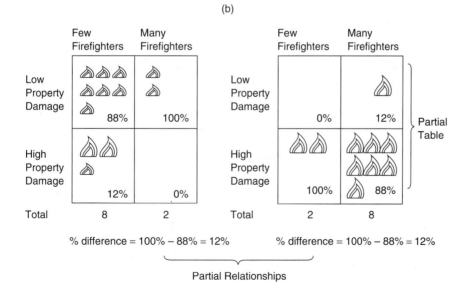

3. Compare the partial relationships with the original relationship: 40% compared with 12%.

We have employed the elaboration procedure to test for a spurious relationship between the number of firefighters and the amount of property damage. Now let's see how elaboration is used to interpret the causal sequence of bivariate relationships by introducing a control variable hypothesized to *intervene* between the independent and dependent variables.

An Intervening Relationship:
Religion and Attitude Toward Abortion

The research on the relationship between religious affiliation and attitudes toward abortion has shown a consistent pattern: religious affiliation is related to the level of support for abortion.[14] In particular, it has been shown that Catholics oppose abortion more than Protestants or Jews.[15]

To test the hypothesis that religion and abortion attitudes are related, we used data from the 1988 to 1991 GSS sample. We limited our analysis to Catholics and Protestants because of the small numbers of respondents with other religious affiliation. Attitudes toward abortion are measured in terms of respondents' approval or disapproval of the following three situations: (1) the woman does not want the baby because the family has a very low income and cannot afford more children; (2) the woman is not married and does not want to marry the father; and (3) the woman does not want to have more children.[16]

The findings are presented in Table 6.10 and illustrated in Figure 6.7. Since, according to the hypothesis, religious affiliation is the independent variable, we use column percentages for our analysis. The results provide some support for the hypothesis that religion is related to attitudes toward abortion. We see that 45 percent of Protestants, compared with 34 percent of Catholics, support a woman's right to an abortion for these cited reasons.

Table 6.10   Religious Affiliation and Support for Abortion

SUPPORT	RELIGIOUS AFFILIATION		
	Catholic	Protestant	Total
Yes	34% (56)	45% (109)	41%
No	66% (107)	55% (131)	59%
Total (N)	100% (163)	100% (240)	100% (403)

Source: General Social Survey, 1988–1991.

---

[14]For example, see Harris Mills, "Religion, Values, and Attitudes Toward Abortion," *Journal for the Scientific Study of Religion* 24, no. 2 (1985): 119–236.

[15]Mario Renzi, "Ideal Family Size as an Intervening Variable Between Religion and Attitudes Toward Abortion," *Journal for the Scientific Study of Religion* 14 (1975): 23–27.

[16]Ibid.

Figure 6.7   Percentage Who Support Abortion by Religious Affiliation

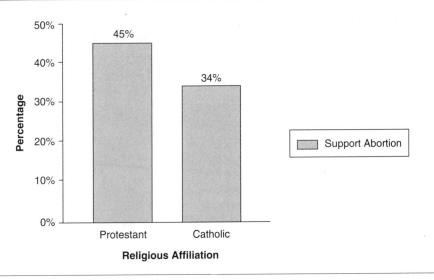

Source: General Social Survey, 1988–1991.

Figure 6.8   The Bivariate Relationship Between Religion and Support for Abortion

These results may suggest the existence of a causal relationship between religion and attitudes toward abortion. According to this interpretation of the relationship, being either Protestant or Catholic leads to a different abortion orientation regardless of other factors. Graphically, this hypothesized relationship is shown in Figure 6.8.

Another body of research findings dealing with religion challenges the conclusion that there is a direct causal link (as suggested by Figure 6.8) between religious affiliation and support for abortion. According to this research literature, some of the difference between Catholics and Protestants can be explained by the variable *preferred family size.*[17] It is argued that religion is systematically related to desired family size: Catholics prefer larger numbers of children than non-Catholics. Similarly, if one conceptualizes abortion as an alternative device to control family size, then support for abortion may also be associated with preferred family size. Therefore, preferred family size operates as an intervening mechanism through which the relationship between religion and abortion attitudes occurs.

To check these ideas, we analyzed the bivariate associations between preferred family size and religion (Table 6.11) and between preferred family size and support for abortion

---

[17]Ibid.

Table 6.11   Religious Affiliation and Preferred Family Size

PREFERRED FAMILY SIZE	RELIGIOUS AFFILIATION		
	Catholic	Protestant	Total
More than 2 children	52% (85)	27% (65)	37%
2 or fewer children	48% (78)	73% (175)	63%
Total (N)	100% (163)	100% (240)	100% (403)

Source: General Social Survey, 1988–1991.

Table 6.12   Preferred Family Size and Support for Abortion

SUPPORT	PREFERRED FAMILY SIZE		
	More than 2 children	2 or fewer children	Total
Yes	25% (38)	50% (127)	41%
No	75% (112)	50% (126)	59%
Total (N)	100% (150)	100% (253)	100% (403)

Source: General Social Survey, 1988–1991.

(Table 6.12).[18] Notice that because the theory suggests that preferred family size operates as an intervening mechanism between religious affiliation and support for abortion, it is analyzed as the dependent variable in Table 6.11 and as the independent variable in Table 6.12.

The data in Tables 6.11 and 6.12 confirm the linkages between preferred family size and religion and preferred family size and support for abortion. First, more Catholics (52%) than Protestants (27%) prefer larger families (Table 6.11). Second, more respondents who prefer smaller families support a woman's right to abortion (50%) compared with those who prefer larger families (25%) (Table 6.12). According to this interpretation of the relationship between religion and abortion attitudes, not only is preferred family size associated with both religious affiliation and support for abortion, but it also intervenes between religious affiliation and

[18]Preferred family size was measured by responses to a question about the ideal number of children for a family. Those respondents who said two or fewer children were ideal were classified as preferring small families; those who answered three or more were classified as preferring large families.

Figure 6.9   Intervening Relationship

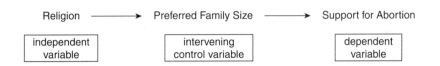

support for abortion. Thus, it is hypothesized that the relation between religion and attitudes toward abortion is *indirect*, and *linked* via the control variable—preferred family size.

The hypothetical causal sequence suggested by this interpretation is shown in Figure 6.9. In this formulation the control variable (preferred family size) is called an *intervening variable*. An **intervening variable** is a control variable that follows an independent variable but precedes the dependent variable in a causal sequence. Because preferred family size follows the independent variable, religion, but precedes the dependent variable, abortion attitudes, it is considered an intervening variable. The relationship between religion and support for abortion shown in Figure 6.9 is called an *intervening relationship*. An **intervening relationship** is one between two variables in which a control variable intervenes between the independent and dependent variables.

Intervening variable   A control variable that follows an independent variable but precedes the dependent variable in a causal sequence.

Intervening relationship   A relationship in which the control variable intervenes between the independent and dependent variables.

We can test the model shown in Figure 6.9 by controlling for preferred family size and repeating the original bivariate analysis between religious affiliation and support for abortion. We control for preferred family size by separating the respondents who indicated that they preferred larger families from those who prefer smaller families. If the causal sequence hypothesized by Figure 6.9 is correct, then the association between religion and abortion attitudes should disappear or diminish considerably once preferred family size has been controlled.

The results presented in Table 6.13 and Figure 6.10 support the notion, as depicted in Figure 6.9, that preferred family size intervenes between religion and abortion attitudes. The associations between religion and abortion attitudes in the two partial tables are smaller than the original bivariate table (Table 6.10). Among respondents who prefer larger families there are smaller differences between Catholics and Protestants regarding a woman's right to an abortion. Twenty-eight percent of Protestants and 24 percent of Catholics support legal abortion. Among those who prefer smaller families, there are also small differences between the two religious groups. Fifty-two percent of Protestants and 46 percent of Catholics are in support of abortion. Thus, we would conclude that Catholics are less favorable to abortion (than Protestants) because they prefer larger families. These findings increase our understanding of the original bivariate relationship between religious affiliation and attitudes toward abortion.

Table 6.13   Religious Affiliation and Support for Abortion After Controlling for Preferred Family Size

Preferred Family Size: 2 or fewer children			
	RELIGIOUS AFFILIATION		
SUPPORT	*Catholic*	*Protestant*	*Total*
Yes	46% (36)	52% (91)	50%
No	54% (42)	48% (84)	50%
Total (*N*)	100% (78)	100% (175)	100% (253)

Preferred Family Size: More than 2 children			
	RELIGIOUS AFFILIATION		
SUPPORT	*Catholic*	*Protestant*	*Total*
Yes	24% (20)	28% (18)	25%
No	76% (65)	72% (47)	75%
Total (*N*)	100% (85)	100% (65)	100% (150)

Source: General Social Survey, 1988–1991.

✔ *Learning Check.*   *You may have noticed that the tests for spuriousness and for an intervening relationship are identical: they both require that the partial associations disappear or diminish considerably! So how can you differentiate between the two? The differentiation is made on theoretical rather than empirical grounds. When a relationship is spurious, there is no causal link between the independent and dependent variables; both are influenced by a causally prior control variable. In an intervening relationship, there is an indirect causal link between the independent and dependent variables; the control variable follows the independent variable but precedes the dependent variable in the causal sequence.*

Conditional Relationships: More on Abortion

What other variables may explain the relationship between religion and attitudes toward abortion? One possible variable is *religious participation*. In their research on abortion attitudes,

Figure 6.10   Percentage Supporting Abortion by Religious Affiliation After Controlling for Preferred Family Size

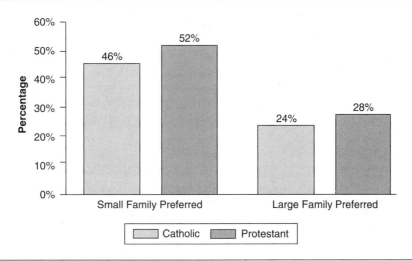

Source: General Social Survey, 1988–1991.

Arney and Trescher[19] found that when religious participation is controlled for, there is little difference in abortion attitudes between Catholics and Protestants who attend church less than once a month. In contrast, among Catholics and Protestants who attend church more than once a month, Catholics were more likely than Protestants to oppose abortion.[20] Other researchers note that age and gender may also influence the relationship between religion and abortion attitudes.

What do these examples have in common? They all specify different conditions under which the relationship between religion and abortion attitudes is expected to hold. For example, Arney and Trescher indicate that the differences in abortion attitudes between Protestants and Catholics might hold under one condition (attend church more than once a month) of the control variable *religious participation* but not under another (attend church less than once a month). Similarly, the relationship may differ for men and women, or for older and younger individuals. When a bivariate relationship differs for different conditions of the control variable, we say that it is a **conditional relationship**. Another way to describe a conditional relationship is to say that there is a *statistical interaction* between the control variable and the independent variable.

Conditional relationship   A relationship in which the control variable's effect on the dependent variable is conditional on its interaction with the independent variable. The relationship between the independent and dependent variables will change according to the different conditions of the control variable.

---

[19]William R. Arney and William H. Trescher, "Trends in Attitudes Toward Abortion, 1972–1975," *Family Planning Perspective* 8 (1976): 117–124.

[20]William V. D'Antonio and Steven Stack, "Religion, Ideal Family Size, and Abortion: Extending Renzi's Hypothesis," *Journal for the Scientific Study of Religion* 19 (1980): 397–408.

Table 6.14   Abortion Morality and Stance on Legal Abortion

ABORTION MORALITY	STANCE ON LEGAL ABORTION		
	Pro-Choice	Pro-Life	Total
Always wrong or depends	37%	98%	57%
Not Wrong	63%	2%	43%
Total	100%	100%	100%
(N)	(337)	(162)	(499)

Source: Adapted from Jacqueline Scott, "Conflicting Belief About Abortion: Legal Approval and Moral Doubts," *Social Psychology Quarterly* 52, no. 4 (1989): 319–326. Copyright © 1989 by the American Sociological Association. Reprinted by permission.

Because conditional relationships are very common, sociology offers many research examples illustrating this pattern of elaboration. One such example comes from a study by Jacqueline Scott on the relationship between stance on legal abortion and opinions about the morality of abortion. The study shows that although nearly all opponents of legal abortion view abortion as morally wrong, not all pro-choice supporters view abortion as morally right. Instead, many pro-choice supporters favor legal abortion despite personal moral reservations.[21] This bivariate relationship between abortion morality and stance on legal abortion is displayed in Table 6.14.

Because stance on legal abortion is the independent variable, percentages are calculated in the columns. The results of this analysis support Scott's hypothesis. Among those who oppose abortion, there is almost unanimous agreement (98%) that abortion is morally wrong. Among those who favor legal abortion, however, the level of incongruence is relatively high: 37 percent support legal abortion despite viewing it as morally wrong or ambiguous.[22]

Although there is little difference between men's and women's attitudes toward the legality of abortion, some argue that women are far more likely to feel that abortion is morally wrong. For example, Carol Gilligan[23] argues that whereas men tend to be more concerned with rights and rules, women are more concerned with caring and relationships. Abortion, therefore, may pose a greater moral dilemma for women than for men. To examine the hypothesis that women are more likely than men to favor legal abortion despite moral reservations, Scott controlled for gender and compared the original relationship between stance on legal abortion and abortion morality among men and women. The cross-tabulation of abortion morality by stance on legal abortion, controlling for gender, is given in Table 6.15. The table shows a marked gender difference in the relationship between abortion morality and stance on legal abortion. Although we can still conclude from Table 6.15 that stance on legal abortion and abortion morality are associated, we need to qualify this conclusion by saying

---

[21]Jacqueline Scott, "Conflicting Belief About Abortion: Legal Approval and Moral Doubts," *Social Psychology Quarterly* 52, no. 4 (1989): 319–326.

[22]Ibid., p. 322.

[23]Carol Gilligan, *In a Different Voice* (Cambridge, MA: Harvard University Press, 1982).

Table 6.15   Abortion Morality and Stance on Legal Abortion After Controlling for Gender

| | *MEN* | | | | *WOMEN* | | |
| | *STANCE ON LEGAL ABORTION* | | | | *STANCE ON LEGAL ABORTION* | | |
ABORTION MORALITY	*Pro-Choice*	*Pro-Life*	*Total*	*ABORTION MORALITY*	*Pro-Choice*	*Pro-Life*	*Total*
Always wrong or depends	29%	96%	50%	Always wrong or depends	46%	100%	64%
Not Wrong	71%	4%	50%	Not Wrong	54%	0%	36%
Total (*N*)	100% (172)	100% (78)	100% (250)	Total (*N*)	100% (165)	100% (84)	100% (249)

Source: Adapted from Jacqueline Scott, "Conflicting Belief About Abortion: Legal Approval and Moral Doubts," *Social Psychology Quarterly* 52, no. 4 (1989): 319-326. Copyright © 1989 by the American Sociological Association. Reprinted by permission.

Figure 6.11   A Conditional Relationship

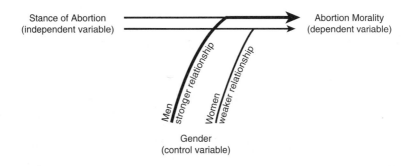

Stance of Abortion (independent variable) → Abortion Morality (dependent variable)

Men stronger relationship / Women weaker relationship

Gender (control variable)

that this association is stronger for men (percentage difference is 96% − 29% = 67%) than for women (percentage difference is 100% − 46% = 54%).

Because the relationship between the independent and dependent variables is different in each of the partial tables, the relationship is said to be a conditional relationship; that is, the original bivariate relationship depends upon the control variable. In our example, the strength of the relationship between abortion morality and stance on legal abortion is conditioned on gender. The conditional relationship between stance on abortion and abortion morality is depicted in Figure 6.11.

The Limitations of Elaboration

The elaboration examples discussed in this section point to the complexity of the social world. We started this chapter by stating that most things around us "go together." It is more accurate to say that most things around us are "tangled," and one of the goals of social science is to "untangle" them. Elaboration is a procedure that helps us "untangle" bivariate relations.

In the illustrations presented in this section, we looked at bivariate relationships that were clarified and reinterpreted when a control variable was introduced. How do we know which variables to control for? In reality, theory provides significant guidance as to the relationships we look for and the sorts of variables that should be introduced as controls. Without theory as a guide, elaboration can become a series of exercises that more closely resembles random shots in the dark than scientific analysis. Even with theory as our guide, the statistical analysis is often more complex than the presentation in this section may suggest. In our examples, when the control variable was introduced, the "real" nature of the relationship "jumped right out at us." It's not always that easy. In fact, most often there is a perilous gap between theory and analysis. This does not mean that you have to abandon your effort to "untangle" bivariate relationships, only that you should be aware of both the importance of theory as a guide to your analysis and the limitations of the statistical analysis.

✔ *Learning Check.* *In this section you have been introduced to a number of important new terms. See if you can write out definitions for the following terms: elaboration, control variable, intervening variable, causally prior variable, spurious relationship, partial relationship, partial table, and conditional relationship. If you cannot provide a definition for each of these terms, you are not clear on the process of elaboration. Go back and review.*

▣ STATISTICS IN PRACTICE: FAMILY SUPPORT FOR THE TRANSITION FROM HIGH SCHOOL

In earlier chapters we saw that statistics helps us analyze how race, class, age, or gender shapes our experiences. However, we focused primarily on how these categories of experience operate separately (we compared men and women, young and old, working class and middle class, and so on). Now we need to think about how these systems interlock in shaping our experience as individuals in society.[24] Everyone has his or her own particular combination of race, social class, age, and gender. These factors act as lenses through which we experience the world. Through analysis of the intersecting effects of these factors, we can understand the ways that others experience the world from their different or similar perspectives.

When we start to see race, class, and gender as intersecting systems of experience, we see, for example, that while white women and women of color may share some common experiences based on their gender, their racial experiences are distinct. Similarly, depending on their race and class, men experience gender differently. For example, we know that the removal of

---

[24]Patricia Hill Collins, "Toward a New Vision: Race, Class, and Gender as Categories of Analysis and Connection," keynote address at Integrating Race and Gender into the College Curriculum, a workshop sponsored by the Center for Research on Women, Memphis State University, Memphis, TN, 1989.

Table 6.16   Race and Class Origin Differences in Percentage of Women Reporting
Family Support for the Transition from High School to College

	BLACK		WHITE	
	Working Class	*Middle Class*	*Working Class*	*Middle Class*
TYPE OF FAMILY SUPPORT	*(N = 50)*	*(N = 50)*	*(N = 50)*	*(N = 50)*
Information				
Entrance exams and colleges	22%	42%	24%	40%
Admission requirements	20%	34%	20%	30%
Financial				
Paid tuition and fees	56%	90%	62%	88%
Emotional				
Emotional support	64%	86%	56%	70%
Encouragement for career	56%	60%	40%	52%

Source: Adapted from Lynn Weber, Elizabeth Higginbotham, and Marianne L. A. Leung, "Race and Class Bias in
Research on Women: A Methodological Note," Research paper 5, presented at the Center for Research on Women,
University of Memphis, 1987.

manufacturing jobs has increased black and Hispanic male job loss. At the same time, many
women of color have found their work opportunities expanded in service and high-tech jobs.[25]

The methods of bivariate analysis and, in particular, the statistical techniques of elaboration are especially suitable for the examination of how race, class, and gender are linked with
social behavior. In Table 6.16 we present the findings of a study that examined the kind of
family support women who are now in professional and managerial positions received when
they made their transition from high school to college. This example illustrates how to analyze the simultaneous operation of race and class using the method of elaboration. The
example also demonstrates the drastic differences in conclusions that would have been drawn
had either or both of these factors been ignored.[26]

Table 6.16 includes five types of family support, representing five dependent variables:
(1) information on entrance examinations and colleges, (2) information on admission requirements, (3) financial support in paying tuition and fees, (4) emotional support, and (5) encouragement for career. Only one category is given for each dependent variable. This category
represents the percentage of women who received family support in each of the specified
areas. For example, 42 percent of black women who were raised in middle-class families

[25]Stanley Eitzen and Maxine Baca Zinn, "Structural Transformation and Systems of Inequality," in
Race, Class, and Gender, ed. Margaret L. Andersen and Patricia Hill Collins (Belmont, CA: Wadsworth,
1998), pp. 233–237.

[26]Lynn Weber Cannon, Elizabeth Higginbotham, and Marianne L. A. Leung, "Race and Class Bias in
Research on Women: A Methodological Note," Research paper 5, presented at the Center for Research on
Women, Memphis, TN, 1987.

($N = 50$) reported that family members had helped them in procuring information on entrance examinations and colleges. The remaining 58 percent of this group (not shown) did not receive family help in this area.

Race and class origin are the two independent variables in this analysis. To estimate the effect of class on family support we compare middle-class–raised women with working-class–raised women among black and white women. The data show that there are large class differences in all types of family support provided to these women by their families. For example, whereas 90 percent of black middle-class families paid tuition for their daughters, only 56 percent of black working-class families did so. A similar pattern is observed among the white women. Similarly, more middle-class families, both black (86%) and white (70%), provided emotional support to their daughters during the transition from high school to college.

The second step involved in looking at a table like this one is to examine whether there are racial differences in family support. To estimate the effect of race we compare black and white women who were raised in working-class and in middle-class families. This comparison reveals virtually no relationship between race and either procurement of information or financial support. For instance, 20 percent of working-class blacks and 20 percent of working-class whites received help with information on college admissions requirements; similarly, 34 percent of middle-class blacks and 30 percent of middle-class whites received support in this category. However, examination of the emotional support category reveals fairly substantial race differences: 86 percent of middle-class–raised black respondents report that their families provided emotional support, compared with only 70 percent of the middle-class–raised whites. Similarly, more working-class black families (64% vs. 56%) provided emotional support to their daughters in the transition from high school to college.

The group that differs most from the others on both emotional support and encouragement for career are black middle-class women, who received the highest degree of family support in each of these categories.

In conclusion, the data reveal a strong relationship between class origin and both information and financial support provided by the family. In addition, the data show relationships between both race and class and emotional encouragement and support. Had the study failed to address both the race and class background of these professional and managerial women, we would have drawn very different conclusions about the role of families in supporting women as they moved from high school to college.[27]

This is another example of the pattern of elaboration examined earlier. In this case, class origin is used as a control variable to elaborate on the relationship between race and family support. In American society, race is associated with class (blacks are more likely to be raised in a working-class family), and class is associated with family support (working-class families are less likely to provide family support). Had we not analyzed the effect of the class background of the women as well as their race, we would have concluded that black women receive far less support in all areas than white women. Such a conclusion could have reinforced a stereotype—that black families are less supportive of their children's education. Such a conclusion would represent a distortion of the real process since it is working-class women, both black and white, who receive less family support.[28]

---

[27]Ibid.

[28]Ibid.

Finally, this example demonstrates the importance of looking at the simultaneous effects of race and class on the lives of women. This is only one among many ways in which race, class, and gender comparisons can be incorporated in a statistical analysis. Moreover, examining the linkages between race, class, and gender cannot be limited to women. While integrating these variables into our analysis introduces complexity to our research, it also suggests new possibilities for thinking that will enrich us all.

MAIN POINTS

- Bivariate analysis is a statistical technique designed to detect and describe the relationship between two variables. A relationship is said to exist when certain values of one variable tend to "go together" with certain values of the other variable.

- A bivariate table displays the distribution of one variable across the categories of another variable. It is obtained by classifying cases based on their joint scores for two variables.

- Percentaging bivariate tables are used to examine the relationship between two variables that have been organized in a bivariate table. The percentages are always calculated within each category of the independent variable.

- Bivariate tables are interpreted by comparing percentages across different categories of the independent variable. A relationship is said to exist if the percentage distributions vary across the categories of the independent variable.

- Variables measured at the ordinal or interval-ratio levels may be positively or negatively associated. With a positive association, higher values of one variable correspond to higher values of the other variable. When there is a negative association between variables, higher values of one variable correspond to lower values of the other variable.

- Elaboration is a technique designed to clarify bivariate associations. It involves the introduction of control variables to interpret the links between the independent and dependent variables.

- In a spurious relationship, both the independent and dependent variables are influenced by a causally prior control variable, and there is no causal link between them.

- In an intervening relationship, the control variable follows the independent variable but precedes the dependent variable in the causal sequence.

- In a conditional relationship, the bivariate relationship between the independent and dependent variables is different in each of the partial tables.

KEY TERMS

bivariate analysis
bivariate table
cell
column variable
conditional relationship
control variable
cross-tabulation
direct causal relationship
elaboration

intervening relationship
intervening variable
marginals
negative relationship
partial relationship
partial tables
positive relationship
row variable
spurious relationship

ON YOUR OWN

Log on to the web-based student study site at http://www.pineforge.com/frankfort-nachmiasstudy4 for additional study questions, quizzes, web resources, and links to social science journal articles reflecting the statistics used in this chapter.

SPSS DEMONSTRATIONS

[GSS02PFP-A]

Demonstration 1: Producing Bivariate Tables

SPSS has a separate procedure designed specifically to produce cross-tabulation tables. It is called the Crosstabs procedure and can be found under *Descriptive Statistics* in the *Analyze* menu. The dialog box for Crosstabs (Figure 6.12) requires us to specify both a variable that will define the rows and one that defines the columns of a table. We will investigate the relationship between support for a legal abortion for a woman who is married but does not want any more children (ABNOMORE) and religious affiliation (RELIG).

Figure 6.12

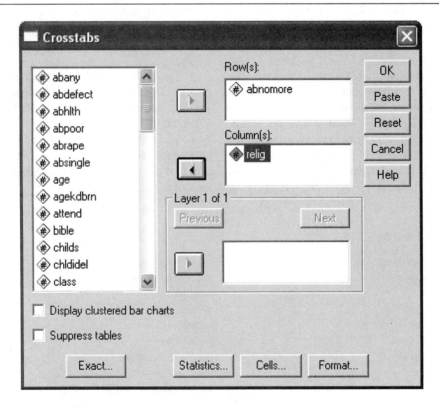

Figure 6.13

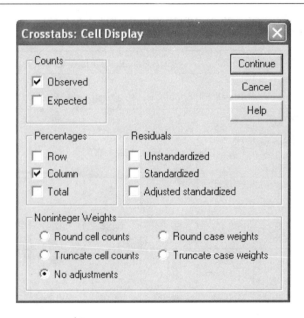

By default, SPSS displays the count in each cell of the table. Normally, then, you should click on the *Cells* button to request percentages (Figure 6.13). As usual, we percentage the table based on the independent or predictor variable, which is religious affiliation. The independent variable is placed in the columns, while the dependent variable is placed in rows. We click on the checkbox for "Column" to percentage the table by religious affiliation. (Note that "Observed" is already checked by default in the Counts section.)

Click on *Continue*, then *OK*, to obtain the table shown in Figure 6.14. SPSS displays both the count and the column percentage in each cell. In the upper left corner of the table, the labels "Count" and "% within RELIG" are displayed as a reminder of what SPSS has placed in each cell. Row totals and column totals are supplied automatically, as is the overall total (487 respondents gave valid responses to both questions). The number of missing responses on one or both variables may also be displayed.

The table shows great differences in support across religious categories for a legal abortion for a married woman who does not want any more children. A majority of Protestants, Catholics, and others (Hindus, Buddhists, Muslims, and others[29]) oppose abortion in this instance, but a majority of Jews and the nonreligious support abortion. Do you find these differences surprising, or are they consistent with your understanding of the social world?

Demonstration 2: Producing Tables with a Control Variable

As we've seen in this chapter, the analysis of data is enhanced when a third variable—a control variable—is added to a bivariate table. In the Crosstabs procedure, the third variable is added in the Layer section of the main dialog box (Figure 6.15). (This box is labeled "Layer 1 of 1" because it is

---

[29]For the purpose of this illustration, category 5 OTHER includes the following religious groups: Buddhism, Hinduism, Moslim/Islam, Orthodox-Christian, Christian, Native American, Inter-Nondenominational. Each of these groups will, however, be listed separately on your output.

Figure 6.14

ABNOMORE MARRIED-WANTS NO MORE CHILDREN * RELIG RS RELIGIOUS PREFERENCE Crosstabulation

			RELIG RS RELIGIOUS PREFERENCE					
			1 Protestant	2 Catholic	3 Jewish	4 None	5 Other (specify)	Total
ABNOMORE MARRIED-WANTS NO MORE CHILDREN	1 YES	Count % within RELIG RS RELIGIOUS PREFERENCE	85 34.6%	51 40.8%	6 85.7%	46 66.7%	24 60.0%	212 43.5%
	2 NO	Count % within RELIG RS RELIGIOUS PREFERENCE	161 65.4%	74 59.2%	1 14.3%	23 33.3%	16 40.0%	275 56.5%
Total		Count % within RELIG RS RELIGIOUS	246 100.0%	125 100.0%	7 100.0%	69 100.0%	40 100.0%	487 100.0%

Figure 6.15

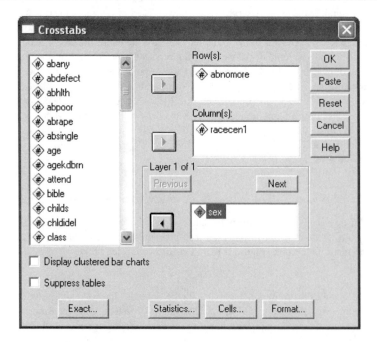

Figure 6.16

ABNOMORE MARRIED-WANTS NO MORE CHILDREN * RACE OF RESPONDENT * SEX RESPONDENTS Crosstabulation

SEX RESPONDENTS SEX				RACE RACE OF RESPONDENT		Total
				1 WHITE	2 BLACK	
1 MALE	ABNOMORE MARRIED-WANTS NO MORE CHILDREN	1 YES	Count	105	6	111
			% within RELIG RS RELIGIOUS PREFERENCE	48.2%	30.0%	46.7%
		2 NO	Count	113	14	127
			% within RELIG RS RELIGIOUS PREFERENCE	51.8%	70.0%	53.3%
	Total		Count	218	20	238
			% within RELIG RS RELIGIOUS PREFERENCE	100.0%	100.0%	100.0%
2 FEMALE	ABNOMORE MARRIED-WANTS NO MORE CHILDREN	1 YES	Count	78	15	93
			% within RELIG RS RELIGIOUS PREFERENCE	40.8%	46.9%	41.7%
		2 NO	Count	113	17	130
			% within RELIG RS RELIGIOUS PREFERENCE	53.2%	53.1%	58.3%
	Total		Count	191	32	223
			% within RELIG RS RELIGIOUS PREFERENCE	100.0%	100.0%	100.0%

possible to have additional levels of control, which are accessed by clicking on the *Next* button.) We will keep the same dependent variable, ABNOMORE, but make RACECEN1 the column variable and SEX the control variable. There is no need to change the numbers displayed in the cells: the count and column percentages are still correct choices. Figure 6.16 shows the bivariate tables of support for legal abortions for married women by race separately for males and females (males are listed first).[30] SPSS labels each table with the value of the control variable for easy reference.

For males, we see that the majority of whites and blacks do not support abortion for a married woman (if she does not want any more children). The largest percentage of men who do not support abortion is found among black males (70%), compared with about 52 percent for white males. In the table for females, too, a majority of both races oppose abortion, but there is no difference between black and white women (53.2% and 53.1%). Finding differences like these is one reason why researchers use control variables in analyses. These tables display a conditional relationship between race and support for abortion for married women when sex is introduced as a control variable.

---

[30]Only blacks and whites are shown.

In later chapters we will use the Statistics button in the Crosstabs dialog box to request additional output to further interpret and evaluate bivariate tables.

SPSS PROBLEMS

[GSS02PFP-A]

1. The data contain responses to questions about the respondent's general happiness (HAPPY) and his or her subjective class identification (CLASS). Analyze the relationship between responses to these two questions with the SPSS Crosstabs procedure, requesting counts and appropriate cell percentages. Click on *Analyze, Descriptive Statistics,* and *Crosstabs* to get started.
   a. What percentage of working-class people responded that they were "very happy"?
   b. What percentage of lower-class people were "very happy"?
   c. What percentage of those who were "pretty happy" were also from the middle and upper classes?
   d. Most of the people who said they were "very happy" were from which two classes?
   e. Is there a relationship between perceived class and perceived happiness? If there is a relationship, describe it. Is it strong or weak? (Hint: Use perceived class as the independent variable.)
   f. Rerun your analysis, this time adding RACECEN1 as a control variable. Is there a difference in the relationship between perceived class and happiness for whites and blacks in the sample? (Because of the large number of racial categories, just compare blacks and whites.)

2. Analyze the relationship between self-reported health condition and general happiness.
   a. Use SPSS to construct a table showing the relationship between health condition (HEALTH) and reported general happiness (HAPPY). (Hint: Use HEALTH as the dependent variable.) Next, use SPSS to construct tables showing the same relationship controlling for sex.
   b. Overall, are women or men more likely to report being "very happy" with their lives?
   c. Do women with higher levels of happiness report poor or excellent health? Is there a relationship between happiness and health? Make sure to support your answer with data from your cross-tabulation.

3. An editorial in your local newspaper reports that people with less education are more likely to support the use of spanking to discipline a child. You decide to use the GSS 2002 data to verify this reported relationship between attitudes toward spanking (SPANKING) and the level of education (use the variable DEGREE, which groups educational attainment into five categories).
   a. Create the appropriate table using SPSS and the Crosstabs procedure.
   b. Create a bar chart grouped by DEGREE to graphically display this same table. In SPSS, use a clustered bar chart. Click on *Graphs* and *Bar*. Select *Clustered* and then *Define*. In the box labeled "Category Axis" insert the variable SPANKING. In the box labeled "Define Clusters By" insert the variable DEGREE. Finally, in the section of the dialog box under "Bars Represent," select *% of cases*.
   c. Is there a relationship between the variables? What is its direction?
   d. Do the data support the newspaper's assertion?
   e. Use the Crosstabs procedure with RACECEN1 as a control variable to create separate tables for SPANKING by DEGREE for blacks and whites.
   f. Is there a difference in support for spanking between blacks with different levels of education?
   g. Is there a difference in attitude between whites with different levels of education?
   h. Use the Crosstabs procedure with SEX as a control variable. Is there a difference in support between women and men?

4. Is there a difference in attitudes about abortion depending on the circumstance of the woman's pregnancy or her reason for an abortion? Separately assess the relationship between SEX and

two abortion items, ABPOOR (Should a woman have an abortion if she can't afford any more children?) and ABHLTH (Should a woman have an abortion if her health is seriously endangered?). What do you conclude?

5. In the GSS 2002, respondents were asked to report which candidate they voted for in the 1996 and 2000 presidential elections (PRES96, PRES00). Does a relationship exist between a respondent's 1996 and 2000 vote? For example, if someone voted for Bill Clinton in 1996, would he/she be likely to vote for Al Gore in 2000?
   a. Which variable should be defined as the dependent variable? Explain your answer.
   b. Using SPSS Crosstabs, create a table with the two variables PRES96 and PRES00. Explain the relationship between the two variables. (Remember, when you discuss your findings you should exclude those respondents who did not vote).
   c. Use the variable SEX as a control variable. Does the relationship between PRES96 and PRES00 hold?
   d. Examine the relationship between PRES96 and PRES00 with a control variable of your choice.

CHAPTER EXERCISES

1. Use the following data on fear, race, and home ownership for this exercise. Variables measure respondent's race, whether the respondent fears walking alone at night, and his/her home ownership.

Respondent	Race	Fear of Walking Alone	Rent/Own
1	W	N	R
2	B	N	R
3	W	Y	R
4	B	N	R
5	W	N	R
6	B	Y	O
7	W	Y	R
8	W	Y	R
9	W	N	O
10	W	N	O
11	W	Y	R
12	W	N	R
13	B	Y	O
14	W	N	R
15	B	N	O
16	B	N	R
17	W	N	O
18	W	N	O
19	B	N	R
20	W	N	O
21	B	Y	R

Race: B = black, W = white; Fear: Y = yes, N = no; R = rent, O = own.

Source: Data based on the 2002 General Social Survey.

a. Construct a bivariate table of frequencies for race and fear of walking alone at night. Which is the independent variable?

b. Calculate percentages for the table based on the independent variable. Describe the relationship between race and fear of walking alone using the table. What sampling issues are involved here?

c. Use the data to construct a bivariate table to compare fear of walking alone at night between people who own their homes and those who rent. Use percentages to show whether there is a difference between homeowners and renters in fear of walking alone.

2. Do women and men have different opinions about welfare spending? Based on a subsample of the 2002 General Social Survey, the output in Figure 6.17 shows respondent's sex (SEX) and attitudes toward welfare spending (NATFARE: Are we spending too little, about right, or too much?). Data are based on individuals 30 years or older.

Figure 6.17

NATFARE WELFARE * SEX RESPONDENTS SEX Crosstabulation

			SEX RESPONDENTS SEX		
			1 MALE	2 FEMALE	Total
NATFARE WELFARE	1 TOO LITTLE	Frequency	115	164	279
	2 ABOUT RIGHT	Frequency	229	273	502
	3 TOO MUCH	Frequency	243	290	533
Total			587	727	1314

a. Which is the independent variable?

b. What are the differences in attitudes between men and women?

c. What might be some other reasons that influence attitudes toward welfare spending? Suggest at least two reasons.

3. Advocates of gay rights often argue that homosexuality is not a "preference" or a choice, but rather an "orientation" that cannot be changed. One of your classmates argues that attitudes about homosexuality often influence political views. Those who think homosexual relations are wrong tend to be more conservative compared to those who do not think that homosexual relations are wrong. Use the following table to answer the questions.

a. What is the dependent variable? The independent variable?
b. What percentage of those polled think that homosexual relations are always wrong?
c. Using the percentages in the table, describe the relationship between views about homosexual relations and political orientation? How strong is the relationship?

		Homosexual Relations		
		Always Wrong	*Not Wrong at All*	*Total*
Political Views	Liberal	84 17.7%	110 38.9%	194 25.6%
	Moderate	187 39.5%	110 38.9%	297 39.2%
	Conservative	203 42.8%	63 22.2%	266 35.2%
Total		474 100.0%	283 100.0%	757 100.0%

Source: 2002 General Social Survey.

Note: Original GSS categories have been recoded for illustration purposes.

4. Refer to Exercise 3. Suppose that a classmate of yours suggests that views about homosexual relations can be explained by the frequency of church attendance. Your classmate shows you the following table taken from the 2002 General Social Survey sample. (Frequencies are shown below.)

		Church Attendance			
		Never	*Several Times a Year*	*Every Week*	*Total*
Homosexual Relations	Always Wrong	43	23	53	119
	Not Wrong at All	42	22	10	74
Total		85	45	63	193

a. Which is the dependent variable in this table? Which is the independent variable?
b. Calculate the percentages using church attendance as the independent variable for each cell in the table. Is there a relationship between church attendance and views about homosexual relations? If so, how strong is it?
c. Suppose that you respond to your classmate by stating that it is not church attendance that explains views about homosexual relations; rather, it is one's opinion about the nature of right and wrong (i.e., morality) that explains attitudes about homosexual relations. Why might there be a potential problem with your argument? Think in terms of assigning variables to the independent and dependent categories.

5. A neighborhood clinic wants to develop a health promotion program aimed at a community with white and black residents. As a part of this outreach program, the clinic staff believes that it is important to assess the level of confidence people have in their medical and health practitioners. The clinic used data from the 2002 General Social Survey to measure levels of confidence by gender and race, adapted and summarized in the following table. (Frequencies are shown below.)

		Race and Sex of Respondent				
		White Males	Black Males	White Females	Black Females	Total
Confidence in	A Great Deal	135	17	136	25	313
Medicine	Only Some	162	23	214	47	446
	Hardly Any	33	6	41	11	91
Total		330	46	391	83	850

Source: General Social Survey 2002.

a. What proportion of respondents report having a "great deal" of confidence in medicine? Hardly any confidence?
b. Is there a difference in levels of confidence between women and men? Is there a difference between blacks and whites?
c. In order to improve levels of confidence in medicine and health practitioners, what group(s) should the clinic reach out to?

6. The educational level of Americans increased throughout the 20th century. The following U.S. census data show the level of education attained by American adults over the age of 25 at three points in time: 1990, 1995, and 2000.

	Educational Level	
Year	High School Graduate	College Graduate
1990	77.6%	21.3%
1995	81.7%	23.0%
2000	84.1%	25.6%

Source: U.S. Bureau of the Census, *Statistical Abstract of the United States,* 2003, Table 227.

a. What is the direction of the relationship between each year and level of education?
b. Use percentage differences to describe the relationship. Why don't the percentages add to 100 percent by year? Is this a problem in analyzing the table?
c. Do these data support the idea that Americans were getting more education in 2000 than 10 years before?

7. Do female and male high school seniors have the same college goals? High school seniors were surveyed about their college plans and asked whether they thought they would graduate from a 4-year program. Data for three time periods are provided in the following table (percentages are displayed).

		1980		1990		1995	
		Females	*Males*	*Females*	*Males*	*Females*	*Males*
College plans for seniors:	Definitely Will	33.6%	35.6%	50.8%	45.8%	59.8%	48.6%
Graduate college	Probably Will	21.3%	23.5%	20.5%	24.0%	20.7%	25.6%
(4-year program)	Definitely/ Probably Won't	45.0%	41.0%	28.8%	30.2%	19.5%	25.8%

Source: U.S. Department of Education, *Trends in Educational Equity of Girls and Women*, NCES 2000–030, Washington, DC: 2000.

    a. Since 1980, how has the pattern of college graduation plans changed for females?

    b. Is the same pattern true for male seniors? Why or why not?

8. In Exercise 1 you found that more blacks than whites are likely to fear walking alone in their neighborhoods. You now wonder if this difference exists because whites are more likely to own their own homes and so live in safer neighborhoods. In other words, you want to try some elaboration.

    a. Use the data from Exercise 1 to construct tables showing the relationship between fear of walking alone and race, controlling for whether the individual rents or owns his or her dwelling.

    b. Does renting versus owning one's dwelling "explain" the difference in fear between whites and blacks? (Use percentage differences to support your answer.)

    c. Has introducing home ownership shown that the relationship between race and fear is spurious, or is home ownership an intervening variable? Explain.

9. Black students in the United States typically attend public schools in which they are a majority (and therefore white students a minority). In the past three decades, in many large cities such as New York, black students have been attending schools with a decreasing proportion of white students. Take a look at the table below and answer the following questions.

A Typical Black Student's Classmates Are	*Elementary School*			*Middle School/Junior High*		
	1980	*1990*	*2001*	*1980*	*1990*	*2001*
White students	89.7%	88.0%	81.8%	91.4%	87.7%	83.8%
Students of color	9.3%	12.0%	18.2%	8.6%	12.3%	16.2%

A Typical Black Student's Classmates Are	*High School*			*College*		
	1980	*1990*	*2001*	*1980*	*1990*	*2001*
White students	93.0%	89.8%	85.7%	95.0%	92.6%	87.3%
Students of color	5.0%	10.2%	14.3%	5.0%	7.4%	12.7%

Source: U.S. Bureau of the Census, *Statistical Abstract of the United States*, 2003, Table 222.

a. Based on the data, is the direction of the trend the same or different at all four educational levels?

b. Is the trend stronger in elementary school or in high school? In high school or in college?

c. Use the column percentages to describe the strength of the relationship between time and student segregation from 1990 to 2001.

10. How much do citizens support the government redistributing wealth as a means of reducing social inequality? Are there differences between Europeans and non-Europeans regarding the government role in redistributing income? In the year 2000, respondents from both European and non-European countries were asked their level of agreement with the following statement: "The government should redistribute income." Their responses are as follows. (Frequencies are shown below.)

		Country	
		European	*Non-European*
The Government	Strongly agree	642	896
Should	Agree	1348	994
Redistribute	Neither agree nor disagree	693	791
Income	Disagree	542	598
	Strongly disagree	251	407
Total		3476	3686

Source: International Social Survey Programme 2000.

a. Of those who disagreed or strongly disagreed, what percent were European?

b. Of those who agreed or strongly agreed, what percent were non-European?

c. Of European respondents, what percent neither agreed nor disagreed?

d. Of non-European respondents, what percent agreed or strongly agreed?

e. Treating whether a country is European or not as the independent variable, calculate the percentages to analyze the relationship between European and non-European countries and attitudes about the government's role in redistributing income.

11. An organization in your state is lobbying to make pornography illegal because its members believe that pornography leads to a breakdown in morals. The following table contains information on the use of pornographic material and attitudes about extramarital sex. (Frequencies are shown below.)

		Seen X-Rated Movie in Past Year		
		Yes	*No*	*Total*
Extramarital Sex	Always wrong	83	268	351
	Almost always wrong	21	49	70
	Sometimes wrong	10	14	24
	Not wrong at all	3	6	9
Total		117	337	454

Source: General Social Survey 2002.

a. Do the GSS data in the table support the organization's beliefs or not? Why? (Calculate the percentages and analyze the relationship.)

b. Your friend argues that there are gender differences in the effect that pornographic materials have on attitudes about extramarital sex. Do the GSS survey data in the following table support her belief? Why or why not? (Frequencies are shown below.)

			Seen X-Rated Movie in Past Year		
			Yes	No	Total
Male	Extramarital Sex	Always wrong	56	132	188
		Almost always wrong	16	20	36
		Sometimes wrong	9	5	14
		Not wrong at all	1	3	4
Total			82	160	242
Female	Extramarital Sex	Always wrong	27	136	163
		Almost always wrong	5	29	34
		Sometimes wrong	1	9	10
		Not wrong at all	2	3	5
Total			35	177	212

Source: General Social Survey 2002.

c. What can you conclude about the relationship between the use of pornographic material and attitudes about extramarital sex when controlling for sex? Is this an example of a conditional relationship?

12. In the previous exercise you examined gender differences in the relationship between the use of pornographic material and attitudes about extramarital sex. Use the following table to describe racial differences in the relationship between the use of pornographic material and attitudes about extramarital sex. (Frequencies are shown below.)

			Seen X-Rated Movie in Past Year		
			Yes	No	Total
White	Extramarital Sex	Always wrong	55	229	284
		Almost always wrong	17	41	58
		Sometimes wrong	9	13	22
		Not wrong at all	2	4	6
Total			83	287	370
Black	Extramarital Sex	Always wrong	22	23	45
		Almost always wrong	3	5	8
		Sometimes wrong	1	1	2
		Not wrong at all	0	1	1
Total			26	30	56

Source: General Social Survey 2002.

a. Describe the relationship between use of pornographic material and attitudes about extramarital sex separately for whites and blacks.

b. Does controlling for race show that the relationship between the use of pornographic material and attitudes about extramarital sex is spurious or conditional? Why or why not?

13. When using crosstabs, it is important to specify which variable is the independent variable and which is the dependent variable. When both variables are attitudes or opinions, it is important to have a reason *why* one variable is the independent variable and why the other variable is the dependent variable.

a. Listed below are two variables with their categories. Discuss why you think one qualifies as the independent variable and one as the dependent variable.

Variable #1 **Science Does More Harm Than Good**	*Variable #2* **Science Solves Environmental Problems**
Strongly agree	Strongly agree
Agree	Agree
Neither agree nor disagree	Neither agree nor disagree
Disagree	Disagree
Strongly disagree	Strongly disagree

b. Looking at the frequencies in the table below, calculate the appropriate percentages according to your classification of the variables in 13a.

		Science Does More Harm Than Good					
		Strongly Agree	Agree	Neither Agree nor Disagree	Disagree	Strongly Disagree	Total
Science Solves	Strongly agree	316	217	162	291	404	1390
Environmental	Agree	289	1728	1088	2585	980	6670
Problems	Neither agree nor disagree	159	877	2253	2120	718	6127
	Disagree	320	1640	1826	4228	1306	9320
	Strongly disagree	308	484	668	1160	998	3618
Total		1392	4946	5997	10384	4406	27125

Source: International Social Survey Programme 2000.

c. What can you conclude about the relationship between these two variables?

14. Does a belief in God influence charitable giving? Assess the table below and discuss whether a relationship exists between a belief in God and giving money to any charitable group over the past five years.

		Belief in God			
		Don't Believe in God	Believe in God Sometimes	God Exists, No Doubt About It	Total
Given Money to a Group in Past 5 Years?	Yes	612	349	1557	2518
	No	2603	1977	9738	14338
Total		3215	2326	11295	16856

Source: International Social Survey Programme 2000.

15. In 2000, respondents in several European countries were asked whether they chose to avoid driving a car for environmental reasons. Their responses are listed below. (Frequencies are shown below.)

		Country					
		Great Britain	Spain	Austria	Ireland	Netherlands	Norway
Avoid Driving	Always	33	15	40	19	47	53
Car for	Often	96	51	165	71	287	206
Environmental	Sometimes	260	122	297	225	541	507
Reasons	Never	343	413	268	655	411	515
Total		732	601	770	970	1286	1281

Source: International Social Survey Programme 2000.

a. Is there a relationship between a respondent's country of residence and whether or not he or she abstains from driving a car for environmental reasons?

b. Approximately 27 percent of Austrian respondents and 26 percent of Dutch respondents say that they always or often avoid driving a car for environmental reasons, whereas for Irish respondents this figure is only 9 percent. Provide some sociological insight as to why these figures are so different.

Chapter 7

Measures of Association for Nominal and Ordinal Variables

▣ PROPORTIONAL REDUCTION OF ERROR: A BRIEF INTRODUCTION

In the previous chapter we focused on one bivariate technique—the method of cross-tabulation—in which the pattern of relationship between two variables was analyzed by making a number of percentage comparisons. In Chapters 7 and 8, we review special **measures of association** for nominal, ordinal, and interval-ratio variables. These measures enable us to use a single summarizing measure or number for analyzing the pattern of relationship between two variables. Such measures of association reflect the strength of the relationship and, at times, its direction (whether it is positive or negative). They also indicate the usefulness of predicting the dependent variable from the independent variable.

In this chapter we discuss two measures of association: lambda (a measure of association for nominal variables) and gamma (a measure of association between ordinal variables). In Chapter 8 we introduce Pearson's correlation coefficient, used for measuring bivariate association between interval-ratio variables.

Measure of association   A single summarizing number that reflects the strength of a relationship, indicates the usefulness of predicting the dependent variable from the independent variable, and often shows the direction of the relationship.

All the measures of association discussed in Chapters 7 and 8 are based on the concept of the **proportional reduction of error**, often abbreviated as **PRE**. According to the concept of PRE, two variables are associated when information about one can help us improve our prediction of the other.

Table 7.1 may help us grasp intuitively the general concept of PRE. Using General Social Survey (GSS) 2002 data for men only, Table 7.1 shows a moderate relationship between the independent variable *number of children* and the dependent variable *support for abortion if the woman is poor and can't afford any more children.* The table shows that 59.1 percent of the men who had two or more children were anti-abortion, compared to only 42.9 percent of men who had one child or no children.

The conceptual formula for all[1] PRE measures of association is

$$PRE = \frac{E1 - E2}{E1} \tag{7.1}$$

where

$E1$ = errors of prediction made when the independent variable is ignored (Prediction 1)

$E2$ = errors of prediction made when the prediction is based on the independent variable (Prediction 2)

All PRE measures are based on comparing predictive error levels that result from each of the two methods of prediction. Let's say that we want to predict a man's position on abortion but we do not know anything about the number of children he has. Based on the row totals in Table 7.1 we could predict that every man in the sample is anti-abortion because this is the modal category of the variable *abortion position*. With this prediction we would make 84

---

[1]Although this general formula provides a framework for all PRE measures of association, only lambda is illustrated with this formula. Gamma, which is discussed in the next section, is calculated with a different formula. Both are interpreted as PRE measures.

Table 7.1   Support for Abortion by Number of Children: Men Only

	Number of Children		*Total*
	None or One Child	*Two or More Children*	
No	36	52	88
	42.9%	59.1%	51.2%
Yes	48	36	84
	57.1%	40.9%	48.8%
Total	84	88	172
	100%	100%	100%

Source: GSS 2002.

errors, because in fact 88 men in this group are anti-abortion but 84 men are pro-choice. (See the row marginals in Table 7.1.) Thus,

$$E1 = 172 - 88 = 84$$

How can we improve this prediction by using the information we have on each man's number of children? For our new prediction, we will use the following rule: If a man has two or more children, we predict that he will be anti-abortion; if a man has one child or none, we predict that he is pro-choice. It makes sense to use this rule because we know, based on Table 7.1, that men with larger families are more likely to be anti-abortion, while men who have one child or none are more likely to be pro-choice. Using this prediction rule, we will make 72 errors (instead of 84) because 36 of the men who have one child or none are actually anti-abortion, whereas 36 of the men who have two children or more are pro-choice (36 + 36 = 72). Thus,

$$E2 = 36 + 36 = 72$$

Our first prediction method, ignoring the independent variable (number of children), resulted in 84 errors. Our second prediction method, using information we have about the independent variable (number of children), resulted in 72 errors. If the variables are *associated*, the second method will result in fewer errors of prediction than the first method. The stronger the relationship is between the variables, the larger will be the reduction in the number of errors of prediction.

Let's calculate the proportional reduction of error for Table 7.1 using Formula 7.1. The proportional reduction of error resulting from using number of children to predict position on abortion is

$$\text{PRE} = \frac{84 - 72}{84} = 0.14$$

A Closer Look 7.1
What Is Strong? What Is Weak? A Guide to Interpretation

The more you work with various measures of association, the better feel you will have for what particular values mean. Until you develop this instinct, though, here are some guidelines regarding what is generally considered a strong relationship and what is considered a weak relationship.

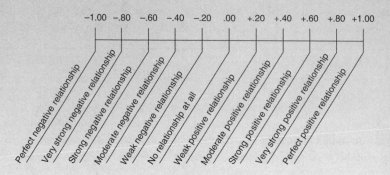

Keep in mind that these are only rough guidelines. Often, the interpretation for a measure of association will depend on the research context. A +.30 in one research field will mean something a little different from a +.30 in another research field. Zero, however, always means the same thing: no relationship.

PRE measures of association can range from 0.0 to ±1.0. A PRE of zero indicates that the two variables are not associated; information about the independent variable will not improve predictions about the dependent variable. A PRE of ±1.0 indicates a perfect positive or negative association between the variables; we can predict the dependent variable without error using information about the independent variable. Intermediate values of PRE will reflect the strength of the association between the two variables and therefore the utility of using one to predict the other. The more the measure of association departs from 0.00 in either direction, the stronger the association. PRE measures of association can be multiplied by 100 to indicate the percentage improvement in prediction.

A PRE of .14 indicates that there is a weak relationship between men's number of children and their position on abortion. (Refer to A Closer Look 7.1 for a discussion of the strength of a relationship.) A PRE of .14 means that we have improved our prediction of men's position on abortion by just 14 percent ($.14 \times 100 = 14\%$) by using information on their number of children.

Proportional reduction of error (PRE)   The concept that underlies the definition and interpretation of several measures of association. PRE measures are derived by comparing the errors made in predicting the dependent variable while ignoring the independent variable with errors made when making predictions that use information about the independent variable.

▣ LAMBDA: A MEASURE OF ASSOCIATION FOR NOMINAL VARIABLES

During the 2000 and 2004 presidential election campaigns, terrorism, the economy, morals, and health care emerged as important issues to voters. Indeed, voting behavior is dependent on a vast set of characteristics and beliefs analyzed by pollsters and social scientists.

In past elections, a candidate's and his party's position on abortion was also important, but in 2004 it became less of an issue in the face of the war with Iraq. How did abortion attitudes influence voters in the 2000 elections? To examine this question, let's look at Table 7.2, which shows 199 respondents of the 2002 GSS classified by their position on abortion (whether they support or do not support) and the candidate they voted for in 2000. We will consider abortion attitudes to be the independent variable and 2000 presidential vote, which may be explained or predicted by the independent variable, to be the dependent variable.

Because abortion attitude and presidential vote are nominal variables, we need to apply a measure of association suitable for calculating relationships between nominal variables. Such a measure will help us determine how strongly associated abortion attitude is with presidential vote. **Lambda** is such a PRE measure.

Table 7.2   2000 Presidential Vote by Abortion Attitudes

2000 Presidential Vote	*Abortion Attitudes (Abortion for Any Reason)*		*Row Total*
	Yes	*No*	
Gore	46	39	85
Bush	41	73	114
Column Total	87	112	199

Source: General Social Survey, 2002.

A Method for Calculating Lambda

Take a look at Table 7.2 and examine the row totals, which show the distribution of the variable *2000 presidential vote.* If we had to predict which candidate people voted for, our best bet would be to guess the mode, which is that everyone voted for President George W. Bush. This prediction will result in the smallest possible error. The number of wrong predictions we make using this method is actually 85, since 114 (the mode) out of 199 voted for Bush (199 − 114 = 85).

Now take another look at Table 7.2, but this time let's consider abortion attitudes when we predict 2000 presidential vote. Again, we can use the mode of 2000 presidential vote, but this time we apply it separately to abortion supporters and nonsupporters. The mode for abortion supporters is "Gore" (46 voters); therefore, we can predict that all abortion supporters are Gore voters. With this method of prediction we make 41 errors, since 46 out of 87 abortion supporters voted for Vice President Al Gore (87 − 46 = 41). Next we look at the group of abortion nonsupporters. The mode for this group is "Bush"; this will be our prediction for the entire group. This method of prediction results in 39 errors ($112 - 73 = 39$). The total number of errors is thus $41 + 39$ or 80 errors.

Let's now put it all together and state the procedure for calculating lambda in more general terms.

1. Find $E1$, the errors of prediction made when the independent variable is ignored. To find $E1$, find the mode of the dependent variable and subtract its frequency from N. For Table 7.2,

   $E1 = N -$ Modal frequency
   $E1 = 199 - 114 = 85$

2. Find $E2$, the errors made when the prediction is based on the independent variable. To find $E2$, find the modal frequency for each category of the independent variable, subtract it from the category total to find the number of errors, then add up all the errors. For Table 7.2,

   Abortion support errors $= 87 - 46 = 41$
   Abortion nonsupport errors $= 112 - 73 = 39$
   $E2 = 41 + 39 = 80$

3. Calculate lambda using Formula 7.1:

$$\text{Lambda} = \frac{E1 - E2}{E1} = \frac{85 - 80}{85} = 0.06$$

Lambda may range in value from 0.0 to 1.0. Zero indicates that there is nothing to be gained by using the independent variable to predict the dependent variable. A lambda of 1.0 indicates that by using the independent variable as a predictor, we are able to predict the dependent variable without any error. In our case, a lambda of .06 is less than one quarter of the distance between 0.0 and 1.0, indicating that for this sample of respondents, abortion attitudes and 2000 presidential vote are only slightly associated.

The proportional reduction of error indicated by lambda, when multiplied by 100, can be interpreted as follows: By using information on respondents' attitudes about abortion to predict 2000 presidential vote, we have reduced our error of prediction by 6.0 percent ($0.06 \times 100 = 6.0\%$).

Lambda An asymmetrical measure of association, lambda is suitable for use with nominal variables and may range from 0.0 to 1.0. It provides us with an indication of the strength of an association between the independent and dependent variables.

Hypothetical Strong Relationship: Stress Level by Dog Ownership
Lambda = .80

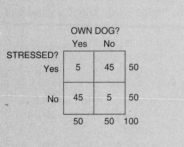
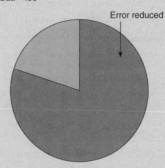

Hypothetical Moderate Relationship: Stress Level by Dog Ownership
Lambda = .40

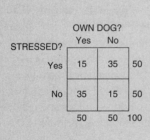
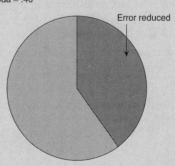

Hypothetical Weak Relationship: Stress Level by Dog Ownership
Lambda = .12

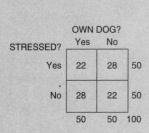
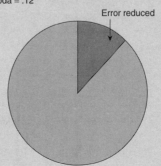

Some Guidelines for Calculating Lambda

Lambda is an **asymmetrical measure of association**. This means that lambda will vary depending on which variable is considered the independent variable and which the dependent variable. In our example, we considered 2000 presidential vote as the dependent variable and abortion attitudes as the independent variable, and not vice versa. Had we instead considered 2000 presidential vote the independent variable and abortion attitudes the dependent variable, the value of lambda would have been slightly larger at 0.08.

Asymmetrical measure of association A measure whose value may vary depending on which variable is considered the independent variable and which the dependent variable.

The method of calculation follows the same guidelines even when the variables are switched. However, exercise caution in calculating lambda, especially when the independent variable is arrayed in the rows rather than in the columns. To avoid confusion it is safer to switch the variables and follow the convention of arraying the independent variable in the columns; then follow the exact guidelines suggested for calculating lambda. Remember, however, that although lambda can be calculated either way, ultimately what guides the decision of which variables to consider as independent or dependent is the theoretical question posed by the researcher.

Lambda is always zero in situations in which the mode for each category of the independent variable falls into the same category of the dependent variable. A problem with interpreting lambda arises in situations in which lambda is zero but other measures of association indicate that the variables are associated. To avoid this potential problem, examine the percentage differences in the table whenever lambda is exactly equal to 0. If the percentage differences are very small (usually 5% or less), lambda is an appropriate measure of association for the table. However, if the percentage differences are larger, indicating that the two variables may be associated, lambda will be a poor choice as a measure of association. In such cases, we may want to discuss the association in terms of the percentage differences or select an alternative measure of association.

✓ *Learning Check.*   *The PRE measure calculated for Table 7.1 was also lambda. Recalculate the lambda for the table, this time assuming that abortion attitudes is the independent variable and number of children is the dependent variable. Has the value of lambda changed?*

▣ GAMMA: AN ORDINAL MEASURE OF ASSOCIATION

In this section, we discuss a way to measure and interpret an association between two *ordinal* variables. If there is an association between the two variables, knowledge of one variable will enable us to make better predictions of the other variable.

Before we illustrate how measures of association between ordinal variables are calculated and interpreted, let's review for a moment the definition of ordinal level variables. Whenever we assign numbers to rank-ordered categories ranging from low to high, we have an ordinal level variable. *Social class* is an example of an ordinal variable. We might classify individuals with respect to their social class status as "upper class," "middle class," or "working class." We can say that a person in the category "upper class" has a higher class position than a person in a "middle class" category (or that a "middle class" position is higher than a "working class" position).

Ordinal variables are very common in social science research. The General Social Survey contains many questions that ask people to indicate their response on an ordinal scale—for example, "very often," "fairly often," "occasionally," or "almost never."

Analyzing the Association Between Ordinal Variables: Job Security and Job Satisfaction

Let's look at a research example in which the association between two ordinal variables is considered. We want to examine the hypothesis that the higher a person's perceived job security, the higher his or her job satisfaction. To examine this hypothesis we selected two questions from the General Social Survey. The following question is a measure of *job satisfaction*:

On the whole, how satisfied are you with the work you do—would you say you are very satisfied, moderately satisfied, a little dissatisfied, or very dissatisfied?[2]

To measure *job security* they asked:

Thinking about the next 12 months, how likely do you think it is that you will lose your job or be laid off—very likely, fairly likely, not too likely, or not at all likely?[3]

Table 7.3 displays the cross-tabulation of these two variables, with *job security* as the independent variable and *job satisfaction* as the dependent variable. We find that 36 percent of those who indicate high job security, but only 11 percent of those who indicate low job security, are "satisfied" with their job. The percentage difference (36% – 11% = 25%) indicates that the variables are related. Several other percentage differences that can be computed on these data (for example, 36% – 22% = 14%; 22% – 11% = 11%) yield smaller percentage differences but lead to the same conclusion—that job satisfaction and job security are associated.

Comparison of Pairs   Let's explore this relationship a bit further. To understand the logic of the ordinal measures of association discussed in this section, we have to restate the relationship not in terms of individual observations but in terms of **paired observations** and their relative position (or rank order) on the two variables. To make the discussion a bit more concrete, let's suppose that we narrow each variable into only two categories, high and low job security and high and low job satisfaction, and that there are only four people involved, one in each cell. For the purpose of this illustration we have assigned names to each individual. Our hypothetical data are presented in Figure 7.1. We now pair four people (six combinations

---

[2]We have recoded the original response categories for this question into the following new categories: very satisfied = high, moderately satisfied = moderate, a little dissatisfied or very dissatisfied = low.

[3]We have recoded responses to this question into three levels of job security: not at all likely = high, not too likely = medium, and fairly likely and very likely = low.

Table 7.3   Job Satisfaction by Job Security*

JOB SATISFACTION (Y)	JOB SECURITY (X)			
	High	Medium	Low	Total
High	a 36.4% (16)	b 22.2% (8)	c 10.8% (14)	18.1% (38)
Moderate	d 43.2% (19)	e 47.2% (17)	f 46.1% (60)	45.7% (96)
Low	g 20.4% (9)	h 30.6% (11)	i 43.1% (56)	36.2% (76)
Total (N)	100% (44)	100% (36)	100% (130)	100% (210)

*The cells in this table have been labeled *a* through *i*.

can be created) and describe their rank order on the question of job security and job satisfaction. These results are presented in Figure 7.2.

Paired observations   Observations compared in terms of their relative rankings on the independent and dependent variables.

Let's consider the first pair, John and Arturo. Notice that John, who is high on job security, is also high on job satisfaction, and that Arturo, who has low job security, is also low on job satisfaction. When we consider the pair, we can say that the person who has higher job security (John) is also the more satisfied of the two. This pair would lead us to conclude that the higher one's job security, the higher one's job satisfaction, or that job satisfaction increases with job security.

Next consider Ruth and May. Ruth, who has low job security, is satisfied with her job. On the other hand, May, who has high job security, is not satisfied with her job. Regarding this pair, we could say that the person who has lower job security is more satisfied. This pair would lead us to conclude that the lower one's job security, the higher one's job satisfaction, or that job satisfaction decreases with job security.

Types of Pairs   Because ordinal variables have direction—that is, their categories can range from low to high—the relationship between ordinal variables also has direction. The direction of the relationship can be *positive* or *negative*. With a positive relationship, if one

Figure 7.1   Job Satisfaction by Job Security of Four People (hypothetical)

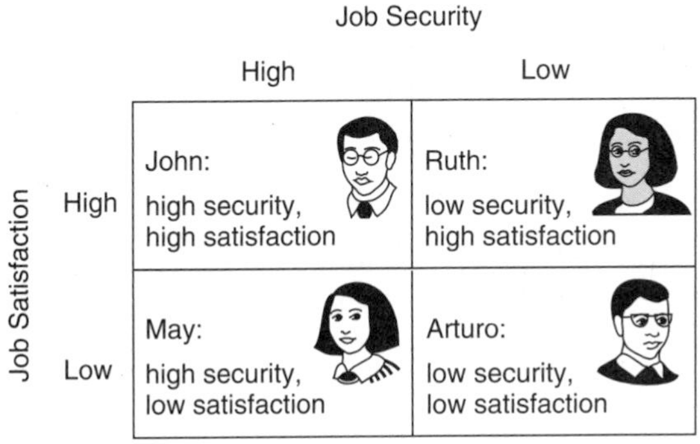

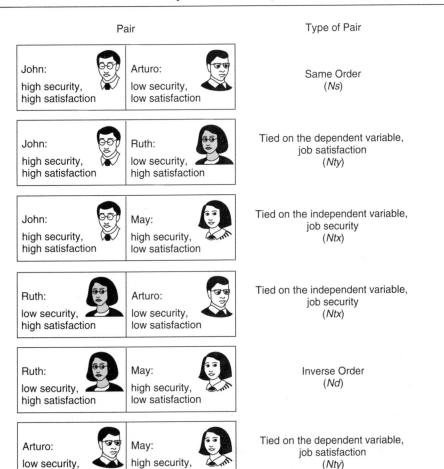

Figure 7.2   Rank Order of Four People on Job Security and Job Satisfaction

person is ranked above another on one variable, he or she would rank above the other person on the second variable. Such a relative ranking of two observations is called a **same order pair.** We label the count of these types of pairs as *Ns*. Same order pairs show a positive association. John and Arturo are a same order pair, displaying a positive association because John, who is higher than Arturo on job security, is also more satisfied than Arturo. For John and Arturo, job satisfaction increases with job security.

Same order pair (Ns)   Paired observations that show a positive association; the member of the pair ranked higher on the independent variable is also ranked higher on the dependent variable.

With a negative relationship, if one person is ranked above another on one variable, he or she ranks below the other person on the second variable. Such relative ranking of a pair of observations is called an **inverse order pair,** and their count is labeled as *Nd*. Inverse order pairs show a negative association. Ruth and May are an inverse order pair. They display a negative association because Ruth, who is lower than May on job security, is higher (more satisfied) than May on job satisfaction. For Ruth and May, job satisfaction decreases with job security.

Inverse order pair (Nd)   Paired observations that show a negative association; the member of the pair ranked higher on the independent variable is ranked lower on the dependent variable.

Note that there are four other pairs in Figure 7.2. These pairs all have the same value on either job satisfaction or job security. For example, Ruth and Arturo have different levels of job satisfaction (Ruth's level of job satisfaction is high while Arturo's is low), but they both have low job security. Pairs that have the same value on a variable are called tied pairs. Tied pairs are used in calculating a measure of association called Somers' *d*, which we will not be covering in this text. Gamma, the most frequently used measure of association for ordinal variables, considers only untied pairs—that is, *same order* and *inverse order* pairs.

Counting Pairs

To calculate gamma we first count the number of same order (*Ns*) pairs and inverse order (*Nd*) pairs that can be obtained from a bivariate table. Once we find *Ns* and *Nd,* the calculation of gamma is straightforward.

Because our illustration is based on only four cases (Figure 7.1), the number of pairs involved is very small and can easily be identified (as in Figure 7.2) and counted. However, because the number of cases in most bivariate tables is considerably larger, the number of pairs that can be generated becomes very large and the process of identifying types of pairs becomes a bit more complicated.

To illustrate the process of identifying and counting all the same order pairs (*Ns*) and inverse order pairs (*Nd*) that can be generated from a bivariate table,[4] let's go back to Table 7.3. Note that the table is constructed so that the cell in the upper left corner represents the highest category on both *X* and *Y* (*high* job security and *high* job satisfaction), with levels of *X* decreasing from left to right and levels of *Y* decreasing from top to bottom. Always make sure the table you are analyzing is arranged in this way before following the procedure outlined here to find *Ns* and *Nd.* If your table is not arranged in this way, you can always rearrange it to follow this format.

Same Order Pairs (Ns) To find the number of same order (*Ns*) pairs, multiply the frequency in each cell in the table by the sum of the frequencies of all the cells that are lower on *both* variables—that is, both below *and* to the right of that cell. Repeat this process for each cell that has cells below it and to its right, and then sum the products. The total of these products is *Ns.*

We begin with the upper left cell, *a,* in Table 7.3. The frequency (16) is multiplied by the sum of the frequencies in cells *e, f, h,* and *i* (17 + 60 + 11 + 56). These are the cells that lie below and to the right of cell *a.* The product equals 2,304 pairs.

This computation is illustrated in Figure 7.3, in which cell *a* and the cells that lie below it and to its right (*e, f, h, i*) are shaded. Move to the next cell, cell *d.* Its frequency, 19, is

Figure 7.3 Counting All Same Order Pairs (*Ns*) from Table 7.5

Job Security

Job Satisfaction		High	Med	Low
	High	*a* 16	*b* 8	*c* 14
	Med	*d* 19	*e* 17	*f* 60
	Low	*g* 9	*h* 11	*i* 56

16 (17 + 60 + 11 + 56) = 2,304

	High	Med	Low
High	*a* 16	*b* 8	*c* 14
Med	*d* 19	*e* 17	*f* 60
Low	*g* 9	*h* 11	*i* 56

19 (11 + 56) = 1,273

Job Security

Job Satisfaction		High	Med	Low
	High	*a* 16	*b* 8	*c* 14
	Med	*d* 19	*e* 17	*f* 60
	Low	*g* 9	*h* 11	*i* 56

8 (60 + 56) = 928

	High	Med	Low
High	*a* 16	*b* 8	*c* 14
Med	*d* 19	*e* 17	*f* 60
Low	*g* 9	*h* 11	*i* 56

17 (56) = 952

Ns = 2,304 + 1,273 + 928 + 952 = 5,457

[4]This process of counting pairs also appears in Chava Frankfort-Nachmias and David Nachmias, *Research Methods in the Social Sciences* (New York: Worth Publishers, 2000), pp. 368–372.

multiplied by $(11 + 56)$, the sum of the frequencies in cells h and i, and the result equals 1,273 pairs. (The direction of the multiplication is illustrated in Figure 7.4.) This procedure is repeated for cells b and e. Because cells g and h have no cells below them and cells c, f, and i have no cells to the right of them, they are not included in the computations of Ns. Now let's sum the total number of same order (Ns) pairs:

$$Ns = 2,304 + 1,273 + 928 + 952 = 5,457$$

Each one of these 5,457 pairs meets the definition of same order (Ns) pairs presented earlier. For instance, take a pair formed from cell a and cell h. The member of the pair from cell a is high on job security and high on job satisfaction, whereas the member of the pair from cell h is medium on job security and low on job satisfaction. Therefore, we can say that the member of the pair that is higher on job security (high versus medium) is also higher on job satisfaction (high versus low).

> ✓ **Learning Check.**   *To convince yourself that each of the 5,457 Ns pairs we counted is indeed a same order pair, try for yourself any of the combinations as depicted in Figure 7.3.*

Inverse Order Pairs (Nd)   To calculate the number of inverse order pairs (Nd) we proceed in exactly the same way, with one difference. Because we are interested in pairs in which the relative ranking of the variables is reversed, we reverse our process and begin with the upper right cell (cell c). We then multiply its frequency by the cells below it and to its left. Repeat this process for each cell that has cells below it and to its left, and then sum the products. The total of these products is Nd. These computations are illustrated in Figure 7.4. Now let's sum to compute Nd:

$$Nd = 784 + 1,200 + 224 + 153 = 2,361$$

All these pairs follow the definition of inverse order. Take a pair formed from cells c and g. The member of the pair from cell c is low on job security and high on job satisfaction. Conversely, the member of the pair from cell g is high on job security and low on job satisfaction. Therefore, we can say that the member of the pair that is lower on job security is higher on job satisfaction.

Before we discuss gamma, let's look again at the meaning of same order and inverse order pairs. *Same order* pairs are all the pairs of observations formed from Table 7.5 that show a positive relationship between the variables: the higher your job security the higher your job satisfaction, or job satisfaction increases with job security. *Inverse order* pairs are all the pairs of observations that show a negative relationship between the variables: the higher your job security the lower your job satisfaction, or job satisfaction decreases with job security. The purpose of our analysis is to determine whether there is an association between job security and job satisfaction. If there is such an association, we want to be able to determine which of the statements best describes the association—that is, whether the association is *positive* or *negative*. Gamma can help us answer these questions.

Figure 7.4   Counting All Inverse Order Pairs (*Nd*) from Table 7.5

Job Security

		High	Med	Low
	High	*a* 16	*b* 8	*c* 14
Job Satisfaction Med		*d* 19	*e* 17	*f* 60
	Low	*g* 9	*h* 11	*i* 56

14 (19 + 17 + 9 + 11) = 784

Job Security

		High	Med	Low
	High	*a* 16	*b* 8	*c* 14
Job Satisfaction Med		*d* 19	*e* 17	*f* 60
	Low	*g* 9	*h* 11	*i* 56

60 (9 + 11) = 1,200

Job Security

		High	Med	Low
	High	*a* 16	*b* 8	*c* 14
Job Satisfaction Med		*d* 19	*e* 17	*f* 60
	Low	*g* 9	*h* 11	*i* 56

8 (19 + 9) = 224

Job Security

		High	Med	Low
	High	*a* 16	*b* 8	*c* 14
Job Satisfaction Med		*d* 19	*e* 17	*f* 60
	Low	*g* 9	*h* 11	*i* 56

17 (9) = 153

Nd = 784 + 1,200 + 224 + 153 = 2,361

▣ CALCULATING GAMMA

Gamma is a symmetrical measure of association suitable for use with ordinal variables or with dichotomous nominal variables. It can vary from 0.0 to ±1.0 and provides us with an indication of the strength and direction of the association between the variables. When there are more same order pairs than inverse order pairs (*Ns* is larger than *Nd*), gamma will be positive. Gamma is calculated using the following formula:

$$\text{Gamma} = \frac{Ns - Nd}{Ns + Nd} \tag{7.2}$$

Using this formula and the pairs calculations we made earlier, let's now find the association between job security and job satisfaction:

$$\text{Gamma} = \frac{5{,}457 - 2{,}361}{5{,}457 + 2{,}361} = 0.40$$

A gamma of 0.40 indicates that there is a moderate positive association between job security and job satisfaction. We can conclude that using information on respondents' job security helps us improve the prediction of their job satisfaction by 40 percent.

Gamma is a **symmetrical measure of association**. This means that the value of gamma will be the same regardless of which variable is the independent variable or the dependent variable. Thus, if we had wanted to predict job security from job satisfaction rather than the opposite, we would have obtained the same gamma.

Gamma A symmetrical measure of association suitable for use with ordinal variables or with dichotomous nominal variables. It can vary from 0.0 to ±1.0 and provides us with an indication of the strength and direction of the association between the variables. When there are more Ns pairs, gamma will be positive; when there are more Nd pairs, gamma will be negative.

Symmetrical measure of association A measure whose value will be the same when either variable is considered the independent variable or the dependent variable.

Positive and Negative Gamma

Note from Formula 7.2 that the size and the direction of gamma (whether positive or negative) are functions of the relative number of same order (Ns) versus inverse order (Nd) pairs. More Ns pairs make gamma positive; more Nd pairs make gamma negative. The larger the difference between Ns and Nd, the larger the size of the coefficient (irrespective of sign). For example, when all the pairs are Ns ($Nd = 0$), gamma equals 1.00:

$$\text{Gamma} = \frac{Ns - 0}{Ns + 0} = 1.0$$

A gamma of 1.0 indicates that the relationship between the variables is positive, and the dependent variable can be predicted without any error based on the independent variable. When Ns is zero, gamma will be −1.0, indicating a perfect and a negative association between the variables:

$$\text{Gamma} = \frac{0 - Nd}{0 + Nd} = -1.0$$

When $Ns = Nd$, gamma will equal zero:

$$\text{Gamma} = \frac{Ns - Nd}{Ns + Nd} = \frac{0}{Ns + Nd} = 0.0$$

A gamma of zero reflects no association between the two variables; hence, there is nothing to be gained by using order on the independent variable to predict order on the dependent variable.

Gamma as a PRE Measure

Like all PRE measures, gamma is based on two methods of prediction. The first method ignores the relative order of pairs on the independent variable (only untied pairs are included in the computation of gamma), whereas the second method considers it. Suppose that we had tried to predict the rank order of each of the 7,818 pairs (the sum of same and inverse order pairs: $Ns + Nd = 5,457 + 2,361 = 7,818$) while ignoring information on their relative rank order on job security. If we had used a random method to make these predictions for each of the pairs, chances are that only about 50 percent of our guesses would have been right. Therefore, we would have made errors about half the time, or $(Ns + Nd) \div 2 = (5,457 + 2,361) \div 2 = 3,909$ errors.

Now let's see if we can improve this prediction by taking the rank order on job security into consideration when predicting the rank order on job satisfaction. The likelihood of improving the prediction depends on the number of Ns versus Nd pairs. When Ns is greater than Nd, as it is in our example, it makes sense to predict that if respondents are higher on job security, they will also be higher on job satisfaction (this is the order displayed by all the Ns pairs)—that is, job satisfaction increases with job security. With this prediction we will be making 2,361 errors—the number of Nd pairs for which this prediction is not correct—or 1,548 fewer errors than before $(3,909 - 2,361 = 1,548)$. When we divide this number by 3,909 (the original number of errors),

$$\frac{1,548}{3,909} = 0.40$$

we obtain a proportional reduction of error of 0.40, which is equal to the gamma coefficient we obtained with Formula 7.2.

Statistics in Practice:
Education and Attendance of Religious Services

When the number of Nd pairs is larger than the number of Ns pairs, gamma is negative and the prediction is that if a person has a higher rank on the independent variable, she or he will have a lower rank on the dependent variable. This order is illustrated in Table 7.4, which displays the cross-tabulation of education and frequency of church attendance of 440 International Social Survey Programme (ISSP) respondents, with education treated as the independent variable. Remember that with a negative relationship, higher values of one variable tend to go together with lower values of the other, and vice versa. In this table, for instance, those with a university degree are more likely to never attend church (37.3%), whereas more respondents with no degree are likely to attend church two to three times per month or more (66.2%).

Table 7.4   Educational Level by Attendance of Religious Services

ATTENDANCE OF RELIGIOUS SERVICES	EDUCATIONAL LEVEL			
	None	Secondary Degree	University Degree	Total
Never	5.2% (4)	32.5% (77)	37.3% (47)	29.1% (128)
Infrequently	28.6% (22)	35.0% (83)	34.9% (44)	33.9% (149)
Two to three times per month or more	66.2% (51)	32.5% (77)	27.8% (35)	37.0% (163)
	100% (77)	100% (237)	100% (126)	100% (440)

Source: ISSP 2000.

To calculate gamma we must first count the number of Ns and Nd pairs. The number of same order (Ns) pairs that can be formed from Table 7.4 is

$$Ns = 4(83 + 44 + 77 + 35) + 22(77 + 35) + 77(44 + 35) + 83(35) = 12{,}408$$

The number of inverse order (Nd) pairs that can be formed from Table 7.4 is

$$Nd = 47(22 + 83 + 51 + 77) + 44(51 + 77) + 77(22 + 51) + 83(51) = 26{,}437$$

Gamma is

$$\frac{12{,}408 - 26{,}437}{12{,}408 + 26{,}437} = -0.36$$

In this example, the number of inverse order pairs ($Nd = 26{,}437$) is larger than the number of same order pairs ($Ns = 12{,}408$). We can predict that if your educational degree is higher than that of the other member of your pair, then your church attendance will be lower—or your church attendance decreases with educational attainment. This is the order displayed by all the inverse order pairs. With this prediction we would make 12,408 errors, the number of same order (Ns) pairs for which this prediction is not correct.

A gamma of –0.36 means that, as expected, the relationship between the variables is negative; that is, as education increases, church attendance decreases. The relationship between these variables is moderate: Using education to predict church attendance results in a proportional reduction of error of 36 percent (0.36×100).

▣ USING ORDINAL MEASURES WITH DICHOTOMOUS VARIABLES

Measures of association for ordinal data are not influenced by the modal category, as is lambda. Consequently, an ordinal measure of association might be preferable for tables when an association cannot be detected by lambda. We can use an ordinal measure for some tables where one or both variables would appear to be measured on a nominal scale. Dichotomous variables (those with only two categories) can be treated as ordinal variables for most purposes. In this chapter we calculated lambda to examine the association between abortion attitudes and number of children (Table 7.1). Although both variables might be considered as nominal variables—because both are dichotomized (yes/no; none or one child/two or more)—they could also be treated as ordinal variables. Thus, the association might also be examined using gamma, an ordinal measure of association.[5]

Let's calculate gamma for Table 7.1. The number of same order pairs that can be formed from Table 7.1 is

$$Ns = 36(36) = 1296$$

The number of inverse order pairs that can be formed from Table 7.1 is

$$Nd = 52(48) = 2496$$

Gamma is

$$\frac{1,296 - 2,496}{1,296 + 2,496} = -0.32$$

A gamma of –0.32 confirms our earlier conclusion that for men, number of children is negatively related to abortion attitudes. Thus, having more children is associated with nonsupport of abortion for men.

▣ READING THE RESEARCH LITERATURE: WORLDVIEW AND ABORTION BELIEFS

Let's conclude this chapter with a typical example of how gamma is presented and interpreted in the social science research literature. The following example is drawn from a study that examines the idea that beliefs about abortion are influenced by a coherent view of the world. Kristin Luker argues that "each side of the abortion debate has an internally coherent and mutually shared view of the world that is tacit, never fully articulated, and most important,

---

[5]For 2 × 2 tables, a measure identical to gamma—Yule's Q—was first introduced by the statistician Udny Yule. However, whereas gamma is suitable for any size table, Yule's Q is appropriate only for 2 × 2 tables. When gamma is calculated for 2 × 2 tables, it is sometimes referred to as Yule's Q.

completely at odds with the world-view held by their opponents."[6] In general, if a person is religious, views the primary role of women as one of taking care of the home and raising the children, thinks that sex should be practiced for procreation only, and has a conservative political viewpoint, then this person also tends to disapprove of abortion under almost any circumstances. Conversely, a person who is not religious, has an egalitarian view of gender roles, has liberal attitudes toward sexuality, and identifies himself or herself as left of center on the political spectrum probably believes in a woman's right to choose whether or not to have an abortion.[7] The major hypothesis of this study is "the more conservative a person's world view, the greater his or her disapproval of abortion."[8]

To test this hypothesis, Daniel Spicer used data from the 1990 General Social Survey, which is based on a representative sample of all the noninstitutionalized adult residents of the continental United States. To measure worldview, Spicer selected six variables: attitudes toward premarital sex, conception of sex roles, religious intensity, political views, fundamentalism, and biblical interpretation. For the purpose of this discussion, we have considered only the first three variables.

Measurement of these variables was based on responses to the following questions:

- *Conception of sex roles* "Do you agree or disagree with this statement? Women should take care of the home and leave running the country up to men."
- *Premarital sex views* "There's been a lot of discussion about the way morals and attitudes about sex are changing in this country. If a man and a woman have sex relations before marriage, do you think it is always wrong, usually wrong, somewhat wrong, or acceptable?"
- *Religious intensity* "Would you call yourself a strong, moderate, or weak (stated religion)?"

The dependent variable *abortion belief* was constructed from responses to four questions dealing with the following circumstances under which the person believed it acceptable or unacceptable for a woman to have an abortion. A person is said to approve of abortion if he or she said yes to all four of the following items:

If she is married and does not want any more children

If the family has a very low income and cannot afford any more children

If she is not married and does not want to marry the man

The woman wants it for any reason

Based on their response to all four questions, respondents were classified into one of two categories: "approve" or "disapprove" of abortion. The bivariate percentage distributions and the gamma for each of the independent variables and abortion belief are presented in Table 7.5.

---

[6]Kristin Luker, *Abortion and the Politics of Motherhood* (Berkeley: University of California Press, 1984), p. 159.

[7]Daniel N. Spicer, "World View and Abortion Beliefs: A Replication of Luker's Implicit Hypothesis," *Sociological Inquiry* 64:1, pp. 115–116 and 120–121. Copyright © 1994 by the University of Texas Press. All rights reserved.

[8]Ibid., pp. 115–116.

Table 7.5   Percentage Approval of Abortion by Selected Independent
Variables (GSS, 1990)

Variable	Category	N	Abortion Approval	Gamma
Premarital sex views	Always wrong	92	15%	−.644
	Usually wrong	42	36%	
	Somewhat wrong	75	51%	
	Acceptable	41	71%	
Sex role views	Traditional	58	21%	−.625
	Liberal	292	53%	
Religious intensity	Strong	261	28%	−.433
	Moderate	97	55%	
	Weak	312	59%	

Source: Adapted from Daniel N. Spicer, "World View and Abortion Beliefs: A Replication of Luker's Implicit Hypothesis," *Sociological Inquiry* 64:1, pp. 115–116 and 120–121. Copyright © 1994 by the University of Texas Press. All rights reserved.

Examining the Data

Begin by examining the structure of the table. Note that it is divided into three parts, one for each independent variable. Each part can be read as a separate table displaying the bivariate percentage and the gamma for each of the independent variables and abortion beliefs. The independent variables and their categories are arrayed in rows of the table; the dependent variable, *abortion approval,* is arrayed in the columns. For each category of the independent variables, the number of people who responded (*N*) and the percentage who approved of abortion are listed. For example, the variable *sex role views* has two categories, "traditional" and "liberal." Of the 58 traditionals, 21 percent approved of abortion, compared with 53 percent of the 292 liberals.

✓ *Learning Check.*   *The percentages that disapproved of abortion are not listed in the table. However, it is very easy to obtain the numbers: simply subtract the percentage that approved from 100 percent for each category. For example, of the 58 traditionals, 79 percent (100% − 21%) disapproved of abortion. Try to complete the table by calculating the percentages that disapproved of abortion for all the variables.*

Interpreting the Data

Next, interpret the data presented in Table 7.5. In reading the table and interpreting the relationship between each of the independent variables and beliefs about abortion, look at both the percentage differences and the value of gamma. Let's begin by comparing the percentages. Following the rules we learned in Chapter 6, when percentages are calculated within rows (as they are in this table), comparisons are made down the column. For instance, to interpret the relationship between religious intensity and abortion beliefs, compare 28 percent with 55 percent and 59 percent. Similarly, to examine the relationship between sex role views and abortion belief, compare 21 percent with 53 percent. Based on the percentage comparisons for each independent variable, the researcher offers the following interpretation of these findings:

> As shown in [Table 7.5], . . . as the idea of premarital sex becomes more acceptable, the approval of the practice of abortion increases . . . people with traditional sex role conceptions tended to disapprove of abortion and those with a modern, liberal conception of the sexes were in favor of abortion rights. Respondents with a high religious intensity disapproved of abortion, while those with weak religious ties were much more liberal on the question of abortion.[9]

The detailed summary of the relationships between the variables is confirmed by the values of gamma displayed in the table. The gamma values range from –.433 for religious intensity to –.644 for attitudes toward premarital sex. These values indicate a moderate to strong relationship between various aspects of one's worldview and abortion beliefs.

Notice that all the gamma values are negative. A negative gamma indicates that low values of one variable "go together" with high values of the other variable, and vice versa.

Often it is tricky to interpret the direction of a relationship between ordinal variables because what is considered "low" or "high" is often a function of arbitrary coding by the researcher. However, a researcher will often specify his or her coding in the text. Spicer makes this statement about the direction of these relationships:

> As high scores on . . . the [independent] variables (*higher* values indicating "*conservative*" social views) were associated with *low* scores (*disapproval*) on the abortion view variable . . . the Gamma values are negative.[10]

In other words, the negative gamma between premarital sex views, sex role views, and religious intensity means that respondents who have more conservative social views tend to disapprove of abortion. Conversely, a large percentage of those who have more liberal views approve of abortion.

MAIN POINTS

- Measures of association are single summarizing numbers that reflect the strength of the relationship between variables, indicate the usefulness of predicting the dependent variable from the independent variable, and often show the direction of the relationship.

- Proportional reduction of error (PRE) underlies the definition and interpretation of several measures of association. PRE

---

[9]Ibid., pp. 120–121.
[10]Ibid., p. 121.

measures are derived by comparing the errors made in predicting the dependent variable while ignoring the independent variable with errors made when making predictions that use information about the independent variable.

- PRE measures may range from 0.0 to ±1.0. A PRE of 0.0 indicates that the two variables are not associated and that information about the independent variable will not improve predictions about the dependent variable. A PRE of ±1.0 means that there is a perfect (positive or negative) association between the variables and that information about the independent variable results in a perfect (without any error) prediction of the dependent variable.

- Measures of association may be symmetrical or asymmetrical. When the measure is symmetrical, its value will be the same regardless of which of the two variables is considered the independent or dependent variable. In contrast, the value of asymmetrical measures of association may vary depending on which variable is considered the independent variable and which the dependent variable.

- Lambda is an asymmetrical measure of association suitable for use with nominal variables. It can range from 0.0 to 1.0 and gives an indication of the strength of an association between the independent and the dependent variables.

- Gamma is a symmetrical measure of association suitable for ordinal variables or for dichotomous nominal variables. It can vary from 0.0 to ±1.0 and reflects both the strength and direction of the association between the variables.

KEY TERMS

asymmetrical measure of association
gamma
inverse order pair (Nd)
lambda

measure of association
proportional reduction of error
 and same order pair (Ns)

ON YOUR OWN

Log on to the web-based student study site at http://www.pineforge.com/frankfort-nachmiasstudy4 for additional study questions, quizzes, web resources, and links to social science journal articles reflecting the statistics used in this chapter.

SPSS DEMONSTRATION

[GSS02PFP-A]

Demonstration 1: Producing Nominal Measures of Association for Bivariate Tables

In Chapter 6 we used the Crosstabs procedure in SPSS to create bivariate tables. That same procedure is used to request measures of association. In this chapter, we'll begin by investigating the relationship between belief in the Bible (BIBLE) and support for legal abortions for women for any reason (ABANY).

Click on *Analyze, Descriptive Statistics*, then *Crosstabs* to get to the Crosstabs dialog box. Put ABANY in the Row(s) box and BIBLE in the Column(s) box. Then click on the *Statistics* button. The Statistics dialog box (Figure 7.5) has about a dozen statistics from which to choose. Notice that four statistics are listed

Figure 7.5

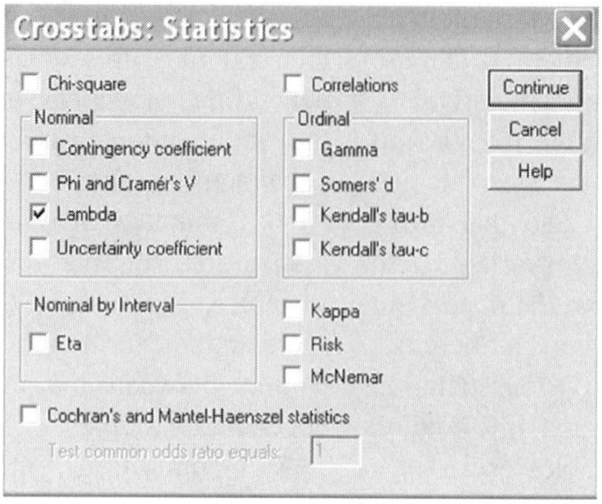

Figure 7.6

Directional Measures

			Value	Asymp. Std. Error[a]	Approx. T[b]	Approx. Sig.
Nominal by Nominal	Lambda	Symmetric	.089	.020	4.171	.000
		abany ABORTION IF WOMAN WANTS FOR ANY REASON Dependent	.187	.041	4.171	.000
		bible FEELINGS ABOUT THE BIBLE Dependent	.000	.000	.[c]	.[c]
	Goodman and Kruskal tau	abany ABORTION IF WOMAN WANTS FOR ANY REASON Dependent	.098	.026		.000[d]
		bible FEELINGS ABOUT THE BIBLE Dependent	.032	.009		.000[d]

in separate categories for "Nominal" and "Ordinal" data. Lambda is listed in the former, and gamma in the latter. The other measures of association, such as Somer's *d*, phi, Cramer's *V*, or Kendall's tau-*b*, will not be discussed in this textbook. The chi-square statistic will be discussed in depth in Chapter 13.

Since both variables are nominal, check the box for lambda. It is critical that we, as users of statistical programs, understand which statistics to select in any procedure. SPSS, like most programs, can't help us select the appropriate statistic for an analysis. Now click on *Continue*, and then *OK* to create the table (Figure 7.6).

The first table should be a bivariate table of our two variables (not shown here). Below this table is a table labeled "Directional Measures." For now we will only concern ourselves with the first two columns. Lambda is listed with three values. We've learned that the value of lambda depends on which variable is considered the dependent variable. In our example, *attitude toward abortion for women* is dependent, so lambda is .187. This indicates a weak relationship between the two variables. We can conclude that knowing the respondent's belief about the Bible increases the ability to predict his or her abortion attitude by just 18.7 percent.

SPSS also calculates a symmetrical lambda for those tables where there is no independent or dependent variable. This calculation goes beyond the scope of this book. (It is not simply an average of the two other values of lambda.) In addition, as a kind of bonus, SPSS provides the Goodman and Kruskal tau statistic, another nominal measure of association, even though it was not requested. These measures will always be produced when lambda is requested.

Demonstration 2: Producing Ordinal Measures of Association for Bivariate Tables

We can also use the same procedures to calculate gamma for ordinal measures. For this demonstration, we'll examine the relationship between educational attainment (DEGREE) and attitudes toward premarital sex (PREMARSX). Respondents were asked to whether premarital sex was wrong—always wrong, almost always wrong, sometimes wrong or not wrong at all. Both variables are ordinal measurements.

Click on *Analyze, Descriptive Statistics*, then *Crosstabs* to get to the Crosstabs dialog box. Put PREMARSX in the Row(s) box and DEGREE in the Column(s) box. Then click on the *Statistics* button. The Statistics dialog box has about a dozen statistics from which to choose. Click on gamma listed in the ordinal box. Notice that gamma is listed in the ordinal box. SPSS produces two separate tables, the first is the bivariate table between PREMARSX and DEGREE and the second is the table of symmetric measures, gamma, which we requested (Figure 7.7).

The gamma statistic is in the first row, under the column labeled "Value." For this bivariate table the gamma is .188, indicating a weak positive relationship between the two variables. About 18.8% of the error in predicting attitudes toward premarital sex would be reduced if we had information about respondent's educational attainment. Notice that given how PREMARSX is coded, the negative gamma indicates that as DEGREE increases, respondents are more likely to indicate that premarital sex is not wrong at all (coded 4).

Figure 7.7

Symmetric Measures

		Value	Asymp. Std. Error[a]	Approx. T[b]	Approx. Sig.
Ordinal by Ordinal	Gamma	.188	.063	2.971	.003
N of Valid Cases		490			

SPSS PROBLEMS

[GSS02PFP-A]

1. Examine the relationship between respondent's health (HEALTH) and social class (CLASS).
   a. Request the appropriate measure of association to describe the relationship.
   b. Add SEX as a control variable and calculate the association measure for each partial table. Is the relationship stronger for women or men? Can you think of reasons why this might be so?
   c. What other control variables may be appropriate? Continue to examine the relationship between HEALTH and CLASS with one additional control variable.

2. GSS respondents were asked whether they believed that women were not suited for politics (FEPOL). Examine how this variable is associated with respondent's sex (SEX) and educational attainment (DEGREE). Calculate the appropriate measure of association for each pair of variables. For the second pair, FEPOL and DEGREE, would gamma be an appropriate measure? Why or why not?

3. Investigate the relationship between the abortion items and various demographic variables (you might begin with gender, age, or race). Examine the relationship of these variables based on the appropriate measures of association. For example, you might examine whether attitude toward each of the abortion items has a similar relationship to gender. That is, if females are supportive of abortion for rape victims, are they also supportive of abortion in other circumstances? Try exploring these relationships further by adding control variables. You might create tables of abortion attitude by race by gender. When you have finished the analysis, write a short report summarizing the findings. Suggest possible causes for the relationships you found.

4. The General Social Survey asked respondents to indicate who they voted for in the 2000 presidential election (PRES00). Examine how respondents' votes are associated with SEX, CLASS, and BIBLE (feelings about the Bible). Calculate the appropriate measure for each pair of variables.

5. ISSP 2000: In this chapter, we examined the relationship between educational attainment and church attendance. Using the ISSP 2000 data set, examine how social class (CLASS) is associated with church attendance (ATTEND). Which do you think is a better predictor of church attendance—education or social class?

CHAPTER EXERCISES

1. Is there a relationship between race of violent offenders and their victims? Data from the U.S. Department of Justice (2003) are presented in Table 7.6.
   a. Let's treat race of offenders as the independent variable and race of victims as the dependent variable. If we first ignore the independent variable and try to predict race of victim, how many errors will we make?
   b. If we now take into account the independent variable, how many errors of prediction will we make for those offenders who are white? Black offenders? Other offenders?
   c. Combine the answers in (a) and (b) to calculate the proportional reduction in error for this table based on the independent variable. How does this statistic improve our understanding of the relationship between the two variables?

2. Let's continue our analysis of offenders and victims of violent crime. In Table 7.7, data for the sex of offenders and the sex of victims are reported.
   a. Treating sex of offender as the independent variable, how many errors of prediction will be made if the independent variable is ignored?
   b. How many fewer errors will be made if the independent variable is taken into account?

Table 7.6

Characteristics of Victim	Characteristics of Offender		
	White	Black	Other
White	3000	483	58
Black	227	2852	11
Other	51	28	109

Table 7.7

Sex of Victim	Sex of Offender	
	Male	*Female*
Male	4326	528
Female	1779	183

c. Combine your answers in (a) and (b) to calculate lambda. Discuss the relationship between these two variables.

d. Which lambda is stronger, the one for sex of offenders/victims or race of offenders/victims?

3. ISSP 2000 respondents were asked whether they believed that science was doing more harm than good to our environment. Their responses are presented in the bivariate table, along with their educational attainment.

	Less Than Secondary Degree	*Secondary Degree*	*University Degree*	*Total*
Strongly agree	14	8	6	28
Agree	55	57	20	132
Neither	74	85	28	187
Disagree	115	151	91	357
Strongly disagree	40	57	56	153
Total	298	358	201	857

Source: ISSP 2000.

a. What proportion of those with less than a secondary degree either "disagree" or "strongly disagree" with the statement? What proportion of those with a university degree either "disagree" or "strongly disagree" with the statement?

b. Calculate the number of N_s and N_d pairs.

c. Using N_s and N_d, calculate gamma for this table. Is gamma positive or negative? Using the value of gamma, interpret the relationship between education and support for the statement.

4. The following table is based on data from the GSS 2002 considering the relationship between social class and health condition. Respondents were asked to assess their current health condition.

	Social Class				
Health	*Lower Class*	*Working Class*	*Middle Class*	*Upper Class*	*Total*
Poor	9	17	11	2	39
Fair	11	58	37	6	112
Good	14	133	150	5	302
Excellent	5	86	110	14	215
Total	39	294	308	27	668

Source: GSS 2002.

    a. Calculate the number of *Ns* and *Nd* pairs.

    b. Using *Ns* and *Nd*, calculate gamma for this table. Is it positive or negative? Make sure you note how each variable has been coded. What is the relationship between social class and one's health?

5. Is there an association between marital status and general happiness? The GSS 2002 asked women and men to rate their happiness on a three-point scale. Table 7.8 displays data on marital status and happiness.

Table 7.8

Happiness Rating	Marital Status					
	Married	*Widowed*	*Divorced*	*Separated*	*Never Married*	*Total*
Very happy	90	7	16	3	38	154
Pretty happy	103	30	50	9	89	281
Not too happy	17	8	20	2	23	70
Total	210	45	86	14	150	505

Source: GSS 2002.

    a. Lambda is the appropriate measure for this table. Why? Without doing the calculation, study this table and offer an estimate of what lambda would be.

    b. Calculate lambda to assess the strength of the relationship between marital status and happiness. Can you explain why lambda has the value that it does?

6. Are you afraid to walk in your neighborhood at night? This question was posed to General Social Survey 2002 respondents. You decide to investigate this variable by sex and race. Do men and women express different fears? Does this change when you consider their race? You obtain the results shown in Tables 7.9a–c.

Use the appropriate PRE measure, plus percentage differences, to summarize the relationship between FEAR and SEX, for all respondents, then for whites and Blacks separately. Can you suggest a reason for any differences you find?

7. In Exercise 5 we learned about what might seem an oddity in the calculation of lambda that caused it to exactly equal zero. Measures of association for ordinal data are not influenced by

Table 7.9a-c

a: Both Whites and Blacks

Fear to Walk Alone at Night	Sex		
	Male	*Female*	*Total*
Yes	27	63	90
No	142	87	229
Total	169	150	319

b: Whites Only

Fear to Walk Alone at Night	Sex		Total
	Male	Female	
Yes	23	54	77
No	135	80	215
Total	158	134	292

c: Blacks Only

Fear to Walk Alone at Night	Sex		Total
	Male	Female	
Yes	4	9	13
No	7	7	14
Total	11	16	27

the modal category, as is lambda. Consequently, an ordinal measure of association might be preferable for tables like Tables 7.9a-c.

   a. Calculate an ordinal measure of association for the relationship between sex and fear of walking at night.

   b. Use this statistic to discuss the strength and direction of the relationship between degree and prediction about fear.

8. In this exercise we will continue to explore how well education helps to explain our social attitudes. This time we'll examine education and its association with political party identification. Calculate the appropriate measure of association for Table 7.10.

Table 7.10

Political Party	Educational Attainment			Total
	Less Than High School	High School	Bachelor's Degree or Higher	
Democrat	66	215	127	408
Independent	37	123	38	198
Republican	31	193	140	364
Total	134	531	305	970

Source: GSS 2002.

9. Continue your exploration of how education relates to various attitudes by investigating how it influences respondents' position on medical testing on animals. Respondents were asked if they agreed to animal testing if it saved lives. The data were taken from the ISSP 2000. What is the appropriate measure of association for this table? Is this a strong or weak relationship? What is its direction?

	Educational Attainment			
Support for Annual Testing	Less Than Secondary Degree	Secondary Degree	University Degree	Total
Strongly agree	47	65	46	158
Agree	136	151	97	384
Neither	49	70	25	144
Disagree	53	50	25	128
Strongly disagree	24	29	8	61
Total	309	365	201	875

Source: ISSP 2000.

10. The ISSP 2002 asked females and males in the former East and West Germany whether a nuclear accident was likely in the future. Bivariate tables for both men and women are presented below.

West Germany

Likelihood of Nuclear Accident	Respondent Sex		
	Males	Females	Total
Very likely	2	3	5
Likely	6	8	14
Unlikely	9	8	17
Very unlikely	0	5	5
Total	17	24	41

East Germany

Likelihood of Nuclear Accident	Respondent Sex		
	Males	Females	Total
Very likely	1	2	3
Likely	4	4	8
Unlikely	5	3	8
Very unlikely	0	1	1
Total	10	10	20

Source: ISSP 2000.

a. What measure of association is appropriate for these tables?

b. Calculate the appropriate measure of association for each table. In which table is there a stronger association between respondent's sex and belief in the likelihood of a nuclear accident—for West Germans or East Germans?

11. In this exercise we'll explore the relationship between feelings about the Bible and attitudes toward three different types of attitudes. The bivariate tables below present the relationship between

Table 7.11a

a: Feelings about the Bible and PREMARSX

Attitudes Toward Premarital Sex	Is the Bible			
	Word of God	Inspired Word	Book of Fables	Total
Always wrong	54	37	4	95
Almost always wrong	6	10	1	17
Sometimes wrong	15	43	5	63
Not wrong at all	19	91	28	138
Total	94	181	38	313

b: Feelings about the Bible and HOMOSEX

Attitudes Toward Homosexuality	Is the Bible			
	Word of God	Inspired Word	Book of Fables	Total
Always wrong	68	86	13	167
Almost always wrong	3	12	0	15
Sometimes wrong	4	14	4	22
Not wrong at all	10	65	27	102
Total	85	177	44	306

c: Feelings about the Bible and PORNLAW

Favor Legalization of Pornography	Is the Bible			
	Word of God	Inspired Word	Book of Fables	Total
Illegal to all	47	67	8	122
Illegal under 18	46	108	27	181
Legal	4	7	6	17
Total	97	182	41	320

feelings about the Bible (word of God, inspired word, or a book of fables) along with PREMARSX (attitudes toward premarital sex), HOMOSEX (attitudes toward homosexuality), and PORNLAW (should pornography be legalized). Calculate the appropriate measure for each table. What can you conclude? Is belief about the Bible associated with these other variables?

12. In a study of adolescent sexual behavior in Ontario, Canada, sexually active teens were asked to report their use of condoms and birth control. The following cross-tabulation of data is presented separately for boys and girls. Results are presented for three age groups.

Males (N = 2918)

Frequency of Use of Protection	Age of Teen			
	12–13	*14–15*	*16–17*	*Total*
Always	145	401	532	1078
Intermittent	81	301	213	595
Never	355	557	333	1245
Total	581	1259	1078	2918

Females (N = 2824)

Frequency of Use of Protection	Age of Teen			
	12-13	*14-15*	*16-17*	*Total*
Always	75	586	726	1387
Intermittent	48	147	284	479
Never	264	446	248	958
Total	387	1179	1258	2824

Source: Adapted from B. Helen Thomas, Alba DiCenso, and Lauren Griffith, "Adolescent Sexual Behaviour: Results from an Ontario Sample: Part II. Adolescent Use of Protection," *Canadian Journal of Public Health* 89, no. 2 (1998): 94–97, Table 1, p. 95. Used by permission.

Calculate the appropriate measure of association for these tables. Interpret the relationship between age and frequency of use of protection for the sample of sexually active Ontario teens. Is there a difference in the level of association for boys and girls?

13. Is there an association between smoking and school performance among teenagers? The following table reports results from the 1990 California Tobacco Survey. School performance is measured on a 4-point scale; smoking status is designated as nonsmoker, former smoker, or current smoker.

School Performance	Nonsmokers	Former Smokers	Current Smokers	Total
Much better than average	753	130	51	934
Better than average	1,439	310	140	1,889
Average	1,365	387	246	1,998
Below average	88	40	58	186
Total	3,645	867	495	5,007

Source: Adapted from Teh-wei Hu, Zihua Lin, and Theodore E. Keeler, "Teenage Smoking: Attempts to Quit and School Performance," *American Journal of Public Health* 88, no. 6 (1998), Table 1, p. 941. Used by permission of The American Public Health Association.

Calculate the appropriate measure of association for this table. Does a relationship exist between smoking status and school performance for California teenagers? Explain.

14. In the beginning of this chapter, we examined the relationship between number of children and attitudes toward abortion (if a woman is poor and cannot afford more children) for a sample of men. In this exercise, we examine the relationship between the same variables for a sample of 154 women from the GSS 2002.

Support Abortion	*Number of Children*		
	None or One Child	*Two or More Children*	*Total*
No	32	58	90
	55.2%	60.4%	58.4%
Yes	26	38	64
	44.8%	39.6%	41.6%
Total	58	96	154
	100%	100%	100%

Calculate the appropriate measure of association for this table. Does a relationship exist between number of children and support for abortion? Explain. How do your findings compare with our calculations for men at the beginning of Chapter 7?

Regression and Correlation

M any research questions require the analysis of relationships between interval-ratio level variables. Environmental studies, for instance, frequently measure opinions and behavior in terms of quantity, percentages, units of production, and dollar amounts. Let's say that we're interested in the relationship between populations' levels of environmental concern and their wealth. Bivariate regression analysis provides us with the tools to express a relationship between two interval-ratio variables in a concise way.[1]

---

[1] Refer to Paul Allison's *Multiple Regression: A Primer* (Thousand Oaks, CA: Pine Forge Press, 1999) for a complete discussion of multiple regression—statistical methods and techniques that consider the relationship between one dependent variable and one or more independent variables.

Since 1900 the world population has more than tripled and the global economy has expanded more than 20 times. This growth has resulted in a tremendous increase both in the consumption of oil and natural gas and in the level of environmental pollution worldwide. The decline in environmental resources combined with the ecological threat to human security has led to a global environmental movement and considerable support for environmental protection.

The 2000 International Social Survey Programme (ISSP) includes a number of questions designed to measure the degree of environmental concern among the public in several countries. We have selected 16 countries for this example. One of the questions asked respondents whether they would be "willing to pay higher prices to protect the environment." The percentage of respondents who indicated that they would be willing to pay higher prices is presented in Table 8.1. Also presented are the mean, variance, and range for these data.

Table 8.1 Percentage of Respondents Willing to Pay Higher Prices to Protect the Environment

Country	Percent Willing to Pay Higher Prices
United States	44.9
Ireland	53.3
Netherlands	61.2
Norway	40.7
Sweden	32.6
Russia	28.2
New Zealand	46.1
Canada	41.8
Japan	53.0
Spain	32.5
Latvia	21.4
Portugal	22.1
Chile	37.5
Switzerland	54.7
Finland	23.1
Mexico	31.6

$$\text{Mean} = \overline{Y} = \frac{\sum Y}{N} = \frac{624.70}{16} = 39.04$$

$$\text{Variance } Y = S_Y^2 = \frac{\sum (Y - \overline{Y})^2}{N - 1} = \frac{2340.78}{15} = 156.05$$

$$\text{Range } Y = 61.2\% - 21.4\% = 39.8\%$$

Source: International Social Survey Programme 2000.

In examining Table 8.1 and the descriptive statistics, notice the variability in the percentage of citizens who are willing to pay higher prices to protect the environment. The percentage of

Table 8.2   GNP per Capita Recorded for 16 Countries in 1998
(in $1,000)

Country	GNP per Capita
United States	29.24
Ireland	18.71
Netherlands	24.78
Norway	34.31
Sweden	25.58
Russia	2.66
New Zealand	14.60
Canada	19.71
Japan	32.35
Spain	14.10
Latvia ·	2.42
Portugal	10.67
Chile	4.99
Switzerland	39.98
Finland	24.28
Mexico	3.84

Mean $= \overline{Y} = \dfrac{\sum Y}{N} = \dfrac{302.22}{16} = 18.89$

Variance $Y = S_Y^2 = \dfrac{\sum (Y - \overline{Y})^2}{N - 1} = \dfrac{2143.35}{15} = 142.89$

Range $Y = \$39.98 - \$2.42 = \$37.56$

Source: The World Bank Group. *Development Education Program Learning Module: Economics, GNP per Capita*. 2004.

respondents willing to pay higher prices ranges from a low of 21.4 percent in Latvia to a high of 61.2 percent in the Netherlands.

One possible explanation for the differences is the economic conditions in these countries.[2] Scholars of the environmental movement have argued that because of limited economic resources, citizens of developing and poorer countries cannot afford to pay for environmental protection. Therefore, there is less support for environmental protection in these countries. One important indicator of economic conditions is the gross national product (GNP) per capita in each country. Table 8.2 displays the GNP per capita in 1998, recorded in thousands of dollars, for each of the 16 countries surveyed. Note that GNP per capita ranges widely, from $2,420 (2.42 × 1,000) in Latvia to $39,980 (39.98 × 1,000) in Switzerland.

---

[2]Steven R. Bechin and Willett Kempton, "Global Environmentalist: A Challenge to the Postmaterialism Thesis?" *Social Science Quarterly* 75, no. 2 (June 1994): 245–266.

▣ THE SCATTER DIAGRAM

Let's examine the possible relationship between the interval-ratio variables *GNP per capita* and *percentage willing to pay higher prices to protect the environment*. One quick visual method used to display such a relationship between two interval-ratio variables[3] is the **scatter diagram** (or **scatterplot**). Often used as a first exploratory step in regression analysis, a scatter diagram can suggest whether two variables are associated.

The scatter diagram showing the relationship between willingness to pay higher prices for environmental protection and per-capita income for the 16 countries is shown in Figure 8.1. In a scatter diagram the scales for the two variables form the vertical and horizontal axes of a graph. Usually the independent variable, *X*, is arrayed along the horizontal axis and the dependent variable, *Y*, along the vertical axis. Because differences in GNP per capita are hypothesized to account for differences in the percentage of those willing to pay higher prices, GNP is assumed as the independent variable and is arrayed along the horizontal axis. Willingness to pay higher prices, the dependent variable, is arrayed along the vertical axis. In Figure 8.1, each dot represents a country; its location lies at the exact intersection of that country's GNP per capita and the percentage willing to pay higher prices.

Figure 8.1   Scatter Diagram of GNP per Capita (in $1,000) and Percentage Willing to Pay Higher Prices to Protect the Environment

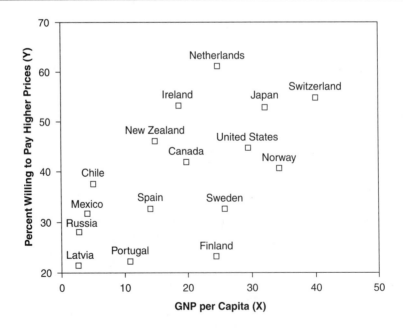

---

[3]We are interested in examining the aggregate relationship between GNP per capita and attitudes toward the environment.

Scatter diagram (scatterplot)   A visual method used to display a relationship between two interval-ratio variables.

Notice that there is an apparent tendency for countries with lower GNP per capita (for example, Latvia and Russia) to also have a lower percentage of people willing to pay higher prices for environmental protection, whereas in countries with a higher GNP per capita (for example, Japan and Switzerland), a higher percentage of people are willing to pay higher prices. In other words, we can say that GNP per capita and willingness to pay higher prices are positively associated. However, there are clearly exceptions to this pattern. For example, Finland has an average GNP per capita ($24,280) but one of the lowest percentages of citizens willing to pay higher prices for environmental protection (23.1%). On the other hand, Chile has one of the lowest GNPs per capita ($4,990), yet close to 40 percent of its citizens are willing to pay higher prices for environmental protection.

Scatter diagrams can also illustrate a negative association between two variables. For example, Figure 8.2 displays the association between GNP per capita and the percentage of respondents in 14 countries who thought that nature and the environment were sacred. Figure 8.2 suggests that low GNP per capita is associated with a higher percentage of citizens who view nature and the environment as sacred. Conversely, high GNP per capita seems to be associated with a lower percentage of citizens who view nature and the environment as sacred. Figure 8.2 illustrates a negative association between GNP per capita and attitudes about the environment.

Figure 8.2   GNP per Capita (in $1,000) and Percentage That View Nature and Environment as Sacred

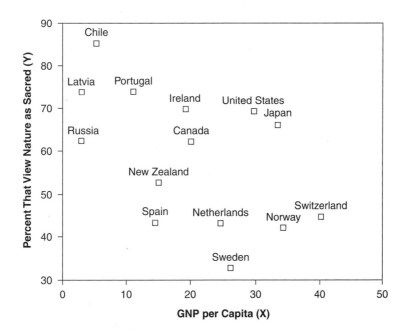

▣ LINEAR RELATIONS AND PREDICTION RULES

Scatter diagrams provide a useful but only preliminary step in exploring a relationship between two interval-ratio variables. We need a more systematic way to express this relationship. Let's examine Figures 8.1 and 8.2 again. Both allow us to see how two sets of measures of environmental concern are related to GNP per capita. The relationships displayed are by no means perfect, but the trends are apparent. In the first case (Figure 8.1), as GNP increases so does the percentage of respondents in most countries who are willing to pay higher prices to protect the environment. In the second case (Figure 8.2), as GNP increases the percentage of respondents who view nature and the environment as sacred decreases.

One way to evaluate these relationships is by expressing them as *linear relationships*. A **linear relationship** allows us to approximate the observations displayed in a scatter diagram with a straight line. In a perfectly linear relationship all the observations (the dots) fall along a straight line (a perfect relationship is sometimes called a **deterministic relationship**), and the line itself provides a predicted value of *Y* (the vertical axis) for any value of *X* (the horizontal axis). For example, in Figure 8.3 we have superimposed a straight line on the scatterplot originally displayed in Figure 8.1. Using this line, we can obtain a predicted value of the percentage willing to pay higher prices for any value of GNP, by reading up to the line from the GNP axis and then over to the percentage axis (indicated by the dotted lines). For example, the predicted value of the percentage willing to pay higher prices for a GNP of $20,000 is 40. Similarly, for a GNP of $37,000 we would get a predicted value of 50 percent willing to pay higher prices.

Figure 8.3   A Straight-Line Graph for GNP per Capita (in $1,000) and Percentage Willing to Pay Higher Prices to Protect the Environment

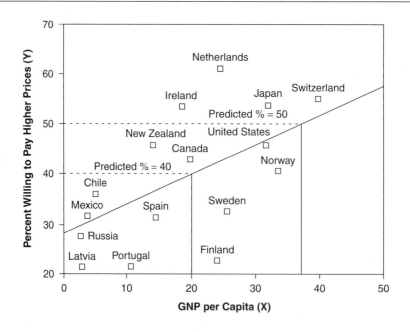

Linear relationship A relationship between two interval-ratio variables in which the observations displayed in a scatter diagram can be approximated with a straight line.

Deterministic (perfect) linear relationship A relationship between two interval-ratio variables in which all the observations (the dots) fall along a straight line. The line provides a predicted value of *Y* (the vertical axis) for any value of *X* (the horizontal axis).

As indicated in Figure 8.3, for the 16 countries surveyed, the actual relationship between GNP per capita and the percentage willing to pay higher prices is not perfectly linear. Although some of the countries lie very close to the line, none falls exactly on the line and some deviate from it considerably. Are there other lines that provide a better description of the relationship between GNP per capita and the percentage willing to pay higher prices?

In Figure 8.4 we have drawn two additional lines that approximate the pattern of relationship shown by the scatter diagram. In each case, notice that even though some of the countries lie close to the line, all fall considerably short of perfect linearity. Is there one line that provides the best linear description of the relationship between GNP per capita and the percentage willing to pay higher prices? How do we choose such a line? What are its characteristics? Before we describe a technique for finding the straight line that most accurately describes the relationship between two variables, we first need to review some basic concepts about how straight-line graphs are constructed.

Figure 8.4 Alternative Straight-Line Graphs for GNP per Capita (in $1,000) and Percentage Willing to Pay Higher Prices to Protect the Environment

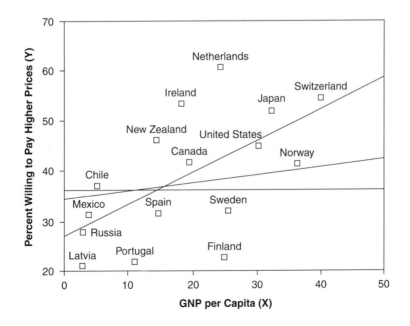

✓ **Learning Check.**   *Use Figure 8.3 to predict the percentage willing to pay higher prices in a country with a GNP of $12,000 and one with a GNP of $27,000.*

Constructing Straight-Line Graphs

To illustrate the fundamentals of straight-line graphs, let's take a simple example. Suppose that in a local school system teachers' salaries are completely determined by seniority. New teachers begin with an annual salary of $12,000, and for each year of seniority their salary increases by $2,000. The seniority and annual salary of six hypothetical teachers are presented in Table 8.3.

Table 8.3   Seniority and Salary of Six Teachers (Hypothetical Data)

Seniority (in years) X	Salary (in dollars) Y
0	12,000
1	14,000
2	16,000
3	18,000
4	20,000
5	22,000

Now let's plot the values of these two variables on a graph (Figure 8.5). Because seniority is assumed to determine salary, let it be our independent variable (X), and let's array it along the horizontal axis. Salary, the dependent variable (Y), is arrayed along the vertical axis. Connecting the six observations in Figure 8.5 gives us a straight-line graph. This graph allows us to obtain a predicted salary value for any value of seniority level simply by reading from the specific seniority level up to the line and then over to the salary axis. For instance, we have marked the lines going up from a seniority of 7 years and then over to the salary axis. We can see that a teacher with 7 years of seniority makes $26,000.

The relationship between salary and seniority, as depicted in Table 8.3 and Figure 8.5, can also be described with the following algebraic equation:

$$Y = 12,000 + 2,000X$$

where

X = seniority (in years)
Y = salary (in dollars)

Figure 8.5   A Perfect Linear Relationship Between Seniority (in Years) and Annual Salary (in $1,000) of Six Teachers (Hypothetical)

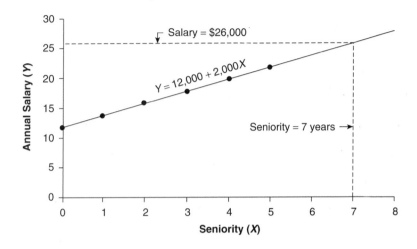

This equation allows us to correctly predict salary (Y) for any value for seniority (X) that we plug into the equation. For example, the salary of a teacher with 5 years of seniority is

$$Y = 12{,}000 + 2{,}000(5) = 12{,}000 + 10{,}000 = 22{,}000$$

Note that we can also plug in values of X that are not shown in Table 8.3. For example, the salary of a teacher with 10 years of seniority is

$$Y = 12{,}000 + 2{,}000(10) = 12{,}000 + 20{,}000 = 32{,}000$$

The equation describing the relation between seniority and salary is an equation for a straight line. The equations for all straight-line graphs have the same general form:

$$Y = a + bX \tag{8.1}$$

where

Y = the predicted score on the dependent variable

X = the score on the independent variable

a = the Y-intercept, or the point where the line crosses the Y-axis; therefore, a is the value of Y when X is 0

b = the slope of the regression line, or the change in Y with a unit change in X. In our example, $a = 12{,}000$ and $b = 2{,}000$. That is, a teacher will make $12,000 with 0 years of seniority, but then her or his salary will go up by $2,000 with each year of seniority.

✔ **Learning Check.**   *For each of these four lines, as X goes up by one unit, what does Y do? Be sure you can answer this question using both the equation and the line.*

Four Lines: Illustrating the Slope and the Y-Intercept

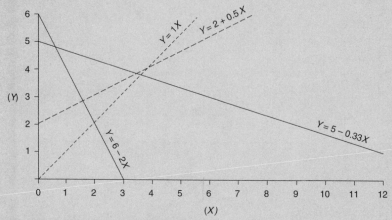

Slope (b) *The change in variable Y (the dependent variable) with a unit change in variable X (the independent variable).*

Y-intercept (a) *The point where the line crosses the Y-axis, or the value of Y when X is 0.*

✔ **Learning Check.**   *Use the linear equation describing the relationship between seniority and salary of teachers to obtain the predicted salary of a teacher with twelve (12) years of seniority.*

Finding the Best-Fitting Line

The straight line displayed in Figure 8.5 and the linear equation representing it $(Y = 12,000 + 2,000X)$ provide a very simple depiction of the relationship between seniority and salary because salary (the Y variable) is completely determined by seniority (the X variable). When each value of Y is completely determined by X, all of the points (observations) lie on the line, and the relationship between the two variables is a deterministic, or perfectly linear, relationship.

However, most relationships we study in the social sciences are not deterministic, and we are not able to come up with a linear equation that allows us to predict Y from X with perfect accuracy. We are much more likely to find relationships approximating linearity, but in which numerous cases don't follow this trend perfectly. For instance, in reality teachers' salaries are not completely determined by seniority, and therefore knowing years of seniority will not provide us with a perfect prediction of their salary level.

When the dependent variable (Y) is not completely determined by the independent variable (X), not all (sometimes none) of the observations will lie exactly on the line. Look back at Figure 8.4, our example of the percentage willing to pay higher prices to protect the environment in relation to GNP per capita. Though each line represents a linear equation showing us how the percentage of citizens willing to pay higher prices rises with a country's GNP per capita, we do not have a perfect prediction in any of the lines. Although all three lines approximate the linear trend suggested by the scatter diagram, very few of the observations lie exactly on any of the lines and some deviate from them considerably.

Given that none of the lines is perfect, our task is to choose one line—the *best-fitting line*. But which is the best-fitting line?

Defining Error   The best-fitting line is the one that generates the least amount of error, also referred to as the **residual**. Let's think about how the residual is defined. Look again at Figure 8.3. For each GNP level, the line (or the equation that this line represents) predicts a value of Y. Canada, for example, with a GNP of nearly \$20,000 gives us a predicted value for Y of 40 percent. But the actual value for Canada is 41.8 percent (see also Table 8.1). Thus, we have two values for Y: (1) a predicted Y, which we symbolize as $\hat{Y}$ and which is generated by the prediction equation, also called the *linear regression equation*

$$\hat{Y} = a + bX$$

and (2) the observed Y, symbolized simply as Y. Thus, for Canada, $\hat{Y} = 40$ percent, whereas $Y = 41.8$ percent.

We can think of the residual as the difference between the observed Y (Y) and the predicted Y ($\hat{Y}$). If we symbolize the residual as e, then

$$e = Y - \hat{Y}$$

The residual for Canada is $41.8 - 40$ percent $= 1.8$ percentage points.

The Residual Sum of Squares $(\sum e^2)$ We want a line or a prediction equation that minimizes e for each individual observation. However, any line we choose will minimize the residual for some observations but may maximize it for others. We want to find a prediction equation that minimizes the residuals over all observations.

There are many mathematical ways of defining the residuals. For example, we may take the algebraic sum of residuals $\sum (Y - \hat{Y})$, the sum of the absolute residuals $\sum (|Y - \hat{Y}|)$, or the sum of the squared residuals $\sum (Y - \hat{Y})^2$. For mathematical reasons, statisticians prefer to work with the third method—squaring and summing the residuals over all observations. The result is the *residual sum of squares*, or $\sum e^2$. Symbolically, $\sum e^2$ is expressed as

$$\sum e^2 = \sum (Y - \hat{Y})^2$$

The Least-Squares Line   The best-fitting regression line is that line where the sum of the squared residuals, or $\sum e^2$, is at a minimum. Such a line is called the **least-squares line** (or **best-fitting line**), and the technique that produces this line is called the **least-squares**

method. The technique involves choosing a and b for the equation $\hat{Y} = a + bX$ such that $\sum e^2$ will have the smallest possible value. In the next section we use the data from the 16 countries to find the least-squares equation. But before we continue, let's review where we are so far.

Least-squares line (best-fitting line)   A line where the residual sum of squares, or $\sum e^2$, is at a minimum.

Least-squares method   The technique that produces the least-squares line.

Review

1. We examined the relationship between GNP per capita and the percentage willing to pay higher prices to protect the environment, using data collected in 16 countries. We used a *scatter diagram* (*scatterplot*) to display the relationship between these variables.

2. The scatter diagram indicated that the relationship between these variables might be *linear*; as GNP per capita increases, so does the percentage of citizens willing to pay higher prices to protect the environment.

3. A more systematic way to analyze the relationship is to develop a *straight-line equation* to predict the percentage willing to pay higher prices based on GNP. We saw that there are a number of straight lines that can approximate the data.

4. The *best-fitting line* is one that minimizes $\sum e^2$. Such a line is called the *least-squares line*, and the technique that produces this line involves choosing the a and b for the equation $\hat{Y} = a + bX$ that minimize $\sum e^2$.

Computing *a* and *b* for the Prediction Equation

Through the use of calculus it can be shown that to figure out the values of a and b in a way that minimizes $\sum e^2$, we need to apply the following formulas:

$$b = \frac{S_{YX}}{S_X^2} \qquad (8.2)$$

$$a = \bar{Y} - b(\bar{X}) \qquad (8.3)$$

where

S_{YX} = the covariance of X and Y

S_X^2 = the variance of X

$\bar{Y}$ = the mean of Y

$\bar{X}$ = the mean of X

a = the Y-intercept

b = the slope of the line

These formulas assume that X is the independent variable and Y is the dependent variable.

Before we compute a and b, let's examine these formulas. The denominator for b is the variance of the variable X. It is defined as follows:

$$\text{Variance }(X) = S_X^2 = \frac{\sum(X - \overline{X})^2}{N - 1}$$

This formula should be familiar to you from Chapter 5. The numerator (S_{YX}), however, is a new term. It is the covariance of X and Y and is defined as

$$\text{Covariance }(X, Y) = S_{YX} = \frac{\sum(X - \overline{X})(Y - \overline{Y})}{N - 1} \tag{8.4}$$

The covariance is a measure of how X and Y vary together. Basically, the covariance tells us to what extent higher values of one variable "go together" with higher values on the second variable (in which case we have a positive covariation) or with lower values on the second variable (which is a negative covariation). Take a look at this formula. It tells us to subtract the mean of X from each X score and the mean of Y from each Y score, and then take the product of the two deviations. The results are then summed for all the cases and divided by $N - 1$.

In Table 8.4 we show the computations necessary to calculate the values of a and b for our 16 countries. The means for GNP per capita and percentage willing to pay higher prices are obtained by summing column 1 and column 2, respectively, and dividing each sum by N. To calculate the covariance we first subtract $\overline{X}$ from each X score (column 3) and $\overline{Y}$ from each Y score (column 5) to obtain the mean deviations. We then multiply these deviations for every observation. The products of the mean deviations are shown in column 7. For example, for the first observation, the United States, the mean deviation for GNP is 10.35 (29.24 − 18.89 = 10.35); for the percentage willing to pay higher prices it is 5.86 (44.9 − 39.04 = 5.86). The product of these deviations, 60.65 (10.35 × 5.86 = 60.65), is shown in column 7. The sum of these products, shown at the bottom of column 7, is 1345.76. Dividing it by 15 ($N - 1$), we get the covariance of 89.72.

The covariance is a measure of the linear relationship between two variables, and its value reflects both the strength and the direction of the relationship. The covariance will be close to zero when X and Y are unrelated; it will be larger than zero when the relationship is positive, and smaller than zero when the relationship is negative.

Now let's substitute the values for the covariance and the variance from Table 8.4 to calculate b:

$$b = \frac{S_{YX}}{S_X^2} = \frac{89.72}{142.89} = 0.63$$

Once b has been calculated, finding a, the intercept, is simple:

$$a = \overline{Y} - b(\overline{X}) = 39.04 - 0.63(18.89) = 27.14$$

Table 8.4   Worksheet for Calculating a and b for the Regression Equation

Country	(1) GNP per Capita (X)	(2) % Willing to Pay (Y)	(3) $(X - \bar{X})$	(4) $(X - \bar{X})^2$	(5) $(Y - \bar{Y})$	(6) $(Y - \bar{Y})^2$	(7) $(X - \bar{X})(Y - \bar{Y})$
United States	29.24	44.9	10.35	107.12	5.86	34.34	60.65
Ireland	18.71	53.3	−0.18	0.03	14.26	203.35	−2.57
Netherlands	24.78	61.2	5.89	34.69	22.16	491.06	130.52
Norway	34.31	40.7	15.42	237.78	1.66	2.76	25.60
Sweden	25.58	32.6	6.69	44.76	−6.44	41.47	−43.08
Russia	2.66	28.2	−16.23	263.41	−10.84	117.51	175.93
New Zealand	14.60	46.1	−4.29	18.40	7.06	49.84	−30.29
Canada	19.71	41.8	0.82	0.67	2.76	7.62	2.26
Japan	32.35	53.0	13.46	181.17	13.96	194.88	187.90
Spain	14.10	32.5	−4.79	22.94	−6.54	42.77	31.33
Latvia	2.42	21.4	−16.47	271.26	−17.64	311.17	290.53
Portugal	10.67	22.1	−8.22	67.57	−16.94	286.96	139.25
Chile	4.99	37.5	−13.90	193.21	−1.54	2.37	21.41
Switzerland	39.98	54.7	21.09	444.79	15.66	245.24	330.27
Finland	24.28	23.1	5.39	29.05	−15.94	254.08	−85.92
Mexico	3.84	31.6	−15.05	226.50	−7.44	55.35	111.97
	$\sum X = 302.22$	$\sum Y = 624.70$	0.00*	2143.35	0.00*	2340.78	1345.76

*Answers may differ due to rounding; however, the exact value of these column totals, properly calculated, will always be equal to zero.

Mean $X = \bar{X} = \dfrac{\sum X}{N} = \dfrac{302.22}{16} = 18.89$

Mean $Y = \bar{Y} = \dfrac{\sum Y}{N} = \dfrac{624.70}{16} = 39.04$

Variance $(Y) = S_Y^2 = \dfrac{\sum(Y - \bar{Y})^2}{N - 1} = \dfrac{2340.78}{15} = 156.05$

Standard deviation $(Y) = S_Y = \sqrt{156.05} = 12.49$

Variance $(X) = S_X^2 = \dfrac{\sum(X - \bar{X})^2}{N - 1} = \dfrac{2143.35}{15} = 142.89$

Standard deviation $(X) = S_X = \sqrt{142.89} = 11.95$

Covariance $(X, Y) = S_{YX} = \dfrac{\sum(X - \bar{X})(Y - \bar{Y})}{N - 1} = \dfrac{1345.76}{15} = 89.72$

The prediction equation is therefore

$$\hat{Y} = 27.14 + 0.63X$$

This equation can be used to obtain a predicted value for the percentage of citizens willing to pay higher prices for environmental protection given a country's GNP per capita. For example, for a country with a GNP per capita of 2 (in $1,000), the predicted percentage is

$$\hat{Y} = 27.14 + 0.63(2) = 28.4$$

Similarly, for a country with a GNP per capita of 12 (in $1,000), the predicted value is

$$\hat{Y} = 27.14 + 0.63(12) = 34.7$$

Now we can plot the straight-line graph corresponding to the regression equation. To plot a straight line we need only two points, where each point corresponds to an *X,Y* value predicted by the equation. We can use the two points we just obtained: (1) $X = 2$, $\hat{Y} = 28.4$ and (2) $X = 12$, $\hat{Y} = 34.7$. In Figure 8.6, the regression line is plotted over the scatter diagram we first displayed in Figure 8.1.

Figure 8.6   The Best-Fitting Line for GNP per Capita and Percentage Willing to Pay More to Protect the Environment

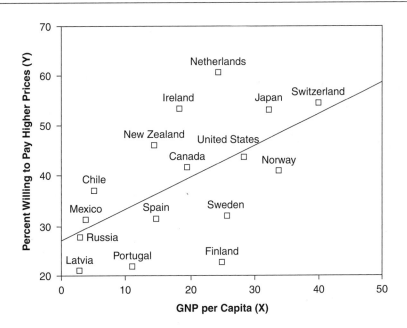

✔ *Learning Check.* *Use the prediction equation to calculate the predicted values of Y for Chile, Canada, and Japan. Verify that the regression line in Figure 8.6 passes through these points.*

Interpreting *a* and *b*

Now let's interpret the coefficients *a* and *b* in our equation. The *b* coefficient is equal to 0.63 percent. This tells us that the percentage of citizens willing to pay higher prices for environmental protection will increase by 0.63 percent for every increment of $1,000 in their country's GNP per capita. Similarly, an increase of $10,000 in a country's GNP corresponds to a 6.3 (0.63 × 10) percent increase in the percentage of citizens willing to pay higher prices for environmental protection.

Note that because the relationships between variables in the social sciences are inexact, we don't expect our regression equation to make perfect predictions for every individual case. However, even though the pattern suggested by the regression equation may not hold for every individual country, it gives us a tool by which to make the best possible guess about how a country's GNP per capita is associated, *on average*, with the willingness of its citizens to pay higher prices for environmental protection. We can say that the slope of 0.63 percent is the estimate of this underlying relationship.

The intercept *a* is the predicted value of *Y* when *X* = 0. Thus, it is the point at which the regression line and the *Y*-axis intersect. With *a* = 27.14, a country with a GNP level equal to zero is predicted to have 27.14 percent of its citizens supporting higher prices for environmental protection. Note, however, that no country has a GNP as low as zero, although Latvia is closest with a GNP of $2,420. As a general rule, be cautious when making predictions for *Y* based on values of *X* that are outside the range of the data. Thus, when the lowest value for *X* is far above zero, the intercept may not have a clear substantive interpretation.

▣ STATISTICS IN PRACTICE: GNP AND VIEWS ABOUT NATURE AND THE ENVIRONMENT

In our ongoing example, we have looked at the association between GNP per capita and concern for the environment as measured by the percentage of people willing to pay higher prices to protect the environment. The regression equation we have estimated from data collected by the World Bank Group and the ISSP in 16 countries shows that as a country's GNP per capita rises, more citizens in that country are willing to pay higher prices to protect the environment.

What do these findings suggest? The conventional wisdom has been that citizens of poorer countries do not or cannot care about the environment. Indeed, our findings seem to suggest that people in wealthy countries hold stronger environmental values than those in poorer countries. But, one may ask, is the only reliable measure of concern for the environment the amount of money one is willing to pay to protect it?

□ A Closer Look 8.1
Understanding the Covariance

Let's say we have a set of eight points for which the mean of *X* is 6 and the mean of *Y* is 3.

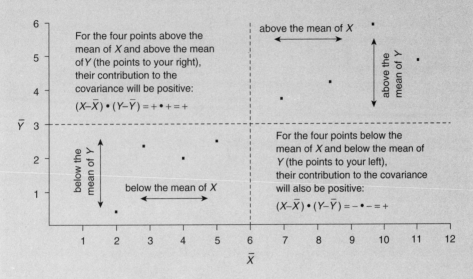

For the four points above the mean of *X* and above the mean of *Y* (the points to your right), their contribution to the covariance will be positive:

$(X - \bar{X}) \cdot (Y - \bar{Y}) = + \cdot + = +$

For the four points below the mean of *X* and below the mean of *Y* (the points to your left), their contribution to the covariance will also be positive:

$(X - \bar{X}) \cdot (Y - \bar{Y}) = - \cdot - = +$

So the covariance in this case will be positive, giving us a positive *b* and a positive *r*.

Now let's say we have a set of eight points that look like this:

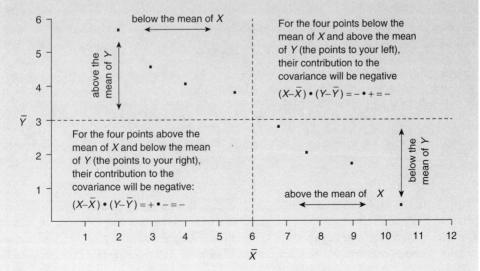

For the four points below the mean of *X* and above the mean of *Y* (the points to your left), their contribution to the covariance will be negative

$(X - \bar{X}) \cdot (Y - \bar{Y}) = - \cdot + = -$

For the four points above the mean of *X* and below the mean of *Y* (the points to your right), their contribution to the covariance will be negative:

$(X - \bar{X}) \cdot (Y - \bar{Y}) = + \cdot - = -$

So the covariance in this case will be negative, giving us a negative *b*.

▣ A Closer Look 8.2
A Note on Nonlinear Relationships

In analyzing the relationship between GNP per capita and percentage willing to pay higher prices for environmental protection, we have assumed that the two variables are linearly related. For the most part, social science relationships can be approximated using a linear equation. It is important to note, however, that sometimes a relationship cannot be approximated by a straight line and is better described by some other, non-linear function. For example, the following scatter diagram shows a nonlinear relationship between age and hours of reading (hypothetical data). Hours of reading increase with age until the twenties, remain stable until the forties, and then tend to decrease with age.

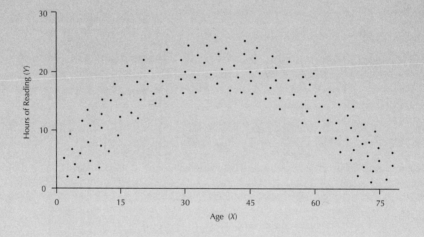

A Nonlinear Relationship Between Age and Hours of Reading per Week

One quick way to find out whether your variables form a linear or a nonlinear pattern of relationship (or whether the variables are related at all!) is to make a scatter diagram of your data. If there is a significant departure from linearity, it would make no sense to fit a straight line to the data. Statistical techniques for analyzing nonlinear relationships between two variables are beyond the scope of this book. Nonetheless, at the very least, you should check for possible departures from linearity when examining your scatter diagram.

In an attempt to challenge the conventional wisdom that people in poor countries lack environmental values, Brechin and Kempton[4] argue that even though few people within the poorest countries would offer monetary payment for anything, even for values they hold highly, they are equally or more likely to agree to commit their labor time. The researchers use the results of a

---

[4]Steven R. Brechin and Willett Kempton, "Global Environmentalism: A Challenge to the Postmaterialism Thesis?" *Social Science Quarterly* 75, no. 2 (June 1994): 245–266.

Table 8.5   Percentage of Citizens That View Nature and Environment as Sacred and GNP per Capita for Fourteen Countries

Country	GNP per Capita (in $1,000) (X)	% w/View That Nature/ Environment Sacred (Y)
United States	29.24	69.7
Ireland	18.71	70.1
Netherlands	24.78	43.6
Norway	34.31	42.4
Sweden	25.58	33.6
Russia	2.66	63.1
New Zealand	14.60	53.1
Canada	19.71	62.8
Japan	32.35	67.9
Spain	14.10	44.0
Latvia	2.42	74.3
Portugal	10.67	74.8
Chile	4.99	85.1
Switzerland	39.98	45.3

survey collected by the Harris organization to examine this argument.[5] In a similar vein, the 2000 ISSP asked respondents whether they thought that nature and the environment were, in some form or another, sacred. The percentage of citizens in each of 14 countries who indicated that nature and the environment were sacred is presented in Table 8.5, together with the GNP per capita for each country. The scatter diagram for these data was displayed earlier, in Figure 8.2.

Let's examine Figure 8.2 once again. The scatter diagram seems to indicate that the two variables—the percentage that view nature and the environment as sacred and GNP per capita—are linearly related. It also illustrates that these variables are negatively associated; that is, as GNP per capita rises, the percentage of citizens that view nature as sacred declines.

For a more systematic analysis of the association we need to estimate the least-squares regression equation for these data. Since we want to predict the percentage of citizens who view nature as sacred, we treat this variable as our dependent variable (Y).

Table 8.6 shows the calculations necessary to find a and b for our data on GNP per capita in relation to the percentage of citizens that view nature and the environment as sacred.

Now let's substitute the values for the covariance and the variance from Table 8.6 to calculate b:

$$b = \frac{S_{YX}}{S_X^2} = \frac{-99.27}{144.71} = -0.69$$

---

[5]Both the Harris and the Gallup survey (discussed earlier in this chapter) examine the relationship between GNP and environmental concerns. However, because the two studies used different samples of countries, we need to be cautious about making generalizations based on looking at them jointly.

Table 8.6   GNP per Capita and Percentage That View Nature and Environment as Sacred for 14 Countries

Country	(1) GNP per Capita (X)	(2) % W/View That Nature Environment Sacred (Y)	(3) $(X - \bar{X})$	(4) $(X - \bar{X})^2$	(5) $(Y - \bar{Y})$	(6) $(Y - \bar{Y})^2$	(7) $(X - \bar{X})(Y - \bar{Y})$
United States	29.24	69.7	9.66	93.32	10.43	108.78	100.75
Ireland	18.71	70.1	−0.87	0.76	10.83	117.29	−9.42
Netherlands	24.78	43.6	5.20	27.04	−15.67	245.55	−81.48
Norway	34.31	42.4	14.73	216.97	−16.87	284.60	−248.50
Sweden	25.58	33.6	6.00	36.00	−25.67	658.95	−154.02
Russia	2.66	63.1	−16.92	286.29	3.83	14.67	−64.80
New Zealand	14.60	53.1	−4.98	24.80	−6.17	38.07	30.73
Canada	19.71	62.8	0.13	0.02	3.53	12.46	0.46
Japan	32.35	67.9	12.77	163.07	8.63	74.48	110.21
Spain	14.10	44.0	−5.48	30.03	−15.27	233.17	83.68
Latvia	2.42	74.3	−17.16	294.47	15.03	225.90	−257.91
Portugal	10.67	74.8	−8.91	79.39	15.53	241.18	−138.37
Chile	4.99	85.1	−14.59	212.87	25.83	667.19	−376.86
Switzerland	39.98	45.3	20.40	416.16	−13.97	195.16	−284.99
	$\sum X = 274.10$	$\sum Y = 829.80$	0.00*	1881.15	0.00*	3117.45	−1290.52

*Answers may differ due to rounding; however, the exact value of these column totals, properly calculated, will always be equal to zero.

$$\text{Mean } X = \bar{X} = \frac{\sum X}{N} = \frac{274.10}{14} = 19.58$$

$$\text{Mean } Y = \bar{Y} = \frac{\sum Y}{N} = \frac{829.80}{14} = 59.27$$

$$\text{Variance } (Y) = S_Y^2 = \frac{\sum (Y - \bar{Y})^2}{N - 1} = \frac{3117.45}{13} = 239.80$$

$$\text{Standard deviation } (Y) = S_Y = \sqrt{239.80} = 15.48$$

$$\text{Variance } (X) = S_X^2 = \frac{\sum (X - \bar{X})^2}{N - 1} = \frac{1881.15}{13} = 144.71$$

$$\text{Standard deviation } (X) = S_X = \sqrt{144.71} = 12.03$$

$$\text{Covariance } (X, Y) = S_{YX} = \frac{\sum (X - \bar{X})(Y - \bar{Y})}{N - 1} = \frac{-1290.52}{13} = -99.27$$

Once b has been calculated, finding a, the intercept, is simple:

$$a = \bar{Y} - b(\bar{X}) = 59.27 - (-0.69)(19.58) = 72.78$$

The prediction equation is therefore

$$\hat{Y} = 72.78 + (-0.69)X$$

This equation can be used to obtain a predicted value for the percentage of citizens who view nature and the environment as sacred given a country's GNP per capita.

Now let's interpret the coefficients a and b in our equation. The b coefficient is equal to -0.69 percent. This tells us that the percentage of citizens who view nature and the environment as sacred will decrease by 0.69 percent for every increment of \$1,000 in their country's GNP per capita. Similarly, an increase of \$10,000 in a country's GNP corresponds to a 6.9 (-0.69×10) percent decrease in the percentage of citizens who view nature and the environment as sacred.

The intercept a is the predicted value of Y when $X = 0$. Thus, it is the point at which the regression line and the Y-axis intersect. With $a = 72.78$, a country with a GNP level equal to zero is predicted to have 72.78 percent of its citizens with the view that nature and the environment are sacred.

⊡ METHODS FOR ASSESSING THE ACCURACY OF PREDICTIONS

So far we have developed two regression equations that are helping us to make predictions about people's views about the environment and their willingness to pay higher prices to protect it. But in both cases our predictions are far from perfect. If we examine Figures 8.6 and 8.7, we can see that we fail to make accurate predictions in every case. Though some of the countries lie pretty close to the regression line, none lies directly on the line—an indication that some error of prediction was made. You must be wondering by now, "Okay, I understand that the model helps us make predictions, but how can I assess the accuracy of these predictions?"

We saw earlier that one way to judge the accuracy of the predictions is to "eyeball" the scatterplot. The closer the observations are to the regression line, the better the "fit" between the predictions and the actual observations. Still we want a more systematic method for making such a judgment. We need a measure that tells us how accurate a prediction the regression model provides. The *coefficient of determination*, or r^2, is such a measure. It tells us how well the bivariate regression model fits the data. Both r^2 and r measure the strength of the association between two interval-ratio variables. Before we discuss these measures, let's first examine the notion of prediction errors.

Prediction Errors

Examine Figure 8.8. It displays the regression line for the variables GNP per capita (X) and the percentage willing to pay higher prices to protect the environment (Y). This regression line and the scatter diagram for the 16 countries were originally presented in Figure 8.6.

In Figure 8.8 we consider the prediction of Y for one country, the Netherlands, out of the 16 countries included in the survey. (The X and Y scores for all 16 countries, including the Netherlands, are presented in Table 8.4.)

Figure 8.7   Regression Line for GNP per Capita (in $1,000) and Percentage That View
Nature and Environment as Sacred

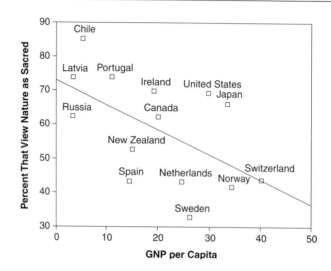

Suppose we didn't know the *actual* Y, the percentage of citizens in the Netherlands who
agreed to pay higher prices for environmental protection. Suppose further that we did not have
knowledge of X, the Netherlands' GNP per capita. Because the mean minimizes the sum of
the squared errors for a set of scores, our best guess for Y would be the mean of Y, or 39.04.
The horizontal line in Figure 8.8 represents this mean. Now let's compare the actual Y, 61.2,
with this prediction:

$$Y - \bar{Y} = 61.2 - 39.04 = 22.16$$

With an error of 22.16, our prediction of the average score for the Netherlands is not very
accurate (this deviation of $Y - \bar{Y}$ is also illustrated in Figure 8.8).

Now let's see if our predictive power can be improved by using our knowledge of X—the
GNP per capita of the Netherlands—and its linear relationship with Y. If we plug the
Netherlands' GNP per capita of 24.78 (per $1,000) into our prediction equation, as follows,

$$\hat{Y} = 27.14 + 0.63(X)$$

$$\hat{Y} = 27.14 + 0.63(24.78) = 42.75$$

we obtain a predicted $\hat{Y}$ of 42.75.

We can now recalculate our new error of prediction by comparing the predicted $\hat{Y}$ with the
actual Y:

$$Y - \hat{Y} = 61.2 - 42.75 = 18.45$$

Although this prediction is by no means perfect, it is an improvement of 3.71 (22.16 −
18.45 = 3.71) over our earlier prediction.

This improvement is illustrated in Figure 8.8. Note that this improvement of 3.71 is equal
to the quantity $\hat{Y} - \bar{Y}$ (42.75 − 39.04 = 3.71). This quantity represents the improvement in the
prediction error resulting from our use of the linear prediction equation.

Figure 8.8   Error Terms for One Observation

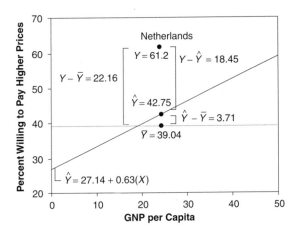

Let's review what we have done. We have two prediction rules and two measures of error. The first prediction rule is in the absence of information on X, predict $\bar{Y}$. The error of prediction is defined as $Y - \bar{Y}$. The second rule of prediction uses X and the regression equation to predict Y. The error of prediction is defined as $Y - \hat{Y}$.

To calculate these two measures of error for all the cases in our sample, we square the deviations and sum them. Thus, for the deviations from the mean of Y we have

$$\Sigma(Y - \bar{Y})^2$$

The sum of the squared deviations from the mean is called the *total sum of squares*, or *SST*:

$$SST = \Sigma(Y - \bar{Y})^2$$

To measure deviation from the regression line, or $\hat{Y}$, we have

$$\Sigma(Y - \hat{Y})^2$$

The sum of squared deviations from the regression line is denoted as the *residual sum of squares*, or *SSE*:

$$SSE = \Sigma(Y - \hat{Y})^2$$

(We discussed this error term, the residual sum of squares, earlier in the chapter.)

The predictive value of the linear regression equations can be assessed by the extent to which the residual sum of squares, or *SSE*, is smaller than the total sum of squares, or *SST*. By subtracting *SSE* from *SST* we obtain the *regression sum of squares*, or *SSR*, which reflects the improvement in the prediction error resulting from our use of the linear prediction equation. *SSR* is defined as

$$SSR = SST - SSE$$

Let's calculate these terms for our data on GNP per capita (X) and the percentage willing to pay higher prices to protect the environment (Y). We already have from Table 8.4 the total sum of squares:

$$SST = \Sigma(Y - \bar{Y})^2 = 2{,}340.78$$

Table 8.7   Worksheet for Calculating Errors Sum of Squares *(SSE)*

Country	(1) GNP per Capita (X)	(2) % Willing to Pay (Y)	(3) Predicted Y $(\hat{Y})$	(4) $(Y - \hat{Y})$	(5) $(Y - \hat{Y})^2$
United States	29.24	44.9	45.56	−.66	.44
Ireland	18.71	53.3	38.93	14.37	206.57
Netherlands	24.78	61.2	42.75	18.45	340.35
Norway	34.31	40.7	48.76	−8.06	64.89
Sweden	25.58	32.6	43.26	−10.66	113.54
Russia	2.66	28.2	28.82	−.62	.38
New Zealand	14.60	46.1	36.34	9.76	95.30
Canada	19.71	41.8	39.56	2.24	5.03
Japan	32.35	53.0	47.52	5.48	30.02
Spain	14.10	32.5	36.02	−3.52	12.41
Latvia	2.42	21.4	28.66	−7.26	52.77
Portugal	10.67	22.1	33.86	−11.76	138.35
Chile	4.99	37.5	30.28	7.22	52.07
Switzerland	39.98	54.7	52.33	2.37	5.63
Finland	24.28	23.1	42.44	−19.34	373.90
Mexico	3.84	31.6	29.56	2.04	4.16
	$\sum X = 302.22$	$\sum Y = 624.70$			$\sum (Y - \hat{Y})^2 = 1495.81$

To calculate the errors sum of squares we will calculate the predicted $\hat{Y}$ for each country, subtract it from the observed Y, square the differences, and sum these for all countries. These calculations are presented in Table 8.7.

The residual sum of squares is thus

$$SSE = \sum(Y - \hat{Y})^2 = 1,495.81$$

The regression sum of squares is then

$$SSR = SST - SSE = 2,340.78 - 1,495.81 = 844.97$$

The Coefficient of Determination (r²) as a PRE Measure   The coefficient of determination, r^2, measures the improvement in the prediction error resulting from our use of the linear prediction equation. The coefficient of determination is a PRE measure of association. We saw in Chapter 7 that all PRE measures adhere to the following formula:

$$PRE = \frac{E1 - E2}{E1}$$

where

$E1$ = prediction errors made when the independent variable is ignored

$E2$ = prediction errors made when the prediction is based on the independent variable

We have all the elements we need to construct a PRE measure. Because the total sum of squares (SST) measures the prediction errors when the independent variable is ignored, we can define

$$E1 = SST$$

Similarly, because the residual sum of squares (SSE) measures the prediction errors resulting from using the independent variable, we can define

$$E2 = SSE$$

We are now ready to define the **coefficient of determination** r^2. It measures the proportional reduction of error associated with using the linear regression equation as a rule for predicting Y:

$$\text{PRE} = r^2 = \frac{E1 - E2}{E1} = \frac{\sum(Y - \overline{Y})^2 - \sum(Y - \hat{Y})^2}{\sum(Y - \overline{Y})^2} \tag{8.5}$$

For our example:

$$r^2 = \frac{2,340.78 - 1,495.81}{2,340.78} = \frac{844.97}{2,340.78} = 0.36$$

The **coefficient of determination** (r^2), reflects the proportion of the total variation in the dependent variable, Y, explained by the independent variable, X. An r^2 of 0.36 means that by using GNP per capita and the linear prediction rule to predict Y—the percentage willing to pay higher prices—we have reduced the error or prediction by 36 percent (0.36×100). We can also say that the independent variable (GNP per capita) explains about 36 percent of the variation in the dependent variable (the percentage willing to pay higher prices), as illustrated in Figure 8.9.

The coefficient of determination ranges from 0.0 to 1.0. An r^2 of 1.0 means that by using the linear regression model we have reduced uncertainty by 100 percent. It also means that the independent variable accounts for 100 percent of the variation in the dependent variable. With an r^2 of 1.0, all the observations fall along the regression line, and the prediction error $[\sum(Y - \hat{Y})^2]$ is equal to 0.0. An r^2 of 0.0 means that using the regression equation to predict Y does not improve the prediction of Y. Figure 8.10 shows r^2 values near 0.0 and near 1.0. In Figure 8.10a, where r^2 is approximately 1.0, the regression model provides a good fit. In contrast, a very poor fit is evident in Figure 8.10b, where r^2 is near zero. An r^2 near zero indicates either poor fit or a well-fitting line with a b of zero.

Figure 8.9   A Pie Graph Approach to r^2

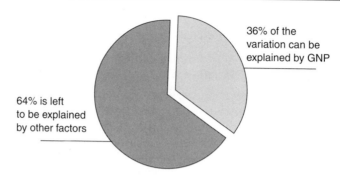

Variation in Percentage Willing to Pay Higher Prices

Figure 8.10   Examples Showing r^2 Near 1.0 and Near 0

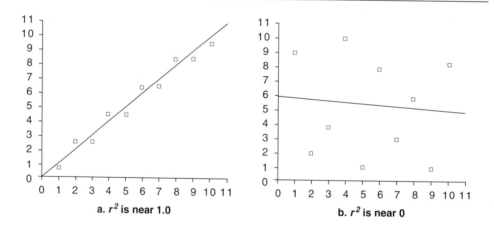

Coefficient of determination (r^2)   A PRE measure reflecting the proportional reduction of error that results from using the linear regression model. It reflects the proportion of the total variation in the dependent variable, Y, explained by the independent variable, X.

Calculating r^2   An easier method for calculating r^2 uses the following equation:

$$r^2 = \frac{[\text{covariance}\,(X, Y)]^2}{[\text{variance}\,(X)][\text{variance}\,(Y)]} = \frac{S^2_{YX}}{S^2_X S^2_Y} \tag{8.6}$$

This formula tells us to divide the square of the covariance of X and Y by the product of the variance of X and the variance of Y.

To calculate r^2 for our example we can go back to Table 8.4, where the covariance and the variances for the two variables have already been calculated:

$S_{YX} = 89.72$

$S_X^2 = 142.89$

$S_Y^2 = 156.05$

Therefore,

$$r^2 = \frac{(89.72)^2}{(142.89)(156.05)} = \frac{8049.68}{22,297.98} = 0.36$$

An r^2 of 0.36 means that by using GNP per capita and the linear prediction rule to predict Y—the percentage willing to pay higher prices—we have reduced uncertainty of prediction by 36 percent (0.36×100). We can also say that the independent variable (GNP per capita) explains 36.1 percent of the variation in the dependent variable (the percentage willing to pay higher prices), as illustrated in Figure 8.9.

Pearson's Correlation Coefficient (*r*)

In the social sciences, it is the square root of r^2, or r —known as **Pearson's correlation coefficient**—that is most often used as a measure of association between two interval-ratio variables:

$$r = \sqrt{r^2}$$

Pearson's r is usually computed directly[6] by using the following definitional formula:

$$r = \frac{[\text{covariance }(X, Y)]}{[\text{standard deviation }(X)][\text{standard deviation }(Y)]} = \frac{S_{YX}}{S_X S_Y} \tag{8.7}$$

Thus, r is defined as the ratio of the covariance of X and Y to the product of the standard deviations of X and Y.

Characteristics of Pearson's r Pearson's r is a measure of relationship or association for interval-ratio variables. Like gamma (introduced in Chapter 7), it ranges from 0.0 to ±1.0, with 0.0 indicating no association between the two variables. An r of +1.0 means that the two variables have a perfect positive association; –1.0 indicates that it is a perfect negative association. The absolute value of r indicates the strength of the linear association between two variables. Thus, a correlation of –0.75 demonstrates a stronger association than a correlation

---

[6]If you obtain r simply by taking the square root of r^2, make sure not to lose the sign of r (r^2 is always positive but r can also be negative), which can be ascertained by looking at the sign of S_{YX}.

Figure 8.11 Scatter Diagrams Illustrating Weak, Moderate, and Strong Relationships as Indicated by the Absolute Value of r

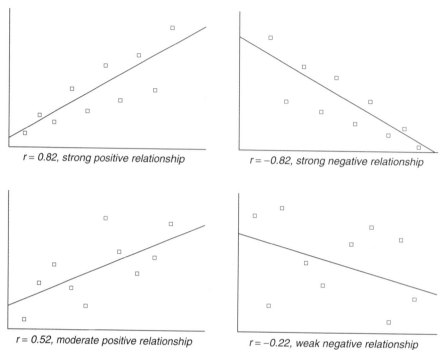

$r = 0.82$, strong positive relationship

$r = -0.82$, strong negative relationship

$r = 0.52$, moderate positive relationship

$r = -0.22$, weak negative relationship

of 0.50. Figure 8.11 illustrates a strong positive relationship, a strong negative relationship, a moderate positive relationship, and a weak negative relationship.

Unlike the b coefficient, r is a symmetrical measure. That is, the correlation between X and Y is identical to the correlation between Y and X. In contrast, b may be different when the variables are switched—for example, when we use Y as the independent variable rather than as the dependent variable.

To calculate r for our example of the relationship between GNP per capita and the percentage of people who support higher prices for environmental protection, let's return to Table 8.4, where the covariance and the standard deviations for X and Y have already been calculated:

$$r = \frac{S_{YX}}{S_X S_Y} = \frac{89.72}{(11.95)(12.49)} = \frac{89.72}{149.26} = 0.60$$

A correlation coefficient of 0.60 indicates that there is a moderate-to-strong positive linear relationship between GNP per capita and the percentage of citizens who support higher prices for environmental protection.

Note that we could have just taken the square root of r^2 to calculate r, because $r = \sqrt{r^2}$, or $\sqrt{0.36} = 0.60$. Similarly, if we first calculate r, we can obtain r^2 simply by squaring r (be careful not to lose the sign of r).

Pearson's correlation coefficient (r)   The square root of r^2; it is a measure of association for interval-ratio variables, reflecting the strength of the linear association between two interval-ratio variables. It can be positive or negative in sign.

▣ STATISTICS IN PRACTICE: TEEN PREGNANCY AND SOCIAL INEQUALITY

The United States has by far the highest rate of teenage pregnancy of any industrialized nation. The pregnancy rate for U.S. teens ages 15 to 19 reached 95.9 pregnancies per 1,000 women in 1990. Although recent data indicate that teen pregnancy rates declined approximately 13 percent, from 1991 to 1997 to 83.7 per 1,000,[7] this rate is still twice as high as in other industrialized nations. These high rates have been attributed, among other factors, to the high rate of poverty and inequality in the United States.

The association between teen pregnancy and poverty and social inequality has been well documented both nationally and internationally. Teen pregnancy rates are higher among people living in poverty, and industrial societies that have done the most to reduce social inequality tend to have the lowest rates of teen pregnancy.[8] The noted sociologist William Wilson has claimed that the disappearance of hundreds of low-skilled jobs in the past 25 years and the resulting increase in unemployment, especially in the inner cities, has led to the increase in teenage pregnancy rates and to welfare dependency.[9] Teenagers living in areas of high unemployment, poverty, and lack of opportunities are six to seven times more likely to become unwed parents.[10]

To examine the degree to which economic factors influence teenage pregnancy rates, we analyze state-by-state data on unemployment rates in 1996 and teenage pregnancy rates in 1997. Using unemployment rate and pregnancy rate, both interval-ratio variables, we can examine the hypothesis that states with higher unemployment rates will tend to have higher teenage pregnancy rates.

Figure 8.12 shows the scatter diagram for unemployment rate and teenage pregnancy rate. Because we are assuming that the unemployment rate in 1996 can predict the teen pregnancy rate in 1997, we are going to treat unemployment rate as our independent variable, *X*. Teen pregnancy rate, then, is our dependent variable, *Y*. The scatter diagram seems to suggest that the two variables are linearly related. It also illustrates that these variables are positively associated; that is, as the state's unemployment rate rises, the teen pregnancy rate rises as well.

---

[7]Centers for Disease Control, 2000.

[8]J. M. Greene and C. L. Ringwalt, "Pregnancy Among Three National Samples of Runaway Homeless Youth," *Journal of Adolescent Health* 23 (1998): 370–377.

[9]William J. Wilson, *The Truly Disadvantaged: The Inner City, the Underclass, and Public Policy* (Chicago: University of Chicago Press, 1987).

[10]Stephanie Coontz, "The Welfare Discussion We Really Need," *Christian Science Monitor* (December 29, 1994): 19.

Figure 8.12   Scatter Diagram for Unemployment Rate and Teenage Pregnancy Rate

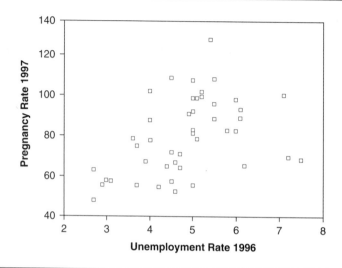

Source: Centers for Disease Control, 2000, and U.S. Census Reports, 2000.

Our bivariate regression equation for 45 states[11] is

$$\hat{Y} = 45.252 + 7.192X$$

In this prediction equation the slope (*b*) is 7.192 and the intercept (*a*) is 45.252. The positive slope, 7.192, confirms our earlier impression, based on the scatter diagram, that the relationship between the unemployment rate and teenage pregnancy rate is positive. In other words, the higher the unemployment rate, the higher the pregnancy rate. A *b* equal to 7.192 means that for every 1 percentage point increase in the unemployment rate the pregnancy rate for teens 15 to 19 will increase by 7.192 pregnancies per 1,000 women. The intercept, *a*, of 45.252 indicates that with full employment (an unemployment rate of 0) the teen pregnancy rate will be 45.252 pregnancies per 1,000 women. The regression line corresponding to this linear regression equation is shown in Figure 8.13.

Based on the linear regression equation, we could predict the teenage pregnancy rate for any state based on its unemployment rate in 1996. For example, with a 1996 unemployment rate of 5 percent, Alabama's predicted 1997 teen pregnancy rate is

$$\hat{Y} = 45.252 + 7.192(5.0) = 81.21$$

With a higher unemployment rate of 7.5 percent, the predicted 1997 teen pregnancy rate for Alaska is

$$\hat{Y} = 45.252 + 7.192(7.5) = 99.19$$

---

[11]Pregnancy rates were not reported for California, Florida, Iowa, New Hampshire, and Oklahoma. The District of Columbia was removed from the analysis due to its extremely high teen pregnancy rate relative to other states.

Figure 8.13   Scatter Diagram Showing Regression Line for Unemployment Rate and
Teenage Pregnancy Rate

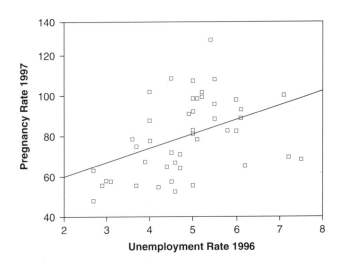

Source: Centers for Disease Control, 2000, and U.S. Census Reports, 2000.

We also calculated *r* and *r²* for these data. We obtained an *r* of 0.43 and an *r²* of
0.43² = 0.18. An *r* of 0.43 indicates that there is a moderate positive relationship between the
unemployment rate in 1996 and the teen pregnancy rate in 1997.

The coefficient of determination, *r²*, measures the proportional reduction of error that
results from using the linear regression model to predict teen pregnancy rates. An *r²* of 0.18
means that by using the regression equation, our prediction of teen pregnancy rates is
improved by 18 percent (0.18 × 100) over the prediction we would make using the mean
pregnancy rate alone. We can also say that the independent variable (unemployment rate)
explains about 18 percent of the variation in the dependent variable (teen pregnancy rate).

This analysis deals with only one factor affecting teen pregnancy rates. Other important
socioeconomic indicators likely to affect teen pregnancy rates—such as poverty rates, welfare
policies, and expenditures on education—would also need to be considered for a complete
analysis of the determinants of teenage pregnancy.

◫ STATISTICS IN PRACTICE:
THE MARRIAGE PENALTY IN EARNINGS

Among factors commonly associated with earnings are human capital variables (for example,
age, education, work experience, and health) and labor market variables (such as the unemployment rate and the structure of occupations). Individual characteristics, such as gender,
race, and ethnicity, also explain disparities in earnings. In addition, marital status has been
linked to differences in earnings. Although marriage is associated with higher earnings for

men, for women it carries a penalty; married women tend to earn less at every educational level than single women.

The lower earnings of married women have been related to differences in labor force experience. Marriage and being the mother of young children tend to limit women's choice of jobs to those that may offer flexible working hours but are generally low paying and offer fewer opportunities for promotion. Moreover, married women tend to be out of the labor market longer and have fewer years on the job than single women. When they reenter the job market or begin their career after their children are grown, they compete with coworkers with considerably more work experience and on-the-job training. (Women who need to become financially independent after divorce or widowhood may share some of the same liabilities as married women.)

All of this suggests that the returns for formal education will be generally lower for married women. Thus, we would expect single women to earn more for each year of formal education than married women. We explore this issue by analyzing the bivariate relationship between level of education and personal income among single and married females (working full-time) that were included in a 2002 GSS sample of 1,500 respondents.[12] We are assuming that level of education (measured in years) can predict personal income, and therefore we treat education as our independent variable, X. Personal income (measured in dollars), then, is the dependent variable, Y. Since both are interval-ratio variables, we can use bivariate regression analysis to examine the difference in returns for education.

Our bivariate regression equation for single females working full-time is

$$\hat{Y} \text{ (single)} = -16,709.40 + 3,054.34X$$

The regression equation tells us that for every unit increase in education—the unit is one year—we can predict an increase of $3,054.34 in the annual income of single women in our sample who work full-time.

The bivariate regression equation for married females working full-time is

$$\hat{Y} \text{ (married)} = -9547.17 + 2,755.81X$$

The regression equation tells us that for every unit increase in education, we can predict an increase of $2,755.81 in the annual income of married women in our sample who work full-time.

This analysis indicates that, as we suggested, the returns for education are lower for married women. For every year of education, the earnings of single women increase by $298.53 ($3,054.34 − $2,755.81) more than those of married women.

Let's use these regression equations to predict the difference in annual income between a single woman and a married woman, both with 16 years of education and working full-time:

$$\hat{Y} \text{ (single)} = -16,709.40 + 3,054.34(16) = 32,160.04$$

$$\hat{Y} \text{ (married)} = -9547.17 + 2,755.81(16) = 34,545.79$$

[12]Analysis is limited to women 40 years and older.

The predicted difference in annual income between a single and a married woman with a college education, both working full-time, is $2,385.75 ($34,545.79 – $32,160.04).

We also calculated the r and r^2 for these data. For single women, $r = 0.42$; for married women $r = 0.334$. These coefficients indicate that for both groups there is a weak-to-moderate (the relationship is slightly stronger for single women) relationship between education and earnings.

To determine how much of the variation in income can be explained by education we need to calculate r^2. For single women,

$$r^2 \text{ (single)} = 0.42^2 = 0.18$$

and for married women,

$$r^2 \text{ (married)} = 0.33^2 = 0.11$$

Using the regression equation, our prediction of income for single women is improved by 18 percent (0.18×100) over the prediction we would make using the mean alone. For married women there is less of an improvement in prediction, 11 percent (0.11×100).

This analysis deals with only one factor affecting earnings—the level of education. Other important factors associated with earnings—including occupation, seniority, race/ethnicity, and age—would need to be considered for a complete analysis of the differences in earnings between single and married women.

▣ MULTIPLE REGRESSION

Thus far we have used examples that involved only two variables: a dependent variable and one independent variable. For example, in the preceding section we attempted to account for variations in teen pregnancy rates on the basis of unemployment rates. We employed a linear bivariate regression equation in which unemployment rate was the independent variable and teen pregnancy rate the dependent variable. Whereas this equation gave us a relatively accurate prediction, it seems reasonable that other variables besides the state's unemployment rate might affect the rate of teenage pregnancy.

For example, we know from theoretical and empirical accounts that lack of educational opportunities also accounts for the rate of teen pregnancy. We might be able to provide a better explanation of teen pregnancy rates if we also included a measure of educational resources in addition to unemployment rates.

Multiple regression is an extension of bivariate regression. It allows us to examine the effect of two or more independent variables on the dependent variable. The calculations involved in multiple regression are quite elaborate, but are easily accomplished using SPSS or other statistical software.

The general form of the multiple regression equation involving two independent variables is

$$\hat{Y} = a + b_1 X_1 + b_2 X_2 \tag{8.8}$$

where

$\hat{Y}$ = the predicted score on the dependent variable
X_1 = the score on independent variable X_1
X_2 = the score on independent variable X_2

a = the Y-intercept, or the value of Y when both X_1 and X_2 are equal to zero

b_1 = the change in Y with a unit change in X_1, when the other independent variable X_2 is controlled

b_2 = the change in Y with a unit change in X_2, when the other independent variable X_1 is controlled

---

Multiple regression   An extension of bivariate regression in which the effects of two or more independent variables on the dependent variable are examined. The general form of the multiple regression equation involving two independent variables is $\hat{Y} = a + b_1X_1 + b_2X_2$

To illustrate, let's expand our investigation of teen pregnancy and add a second independent variable—the state's 1996 expenditure per pupil in elementary and secondary schools. We are hypothesizing that the higher the state's expenditure, the lower the teen pregnancy rate. We are also hypothesizing, as we did earlier, that the higher the state's unemployment rate, the higher the teen pregnancy rate. The multiple linear equation that incorporates both the unemployment rate and the level of educational expenditures as predictors of teen pregnancy rates is

$$\hat{Y} = 49.813 + 9.736X_1 - .007X_2$$

where

$\hat{Y}$ = predicted pregnancy rate, 1997
X_1 = state unemployment rate, 1996
X_2 = expenditure per pupil, 1996

This equation tells us that a state's pregnancy rate goes up by 9.736 per 1,000 women for each 1 percent increase in the unemployment rate (X_1), holding expenditure per pupil (X_2) constant. Similarly, the state's pregnancy rate goes down by 0.007 with each \$1 increase in the state's expenditure per pupil (X_2), holding the unemployment rate (X_1) constant.

Controlling for the effect of one variable while examining the effect of the other allows us to separate out the effects of each predictor independently of the other. For example, given two states with the same unemployment rate, the state with \$1 additional in expenditure per pupil is expected to have a teen pregnancy rate that is 0.007 lower. Similarly, given two states with the same level of expenditure per pupil, the state where the unemployment rate is 1 percentage point higher will have 9.736 more pregnancies per 1,000 women. Finally, the value of a (49.813) reflects the state's teen pregnancy rate when both the unemployment rate and the state's expenditure per pupil are equal to zero. Although it seems unlikely that any state will ever have rates as low as zero, the value of a is a baseline that must be added to the equation for pregnancy rate to be properly estimated.

Using our prediction equation for rates of teen pregnancy, we can predict the teen pregnancy rates for states with given levels of unemployment and expenditures per pupil. For example, with an unemployment rate of 5.20 percent and an expenditure per pupil of $4,012, Arizona's predicted pregnancy rate for 1997 is

$$\hat{Y} = 49.813 + 9.736(5.20) - 0.007(4012) = 72.36$$

Like bivariate regression, multiple regression analysis yields a **coefficient of determination**, symbolized as R^2 (corresponding to r^2 in the bivariate case). R^2 measures the proportional reduction of error that results from using the linear regression model. It reflects the proportion of the total variation in the dependent variable that is explained jointly by two or more independent variables. We obtained an R^2 of 0.27. This means that by using states' unemployment rates and expenditures per pupil to predict pregnancy rates we have reduced error of prediction by 27 percent (0.27×100). We can also say that the independent variables *unemployment rates* and *expenditures per pupil* in combination explain almost 27 percent of the variation in states' pregnancy rates.

The inclusion of expenditure per pupil did not improve our prediction of teen pregnancy rate by much. As we saw earlier, unemployment rate accounted for 18 percent of the variation in teen pregnancy rate. The addition of expenditure per pupil to the prediction equation resulted in an increase of only 9 percent ($27\% - 18\% = 9\%$) in the percentage of explained variation in teen pregnancy rate.

As in the bivariate case, the square root of R^2, or R, is **Pearson's multiple correlation coefficient.** It measures the linear relationship between the dependent variable and the combined effect of two or more independent variables. For our example, $R = 0.52$. It indicates that there is a moderate relationship between the dependent variable *teen pregnancy rate* and both independent variables, *unemployment rate* and *expenditure per pupil.*

✔ ***Learning Check.*** *Use the prediction equation describing the relationship between teen pregnancy and both unemployment and expenditures on education to calculate the 1997 predicted teen pregnancy rate for a state with an unemployment rate of 3 percent and an expenditure per pupil of $6,000.*

Multiple coefficient of determination (R^2) Measure that reflects the proportion of the total variation in the dependent variable that is explained jointly by two or more independent variables.

Pearson's multiple correlation (R) Measure of the linear relationship between the independent variable and the combined effect of two or more independent variables.

MAIN POINTS

- A scatter diagram (also called scatter-plot) is a quick visual method used to display relationships between two interval-ratio variables. It is used as a first exploratory step in regression analysis and can suggest to us whether two variables are associated.

- Equations for all straight lines have the same general form:

$$\hat{Y} = a + bX$$

where

$\hat{Y}$ = the predicted score on the dependent variable

X = the score on the independent variable

a = the Y-intercept, or the point where the line crosses the Y-axis; therefore, a is the value of Y when X is 0

b = the slope of the line, or the change in Y with a unit change in X

- The best-fitting regression line is that line where the residual sum of squares, or Σe^2, is at a minimum. Such a line is called the least-squares line, and the technique that produces this line is called the least-squares method.

- The coefficient of determination (r^2) and Pearson's correlation coefficient (r) measure how well the regression model fits the data. Pearson's r also measures the strength of the association between the two variables. The coefficient of determination, r^2, can be interpreted as a PRE measure. It reflects the proportional reduction of error resulting from use of the linear regression model.

- The general form of the multiple regression equation involving two independent variables is

$$\hat{Y} = a + b_1X_1 + b_2X_2$$

where

$\hat{Y}$ = the predicted score on the dependent variable

X_1 = the score on independent variable X_1

X_2 = the score on independent variable X_2

a = the Y-intercept, or the value of Y when both X_1 and X_2 are equal to zero

b_1 = the change in Y with a unit change in X_1, when the other independent variable X_2 is controlled

b_2 = the change in Y with a unit change in X_2, when the other independent variable X_1 is controlled

- The multiple coefficient of determination (R^2) and Pearson's multiple correlation coefficient (R) measure how well the regression model fits the data. Pearson's R also measures the strength of the association between the dependent variable and the independent variables. The coefficient of determination, R^2, can be interpreted as a PRE measure. It reflects the proportional reduction of error resulting from use of the linear regression model.

KEY TERMS

coefficient of determination (r^2)
deterministic (perfect) linear relationship
linear relationship
least-squares method
least-squares line (best-fitting line)
multiple coefficient of determination (R^2)

multiple regression
Pearson's correlation coefficient (r)
Pearson's multiple correlation coefficient (R)
scatter diagram (scatterplot)
slope (b)
Y-intercept (a)

ON YOUR OWN

Log on to the web-based student study site at http://www.pineforge.com/frankfort-nachmiasstudy4 for additional study questions, quizzes, web resources, and links to social science journal articles reflecting the statistics used in this chapter.

SPSS DEMONSTRATIONS

[ISSP00PFP]

Demonstration 1: Producing Scatterplots (Scatter Diagrams)

Do people with more education work longer hours at their jobs? This question can be explored with SPSS using the techniques discussed in this chapter for interval-ratio data because *hours worked per week* (WRKHRS) and *number of years of education* (EDUCYRS) are both coded at an interval-ratio level in the ISSP00PFP file.

We begin by looking at a scatterplot of these two variables. The Scatter procedure can be found under the *Graphs* menu choice. In the opening dialog box, click on *Scatter/Dot* (which means we want to produce a standard scatterplot with two variables), select the icon for *Simple Scatter*, and then click on *Define*.

The Scatterplot dialog box (Figure 8.14) requires that we specify a variable for both the *X*- and *Y*-axes. We place EDUCYRS (number of years of education) on the *X*-axis because we consider it the

Figure 8.14

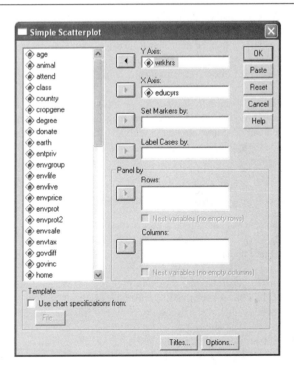

Figure 8.15

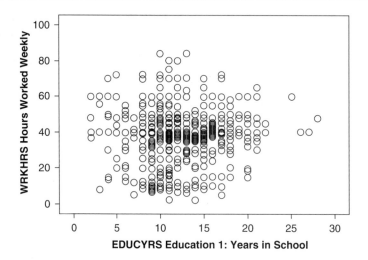

independent variable and WRKHRS (number of hours worked per week) on the *Y*-axis because it is the dependent variable. Then click on *OK*.

SPSS creates the requested graph (Figure 8.15). You can edit it to change its appearance by double-clicking on the chart in the viewer. This displays the chart in a chart window. You can edit the chart from the menus, from the toolbar, or by double-clicking on the object you want to edit.

It is difficult to tell by eye whether or not there is a relationship between the two variables, so we will ask SPSS to place the regression line on the plot. To add a regression line to the plot, we start by double-clicking on the scatterplot to open the Chart Editor. Click on *Elements* from the main menu, then *Fit Line at Total*. In the section of the dialog box headed "Fit Method," select *Linear*. Click *Apply* and then *Close*. Finally, in the Chart Editor, click on *File* and then *Close*. The result of these actions is shown in Figure 8.16.

Since the regression line is essentially horizontal, we observe that there is a slight positive relationship between education and number of hours worked each week. The predicted value for those with 20 years of education is about 40 hours, compared with less than 40 hours per week for those with less than 10 years of education. However, because there is a lot of scatter around the line, the predictive power of the model is weak.

Demonstration 2: Producing Correlation Coefficients

To further quantify the effect of education on hours worked, we request a correlation coefficient. This statistic is available in the Bivariate procedure, which is located by clicking on *Analyze*, *Correlate*, then *Bivariate* (Figure 8.17). Place the variables you are interested in correlating, EDUCYRS and WRKHRS, in the Variable(s) box, then click on *OK*.

SPSS produces a matrix of correlations, shown in Figure 8.18. We are interested in the correlation in the bottom left-hand cell, .028. We see that this correlation is closer to 0 than to 1, which tells us that education is not a very good predictor of hours worked, even if it is true that those with more education work longer hours at their job. The number under the correlation coefficient, 730, is the number of valid cases (*N*)—those respondents who gave a valid response to both questions. The number is reduced because not everyone in the sample is working.

Figure 8.16

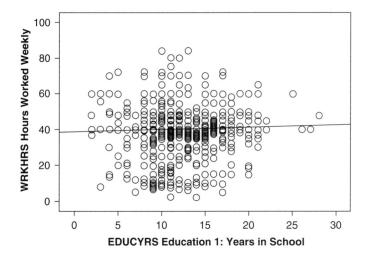

Figure 8.17

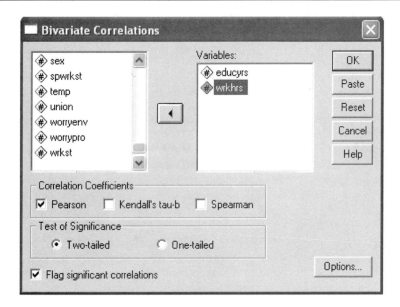

Demonstration 3: Producing a Regression Equation

Next, we will use SPSS to calculate the best-fitting regression line and the coefficient of determination. This procedure is located by clicking on *Analyze*, *Regression*, then *Linear*. The Linear Regression dialog box (Figure 8.19) provides boxes in which to enter the dependent variable, WRKHRS, and one or

Figure 8.18

Correlations

		educyrs	wrkhrs
EDUCYRS	Pearson Correlation	1	.028
	Sig. (2-tailed)		.449
	N	1350	730
WRKHRS	Pearson Correlation	.028	1
	Sig. (2-tailed)	.449	
	N	730	777

Figure 8.19

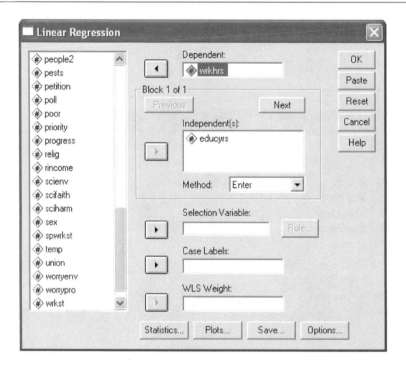

more independent variables, EDUCYRS (regression allows more than one). The Linear Regression dialog box offers many other choices, but the default output from the procedure contains all that we need.

SPSS produces a great deal of output, which is typical for many of the more advanced statistical procedures in the program. We've selected two portions of the output to review here (Figure 8.20). Under the Model Summary, the coefficient of determination is labeled "R Square." Its value is .001, which is very weak. Educational attainment explains little of the variation in hours worked, only about

Figure 8.20

Model Summary

Model	R	R Square	Adjusted R Square	Std. Error of the Estimate
1	.028[a]	.001	−.001	13.229

Coefficients[a]

Model		Unstandardized Coefficients		Standardized Coefficients	t	Sig.
		B	Std. Error	Beta		
1	(Constant)	38.378	1.712		22.424	.000
	educyrs	.099	.131	.028	.757	.449

0.1 percent. This is probably not too surprising; for example, people who own a small business may have no more than a high school education but work very long hours in their business.

The regression equation results are in the Coefficients section. The regression equation coefficients are listed in the column headed "B." The coefficient for EDUCYRS, or b, is about .099; the intercept term, or a, identified in the "(Constant)" row, is 38.378. Thus, we would predict that every additional year of education increases the number of hours worked each week by about one tenth of an hour, or just over 5 minutes. Or we could predict that those with a high school education work, on average, $38.378 + 12(.099)$ hours, or about 40 hours.

Demonstration 4: Producing a Multiple Regression Equation

What other variables, in addition to education, affect the number of hours worked? One answer to this question is that age (AGE) has something to do with the number of hours worked per week. To answer this question, we will use SPSS to calculate a multiple regression equation and a multiple coefficient of determination. This procedure is similar to the one used to generate the bivariate regression equation. Click on *Analyze, Regression*, then *Linear*. The Linear Regression dialog box (Figure 8.21) provides boxes in which to enter the dependent variable, WRKHRS, and the independent variables, EDUCYRS and AGE. We place EDUCYRS (number of years of education) and AGE (age in years) in the box for the independent variables and WRKHRS (the number of hours worked per week) in the box for the dependent variable, and click on *OK*.

SPSS produces a great deal of output. We've selected two portions of the output to review here (Figure 8.22).

Under the Model Summary, the multiple correlation coefficient labeled "*R*" is .028. This tells us that education and age are weakly associated with hours worked. The coefficient of determination is labeled "R Square." Its value is .001. In addition, SPSS provides an "Adjusted R Square" which is −.002 (i.e., 0.0). The "Adjusted R Square" adjusts the R^2 coefficient for the number of predictors in the equation. Generally, the adjusted R^2 will be lower, relative to R^2, the larger the number of predictors. An R^2 of .001 means that educational attainment and age jointly explain about 0.1 percent of the variation in hours worked.

Figure 8.21

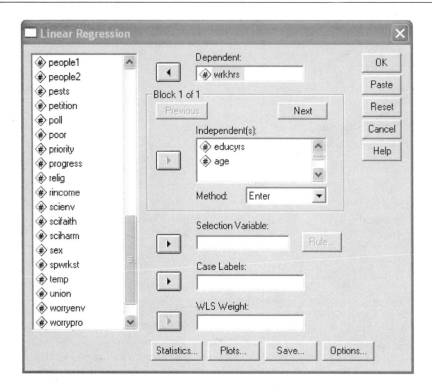

Figure 8.22

Model Summary

Model	R	R Square	Adjusted R Square	Std. Error of the Estimate
1	.028[a]	.001	−.002	13.220

Coefficients[a]

Model		Unstandardized Coefficients		Standardized Coefficients	t	Sig.
		B	Std. Error	Beta		
1	(Constant)	38.643	2.476		15.606	.000
	EDUCYRS	.097	.132	.027	.734	.463
	AGE	−.005	.039	−.005	−.129	.897

The regression equation results are in the Coefficients section. The regression equation coefficients are listed in the column headed "B." The coefficient for EDUCYRS is about .097, and for AGE it is −.005. The intercept term, or a, identified in the "(Constant)" row, is 38.643. Thus, we would predict that, holding age constant, every additional year of education increases the number of hours worked each week by a little over one tenth of an hour. Similarly, holding education constant, each year of age decreases the number of hours worked by less than one-hundredth of an hour. We could predict that those with a high school education and 20 years of age work, on average, $38.643 + 12(.097) + 20(−.005)$ hours, or about 38 hours.

SPSS PROBLEMS

[ISSP00PFP]

1. Use the 2000 ISSP data file to study the relationship between the number of persons in a respondent's household (HOMPOP) and number of hours worked per week (WRKHRS).
   a. Construct a scatterplot of these two variables in SPSS and place the best-fit linear regression line on the scatterplot. Describe the relationship between the number of persons in a respondent's household (HOMPOP) and the number of hours worked per week (WRKHRS).
   b. Have SPSS calculate the regression equation predicting WRKHRS with HOMPOP. What are the intercept and the slope? What are the coefficient of determination and the correlation coefficient?
   c. What is the predicted number of hours worked per week for someone with a household size of three persons?
   d. What is the predicted number of hours worked per week for someone with a household size of eight persons?
   e. Can you find a way for SPSS to calculate the error of prediction and predicted value for each respondent and save them as new variables?

2. Use the same variables as in Exercise 1, but do the analysis separately for Americans and Russians. Begin by locating the variable COUNTRY. Click *Data*, *Split File*, and then select *Organize Output by Groups*. Insert COUNTRY into the box and click *OK*. Now SPSS will split your results by country.
   a. Have SPSS calculate the regression equation for Americans and Russians. (Note: You will need to scroll down through your output to find the results for Americans and Russians.) How similar are they?
   b. What is the predicted number of hours worked per week for a Russian respondent with a household size of two persons? Six persons? For an American respondent with the same household sizes? Which is greater?

3. Use the same variables as in Exercise 1, but do the analysis separately for women and men. Begin by locating the variable SEX. Click *Data*, *Split File*, and then select *Organize Output by Groups*. Insert SEX into the box and click *OK*. (Note: Be sure to remove COUNTRY from the box if it is still there from the previous exercise.) Now SPSS will split your results by sex.
   a. Is there any difference between the regression equations for men and women?
   b. What is the predicted number of hours worked per week for women and men with the same household sizes: one person, four persons, seven persons?

4. Use the same variables as in Exercise 1, but do the analysis separately for married and divorced respondents. Begin by locating the variable MARITAL. Click *Data*, *Split File*, and then select *Organize Output by Groups*. Insert MARITAL into the box and click *OK*. (Note: Be sure to remove SEX and/or COUNTRY from the box if they are still there from the previous exercises.) Now SPSS will split your results by marital status.

   a. Is there any difference between the regression equations for married and divorced respondents?

   b. What is the predicted number of hours worked per week for married and divorced respondents with household sizes: one person, four persons, seven persons?

   c. What differences, if any, do you find? Can the hours worked per week be predicted better for married or divorced respondents?

5. Use the 2002 GSS file [GSS02PFP-A] to investigate the relationship between the respondent's education (EDUC) and the education received by his or her father and mother (PAEDUC and MAEDUC, respectively).

   a. Use SPSS to find the correlation coefficient, the coefficient of determination, and the regression equation predicting the respondent's education with father's education only. Interpret your results.

   b. Use SPSS to find the multiple correlation coefficient, the multiple coefficient of determination, and the regression equation predicting the respondent's education with father's *and* mother's education. Interpret your results.

   c. Did taking into account the respondent's mother's education improve our prediction? Discuss this on the basis of the results from 5b.

   d. Using the regression equation from 5a, calculate the predicted number of years of education for a person with a father with 12 years of education. Then repeat this procedure, adding in a mother's 12 years of education and using the regression equation from 5b.

CHAPTER EXERCISES

1. For a variety of reasons, a larger percentage of people are concerned about the state of the environment today than in years past. This has led to the formation of environmental action groups that attempt to alter environmental policies nationally and around the globe. A large number of environmental action groups subsist on the donations of concerned citizens. Based on the following eight countries, examine the data to determine the extent of the relationship between simply being concerned about the environment and actually giving money to environmental groups.

Country	Percent Concerned	Percent Donating Money
Austria	35.5	27.8
Denmark	27.2	22.3
Netherlands	30.1	44.8
Philippines	50.1	6.8
Russia	29.0	1.6
Slovenia	50.3	10.7
Spain	35.9	7.4
United States	33.8	22.8

Source: International Social Survey Programme 2000.

a. Construct a scatterplot of the two variables, placing percent concerned about the environment on the horizontal or *X*-axis and the percent donating money to environmental groups on the vertical or *Y*-axis.

b. Does the relationship between the two variables seem linear? Describe the relationship.

c. Find the value of the Pearson correlation coefficient that measures the association between the two variables, and offer an interpretation.

2. There is often thought to be a relationship between a person's educational attainment and the number of children he or she has. The hypothesis is that as one's educational level increases, he or she has fewer children. Investigate this conjecture with 25 cases drawn from the 2002 GSS file. The following table displays educational attainment, in years, and the number of children for each respondent.

Education	Children	Education	Children
16	0	12	2
12	1	12	3
12	3	11	1
6	6	12	2
14	2	11	2
14	2	12	0
16	2	12	2
12	2	12	3
17	2	12	4
12	3	12	1
14	4	14	0
13	0	12	3
12	1		

a. Calculate the Pearson correlation coefficient for these two variables. Does its value support the hypothesized relationship?

b. Calculate the least-squares regression equation using education as a predictor variable. What is the value of the slope, *b*? What is the value of the intercept, *a*?

c. What is the predicted number of children for a person with a college degree (16 years of education)?

d. Does any respondent actually have this number of children? If so, what is his or her level of education? If not, is this a problem or an indication that the regression equation you calculated is incorrect? Why or why not?

3. Births out of wedlock have been on the rise in the United States for many years. Discussions of this social phenomenon often focus on the greater number of births to unwed minority mothers. However, births to unwed white mothers have also increased. The following data show the percentage of births to mothers who were not married, separately for whites and nonwhites, over a 50-year period.

a. Calculate the correlation coefficient between the percentage of unwed births for whites and minorities. What is its value?

b. Provide an interpretation for the coefficient. Substantively, what does the value of the correlation coefficient imply about the similarity in the rate of increase of births out of wedlock for whites and nonwhites?

Percentage of Unwed Births by Year and Race

Year	Race	
	White	Nonwhite
1950	1.8	18.0
1955	1.9	20.2
1960	2.3	21.6
1965	4.0	26.3
1970	5.7	34.9
1975	7.3	44.2
1980	11.0	49.4
1985	14.7	52.9
1990	20.4	57.1
1995	25.3	58.6
1999	26.7	56.6

Source: Data from the National Center for Health Statistics, 2000.

4. In Chapter 5, Exercise 9, we studied the variability of crime rates and police expenditures in the eastern and midwestern United States. We've now been asked to investigate the hypothesis that the number of crimes is related to police expenditures per capita because states with higher crime rates are likely to increase their police force, thereby spending more on the number of officers on the street.

State	Number of Crimes	Police Protection Expenditures per Capita (Dollars), per 10,000 People
Maine	2,875	122.5
New Hampshire	2,282	141.8
Vermont	2,819	102.8
Massachusetts	3,262	218.7
Rhode Island	3,583	179.2
Connecticut	3,389	193.6
New York	3,279	292.4
New Jersey	3,400	236.6
Pennsylvania	3,114	171.2
Ohio	3,997	179.4
Indiana	3,766	124.5
Illinois	4,515	224.4
Michigan	4,325	172.3
Wisconsin	3,296	196.6
Minnesota	3,598	166.8
Iowa	3,224	135.8
Missouri	4,578	153.9
North Dakota	2,394	102.9
South Dakota	2,644	115.3
Nebraska	4,108	128.8
Kansas	4,439	161.6

Source: U.S. Bureau of the Census, *Statistical Abstract of the United States*, 2003, Tables 307 and 341.

a. Construct a scatter diagram of the number of crimes and police expenditures per capita, with number of crimes as the predictor variable. What can you say about the relationship between these two variables based on the scatterplot?

b. Find the least-squares regression equation that predicts police expenditures per capita from the number of crimes. What is the slope? What is the intercept?

c. Calculate the coefficient of determination (r^2), and provide an interpretation.

d. If the number of crimes increased by 100 for a state, by how much would you predict police expenditures per capita to increase?

e. Does it make sense to predict police expenditures per capita when the number of crimes is equal to zero? Why or why not?

5. Before calculating a correlation coefficient or a regression equation, it is always important to examine a scatter diagram between two variables to see how well a straight line fits the data. If a straight line does not appear to fit, other curves can be used to describe the relationship (this subject is not discussed in our text).

The SPSS scatterplot in Figure 8.23 and output shown in Figure 8.24 display the relationship between education (measured in years) and television viewing (measured in hours) based on 2002 GSS data. We can hypothesize that as educational attainment increases, hours of television viewing will decrease, indicating a negative relationship between the two variables.

a. Assess the relationship between the two variables based on the scatterplot and output for the *b* coefficient. Is there a relationship between these two variables as hypothesized? Is it a negative or a positive relationship?

b. Describe the relationship between these two variables using representative values of years of education and hours of television viewing. For example, if an individual has 16 years of education, what are the predicted hours of television viewing? How can you determine this?

c. Does a straight line adequately represent (visually) the relationship between these two variables? Why or why not?

Figure 8.23

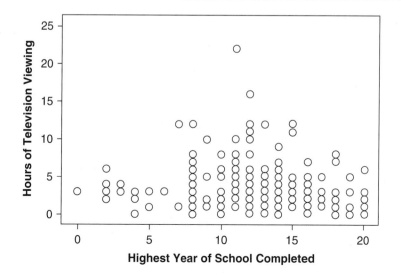

Figure 8.24

Coefficients[a]

Model		Unstandardized Coefficients		Standardized Coefficients	t	Sig.
		B	Std. Error	Beta		
1	(Constant)	4.918	.473		10.407	.000
	EDUC	−.141	.035	−.177	−4.020	.000

Model Summary

Model	R	R Square	Adjusted R Square	Std. Error of the Estimate
1	.177[a]	.031	.029	2.351

6. Using the countries in South America, let's analyze the relationship between GNP and infant mortality rate (IMR).

Country	GNP per Capita in 1997 (Dollars)	Infant Mortality in 1998 (per 1,000 Births)
Argentina	8,030	19.0
Bolivia	1,010	60.0
Brazil	4,630	33.0
Chile	4,990	10.0
Colombia	2,740	23.0
Ecuador	1,520	32.0
Paraguay	1,760	24.0
Peru	2,440	40.0
Uruguay	6,070	16.0
Venezuela	3,530	21.0

Source: Data from the Population Reference Bureau, 2000, and the World Bank, 1999.

a. Construct a scatterplot from the following data, predicting IMR from GNP. What is the relationship between GNP and IMR for these 10 countries in South America?
b. Does it appear (visually) that a straight line fits these data? Why or why not?
c. Calculate the correlation coefficient and coefficient of determination. Do these values offer further support for your answer to (b)? How?

7. Social scientists have long been interested in the aspirations and achievements of people in the United States. Research on social mobility, status, and educational attainment has provided convincing evidence on the relationship between parents' and children's socioeconomic achievement. The GSS 2002 data set has information on the educational level of respondents and their mothers. Use this information for the following selected nonrandom subsample of respondents to see whether those whose mothers had more education are more likely to have more education themselves.

Mother's Highest School Year Completed	Respondent's Highest School Year Completed
0	12
15	13
6	9
9	12
16	16
12	12
6	16
18	14
12	13
14	12
14	18
7	12

a. Construct a scatterplot, predicting the highest year of respondent's schooling with the highest year of the mother's schooling.

b. Calculate the regression equation with mother's education as the predictor variable, and draw the regression line on the scatterplot. What is the slope? What is the intercept? Describe how the straight line "fits" the data.

c. What is the error of prediction for the second case (the person with 13 years of education and mother's education 15 years)? What is the error of prediction for the person with 18 years of education and mother's education 14 years?

d. What is the predicted years of education for someone whose mother received 4 years of education? How about for someone whose mother received 12 years of education?

e. Calculate the mean number of years of education for respondents and for respondents' mothers. Plot this point on the scatterplot. Where does it fall? Can you think of a reason why this should be true?

8. In Exercise 6, we investigated the relationship between infant mortality rate and GNP in South America. The birth rates (number of live births per 1,000 inhabitants) in these same countries are shown in the following table:

Country	Birth Rate in 1999
Argentina	19
Bolivia	32
Brazil	20
Chile	18
Colombia	24
Ecuador	24
Paraguay	30
Peru	25
Uruguay	17
Venezuela	25

Source: Data from World Bank, 2000.

a. Construct a scatterplot for GNP and birth rate and one for infant mortality rate and birth rate. Do you think each can be characterized by a linear relationship?

b. Calculate the coefficient of determination and correlation coefficient for each relationship.

c. Use this information to describe the relationship between the variables.

9. In 2004 a U.S. Census Bureau report revealed that approximately 12.5 percent of all Americans were living below the poverty line in 2003. This figure is higher than in 2002, when the poverty rate was 12.1 percent. This translates to an increase of 1.3 million Americans living below the poverty line. Individuals and families living below the poverty line face many obstacles, the least of which is access to health care. In many cases, those living below the poverty line are without any form of health insurance. Using data from the U.S. Census Bureau, analyze the relationship between living below the poverty line and access to health care.

State	% Below Poverty Line (Average from 2001 to 2003)	% Without Health Insurance (Average from 2001 to 2003)
Alabama	15.1	13.3
California	12.9	18.7
Idaho	11.0	17.5
Louisiana	16.9	19.4
New Jersey	8.2	13.4
New York	14.2	15.5
Pennsylvania	9.9	10.7
Rhode Island	10.7	9.3
South Carolina	14.0	13.1
Texas	15.8	24.6
Washington	11.4	14.3
Wisconsin	8.8	9.5

Source: U.S. Bureau of the Census. Current Population Reports P60–226, *Income, Poverty and Health Insurance Coverage in the United States*: 2003.

a. Construct a scatterplot, predicting the percentage without health insurance with the percentage living below the poverty level. Does it appear that a straight-line relationship will fit the data?

b. Calculate the regression equation with percent without health insurance as the dependent variable, and draw the regression line on the scatterplot. What is its slope? What is the intercept? Has your opinion changed about whether a straight line seems to fit the data? Are there any states that fall far from the regression line? Which one(s)?

c. What percentage of the population must be living below the poverty line to obtain a predicted value of 5 percent without health insurance?

d. Predicting a value that falls beyond the observed range of the two variables in a regression is problematic at best, so your answer in (c) isn't necessarily statistically believable. However, what is a nonstatistical, or substantive, reason why making such a prediction might be important?

10. In Table 8.4 of this chapter, we calculated a regression equation, thereby expressing the linear relationship between GNP per capita and the percentage of respondents from several countries

that are willing to pay higher prices to protect the environment. We discovered a positive relationship between the two variables—as one increases, so does the other. Now, let's take our analysis one step further by looking into the relationship between GNP per capita and the percentage of respondents willing to pay more in taxes.

Country	*GNP per Capita*	*% Willing to Pay Higher Taxes*
Canada	19.71	24.0
Chile	4.99	29.1
Finland	24.28	12.0
Ireland	18.71	34.3
Japan	32.35	37.2
Latvia	2.42	17.3
Mexico	3.84	34.7
Netherlands	24.78	51.9
New Zealand	14.60	31.1
Norway	34.31	22.8
Portugal	10.67	17.1
Russia	2.66	29.9
Spain	14.10	22.2
Sweden	25.58	19.5
Switzerland	39.98	33.5
United States	29.24	31.6

Sources: The World Bank Group. Development Education Program Learning Module: *Economics, GNP per Capita.* 2004.
ISSP 2000.

a. In the chapter, we used Table 8.4 to illustrate how to calculate the slope and intercept used in the regression table. Using Table 8.4 as a model, create a similar table using the data above for GNP per capita and the percent willing to pay higher taxes.

b. From the table that you created in 10a, calculate a and b and write out the regression equation (i.e., prediction equation).

c. Calculate and interpret error type, $E2$.

d. Using your answer from 10c, calculate the PRE measure, r^2. Interpret.

e. About what percent of citizens are willing to pay higher taxes for a country with a GNP per capita of 3.0 (i.e., \$3,000)? For a GNP per capita of 30.0 (i.e., \$30,000)?

11. In Exercise 5 we examined the relationship between years of education and hours of television watched per day. We saw that as education increases, hours of television viewing decreases. The number of children a family has could also affect how much television is viewed per day. Having children may lead to more shared and supervised viewing and thus increase the number of viewing hours. The SPSS output in Figure 8.25, based on 2002 GSS data, displays the relationship between television viewing (measured in hours per day) and both education (measured in years) and number of children. We hypothesize that whereas more education may lead to less viewing, the number of children has the opposite effect: having more children will result in more hours of viewing per day.

Figure 8.25

Coefficients[a]

Model		Unstandardized Coefficients		Standardized Coefficients	t	Sig.
		B	Std. Error	Beta		
1	(Constant)	4.918	.473		10.407	.000
	EDUC	−.141	.035	−.177	−4.020	.000
	CHILDS	.077	.062	.056	1.247	.213

Model Summary

Model	R	R Square	Adjusted R Square	Std. Error of the Estimate
1	.186[a]	.034	.031	2.350

a. What is the b coefficient for education? For number of children? Interpret each coefficient. Is the relationship between education and hours of viewing as hypothesized? How about number of children and television viewing?

b. Using the multiple regression equation with both education and number of children as independent variables, calculate the number of hours of television viewing for a person with 16 years of education and two children. Compare this with the predicted value using the equation in Exercise 5.

c. Compare the r^2 value from Exercise 5 with the R^2 value from this regression. Does using education and number of children jointly reduce the amount of error involved in predicting hours of television viewed per day?

12. In 2004, the U.S. Census published a report saying that the number of Americans living below the federal poverty line was at an all-time high. We want to know if the percentage of residents in each state living below the federal poverty line can be predicted by taking into account both states' racial composition and residents' educational attainment. Figure 8.26 displays the results of multivariate regression ($N = 50$ states), predicting the percentage of a state's residents living below the federal poverty line between 2002 and 2003 using the percentage of black residents in each state in 2002 and percentage residents in each state with at least a high school diploma in 2002. Use these results to answer the questions below.

a. What is the b coefficient for the percentage of black residents in each state? For the percentage of states' residents with at least a high school diploma? Interpret each coefficient. Do these results support the idea that poverty can be explained, at least in part, by considering the racial composition and education level of states' residents? Why or why not? Use the appropriate statistics to make your argument.

b. Utilize the regression results to predict the percentage of a state's residents living below the federal poverty line. Use the 2002 mean value of 10.2 percent for the percentage of black residents in each state and the 2002 mean of 85.6 percent for the percentage of states' residents with at least a high school diploma. Is the predicted value below or above the mean value of 11.7 percent living below the federal poverty line between 2002 and 2003?

Figure 8.26

Coefficients[a]

Model		Unstandardized Coefficients		Standardized Coefficients	t	Sig.
		B	Std. Error	Beta		
1	(Constant)	66.477	7.242		9.180	.000
	% BLACK RESIDENTS	−.045	.034	−.144	−1.304	.199
	% W/ HS DIPLOMA	−.635	.082	−.851	−7.718	.000

Model Summary

Model	R	R Square	Adjusted R Square	Std. Error of the Estimate
1	.779[a]	.607	.591	1.91239

    c. What is the coefficient of determination? By how much has our prediction of the percent living below the federal poverty line improved by employing the multiple regression equation?

13. Before leaving this chapter, make sure that you can correctly answer the following questions.
    a. True or false: It is possible, in fact it often is the case, that your slope, *b*, will be a positive value and your correlation coefficient, *r*, will be a negative value.
    b. Both *a* and *b* refer to changes in which variable, the independent or dependent?
    c. The coefficient of determination, r^2, is a PRE measure. What does this mean?
    d. True or false: All regression equations reflect *causal* relationships expressed as linear functions.

The Normal Distribution

I n the preceding chapters we have learned some important things about distributions: how
to organize them into frequency distributions; how to display them using graphs; and how
to describe their central tendencies and variation using measures such as the mean and the
standard deviation. We have also learned that distributions can have different shapes. Some

distributions are symmetrical; others are negatively or positively skewed. The distributions we have described so far are all *empirical distributions*; that is, they are all based on real data.

The distribution we describe in this chapter—known as the *normal curve* or the **normal distribution**—is a theoretical distribution. A *theoretical distribution* is similar to an empirical distribution in that it can be organized into frequency distributions, displayed using graphs, and described by its central tendency and variation using measures such as the mean and the standard deviation. However, unlike an empirical distribution, a theoretical distribution is based on theory rather than on real data. The value of the theoretical normal distribution lies in the fact that many empirical distributions we study seem to approximate it. We can often learn a lot about the characteristics of these empirical distributions based on our knowledge of the theoretical normal distribution.

◙ PROPERTIES OF THE NORMAL DISTRIBUTION

The normal curve (Figure 9.1) looks like a bell-shaped frequency polygon. Because of this property it is sometimes called the *bell-shaped curve*. One of the most striking characteristics of the normal distribution is its perfect symmetry. Notice that if you fold Figure 9.1 exactly in the middle, you have two equal halves, each the mirror image of the other. This means that precisely half the observations fall on each side of the middle of the distribution. In addition, the midpoint of the normal curve is the point having the maximum frequency. This is also the point at which three measures coincide: the mode (the point of the highest frequency), the median (the point that divides the distribution into two equal halves), and the mean (the average of all the scores). Notice also that most of the observations are clustered around the middle, with the frequencies gradually decreasing at both ends of the distribution.

Normal distribution A bell-shaped and symmetrical theoretical distribution with the mean, the median, and the mode all coinciding at its peak and with the frequencies gradually decreasing at both ends of the curve.

Figure 9.1   The Normal Curve

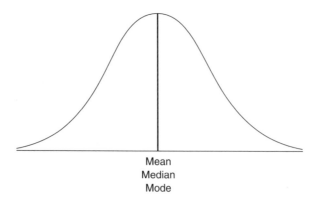

Mean
Median
Mode

Empirical Distributions Approximating the Normal Distribution

The normal curve is a theoretical ideal, and real-life distributions never match this model perfectly. However, researchers study many variables (for example, standardized tests such as the SAT, ACT, or GRE; height; athletic ability; and numerous social and political attitudes) that closely resemble this theoretical model. When we say that a variable is "normally distributed," we mean that a graphic display will reveal an approximately bell-shaped and symmetrical distribution closely resembling the idealized model shown in Figure 9.1. This property makes it possible for us to describe many empirical distributions based on our knowledge of the normal curve.

An Example: Final Grades in Statistics

It is easier to understand the properties of a normal curve if we think in terms of a real distribution that is near normal. Let's examine the frequencies and the bar chart presented in Table 9.1. These data are the final scores of 1,200 students who took Professor Frankfort-Nachmias's social statistics class at the University of Wisconsin–Milwaukee between 1983 and 1993. To convince you that the variable *final score in statistics* is normally distributed, we overlaid a normal curve on the distribution shown in Table 9.1. Notice how closely our empirical distribution of statistics scores approximates the normal curve!

Table 9.1   Final Grades in Social Statistics of 1,200 Students (1983–1993): A Near Normal Distribution

Midpoint Score	Frequency Bar Chart	Freq	Cum Freq	%	Cum %
40	*	4	4	0.33	0.33
50	*******	78	82	6.50	6.83
60	**************	275	357	22.92	29.75
70	************************	483	840	40.25	70.00
80	**************	274	1,114	22.83	92.83
90	*******	81	1,195	6.75	99.58
100	*	5	1,200	0.42	100.00
	10 50 100 200 300 400 500				

Mean $(\bar{Y}) = 70.07$ Median = 70.00   Mode = 70.00

Standard deviation $(S_Y) = 10.27$

Figure 9.2   Two Normal Distributions with Equal Means but Different Standard
Deviations

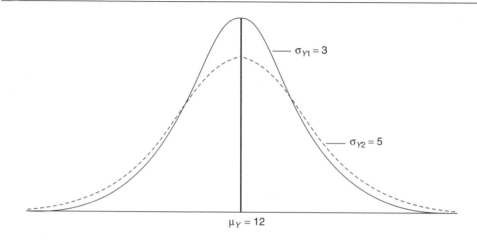

$\mu_Y = 12$

Notice that 70 is the most frequent score obtained by the students, and therefore it is the mode of the distribution. Because about half the students are either above (49.99%) or below (50.01%) this score (based on raw frequencies), both the mean (70.07) and the median (70) are approximately 70. Also shown in Table 9.1 is the gradual decrease in the number of students who scored either above or below 70. Very few students scored higher than 90 or lower than 50.

When we use the term *normal curve*, we are not referring to identical distributions. The shape of a normal distribution varies, depending on the mean and standard deviation of the particular distribution. For example, in Figure 9.2 we present two normally shaped distributions with identical means ($\mu_Y = 12$) but with different standard deviations ($\sigma_{Y_1} = 3$, and $\sigma_{Y_2} = 5$). Notice that the distribution with the larger standard deviation appears relatively wider and flatter.

Areas Under the Normal Curve

Regardless of the precise shape of the distribution, in all normal or nearly normal curves we find a constant proportion of the area under the curve lying between the mean and any given distance from the mean when measured in standard deviation units. The area under the normal curve may be conceptualized as a proportion or percentage of the number of observations in the sample. Thus, the entire area under the curve is equal to 1.00, or 100 percent (1.00 × 100) of the observations. Because the normal curve is perfectly symmetrical, exactly 0.50 or 50 percent of the observations lie above or to the right of the center, which is the mean of the distribution, and 50 percent lie below or to the left of the mean.

In Figure 9.3, note the percentage of cases that will be included between the mean and 1, 2, and 3 standard deviations above and below the mean. The mean of the distribution divides it exactly in half: 34.13 percent is included between the mean and 1 standard deviation to the right of the mean; the same percentage is included between the mean and 1 standard deviation to the left of the mean. The plus signs indicate standard deviations above the mean; the minus signs denote standard deviations below the mean. Thus, between the mean and ±1

Figure 9.3   Percentages Under the Normal Curve

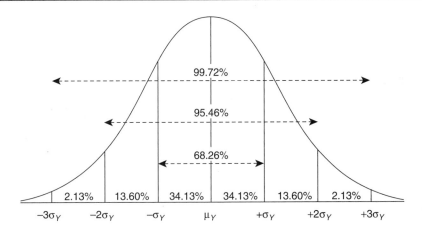

standard deviation, 68.26 percent of all the observations in the distribution occur; between the mean and ±2 standard deviations, 95.46 percent of all observations in the distribution occur; and between the mean and ±3 standard deviations, 99.72 percent of the observations occur.

✔ **Learning Check.**   *Review and confirm the properties of the normal curve. What is the area underneath the curve equal to? What percentage of the distribution is within 1 standard deviation? Within 2 and 3 standard deviations? Verify the percentage of cases by summing the percentages in Figure 9.3.*

Interpreting the Standard Deviation

The fixed relationship between the distance from the mean and the areas under the curve represents a property of the normal curve that has highly practical applications. As long as a distribution is normal and we know the mean and the standard deviation, we can determine the relative frequency (proportion or percentage) of cases that fall between any score and the mean.

This property provides an important interpretation for the standard deviation of empirical distributions that are approximately normal. For such distributions, when we know the mean and the standard deviation, we can determine the percentage of scores that are within any distance, measured in standard deviation units, from that distribution's mean. For example, we know that college entrance tests such as the SAT and ACT are normally distributed. The SAT, for instance, has a mean of 500 and a standard deviation of 100. This means that approximately 68 percent of the students who take the test obtain a score between 400 (1 standard deviation below the mean) and 600 (1 standard deviation above the mean). We can also anticipate that approximately 95 percent of the students who take the test will score between 300 (2 standard deviations below the mean) and 700 (2 standard deviations above the mean).

Not every empirical distribution is normal. We've learned that the distributions of some common variables, such as income, are skewed and therefore not normal. The fixed relationship between the distance from the mean and the areas under the curve applies only to distributions that are normal or approximately normal.

▣ STANDARD (Z) SCORES

We can express the difference between any score in a distribution and the mean in terms of *standard scores*, also known as *Z scores*. A **standard (Z) score** is the number of standard deviations that a given raw score (or the observed score) is above or below the mean. A raw score can be transformed into a Z score to find how many standard deviations it is above or below the mean.

Standard (Z) score   The number of standard deviations that a given raw score is above or below the mean.

Transforming a Raw Score into a Z Score

To transform a raw score into a Z score, we divide the difference between the score and the mean by the standard deviation. For instance, to transform a final score in the statistics class into a Z score, we subtract the mean of 70.07 from that score and divide the difference by the standard deviation of 10.27. Thus, the Z score of 80 is

$$\frac{80 - 70.07}{10.27} = 0.97$$

or 0.97 standard deviations above the mean. Similarly, the Z score of 60 is

$$\frac{60 - 70.07}{10.27} = -0.98$$

or 0.98 standard deviations below the mean; the negative sign indicates that this score is below the mean.

This calculation, in which the difference between a raw score and the mean is divided by the standard deviation, gives us a method of standardization known as *transforming a raw score into a Z score* (also known as a standard score). The Z score formula is

$$Z = \frac{Y - \overline{Y}}{s_Y} \qquad (9.1)$$

Table 9.2   Final Social Science Statistics Scores Converted to *Z* Scores

Final Score	Z Score
40	$Z = \dfrac{40 - 70.07}{10.27} = \dfrac{-30.07}{10.27} = -2.93$
50	$Z = \dfrac{50 - 70.07}{10.27} = \dfrac{-20.07}{10.27} = -1.95$
60	$Z = \dfrac{60 - 70.07}{10.27} = \dfrac{-10.07}{10.27} = -0.98$
70	$Z = \dfrac{70 - 70.07}{10.27} = \dfrac{-0.07}{10.27} = -0.01$
80	$Z = \dfrac{80 - 70.07}{10.27} = \dfrac{9.93}{10.27} = 0.97$
90	$Z = \dfrac{90 - 70.07}{10.27} = \dfrac{19.93}{10.27} = 1.94$
100	$Z = \dfrac{100 - 70.07}{10.27} = \dfrac{29.93}{10.27} = 2.91$
$\overline{Y} = 70.07$	$S_T = 10.27$

A *Z* score allows us to represent a raw score in terms of its relationship to the mean and to the standard deviation of the distribution. It represents how far a given raw score is from the mean in standard deviation units. A positive *Z* indicates that a score is larger than the mean, and a negative *Z* indicates that it is smaller than the mean. The larger the *Z* score, the larger the difference between the score and the mean.

Transforming a Z Score into a Raw Score

For some normal curve applications, we need to reverse the process, transforming a *Z* score into a raw score instead of transforming a raw score into a *Z* score. A *Z* score can be converted to a raw score to find the score associated with a particular distance from the mean when this distance is expressed in standard deviation units. For example, suppose we are interested in finding out the final score in the statistics class that lies 1 standard deviation above the mean. To solve this problem we begin with the *Z*-score formula:

$$Z = \frac{Y - \overline{Y}}{S_Y}$$

Note that for this problem we have the values for Z ($Z = 1$), the mean ($Y = 70.07$), and the standard deviation ($S_T = 10.27$), but we need to determine the value of Y:

$$1 = \frac{Y - 70.07}{10.27}$$

Through simple algebra we solve for Y:

$$Y = 70.07 + 1(10.27) = 70.07 + 10.27 = 80.34$$

The score of 80.34 lies 1 standard deviation (or 1 Z score) above the mean of 70.07.
The general formula for transforming a Z score into a raw score is

$$Y = \bar{Y} + Z(S_Y) \tag{9.2}$$

Thus, to transform a Z score into a raw score, multiply the Z score by the standard deviation and add this product to the mean.

Now, what statistics score lies 1.5 standard deviations below the mean? Because the score lies below the mean, the Z score is negative. Thus,

$$Y = 70.07 + (-1.5)(10.27) = 70.07 - 15.41 = 54.66$$

The score of 54.66 lies 1.5 standard deviations below the mean of 70.07.

✓ **Learning Check.** *Transform the Z scores in Table 9.2 back into raw scores. Your answers should agree with the raw scores listed in the table.*

▣ THE STANDARD NORMAL DISTRIBUTION

When a normal distribution is represented in standard scores (Z scores), we call it the **standard normal distribution**. Standard scores, or Z scores, are numbers that tell us the distance between an actual score and the mean in terms of standard deviation units. The standard normal distribution has a mean of 0.0 and a standard deviation of 1.0.

> ***Standard normal distribution*** A normal distribution represented in standard (Z) scores.

Figure 9.4 shows a standard normal distribution with areas under the curve associated with 1, 2, and 3 standard scores above and below the mean. To help you understand the

Figure 9.4   The Standard Normal Distribution

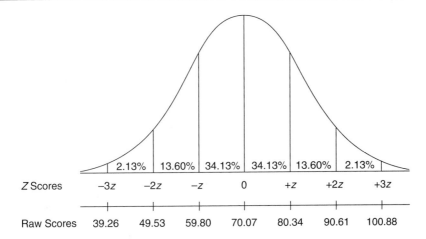

relationship between raw scores of a distribution and standard Z scores, we also show the raw scores in the statistics class that correspond to these standard scores. For example, notice that the mean for the statistics score distribution is 70.07; the corresponding Z score—the mean of the standard normal distribution—is 0. The score of 80.34 is 1 standard deviation above the mean $(70.07 + 10.27 = 80.34)$; therefore, its corresponding Z score is +1. Similarly, the score of 59.80 is 1 standard deviation below the mean $(70.07 - 10.27 = 59.80)$, and its Z-score equivalent is -1.

> ✓ **Learning Check.**   *Can you explain why the mean of the standard normal curve is 0 and the standard deviation is equal to 1?*

▣ THE STANDARD NORMAL TABLE

We can use Z scores to determine the proportion of cases that are included between the mean and any Z score in a normal distribution. The areas or proportions under the standard normal curve, corresponding to any Z score or its fraction, are organized into a special table called the **standard normal table**. The table is presented in Appendix B. In this section we discuss how to use this table.

The Structure of the Standard Normal Table

Table 9.3 reproduces a small part of the standard normal table. Note that the table consists of three columns (rather than having one long table, we have moved the second half of the table next to the first half, so the three columns are presented in a six-column format).

Table 9.3   The Standard Normal Table

(A) Z	(B) Area Between Mean and Z	(C) Area Beyond Z	(A) Z	(B) Area Between Mean and Z	(C) Area Beyond Z
0.00	0.0000	0.5000	0.21	0.0832	0.4168
0.01	0.0040	0.4960	0.22	0.0871	0.4129
0.02	0.0080	0.4920	0.23	0.0910	0.4090
0.03	0.0120	0.4880	0.24	0.0948	0.4052
0.04	0.0160	0.4840	0.25	0.0987	0.4013
0.05	0.0199	0.4801	0.26	0.1026	0.3974
0.06	0.0239	0.4761	0.27	0.1064	0.3936
0.07	0.0279	0.4721	0.28	0.1103	0.3897
0.08	0.0319	0.4681	0.29	0.1141	0.3859
0.09	0.0359	0.4641	0.30	0.1179	0.3821
0.10	0.0398	0.4602	0.31	0.1217	0.3783
0.11	0.0438	0.4562	0.32	0.1255	0.3745
0.12	0.0478	0.4522	0.33	0.1293	0.3707
0.13	0.0517	0.4483	0.34	0.1331	0.3669
0.14	0.0557	0.4443	0.35	0.1368	0.3632
0.15	0.0596	0.4404	0.36	0.1406	0.3594
0.16	0.0636	0.4364	0.37	0.1443	0.3557
0.17	0.0675	0.4325	0.38	0.1480	0.3520
0.18	0.0714	0.4286	0.39	0.1517	0.3483
0.19	0.0753	0.4247	0.40	0.1554	0.3446
0.20	0.0793	0.4207			

Column A lists positive Z scores. Because the normal curve is symmetrical, the proportions that correspond to positive Z scores are identical to the proportions corresponding to negative Z scores.

Column B shows the area included between the mean and the Z score listed in column A. Note that when Z is positive the area is located on the right side of the mean (see Figure 9.5a), whereas for a negative Z score the same area is located left of the mean (Figure 9.5b).

Column C shows the proportion of the area that is beyond the Z score listed in column A. Areas corresponding to positive Zs are on the right side of the curve (see Figure 9.5a). Areas corresponding to negative Z scores are identical except that they are on the left side of the curve (Figure 9.5b).

Standard normal table   A table showing the area (as a proportion, which can be translated into a percentage) under the standard normal curve corresponding to any Z score or its fraction.

Figure 9.5   Areas Between Mean and Z (B) and Beyond Z (C)

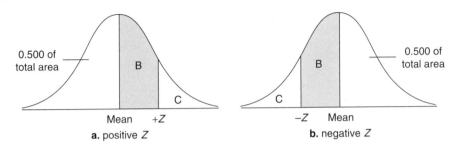

Transforming Z Scores into Proportions (or Percentages)

We illustrate how to use Appendix B with some simple examples, using our data on students' final statistics scores (see Table 9.1). The examples in this section are applications that require the transformation of Z scores into proportions (or percentages).

Finding the Area Between the Mean and a Specified Positive Z Score

Use the standard normal table to find the area between the mean and a specified positive Z score. To find the percentage of students whose scores range between the mean (70.07) and 85, follow these steps.

1. Convert 85 to a Z score:

$$Z = \frac{85 - 70.07}{10.27} = 1.45$$

2. Look up 1.45 in column A (in Appendix B) and find the corresponding area in column B, 0.4265. We can translate this proportion into a percentage ($0.4265 \times 100 = 42.65\%$) of the area under the curve included between the mean and a Z of 1.45 (Figure 9.6).

3. Thus, 42.65 percent of the students scored between 70.07 and 85.

To find the actual number of students who scored between 70.07 and 85, multiply the proportion 0.4265 by the total number of students. Thus, approximately 512 students ($0.4265 \times 1,200 = 512$) obtained a score between 70.07 and 85.

Finding the Area Between the Mean and a Specified Negative Z Score

What is the percentage of students whose scores ranged between 65 and 70.07? We can use the standard normal table and the following steps to find out.

Figure 9.6   Finding the Area Between the Mean and a Specified Positive Z Score

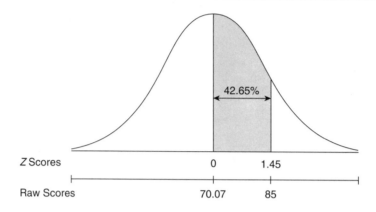

Figure 9.7   Finding the Area Between the Mean and a Specified Negative Z Score

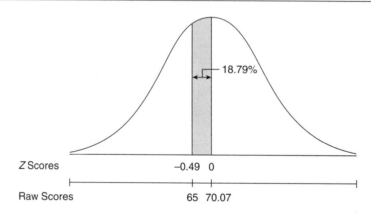

1. Convert 65 to a Z score:

$$Z = \frac{65 - 70.07}{10.27} = -0.49$$

2. Because the proportions that correspond to positive Z scores are identical to the proportions corresponding to negative Z scores, we ignore the negative sign of Z and look up 0.49 in column A. The area corresponding to a Z score of 0.49 is 0.1879. This indicates that 0.1879 of the area under the curve is included between the mean and a Z of −0.49 (Figure 9.7). We convert this proportion to 18.79 percent ($0.1879 \times 100 = 18.79\%$).

3. Thus, approximately 225 ($0.1879 \times 1,200 = 225$) students obtained a score between 65 and 70.07.

Finding the Area Between Two Z Scores on Opposite Sides of the Mean

When the scores we are interested in lie on opposite sides of the mean, we add the areas together. For example, suppose we want to find the number of students who scored between 62 and 72.

1. First, find the Z scores corresponding to 62 and 72:

$$Z = \frac{72 - 70.07}{10.27} = 0.19 \quad Z = \frac{62 - 70.07}{10.27} = -0.79$$

2. Look up the areas corresponding to these Z scores. We find that 0.19 corresponds to an area of 0.0753 (or 7.53%) and –0.79 corresponds to an area of 0.2852 (28.52%) (Figure 9.8).

3. Because the scores are on opposite sides of the mean, add the two areas obtained in step 2. The total area between these scores is 7.53% + 28.52% = 36.05%.

4. The number of students who scored between 62 and 72 is 433 (1,200 × 0.3605).

Figure 9.8   Finding the Area Between Two Z Scores on Opposite Sides of the Mean

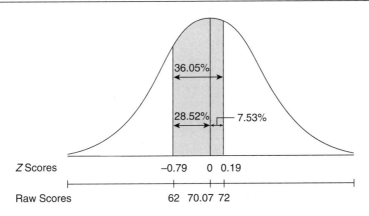

Finding the Area Above a Positive Z Score or Below a Negative Z Score

We can compare students who have done very well or very poorly to get a better idea of how they compare with other students in the class.

To identify students who did very well, we selected all students who scored above 85. To find how many students scored above 85, first convert 85 to a Z score:

$$Z = \frac{85 - 70.07}{10.27} = 1.45$$

Figure 9.9   Finding the Area Above a Positive Z Score or Below a Negative Z Score

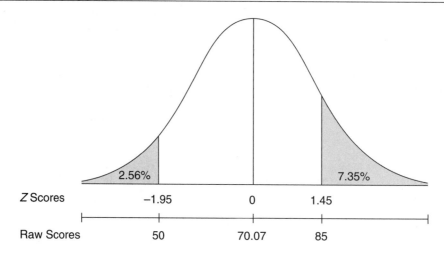

Thus, the Z score corresponding to a final score of 85 in statistics is equal to 1.45.

The area beyond a Z of 1.45 includes all students who scored above 85. This area is shown in Figure 9.9. To find the proportion of students whose scores fall into this area, refer to the entry in column C that corresponds to a Z of 1.45, 0.0735. This means that 7.35 percent ($0.0735 \times 100 = 7.35\%$) of the students scored above 85. To find the actual number of students in this group, multiply the proportion 0.0735 by the total number of students. Thus, there were $1,200 \times 0.0735$, or about 88 students, who scored above 85 over the 10-year period.

A similar procedure can be applied to identify the number of students who did not do well in the class. The cutoff point for poor performance in this class was the score of 50. To determine how many students did poorly, we first converted 50 to a Z score:

$$Z = \frac{50 - 70.07}{10.27} = -1.95$$

The Z score corresponding to a final score of 50 is equal to –1.95. The area beyond a Z of –1.95 includes all students who scored below 50. This area is also shown in Figure 9.9. Locate the proportion of students in this area in column C, in the entry corresponding to a Z of 1.95. (Remember the proportions corresponding to positive or negative Zs are identical.) This proportion is equal to 0.0256. Thus, 2.56 percent ($0.0256 \times 100 = 2.56\%$) of the group, or about 31 ($0.0256 \times 1,200$) students, performed poorly in statistics.

Transforming Proportions (or Percentages) into Z Scores

The examples in this section are applications that require transforming proportions (or percentages) into Z scores.

Figure 9.10   Finding a Z Score Bounding an Area Above It

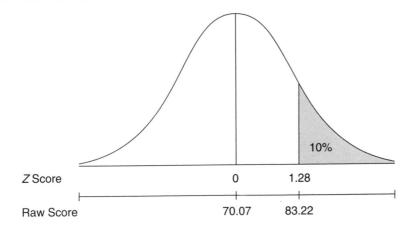

Finding a Z Score Bounding an Area Above It

Assuming a grade of A is assigned to the top 10 percent of the students, what would it take to get an A in the class? To answer this question we need to identify the cutoff point for the top 10 percent of the class. This problem involves two steps:

1. Find the Z score that bounds the top 10 percent, or 0.1000 (0.1000 × 100 = 10%), of all the students who took statistics (Figure 9.10).

   Refer to the areas under the normal curve, shown in Appendix B. First, look for an entry of 0.1000 (or the value closest to it) in column C. The entry closest to 0.1000 is 0.1003. Then locate the Z in column A that corresponds to this proportion. The Z score associated with the proportion 0.1003 is 1.28.

2. Find the final score associated with a Z of 1.28.

   This step involves transforming the Z score into a raw score. We learned earlier in this chapter (Formula 9.2) that to transform a Z score into a raw score we multiply the score by the standard deviation and add that product to the mean. Thus,

$$Y = 70.07 + 1.28(10.27) = 70.07 + 13.15 = 83.22$$

The cutoff point for the top 10 percent of the class is a score of 83.22.

Finding a Z Score Bounding an Area Below It

Now let's assume that a grade of F was assigned to the bottom 5 percent of the class. What would be the cutoff point for a failing score in statistics? Again, this problem involves two steps:

1. Find the Z score that bounds the lowest 5 percent, or 0.0500, of all the students who took the class (Figure 9.11).

Figure 9.11   Finding a Z Score Bounding an Area Below It

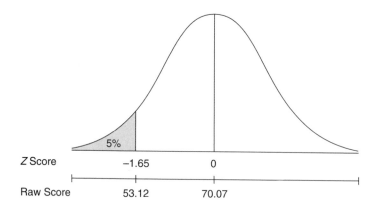

Refer to the areas under the normal curve, and look for an entry of 0.0500 (or the value closest to it) in column C. The entry closest to 0.0500 is 0.0495. Then locate the Z in column A that corresponds to this proportion, 1.65. Because the area we are looking for is on the left side of the curve—that is, below the mean—the Z score is negative. Thus, the Z associated with the lowest 0.0500 (or 0.0495) is –1.65.

2. To find the final score associated with a Z of –1.65, convert the Z score to a raw score:

$$Y = 70.07 + (-1.65)(10.27) = 70.07 - 16.95 = 53.12$$

The cutoff for a failing score in statistics is 53.12.

> ✔ **Learning Check.**   *Can you find the number of students who got a score of at least 90 in the statistics course? How many students got a score below 60?*

Working with Percentiles in a Normal Distribution

In Chapter 4 we defined percentiles as scores below which a specific percentage of the distribution falls. For example, the 95th percentile is a score that divides the distribution so that 95 percent of the cases are below it and 5 percent are above it. How are percentile ranks determined? How do you convert a percentile rank to a raw score? To determine the percentile rank of a raw score requires transforming Z scores into proportions or percentages. Converting percentile ranks to raw scores is based on transforming proportions or percentages into Z scores. In the following examples, we illustrate both procedures based on our statistics scores example.

Figure 9.12 Finding the Percentile Rank of a Score Higher Than the Mean

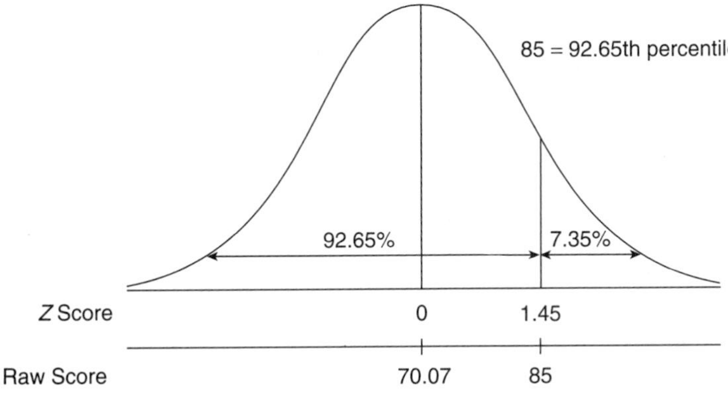

Finding the Percentile Rank of a Score Higher Than the Mean

Suppose you are one of the 1,200 students who took the statistics course. Your final score in the course was 85. How well did you do relative to the other students who took the class? To evaluate your performance you must translate your raw score into a percentile rank. Figure 9.12 illustrates this problem.

To find the percentile rank of a score higher than the mean, follow these steps.

1. Convert the raw score to a Z score:

$$Z = \frac{85 - 70.07}{10.27} = 1.45$$

The Z score corresponding to a raw score of 85 is 1.45.

2. Find the area beyond Z in Appendix B, column C. The area beyond a Z score of 1.45 is 0.0735.

3. Subtract the area from 1.00 and multiply by 100 to obtain the percentile rank:

$$\text{percentile rank} = (1.0000 - 0.0735 = 0.9265)100 = 92.65\%$$

Being in the 92.65th percentile means that 92.65 percent of all the students enrolled in social statistics scored lower than 85 and 7.35 percent scored higher than 85.

Finding the Percentile Rank of a Score Lower Than the Mean

Now let's say that you were unfortunate enough to obtain a score of 65 in the class. What is your percentile rank? Again, to evaluate your performance you must translate your raw score into a percentile rank. Figure 9.13 illustrates this problem.

Figure 9.13   Finding the Percentile Rank of a Score Lower Than the Mean

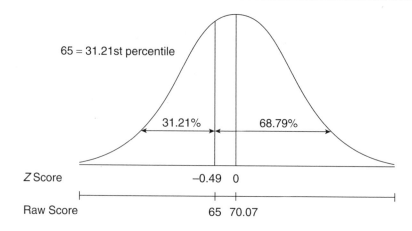

To find the percentile rank of a score lower than the mean, follow these steps.

1. Convert the raw score to a Z score:

$$Z = \frac{65 - 70.07}{10.27} = -0.49$$

The Z score corresponding to a raw score of 65 is –0.49.

2. Find the area beyond Z in Appendix B, column C. The area beyond a Z score of –0.49 is 0.3121.

3. Multiply the area by 100 to obtain the percentile rank:

$$\text{percentile rank} = 0.3121(100) = 31.21\%$$

The 31.21st percentile rank means that 31.21 percent of all the students enrolled in social statistics did worse than you (that is, 31.21% scored lower than 65 but 68.79% scored higher than 65).

✔ *Learning Check.*   *In Chapter 4 we learned to identify percentiles using cumulative percentages in a distribution. Examine Table 9.1 and find the 92nd percentile. Does your answer differ from the results we obtained earlier (finding the percentile rank of a score higher than the mean)? If it does, explain why.*

Figure 9.14 Finding the Raw Score Associated with a Percentile Higher Than 50

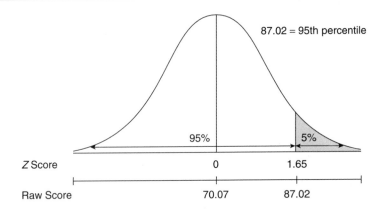

Finding the Raw Score Associated with a Percentile Higher Than 50

Now let's assume that our graduate program in sociology will accept only students who scored at the 95th percentile. What is the cutoff point required for admission? Figure 9.14 illustrates this problem.

To find the score associated with a percentile higher than 50, follow these steps.

1. Divide the percentile by 100 to find the area below the percentile rank:

$$\frac{95}{100} = 0.95$$

2. Subtract the area below the percentile rank from 1.00 to find the area above the percentile rank:

$$1.00 - 0.95 = 0.05$$

3. Find the Z score associated with the area above the percentile rank.

   Refer to the area under the normal curve, shown in Appendix B. First, look for an entry of 0.0500 (or the value closest to it) in column C. The entry closest to 0.0500 is 0.0495. Now locate the Z in column A that corresponds to this proportion, 1.65.

4. Convert the Z score to a raw score:

$$Y = 70.07 + 1.65(10.27) = 70.07 + 16.95 = 87.02$$

The final statistics score associated with the 95th percentile is 87.02. This means that you will need a score of 87.02 or higher to be admitted to the graduate program in sociology.

✓ *Learning Check.*   *In a normal distribution, how many standard deviations from the mean is the 95th percentile? If you can't answer this question, review the material in this section.*

Figure 9.15   Finding the Raw Score Associated with a Percentile Lower Than 50

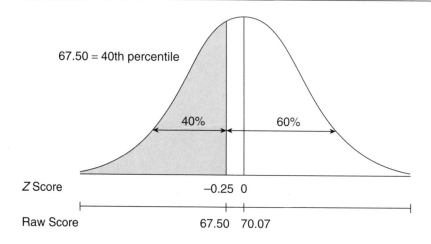

Finding the Raw Score Associated with a Percentile Lower Than 50

Finally, what is the score associated with the 40th percentile? To find the percentile rank of a score lower than 50 follow these steps (Figure 9.15).

1. Divide the percentile by 100 to find the area below the percentile rank:

$$\frac{40}{100} = 0.40$$

2. Find the Z score associated with this area.

   Refer to the area under the normal curve, shown in Appendix B. First, look for an entry of 0.4000 (or the value closest to it) in column C. The entry closest to 0.4000 is 0.4013. Now locate the Z in column A that corresponds to this proportion. The Z score associated with the proportion 0.4013 is –0.25.

3. Convert the Z score to a raw score:

$$Y = 70.07 + (-0.25)(10.27) = 70.07 - 2.578 = 67.50$$

The final statistics score associated with the 40th percentile is 67.50. This means that 40 percent of the students scored below 67.50 and 60 percent scored above it.

✔ **Learning Check.**   *What is the raw score in statistics associated with the 50th percentile?*

▣ A FINAL NOTE

In this chapter we have learned how the properties of the theoretical normal curve can be applied to describe important characteristics of empirical distributions that are approximately normal. The normal curve has other practical applications as well, however, beyond the description of "real life" distributions. In subsequent chapters we will see that the normal distribution also enables us to describe the characteristics of a theoretical distribution of great significance in inferential statistics: the sampling distribution. The techniques learned in this chapter—transforming scores and finding areas under the normal curve—will be used in many of the procedures described in subsequent chapters. Make sure you understand these techniques before you proceed to the next chapter.

MAIN POINTS

- The normal distribution is central to the theory of inferential statistics. It also provides a model for many empirical distributions that approximate normality.

- In all normal or nearly normal curves, we find a constant proportion of the area under the curve lying between the mean and any given distance from the mean when measured in standard deviation units.

- The standard normal distribution is a normal distribution represented in standard scores, or Z scores. Z scores express the number of standard deviations that a given score is above or below the mean. The proportions corresponding to any Z score or its fraction are organized into a special table called the standard normal table.

KEY TERMS

normal distribution
standard normal distribution

standard normal table
standard (Z) score

ON YOUR OWN

Log on to the web-based student study site at http://www.pineforge.com/frankfort-nachmiasstudy4 for additional study questions, quizzes, web resources, and links to social science journal articles reflecting the statistics used in this chapter.

SPSS DEMONSTRATION

[GSS02PFP-A]

Producing Z Scores with SPSS

In this chapter we have discussed the theoretical normal curve, Z scores, and the relationship between raw scores and Z scores. The SPSS Descriptives procedure can calculate Z scores for any distribution. We'll use it to study the distribution of occupational prestige in the 2002 GSS file. Locate the Descriptives procedure in the *Analyze* menu, under *Descriptive Statistics*, then *Descriptives*. We can select one or more variables to place in the Variable(s) box; for now, we'll just place PRESTG80 in this box (Figure 9.16). A checkbox in the bottom left corner tells SPSS to create standardized values, or Z scores, as new variables. Any new variable is placed in a new column in the Data View window and will then be available for additional analyses. Click on *OK* to run the procedure.

The output from Descriptives (Figure 9.17) is brief, listing the mean and standard deviation for PRESTG80, plus the minimum and maximum values and the number of valid cases.

Though not indicated in the Output window, SPSS has created a new Z score variable for PRESTG80. To see this new variable, switch to the Data View screen. Then go to the last column by

Figure 9.16

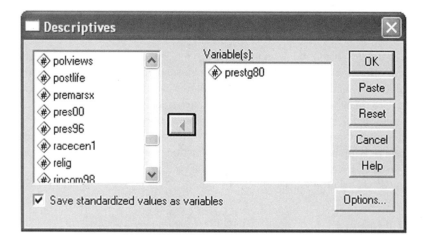

Figure 9.17

	N	Minimum	Maximum	Mean	Std. Deviation
RS OCCUPATIONAL PRESTIGE (1980)	1430	17	86	43.77	13.784
Valid N (listwise)	1430				

Figure 9.18

	sppres80	tvhours	wwwhr	Zprestg80
1	0	1	1	.95989
2	47	-1	3	.23443
3	0	-1	2	1.61280
4	0	-1	-1	.52461
5	49	-1	-1	1.46771
6	24	-1	-1	-.27339
7	0	-1	10	.52461
8	0	-1	4	-.99885
9	51	-1	-1	-.85376
10	59	-1	5	.52461

pressing the *End* key (Figure 9.18). By default, SPSS appends a Z to the variable name, so the new variable is called ZPRESTG80. The first case in the file has a Z score of .95989, so the prestige score for this person must be above the mean of 43.77. If we locate the respondent's PRESTG80 score, we see that the score for this person was 57 (not pictured), above the mean as we expected.

If the data file is saved, the new Z score variable will be saved along with the original data and then can be used in analyses. In addition, if we have SPSS calculate the mean and standard deviation of ZPRESTG80, we find that they are equal to 0 and 1.00, respectively.

SPSS PROBLEMS

[GSS02PFP-A]

1. The majority of variables that social scientists study are not normally distributed. This doesn't typically cause problems in analysis when the goal of a study is to calculate means and standard deviations—as long as sample sizes are greater than about 50. (This will be discussed in later chapters.) However, when characterizing the distribution of scores in one sample, or in a complete population (if this information is available), a nonnormal distribution can cause complications. We can illustrate this point by examining the distribution of age in the GSS data file.

   a. Create a histogram for AGE (click on *Graphs, Histogram*, insert the variable AGE) with a superimposed normal curve (click on the option, "Display Normal Curve"). How does the distribution of AGE deviate from the theoretical normal curve?

   b. Calculate the mean and standard deviation for AGE in this sample, using either the Frequencies or Descriptives procedure.

   c. Assuming the distribution of AGE is normal, calculate the number of people who should be 25 years of age or less.

   d. Use the Frequencies procedure to construct a table of the percentage of cases at each value of AGE. Compare the theoretical calculation in (c) with the actual distribution of age in the sample. What percentage of people in the sample are 25 years old or less? Is this value close to what you calculated? Why might there be a discrepancy?

2. SPSS will calculate standard scores for any distribution. Examine the distribution of EDUC (years of school completed).
   a. Have SPSS calculate Z scores for EDUC. (See SPSS Demonstration if this is unclear.)
   b. What is the equivalent Z score for someone who completed 18 years of education?
   c. Use the Frequencies procedure to find the percentile rank for a score of 18.
   d. Does the percentile rank you found from Frequencies correspond to the Z score for a value of 18? In other words, is the distribution for years of education normal? If so, then the Z score SPSS calculates should be very close, after transforming it into an appropriate area, to the percentile rank for that same score.
   e. Create histograms for EDUC and the new variable ZEDUC. Explain why they have the same shape.

3. Repeat the procedure in Exercise 2, this time running separate analyses for men versus women (SEX) and blacks versus whites (RACECEN1) based on the variable EDUC. Remember, you can run separate analyses using the Data–Split File command. Click on *Data*, *Split File*, *Organize Output by Groups*, and select either SEX or RACECEN1. Is there a difference in EDUC among men/women and blacks/whites in the GSS sample? How would you describe the distribution of EDUC for the four groups?

CHAPTER EXERCISES

1. We discovered that 905 GSS respondents in 2002 watched television for an average of 3 hours a day, with a standard deviation of 2.4 hours. Answer the following questions, assuming the distribution of the number of television hours is normal.
   a. What is the Z score for a person who watches more than 8 hours per day?
   b. What proportion of people watch television less than 5 hours a day? How many does this correspond to in the sample?
   c. What number of television hours per day corresponds to a Z score of +1?
   d. What is the percentage of people who watch between 1 and 6 hours of television per day?

2. If a particular distribution you are studying is not normal, it may be difficult to determine the area under the curve of the distribution or translate a raw score into a Z value. Is this statement true? Why or why not?

3. In 2002 the average population of all states in the United States (excluding Washington, DC) was 5,755,980 with a standard deviation of 6,386,850.[1]
   a. California's population in 2002 was 35,116,000. If the population distribution is approximately normal, convert the value of California's population to a Z score.
   b. For a normal distribution, what percentage of cases should fall less than 1 standard deviation below the mean? For a normal distribution, how many states would fall below this value in 2002? (Hint: You don't need a listing of each state's population to answer this question.)
   c. What does your answer in (b) imply about the shape of the distribution of population for the 50 states? When a distribution isn't normal, what statistic is a better measure of central tendency than the mean?

---

[1]*Source*: U.S. Bureau of the Census, *Statistical Abstract of the United States*, 2003, Table 17.

4. A social psychologist has developed a test to measure gregariousness. The test is normed so that it has a mean of 70 and a standard deviation of 20, and the gregariousness scores are normally distributed in the population of college students used to develop the test.
   a. What is the percentile rank of a score of 40?
   b. What percentage of scores falls between 35 and 90?
   c. What is the standard score for a test score of 65?
   d. What proportion of students should score above 115?
   e. What is the cutoff score below which 87 percent of all scores fall?

5. The 2002 General Social Survey provides the following statistics for the average years of education for lower, working, middle, and upper class respondents, and their associated standard deviations.

	Mean	*Standard Deviation*	*N*
Lower class	10.34	3.18	89
Working class	12.64	2.50	675
Middle class	14.19	2.39	675
Upper class	15.00	3.27	49

   a. Assuming that years of education is normally distributed in the population, what proportion of working class respondents have 12 to 16 years of education? What proportion of upper class respondents have 12 to 16 years of education?
   b. What is the probability that a working class respondent, drawn at random from the population, will have more than 16 years of education? What is the equivalent probability for a middle class respondent drawn at random?
   c. What is the probability that a lower or upper class respondent will have less than 12 years of education?
   d. Find the upper and lower limits, centered around the mean, that will include 50 percent of all working class respondents.
   e. If years of education is actually positively skewed in the population, how would that change your other answers?

6. The following table displays information on the abortion rate—the number of abortions per 1,000 women—for each U.S. state and the District of Columbia in 2000.
   a. What are the mean and the standard deviation for the abortion rate for all states?
   b. Using the information from (a), how many states fall more than 1 standard deviation above the mean? How does this number compare with the number expected from the theoretical normal curve distribution? Can you suggest anything these states have in common that might cause them to have higher abortion rates?
   c. How many states fall more than 1 standard deviation below the mean? Is this number greater or lower than the expected value from the theoretical normal curve? Again, can you suggest any characteristics these states have in common that might cause them to have lower abortion rates?
   d. Create a histogram of abortion rates for all 50 states and Washington, D.C. Does the distribution appear to be normal? Use this information to further explain why the number of states falling more than 1 standard deviation below the mean differs from the expected value.

Abortion Rate (per 1,000 women) in the United States by State: 2000

Alabama	14.3	Kentucky	5.3	North Dakota	9.9
Alaska	11.7	Louisiana	13.0	Ohio	16.5
Arizona	16.5	Maine	9.9	Oklahoma	10.1
Arkansas	9.8	Maryland	29.0	Oregon	23.5
California	31.2	Massachusetts	21.4	Pennsylvania	14.3
Colorado	15.9	Michigan	21.6	Rhode Island	24.1
Connecticut	21.1	Minnesota	13.5	South Carolina	9.3
Delaware	31.9	Mississippi	5.9	South Dakota	5.5
D.C.	68.1	Missouri	6.0	Tennessee	15.2
Florida	31.9	Montana	13.5	Texas	18.8
Georgia	16.9	Nebraska	11.6	Utah	6.6
Hawaii	22.1	Nevada	32.2	Vermont	12.7
Idaho	7.0	New Hampshire	11.2	Virginia	18.1
Illinois	23.2	New Jersey	36.3	Washington	20.3
Indiana	9.4	New Mexico	14.7	West Virginia	6.8
Iowa	9.8	New York	39.1	Wisconsin	9.6
Kansas	21.4	North Carolina	21.0	Wyoming	0.9

Source: U.S. Bureau of the Census, *Statistical Abstract of the United States*, 2003, Table 104.

7. Information on the occupational prestige scores for blacks and whites is given below.

	Mean	*Standard Deviation*	*N*
Blacks	36.81	13.81	383
Whites	44.75	13.75	2082

Source: GSS 2002.

a. What percentage of whites should have occupational prestige scores above 60? How many whites in the sample should have occupational prestige scores above 60?

b. What percentage of blacks should have occupational prestige scores above 60? How many blacks should have occupational prestige scores above 60?

c. What proportion of whites have prestige scores between 30 and 70? How many whites have prestige scores between 30 and 70?

d. Given that the black sample size is 383, how many blacks in the sample should have an occupational prestige score between 30 and 60?

8. SAT scores are normed so that, in any year, the mean of the verbal or math test should be 500 and the standard deviation 100. Assuming this is true (it is only approximately true, both because of variation from year to year and because scores have decreased since the SAT tests were first developed), answer the following questions.

a. What percentage of students score above 625 on the math SAT in any given year?

b. What percentage of students score between 400 and 600 on the verbal SAT?

c. A college decides to liberalize its admission policy. As a first step, the admissions committee decides to exclude only those applicants scoring below the 20th percentile on the verbal SAT. Translate this percentile into a Z score. Then calculate the equivalent SAT verbal test score.

9. The Chicago police department was asked by the mayor's office to estimate the cost of crime to citizens of Chicago. The police began their study with the crime of burglary, relying on a random sample of 500 files (there is too much crime to calculate statistics for all the crimes committed). They found the average dollar loss in a burglary was $678, with a standard deviation of $560, and that the dollar loss was normally distributed.
   a. What proportion of burglaries had dollar losses above $1,000?
   b. What is the probability that any one burglary had a dollar loss above $400?
   c. What proportion of burglaries had dollar losses below $500?

10. The number of hours people work each week varies widely for many reasons. Using the 2002 General Social Survey, you find that the mean number of hours worked last week was 34.9 with a standard deviation of 15.6 hours, based on a sample size of 2715.
   a. Assume that hours worked is approximately normally distributed in the sample. What is the probability that someone in the sample will work 60 or more hours in a week? How many people in the sample of 2715 should have worked 60 or more hours?
   b. What is the probability that someone will work 30 or fewer hours in a week (that is, work part-time)? How many people does this represent in the sample?
   c. What number of hours worked per week corresponds to the 60th percentile?

11. It is increasingly true that government agencies, on all levels, use examinations to screen applicants and remove bias from the hiring process. Consider a police department in a large midwestern city that uses such an examination to hire new officers. The mean score for all applicants this year on the exam is 98, with a standard deviation of 13. The distribution of scores for the applicants is approximately normal.
   a. Assume that only 12 percent of all applicants can be accepted this year. Will an applicant be accepted if his or her score on the exam is 115?
   b. What is the cutoff score of this year's test? In other words, what score is above 88 percent of all scores in the distribution?
   c. What is the Z value for this score?

12. A company tests applicants for a job by giving writing and software proficiency tests. The means and standard deviations for each exam follow, along with the scores for two applicants, Bill and Ted. Assume test scores are normally distributed.

Exam	Mean	Standard Deviation	Bill	Ted
Writing	56.4	9.3	65	67
Software use	68.7	5.6	70	75

   a. On which test did Bill do better, relative to the other applicants? Calculate appropriate statistics to answer this question.
   b. On which test did Ted do better, relative to the other applicants? Calculate statistics to answer this question.
   c. What proportion of applicants scored below Bill's software test score?
   d. What is the percentile rank of Ted's writing score?

13. What is the value of the mean for any standard normal distribution? What is the value of the standard deviation for any standard normal distribution? Explain why this is true for any standard normal distribution.

14. You are asked to do a study of shelters for abused and battered women to determine the necessary capacity in your city to provide housing for most of these women. After recording data for a whole year, you find that the mean number of women in shelters each night is 250, with a standard deviation of 75. Fortunately, the distribution of the number of women in the shelters each night is normal, so you can answer the following questions posed by the city council.
   a. If the city's shelters have a capacity of 350, will that be enough places for abused women on 95 percent of all nights? If not, what number of shelter openings will be needed?
   b. The current capacity is only 220 openings because some shelters have closed. What is the percentage of nights that the number of abused women seeking shelter will exceed current capacity?

15. Based on the chapter discussion:
   a. What are the properties of the normal distribution? Why is it called "normal"?
   b. What is the meaning of a positive (+) Z score? What is the meaning of a negative (−) Z score?

Chapter 10

Sampling and Sampling Distributions

U ntil now we have ignored the question of who or what should be observed when we collect data or whether the conclusions based on our observations can be generalized to a larger group of observations. The truth is that we are rarely able to study or observe everyone or everything we are interested in. Though we have learned about various methods to analyze observations, remember that these observations represent only a tiny fraction of all the possible observations we might have chosen. Consider the following examples.

Example 1 The student union on your campus is trying to find out how it can better address the needs of commuter students and has commissioned you to conduct a needs assessment survey. You have been given enough money to survey about 500 students. Given that there are nearly 15,000 commuters on your campus, is this an impossible task?

Example 2 Your chancellor has appointed a task force to investigate issues of concern to the lesbian, gay, and bisexual community at the university. The task force has been charged with assessing the campus climate for members of these university communities and studying the coverage of lesbian, gay, and bisexual subjects in the curriculum. There are about 30,000 students, faculty, and staff on your campus and about 2,000 courses offered every year. How should the task force proceed?

What do these problems have in common? In both situations the major problem is that there is too much information and not enough resources to collect and analyze all of it.

▣ AIMS OF SAMPLING[1]

Researchers in the social sciences almost never have enough time or money to collect information about the entire group that interests them. Known as the **population**, this group includes all the cases (individuals, objects, or groups) in which the researcher is interested. For example, in our first illustration the population is all 15,000 commuter students; the population in the second illustration consists of all 30,000 faculty, staff, and students and 2,000 courses.

Fortunately, we can learn a lot about a population if we carefully select a subset of it. This subset is called a **sample**. Through the process of *sampling*—selecting a subset of observations from the population of interest—we attempt to generalize to the characteristics of the larger group (population) based on what we learn from the smaller group (the sample). This is the basis of *inferential statistics*—making predictions or inferences about a population from observations based on a sample.

The term **parameter**, associated with the population, refers to measures used to describe the distribution of the population we are interested in. For instance, the average commuting time for *all* 15,000 students on your campus is a population parameter because it refers to a population characteristic. In previous chapters we have learned many ways of describing a distribution, such as a proportion, a mean, or a standard deviation. When used to describe the population distribution, these measures are referred to as parameters. Thus, a population mean, a population proportion, and a population standard deviation are all parameters.

Population   A group that includes all the cases (individuals, objects, or groups) in which the researcher is interested.

Sample   A relatively small subset selected from a population.

---

[1]This discussion has benefited from a more extensive presentation on the aims of sampling in Richard Maisel and Caroline Hodges Persell, *How Sampling Works* (Thousand Oaks, CA: Pine Forge Press, 1996).

Table 10.1   Sample and Population Notations

Measure	Sample Notation	Population Notation
Mean	$\bar{Y}$	μ_Y
Proportion	p	π
Standard deviation	S_Y	σ_Y
Variance	S_Y^2	σ_Y^2

We use the term **statistic** when referring to a corresponding characteristic calculated for the sample. For example, the average commuting time for a *sample* of commuter students is a sample statistic. Similarly, a sample mean, a sample proportion, and a sample standard deviation are all statistics.

In this and the following chapter we discuss some of the principles involved in generalizing results from samples to the population. In our discussion we will use different notations when referring to sample statistics and population parameters. Table 10.1 presents the sample notation and the corresponding population notation.

The distinctions between a sample and a population and between a parameter and a statistic are illustrated in Figure 10.1. We've included for illustration the population parameter of .60—the proportion of white respondents in the population. However, since we almost never have enough resources to collect information about the population, it is rare that we know the value of a parameter. The goal of most research is to find the population parameter. Researchers usually select a sample from the population to obtain an estimate of the population parameter. Thus, the major objective of sampling theory and statistical inference is to provide estimates of unknown parameters from sample statistics that can be easily obtained and calculated.

Parameter   A measure (for example, mean or standard deviation) used to describe the population distribution.

Statistic   A measure (for example, mean or standard deviation) used to describe the sample distribution.

✔ *Learning Check.*   *It is important that you understand what the terms population, sample, parameter, and statistic mean. Use your own words so the meaning makes sense to you. If you cannot clearly define these terms, review the preceding material. You will see these sample and population notations over and over again. If you memorize them, you will find it much easier to understand the formulas used in inferential statistics.*

Figure 10.1   The Proportion of White Respondents in a Population and in a Sample

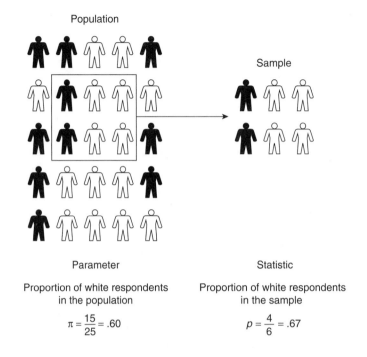

Parameter

Proportion of white respondents
in the population

$$\pi = \frac{15}{25} = .60$$

Statistic

Proportion of white respondents
in the sample

$$p = \frac{4}{6} = .67$$

◙ SOME BASIC PRINCIPLES OF PROBABILITY

We all use the concept of probability in everyday conversation. We might ask, "What is the probability that it will rain tomorrow?" or "What is the likelihood that we will do well on a test?" In everyday conversations our answers to these questions are rarely systematic, but in the study of statistics *probability* has a far more precise meaning.

In the following sections we will discuss a variety of techniques adopted by social scientists to select samples from populations. The techniques all follow a general approach called *probability sampling*. Before we discuss these techniques, we will review some theories and principles of probability.

Generally speaking, probability can be understood as taking one of three theoretical forms: classical theory, relative frequency theory, or subjectivist theory.[2] Classical probability theory applies to situations in which we attempt to specify the number of times a desired outcome will occur relative to the set of all possible and equally likely outcomes. Consider, for example, the outcome of rolling a 3 with a six-sided, equally weighted, die. The probability of rolling a 3 is 1 ÷ 6 because this outcome can occur only once out of a total of six possible outcomes: 1, 2, 3, 4, 5, 6.

---

[2]This discussion is based on C. Stephen Layman's, *The Power of Logic* (Mountain View, CA: Mayfield Publishing, 1999).

Relative frequency theory applies to situations when we have a prior "track record" to work with. For example, a community group may want to know the probability that a convicted sex offender will re-offend if released from prison. Since we know that this individual has offended in the past, we should consider these instances when stating the probability of re-offense. In other words, our knowledge of prior events informs the probabilities that we specify.

Subjectivist probability theory applies to situations in which the possible outcomes are specified on the basis of a belief or opinion. Consider, for example, the U.S.-led war in Iraq that began in 2003. Because of the multitude of factors involved, the specified probability of winning the war in Iraq depends on the individual or group that specifies it. For example, President George W. Bush and his administration may offer very optimistic odds, whereas members of the opposing political party may not.

As we will discuss, all three forms of probability theory are "correct" insofar as each informs our sampling techniques.[3] What is important is that we are consistent in our method of assigning probabilities to specific outcomes, as well as with the selection of samples from the population.

Probabilities are usually measured in terms of proportions that range in value from 0 to 1. The closer the proportion is to 1, the more likely it is that the desired outcome will occur. A proportion that is close to 0 means that the desired outcome is highly unlikely. Let's consider, for example, a sample of 1,705 Russian respondents from the 2000 International Social Survey Programme. Respondents were asked their opinions regarding what they saw as Russia's highest national priority; their responses are summarized in Table 10.2.

The ratio of respondents with the opinion that Russia's highest national priority is to protect freedom of speech is approximately 94:1705, or when reduced, 1:18. To convert a ratio to a proportion, we divide the numerator (94) by the denominator (1705), as shown in Table 10.2. Thus, the probability 94:1705 is equivalent to 0.06 (94 ÷ 1705 = 0.06). Now, imagine that we wrote down each of the 1,705 respondents' names and placed them in a hat. Because the proportion of 0.056 is closer to 0 than it is to 1, it is unlikely that we would select a respondent with the opinion that Russia's highest national priority is to protect freedom of speech.

Table 10.2   Views on Russia's Highest National Priority

	Frequency	*Proportion*
Establish order in nation	244	0.14
Give citizens more say	393	0.23
Fight rising prices	889	0.52
Protect freedom of speech	94	0.06
Undecided/can't choose	85	0.05
Total	1705	1.00

---

[3]This was suggested by David C. Howell in *Statistical Methods for Psychology* (Duxbury: Wadsworth Publishing, 2002). Howell also notes that those who adopt a subjectivist view of probability theory tend to disagree with a hypothesis testing orientation in general, including the use of random sampling techniques.

✓ *Learning Check.*   *What is the probability of drawing an ace out of a normal deck of 52 playing cards? It's not 1 ÷ 52. There are four aces, so the probability is 4 ÷ 52 or 1 ÷ 13. The proportion is 0.08. The probability of drawing the ace of spades is 1 ÷ 52.*

▣ PROBABILITY SAMPLING

Social researchers are usually more systematic in their effort to obtain samples that are representative of the population than we are when we gather information in our everyday life. Such researchers have adopted a number of approaches for selecting samples from popula-tions. Only one general approach, *probability sampling*, allows the researcher to use the prin-ciples of statistical inference to generalize from the sample to the population.

Probability sampling is a method that enables the researcher to specify for each case in the population the probability of its inclusion in the sample. The purpose of probability sampling is to select a sample that is as representative as possible of the population. The sample is selected in such a way as to allow use of the principles of probability to evaluate the generalizations made from the sample to the population. A probability sample design enables the researcher to estimate the extent to which the findings based on one sample are likely to differ from what would be found by studying the entire population.

Probability sampling   A method of sampling that enables the researcher to spec-ify for each case in the population the probability of its inclusion in the sample.

Although accurate estimates of sampling error can be made only from probability samples, social scientists often use nonprobability samples because they are more convenient and cheaper to collect. Nonprobability samples are useful under many circumstances for a variety of research purposes. Their main limitation is that they do not allow the use of the method of inferential statistics to generalize from the sample to the population. Because in this and the next chapter we deal only with inferential statistics, we do not discuss nonprobability sam-pling. In the following sections we will learn about three sampling designs that follow the principles of probability sampling: the simple random sample, the systematic random sample, and the stratified random sample.[4]

The Simple Random Sample

The *simple random sample* is the most basic probability sampling design, and it is incor-porated into even more elaborate probability sampling designs. A **simple random sample** is

---

[4]The discussion in these sections is based on Chava Frankfort-Nachmias and David Nachmias, *Research Methods in the Social Sciences* (New York: Worth Publishers, 2000), pp. 167–177.

a sample design chosen in such a way as to ensure that (1) every member of the population has an equal chance of being chosen and (2) every combination of N members has an equal chance of being chosen.

Let's take a very simple example to illustrate. Suppose we are conducting a cost-containment study of the 10 hospitals in our region, and we want to draw a sample of two hospitals to study intensively. We can put into a hat 10 slips of paper, each representing one of the 10 hospitals and mix the slips carefully. We select one slip out of the hat and identify the hospital it represents. We then make the second draw and select another slip out of the hat and identify it. The two hospitals we identified on the two draws become the two members of our sample: (1) Assuming we made sure the slips were really well mixed, pure chance determined which hospital was selected on each draw. The sample is a simple random sample because every hospital had the same chance of being selected as a member of our sample of two, and (2) every combination of ($N = 2$) hospitals was equally likely to be chosen.

Simple random sample   A sample designed in such a way as to ensure that (1) every member of the population has an equal chance of being chosen and (2) every combination of N members has an equal chance of being chosen.

Researchers usually use computer programs or tables of random numbers in selecting random samples. An abridged table of random numbers is reproduced in Appendix A. To use a random number table, list each member of the population and assign the member a number. Begin anywhere on the table and read each digit that appears in the table in order—up, down, or sideways; the direction does not matter, as long as it follows a consistent path. Whenever we come across a digit in the table of random digits that corresponds to the number of a member in the population of interest, that member is selected for the sample. Continue this process until the desired sample size is reached.

Suppose now that, in your job as a hospital administrator, you are planning to conduct a cost-containment study by examining patients' records. Out of a total of 300 patients' records, you want to draw a simple random sample of five. You follow these steps:

1. Number the patient accounts, beginning with 001 for the first account and ending with 300, which represents the 300th account.

2. Use some random process to enter Appendix A (you might close your eyes and point a pencil). For our illustration, let's start with the first column of numbers. Notice that each column lists five-digit numbers. Because your population contains only three-digit numbers (001–300), drop the last two digits of each number and read only the first three digits in each group of numbers. (Alternatively, you could choose any other group of three-digit numbers in this block—for example, the last three digits in the block.)

3. Dropping the last two digits of each five-digit block and proceeding down the column, you obtain the following three-digit numbers:

104*	375	963	289*
223*	779	895	635
241*	995	854	094*
421			

Among these numbers, five correspond to numbers within the range of numbers assigned to the patient records. They are marked with an asterisk. The last number listed is 094 from line 13. You do not need to list more numbers because you already have five different numbers that qualify for inclusion in the sample. The asterisked numbers represent the records you will choose for your sample because these are the only ones that fall between 001 and 300, the range you specified.

4. We now have five records in our simple random sample. Let's list them: 104, 223, 241, 289, and 094.

The Systematic Random Sample

Now let's look at a sampling method that is easier to implement than a simple random sample. The *systematic random sample*, although not a true probability sample, provides results very similar to those obtained with a simple random sample. It uses a ratio, K, obtained by dividing the population size by the desired sample size:

$$K = \frac{\text{population size}}{\text{sample size}}$$

Systematic random sampling   is a method of sampling in which every Kth member in the total population is chosen for inclusion in the sample after the first member of the sample is selected at random from among the first K members in the population.

Recall our example in which we had a population of 15,000 commuting students and our sample was limited to 500. In this example,

$$K = \frac{15,000}{500} = 30$$

Using a systematic random sampling method, we first choose any one student at random from among the first 30 students on the list of commuting students. Then we select every 30th student after that until we reach 500, our desired sample size. Suppose that our first student selected at random happens to be the eighth student on the list. The second student in our sample is then 38th on the list ($8 + 30 = 38$). The third would be $38 + 30 = 68$, the fourth, $68 + 30 = 98$, and so on. The systematic random sample is illustrated in Figure 10.2.

Figure 10.2   Systematic Random Sampling

From a population of 40 students, let's select a systematic random sample of 8 students. Our skip interval will be 5 (40 ÷ 8 = 5). Using a random number table, we choose a number between 1 and 5. Let's say we choose 4. We then start with student 4 and pick every 5th student:

Our trip to the random number table could have just as easily given us a 1 or a 5, so all the students do have a chance to end up in our sample.

Systematic random sampling   A method of sampling in which every *K*th member (*K* is a ratio obtained by dividing the population size by the desired sample size) in the total population is chosen for inclusion in the sample after the first member of the sample is selected at random from among the first *K* members in the population.

✔ *Learning Check.*   *How does a systematic random sample differ from a simple random sample?*

The Stratified Random Sample

A third type of probability sampling is the *stratified random sample*. We obtain a **stratified random sample** by (1) dividing the population into subgroups based on one or more variables central to our analysis and then (2) drawing a simple random sample from each of

A Closer Look 10.1
Disproportionate Stratified Samples and Diversity

Disproportionate stratified sampling is especially useful given the increasing diversity of Justify both ends American society. In a diverse society factors such as race, ethnicity, class, and gender, as well as other categories of experience such as age, religion, and sexual orientation, become central in shaping our experiences and defining the differences among us. These factors are an important dimension of the social structure, and they not only operate independently but also are experienced simultaneously by all of us.* For example, if you are a white woman, you may share some common experiences with a woman of color based on your gender, but your racial experiences are going to be different. Moreover, your experiences within the race/gender system are further conditioned by your social class. Similarly, if you are a man, your experiences are shaped as much by your class, race, and sexual orientation as they are by your gender. If you are a black gay man, for instance, you might not benefit equally from patriarchy compared with a classmate who is a white heterosexual male.

What are the research implications of an inclusive approach that emphasizes social differences? Such an approach will include women and men in a study of race, Latinos and people of color when considering class, and women and men of color when studying gender. Such an approach makes the experience of previously excluded groups more visible and central because it puts those who have been excluded at the center of the analysis so that we can better understand the experience of all groups, including those with privilege and power.

What are the sampling implications of such an approach? Let's think of an example. Suppose you are looking at the labor force experiences of black and Latina women who are over 50 years of age, and you want to compare these experiences with those of white women in the same age group. Both Latina and black women comprise a small proportion of the population. A proportional sample probably would not include enough Latina or black women to provide an adequate basis for comparison with white women. To make such comparisons, it would be desirable to draw a disproportionate stratified sample that deliberately overrepresents both Latina and black women so that these subsamples will be of sufficient size (Figure 10.3).

*Margaret L. Andersen and Patricia Hill Collins, *Race, Class, and Gender* (Belmont, CA: Wadsworth, 2003).

the subgroups. We could stratify by race/ethnicity, for example, by dividing the population into different racial/ethnic groups and then drawing a simple random sample from each group. For instance, suppose we want to compare the attitudes of Latinos toward abortion with the attitudes of white and black respondents. Our population of interest consists of 1,000 individuals, with 700 (or 70%) whites, 200 (or 20%) blacks, and 100 (10%) Latinos. In such a **proportionate stratified sample**, the size of the sample selected from each subgroup is proportional to the size of that subgroup in the entire population.

In a **disproportionate stratified sample**, the size of the sample selected from each subgroup is deliberately made disproportional to the size of that subgroup in the population.

Figure 10.3   A Random Sample Stratified by Race/Ethnicity

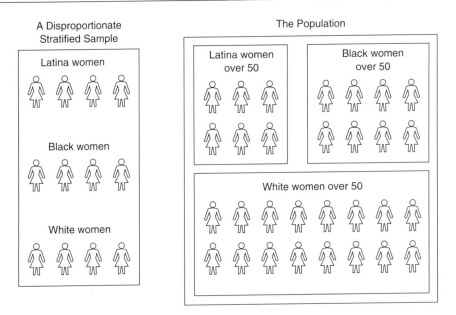

For instance, for our example we could select a sample ($N = 180$) consisting of 90 whites (50%), 45 blacks (25%), and 45 Latinas (25%). In such a sampling design, although the sampling probabilities for each population member are not equal (they vary between groups), they are *known*, and therefore we can make accurate estimates of error in the inference process.[5] Disproportionate stratified sampling is especially useful when we want to compare subgroups with each other, and when the size of some of the subgroups in the population is relatively small. Proportionate sampling can result in the sample having too few members from a small subgroup to yield reliable information about them.

Stratified random sample   A method of sampling obtained by (1) dividing the population into subgroups based on one or more variables central to our analysis and (2) then drawing a simple random sample from each of the subgroups.

Proportionate stratified sample   The size of the sample selected from each subgroup is proportional to the size of that subgroup in the entire population.

Disproportionate stratified sample   The size of the sample selected from each subgroup is disproportional to the size of that subgroup in the population.

---

[5]We discuss more on sampling error in the next section.

✓ *Learning Check.*   *Can you think of some research questions that could best be studied using a disproportionate stratified random sample? When might it be important to use a proportionate stratified random sample?*

▣ THE CONCEPT OF THE SAMPLING DISTRIBUTION

We began this chapter with a few examples illustrating why researchers in the social sciences almost never collect information on the entire population that interests them. Instead, they usually select a sample from that population and use the principles of statistical inference to estimate the characteristics, or parameters, of that population based on the characteristics, or statistics, of the sample. In this section we describe one of the most important concepts in statistical inference—*sampling distribution.* The sampling distribution helps estimate the likelihood of our sample statistics and, therefore, enables us to generalize from the sample to the population.

The Population

To illustrate the concept of the sampling distribution, let's consider as our population the 20 individuals listed in Table 10.3.[6] Our variable, Y, is the income (in dollars) of these 20 individuals, and the parameter we are trying to estimate is the mean income.

We use the symbol μ_Y to represent the population mean; the Greek letter mu (μ) stands for the mean, and the subscript Y identifies the specific variable, income. Using Formula 4.1, we can calculate the population mean:

$$\mu_Y = \frac{\sum Y}{N} = \frac{Y_1 + Y_2 + Y_3 + Y_4 + Y_5 + \cdots + Y_{20}}{20}$$

$$= \frac{11,350 + 7,859 + 41,654 + 13,445 + 17,458 + \cdots + 25,671}{20}$$

$$= 22,766$$

---

[6]The population of the 20 individuals presented in Table 10.3 is considered a finite population. A finite population consists of a finite (countable) number of elements (observations). Other examples of finite populations include all women in the labor force in 2005 and all public hospitals in New York City. A population is considered infinite when there is no limit to the number of elements it can include. Examples of infinite populations include all women in the labor force, in the past or future. Most samples studied by social scientists come from finite populations. However, it is also possible to sample from an infinite population.

Table 10.3   The Population: Personal Income for 20 Individuals
(Hypothetical Data)

Individzual	*Income (Y)*
Case 1	11,350 (Y_1)
Case 2	7,859 (Y_2)
Case 3	41,654 (Y_3)
Case 4	13,445 (Y_4)
Case 5	17,458 (Y_5)
Case 6	8,451 (Y_6)
Case 7	15,436 (Y_7)
Case 8	18,342 (Y_8)
Case 9	19,354 (Y_9)
Case 10	22,545 (Y_{10})
Case 11	25,345 (Y_{11})
Case 12	68,100 (Y_{12})
Case 13	9,368 (Y_{13})
Case 14	47,567 (Y_{14})
Case 15	18,923 (Y_{15})
Case 16	16,456 (Y_{16})
Case 17	27,654 (Y_{17})
Case 18	16,452 (Y_{18})
Case 19	23,890 (Y_{19})
Case 20	25,671 (Y_{20})
Mean (μ_Y) = 22,766	Standard deviation (σ_Y) = 14,687

Using Formula 5.3, we can also calculate the standard deviation for this population distribution. We use the Greek symbol sigma (σ) to represent the population's standard deviation and the subscript Y to stand for our variable, income:

$$\sigma_Y = 14,687$$

Of course, most of the time we do not have access to the population. So instead we draw one sample, compute the mean—the statistic—for that sample, and use it to estimate the population mean—the parameter.

The Sample

Let's pretend that μ_Y is unknown and that we estimate its value by drawing a random sample of three individuals ($N = 3$) from the population of 20 individuals and calculate the mean income for that sample. The incomes included in that sample are as follows:

Case 8	18,342
Case 16	16,456
Case 17	27,654

Now let's calculate the mean for that sample:

$$\bar{Y} = \frac{18{,}342 + 16{,}456 + 27{,}654}{3} = 20{,}817$$

Notice that our sample mean ($\bar{Y}$), \$20,817, differs from the actual population parameter, \$22,766. This discrepancy is due to sampling error. **Sampling error** is the discrepancy between a sample estimate of a population parameter and the real population parameter. By comparing the sample statistic with the population parameter, we can determine the sampling error. The sampling error for our example is 1,949 (22,766 − 20,817 = 1,949).

Now let's select another random sample of three individuals. This time the incomes included are

Case 15	18,923
Case 5	17,458
Case 17	27,654

The mean for this sample is

$$\bar{Y} = \frac{18{,}923 + 17{,}458 + 27{,}654}{3} = 21{,}345$$

The sampling error for this sample is 1,421 (22,766 − 21,345 = 1,421), somewhat less than the error for the first sample we selected.

The Dilemma

Although comparing the sample estimates of the average income with the actual population average is a perfect way to evaluate the accuracy of our estimate, in practice we rarely have information about the actual population parameter. If we did, we would not need to conduct a study! Moreover, few, if any, sample estimates correspond exactly to the actual population parameter. This, then, is our dilemma: If sample estimates vary and if most estimates result in some sort of sampling error, how much confidence can we place in the estimate? On what basis can we infer from the sample to the population?

Sampling error   The discrepancy between a sample estimate of a population parameter and the real population parameter.

The Sampling Distribution

The answer to this dilemma is to use a device known as the sampling distribution. The **sampling distribution** is a theoretical probability distribution of all possible sample values for the statistic in which we are interested. If we were to draw all possible random samples of the same size from our population of interest, compute the statistic for each sample, and plot the frequency distribution for that statistic, we would obtain an approximation of the sampling distribution. Every statistic—for example, a proportion, a mean, or a variance—has a sampling distribution. Because it includes all possible sample values, the sampling distribution enables us to compare our sample result with other sample values and determine the likelihood associated with that result.[7]

▣ THE SAMPLING DISTRIBUTION OF THE MEAN

Sampling distributions are theoretical distributions, which means that they are never really observed. Constructing an actual sampling distribution would involve taking all possible random samples of a fixed size from the population. This process would be very tedious because it would involve a very large number of samples. However, to help grasp the concept of the sampling distribution, let's illustrate how one could be generated from a limited number of samples.

An Illustration

For our illustration, we use one of the most common sampling distributions—the sampling distribution of the mean. The **sampling distribution of the mean** is a theoretical distribution of sample means that would be obtained by drawing from the population all possible samples of the same size.

Let's go back to our example in which our population is made up of 20 individuals and their incomes. From that population (Table 10.3) we now randomly draw 50 possible samples of size 3, computing the mean income for each sample and replacing it before drawing another.

In our first sample of size 3 we draw three incomes: $8,451, $41,654, and $18,923. The mean income for this sample is

$$\overline{Y} = \frac{8,451 + 41,654 + 18,923}{3} = 23,009$$

Now we restore these individuals to the original list and select a second sample of three individuals. The mean income for this sample is

$$\overline{Y} = \frac{15,436 + 25,345 + 16,456}{3} = 19,079$$

---

[7]Here we are using an idealized example in which the sampling distribution is actually computed. However, please bear in mind that in practice one never computes a sampling distribution because it is also infinite.

Table 10.4   Mean Income of 50 Samples of Size 3

Sample	Mean ($\bar{Y}$)
First	23,009
Second	19,079
Third	18,873
Fourth	26,885
Fifth	21,847
.	.
.	.
.	.
Fiftieth	26,645
Total (M) = 50	$\sum \bar{Y}$ = 1,237,482

We repeat this process 48 more times, each time computing the sample mean and restoring the sample to the original list. Table 10.4 lists the means of the first five and the 50th samples of $N = 3$ that were drawn from the population of 20 individuals. (Note that $\sum \bar{Y}$ refers to the sum of all the means computed for each of the samples and M refers to the total number of samples that were drawn.)

The grouped frequency distribution for all 50 sample means ($M = 50$) is displayed in Table 10.5; Figure 10.4 is a histogram of this distribution. This distribution is an example of a sampling distribution of the mean. Notice that in its structure the sampling distribution resembles a frequency distribution of raw scores, except that here each score is a sample mean and the corresponding frequencies are the number of samples with that particular mean value. For example, the third interval in Table 10.5 ranges from $19,500 to $23,500, with a corresponding frequency of 14, or 28 percent. This means that we drew 14 samples (28%) with means ranging between $19,500 and $23,500.

Remember the distribution depicted in Table 10.5 and Figure 10.4 is an empirical distribution, whereas the sampling distribution is a theoretical distribution. In reality, we never really construct the sampling distribution. However, even this simple empirical example serves to illustrate some of the most important characteristics of the sampling distribution.

Sampling distribution of the mean   A theoretical probability distribution of sample means that would be obtained by drawing from the population all possible samples of the same size.

Review

Before we continue, let's take a moment to review the three distinct types of distribution.

The Population   We began with the *population distribution* of 20 individuals. This distribution actually exists. It is an empirical distribution that is usually unknown to us. We are interested in estimating the mean income for this population.

The Sample   We drew a sample from that population. The *sample distribution* is an empirical distribution that is known to us and is used to help us estimate the mean of the

Table 10.5 Sampling Distribution of Sample Means for Sample Size
$N = 3$ Drawn from the Population of 20 Individuals'
Incomes

Sample Mean Intervals	Frequency	Percentage (%)
11,500–15,500	6	12
15,500–19,500	7	14
19,500–23,500	14	28
23,500–27,500	4	8
27,500–31,500	9	18
31,500–35,500	7	14
35,500–39,500	1	2
39,500–43,500	2	4
Total (*M*)	50	100

Figure 10.4 Sampling Distribution of Sample Means for Sample Size $N = 3$ Drawn from
the Population of 20 Individuals' Incomes

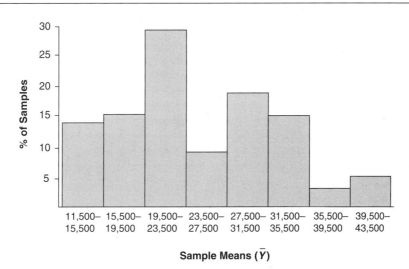

Sample Means ($\bar{Y}$)

population. We selected 50 samples of $N = 3$ and calculated the mean income. We usually use
the sample mean ($\bar{Y}$) as an estimate of the population mean (μ_y).

The Sampling Distribution of the Mean For illustration, we generated an approximation
of the sampling distribution of the mean, consisting of 50 samples of $N = 3$. *The sampling distribution of the mean* does not really exist. It is a theoretical distribution.

To help you understand the relationship among the population, the sample, and the sampling distribution, we have illustrated in Figure 10.5 the process of generating an empirical
sampling distribution of the mean. From a population of raw scores (*Y*s), we draw *M* samples
of size *N* and calculate the mean of each sample. The resulting sampling distribution of the
mean, based on *M* samples of size *N*, shows the values that the mean could take and the

Figure 10.5   Generating the Sampling Distribution of the Mean

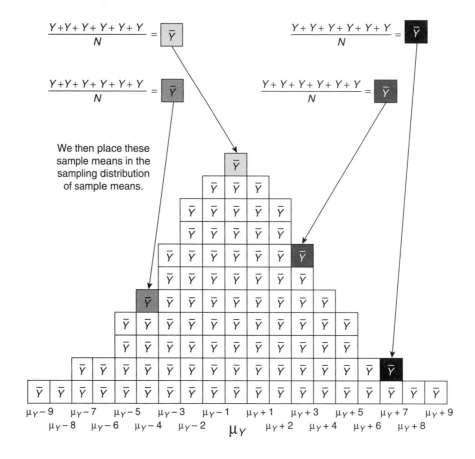

From a population (with a population mean of μ_Y) we start drawing samples and calculating the means for those samples:

$$\frac{Y+Y+Y+Y+Y+Y}{N} = \bar{Y}$$

$$\frac{Y+Y+Y+Y+Y+Y}{N} = \bar{Y}$$

$$\frac{Y+Y+Y+Y+Y+Y}{N} = \bar{Y}$$

$$\frac{Y+Y+Y+Y+Y+Y}{N} = \bar{Y}$$

We then place these sample means in the sampling distribution of sample means.

frequency (number of samples) associated with each value. Make sure you understand these relationships. The concept of the sampling distribution is crucial to understanding statistical inference. In this and the next chapter, we learn how to employ the sampling distribution to draw inferences about the population on the basis of sample statistics.

The Mean of the Sampling Distribution

Like the sample and population distributions, the sampling distribution can be described in terms of its mean and standard deviation. We use the symbol $\mu_{\bar{Y}}$ to represent the mean of the sampling distribution. The subscript $\bar{Y}$ indicates that the variable of this distribution is the mean. To obtain the mean of the sampling distribution, add all the individual sample means

($\sum \overline{Y}$ = 1,237,482) and divide by the number of samples (M = 50). Thus, the mean of the sampling distribution of the mean is actually the mean of means:

$$\mu_{\overline{Y}} = \frac{\sum \overline{Y}}{M} = \frac{1,237,482}{50} = 24,750$$

The Standard Error of the Mean

The standard deviation of the sampling distribution is also called the **standard error of the mean**. The standard error of the mean, $\sigma_{\overline{Y}}$, describes how much dispersion there is in the sampling distribution, or how much variability there is in the value of the mean from sample to sample:

$$\sigma_{\overline{Y}} = \frac{\sigma_Y}{\sqrt{N}}$$

This formula tells us that the standard error of the mean is equal to the standard deviation of the population (σ_Y) divided by the square root of the sample size (N). For our example, because the population standard deviation is 14,687 and our sample size is 3, the standard error of the mean is

$$\sigma_{\overline{Y}} = \frac{14,687}{\sqrt{3}} = 8,480$$

Standard error of the mean The standard deviation of the sampling distribution of the mean. It describes how much dispersion there is in the sampling distribution of the mean.

◎ THE CENTRAL LIMIT THEOREM

In Figures 10.6a and 10.6b, we compare the histograms for the population and sampling distributions of Tables 10.2 and 10.4. Figure 10.6a shows the population distribution of 20 incomes, with a mean μ_Y = 22,766 and a standard deviation σ_Y = 14,687. Figure 10.6b shows the sampling distribution of the means from 50 samples of N = 3 with a mean $\mu_{\overline{Y}}$ = 24,749 and a standard deviation (the standard error of the mean) $\sigma_{\overline{Y}}$ = 8,479. These two figures illustrate some of the basic properties of sampling distributions in general and the sampling distribution of the mean in particular.

First, as can be seen from Figures 10.6a and 10.6b, the shapes of the two distributions differ considerably. Whereas the population distribution is skewed to the right, the sampling distribution of the mean is less skewed—that is, closer to symmetry and a normal distribution.

Second, whereas only a few of the sample means coincide exactly with the population mean, $22,766, the sampling distribution centers around this value. The mean of the sampling distribution is a pretty good approximation of the population mean.

◼ A Closer Look 10.2
Population, Sample, and Sampling Distribution Symbols

In the discussions that follow we make frequent references to the mean and standard deviation of the three distributions. To distinguish among the different distributions, we use certain conventional symbols to refer to the means and standard deviations of the sample, the population, and the sampling distribution. Notice that we use Greek letters to refer to both the sampling and the population distributions.

	Mean	Standard Deviation
Sample distribution	$\bar{Y}$	S_Y
Population distribution	μ_Y	σ_Y
Sampling distribution of $\bar{Y}$	$\mu_{\bar{Y}}$	$\sigma_{\bar{Y}}$

Third, the variability of the sampling distribution is considerably smaller than the variability of the population distribution. Notice that the standard deviation for the sampling distribution ($\sigma_{\bar{Y}} = 8,479$) is almost half that for the population ($\sigma_Y = 14,687$).

These properties of the sampling distribution are even more striking as the sample size increases. To illustrate the effect of a larger sample on the shape and properties of the sampling distribution, we went back to our population of 20 individual incomes and drew 50 additional samples of $N = 6$. We calculated the mean for each sample and constructed another sampling distribution. This sampling distribution is shown in Figure 10.6c. It has a mean $\mu_{\bar{Y}} = 24,064$, and a standard deviation $\sigma_{\bar{Y}} = 5,995$. Notice that as the sample size increased, the sampling distribution became more compact. This decrease in the variability of the sampling distribution is reflected in a smaller standard deviation: with an increase in sample size from $N = 3$ to $N = 6$ the standard deviation of the sampling distribution decreased from 8,479 to 5,995. Furthermore, with a larger sample size the sampling distribution of the mean is an even better approximation of the normal curve.

These properties of the sampling distribution of the mean are summarized more systematically in one of the most important statistical principles underlying statistical inference. It is called the **Central Limit Theorem**, and it states: If all possible random samples of size N are drawn from a population with a mean μ_Y and a standard deviation σ_Y, then as N becomes larger, the sampling distribution of sample means becomes approximately normal, with mean $\mu_{\bar{Y}}$ and standard deviation

$$\sigma_{\bar{Y}} = \frac{\sigma_Y}{\sqrt{N}}.$$

The significance of the Central Limit Theorem is that it tells us that with *sufficient sample size* the sampling distribution of the mean will be normal regardless of the shape of the population distribution. Therefore, even when the population distribution is skewed, we can still assume that

Figure 10.6   Three Income Distributions

a. Population distribution of personal income for 10 individuals (hypothetical data)

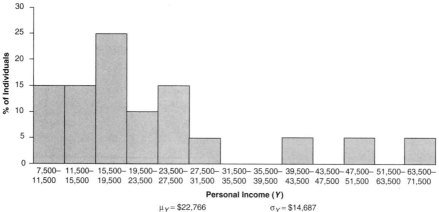

$\mu_Y = \$22,766$ $\sigma_Y = \$14,687$

b. Sampling distribution of sample means for sample size $N = 3$ drawn from the population of 20 individuals' incomes

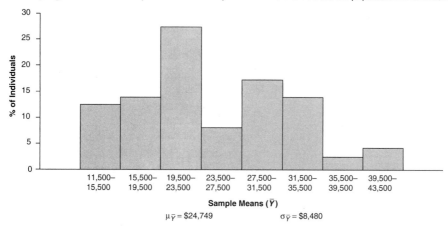

$\mu_{\bar{Y}} = \$24,749$ $\sigma_{\bar{Y}} = \$8,480$

c. Sampling distribution of sample means for sample size $N = 6$ drawn from the population of 20 individuals' incomes

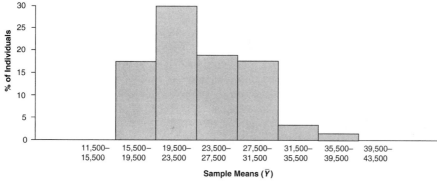

$\mu_{\bar{Y}} = \$24,064$ $\sigma_{\bar{Y}} = \$5,995$

the sampling distribution of the mean is normal, given random samples of large enough size. Furthermore, the Central Limit Theorem also assures us that (1) as the sample size gets larger, the mean of the sampling distribution becomes equal to the population mean; and (2) as the sample size gets larger, the standard error of the mean (the standard deviation of the sampling distribution of the mean) decreases in size. The standard error of the mean tells how much variability in the sample estimates there is from sample to sample. The smaller the standard error of the mean, the closer (on average) the sample means will be to the population mean. Thus, the larger the sample, the more closely the sample statistic clusters around the population parameter.

Central Limit Theorem   If all possible random samples of size N are drawn from a population with a mean μ_Y and a standard deviation σ_Y, then as N becomes larger, the sampling distribution of sample means becomes approximately normal, with mean $\mu_{\bar{Y}}$ and standard deviation

$$\sigma_{\bar{Y}} = \frac{\sigma_Y}{\sqrt{N}}.$$

✓ *Learning Check.* *Make sure you understand the difference between the number of samples that can be drawn from a population and the sample size. Whereas the number of samples is infinite in theory, the sample size is under the control of the investigator.*

The Size of the Sample

Although there is no hard-and-fast rule, a general rule of thumb is that when N is 50 or more, the sampling distribution of the mean will be approximately normal regardless of the shape of the distribution. However, we can assume that the sampling distribution will be normal even with samples as small as 30 if we know that the population distribution approximates normality.

✓ *Learning Check.* *What is a normal population distribution? If you can't answer this question, go back to Chapter 9. You must understand the concept of a normal distribution before you can understand the techniques involved in inferential statistics.*

The Significance of the Sampling Distribution and the Central Limit Theorem

In the preceding sections we have covered a lot of abstract material. You may have a number of questions at this time. Why is the concept of the sampling distribution so important? What is the significance of the Central Limit Theorem? To answer these questions, let's go back and review our 20 incomes example.

In order to estimate the mean income of a population of 20 individuals, we drew a sample of three cases and calculated the mean income for that sample. Our sample mean, $\overline{Y} = 20{,}817$, differs from the actual population parameter, $\mu_Y = 22{,}766$. When we selected different samples, we found each time that the sample mean differed from the population mean. These discrepancies are due to sampling errors. Had we taken a number of additional samples, we probably would have found that the mean was different each time because every sample differs slightly. Few, if any, sample means would correspond exactly to the actual population mean. Usually we have only one sample statistic as our best estimate of the population parameter.

So now let's restate our dilemma: If sample estimates vary and if most result in some sort of sampling error, how much confidence can we place in the estimate? On what basis can we infer from the sample to the population?

The solution lies in the sampling distribution and its properties. Because the sampling distribution is a theoretical distribution that includes all possible sample outcomes, we can compare our sample outcome with it and estimate the likelihood of its occurrence.

Since the sampling distribution is theoretical, how can we know its shape and properties so that we can make these comparisons? Our knowledge is based on what the Central Limit Theorem tells us about the properties of the sampling distribution of the mean. We know that if our sample size is large enough (at least 50 cases), most sample means will be quite close to the true population mean. It is highly unlikely that our sample mean would deviate much from the actual population mean.

In Chapter 9 we saw that in all normal curves, a constant proportion of the area under the curve lies between the mean and any given distance from the mean when measured in standard deviation units, or Z scores. We can find this proportion in the standard normal table (Appendix B).

Knowing that the sampling distribution of the means is approximately normal, with a mean $\mu_{\overline{Y}}$ and a standard deviation $\sigma_Y/\sqrt{N}$ (the standard error of the mean), we can use Appendix B to determine the probability that a sample mean will fall within a certain distance—measured in standard deviation units, or Z scores—of $\mu_{\overline{Y}}$ or μ_Y. For example, we can expect approximately 68 percent (or we can say the probability is approximately 0.68) of all sample means to fall within ± 1 standard error ($\sigma_{\overline{Y}} = \frac{\sigma_Y}{\sqrt{N}}$, or the standard deviation of the sampling distribution of the mean) of $\mu_{\overline{Y}}$ or μ_Y. Similarly, the probability is about 0.95 that the sample mean will fall within ± 2 standard errors of $\mu_{\overline{Y}}$ or μ_Y. In the next chapter we will see how this information helps us evaluate the accuracy of our sample estimates.

✔ *Learning Check.*   *Suppose a population distribution has a mean $\mu_Y = 150$ and a standard deviation $\sigma_Y = 30$ and you draw a simple random sample of $N = 100$ cases. What is the probability that the mean is between 147 and 153? What is the probability that the sample mean exceeds 153? Would you be surprised to find a mean score of 159? Why? (Hint: To answer these questions you need to apply what you learned in Chapter 9 about Z scores and areas under the normal curve [Appendix B].) Remember, to translate a raw score into a Z score we used this formula:*

$$Z = \frac{Y - \overline{Y}}{S_Y}$$

However, because here we are dealing with a sampling distribution, replace Y with the sample mean $\overline{Y}$, $\overline{Y}$ with the sampling distribution's mean $\mu_{\overline{Y}}$, and S_Y with the standard error of the mean $\frac{\sigma_Y}{\sqrt{N}}$.

$$Z = \frac{\overline{Y} - \mu_{\overline{Y}}}{\sigma_Y / \sqrt{N}}$$

MAIN POINTS

• Through the process of sampling, researchers attempt to generalize to the characteristics of a large group (the population) from a subset (sample) selected from that group. The term *parameter,* associated with the population, refers to the information we are interested in finding out. *Statistic* refers to a corresponding calculated sample statistic.

• A probability sample design allows us to estimate the extent to which the findings based on one sample are likely to differ from what we would find by studying the entire population.

• A simple random sample is chosen in such a way as to ensure that every member of the population and every combination of N members have an equal chance of being chosen.

• In systematic sampling, every Kth member in the total population is chosen for

inclusion in the sample after the first member of the sample is selected at random from the first K members in the population.

• A stratified random sample is obtained by (1) dividing the population into subgroups based on one or more variables central to our analysis and (2) then drawing a simple random sample from each of the subgroups.

• The sampling distribution is a theoretical probability distribution of all possible sample values for the statistic in which we are interested. The sampling distribution of the mean is a frequency distribution of all possible sample means of the same size that can be drawn from the population of interest.

• According to the Central Limit Theorem, if all possible random samples of size N are drawn from a population with a mean μ_Y and a standard deviation σ_Y, then as N becomes larger, the sampling distribution of sample means becomes approximately normal, with mean $\mu_{\bar{Y}}$ and standard deviation $\frac{\sigma_Y}{\sqrt{N}}$.

- The Central Limit Theorem tells us that with sufficient sample size, the sampling distribution of the mean will be normal regardless of the shape of the population distribution. Therefore, even when the population distribution is skewed, we can still assume that the sampling distribution of the mean is normal, given a large enough randomly selected sample size.

KEY TERMS

Central Limit Theorem
disproportionate stratified sample
parameter
population
probability sampling
proportionate stratified sample
sample

sampling distribution
sampling distribution of the mean
sampling error
simple random sample
standard error of the mean
statistic
stratified random sample
systematic random sampling

ON YOUR OWN

Log on to the web-based student study site at http://www.pineforge.com/frankfort-nachmiasstudy4 for additional study questions, quizzes, web resources, and links to social science journal articles reflecting the statistics used in this chapter.

SPSS DEMONSTRATION

[ISSP00PFP]

Selecting a Random Sample

In this chapter we've discussed various types of samples and the definition of the standard error of the mean. Usually, data entered into SPSS have already been sampled from some larger population. However, SPSS does have a sampling procedure that can take random samples of data. Systematic samples and stratified samples can also be drawn with SPSS, but they require the use of the SPSS command language.

When might it be worthwhile to use the SPSS Sample procedure? One instance is when doing preliminary analysis of a very large data set. For example, if you worked for your local hospital and had complete data records for all patients (tens of thousands), there would be no need to use *all* the data during initial analysis. You could select a random sample of individuals and use the subset of data for preliminary analysis. Later, the complete patient data set could be used for completing your final analyses.

To use the Sample procedure, click on *Data* from the main menu, then on *Select Cases*. The opening dialog box (Figure 10.7) has four choices that will select a subset of cases via various methods. By default, the *All cases* button is checked. We click on the *Random sample of cases* button, then on the *Sample* button to give SPSS our specification.

The next dialog box (Figure 10.8) provides two options to create a random sample. The most convenient is normally the first, where we tell SPSS what percentage of cases to select from the larger file. Alternatively, we can tell SPSS to take an exact number of cases. The second option is available because SPSS will only take approximately the percentage specified in the first option.

We type "10" in the box to ask for 10 percent of the original sample of 1,500 respondents from the ISSP. Then click on *Continue* and *OK*, as usual, to process the request.

Figure 10.7

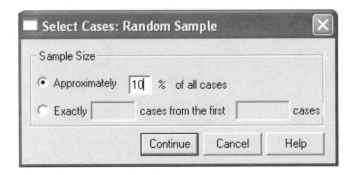

Figure 10.8

SPSS does not delete the cases from the active data file that aren't selected for the sample. Instead, they are filtered out (you can identify them in the Data Editor window by the slash across their row number). This means that we can always return to the full data file by going back to the Select Cases dialog box and selecting the *All cases* button.

When SPSS processes our request, it tells us that the data have been filtered by putting the words "Filter On" in the status area at the bottom of the SPSS window (the status area has many helpful messages from SPSS).

Figure 10.9

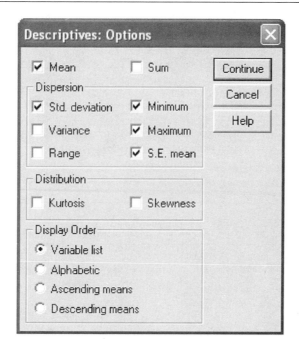

To demonstrate the effect of sampling, we ask for univariate statistics for the variable EDUCYRS, measuring the number of years of education that a respondent has. Click on *Analyze, Descriptive Statistics,* and then *Descriptives* to open this dialog box. Place EDUCYRS in the variable list. Click on the *Options* button to obtain the box shown in Figure 10.9. Select the mean, standard deviation, minimum, and maximum values. In addition, we'll add the standard error of the mean by clicking the *S.E. mean* box. Then click *Continue* and *OK* to put SPSS to work.

The results (Figure 10.10) show that the number of valid cases is exactly 140, or 10 percent of the whole file. The mean of EDUCYRS is 11.64, and the standard error of the mean is .277.

If we repeat the process, this time asking for a 25 percent sample, we obtain the results shown in Figure 10.11.

How closely does the mean for EDUCYRS from these two random samples match that of the full file? The mean for all 1,350 respondents (the other 150 respondents did not have valid responses) is 11.57 years, so it certainly appears that SPSS did take a random sample of this larger file. Both samples produced means and standard deviations that are within the range of the population parameters.

SPSS PROBLEM

Using ISSP00PFP, repeat the SPSS demonstration, selecting 25 percent, 50 percent, and 75 percent samples and requesting descriptives for WRKHRS and HOMPOP. Compare your descriptive statistics with descriptives for the entire sample. What can you say about the accuracy of your random samples?

Figure 10.10

	N	Minimum	Maximum	Mean		Std.
	Statistic	Statistic	Statistic	Statistic	Std. Error	Statistic
EDUCYRS	140	1	30	11.64	.277	3.272
Valid N (listwise)	140					

Your results may differ from the results presented here. We are asking SPSS to generate a random selection of cases and you may not get the same selection of cases as we did.

Figure 10.11

	N	Minimum	Maximum	Mean		Std.
	Statistic	Statistic	Statistic	Statistic	Std. Error	Statistic
EDUCYRS	352	1	30	11.43	.21	3.922
Valid N (listwise)	352					

Your results may differ from the results presented here. We are asking SPSS to generate a random selection of cases and you may not get the same selection of cases as we did.

CHAPTER EXERCISES

1. Explain which of the following is a statistic and which is a parameter.
   a. The mean age of Americans from the 2000 census
   b. The unemployment rate for the population of U.S. adults, estimated by the government from a large sample
   c. The percentage of Texans opposed to abortion from a poll of 1,000 residents
   d. The mean salaries of employees at your school (for example, administrators, faculty, maintenance, etc.)
   e. The percentage of students at your school who receive financial aid

2. The mayor of your city has been talking about the need for a tax hike. The city's newspaper uses letters sent to the editor to judge public opinion about this possible hike, reporting on their results in an article.
   a. Do you think that these letters represent a random sample? Why or why not?
   b. What alternative sampling method would you recommend to the mayor?

3. The following four common situation scenarios involve selecting a sample and understanding how a sample relates to a population.
   a. A friend interviews every 10th shopper that passes by her as she stands outside one entrance of a major department store in a shopping mall. What type of sample is she selecting? How might you define the population from which she is selecting the sample?

b. A political polling firm samples 50 potential voters from a list of registered voters in each county in a state to interview for an upcoming election. What type of sample is this? Do you have enough information to tell?

c. Another political polling firm in the same state selects potential voters from the same list of registered voters with a very different method. First, they alphabetize the list of last names, then pick the first 20 names that begin with an A, the first 20 that begin with a B, and so on until Z (the sample size is thus 20 × 26, or 520). Is this a probability sample?

d. A social scientist gathers a carefully chosen group of 20 people whom she has selected to represent a broad cross-section of the population in New York City. She interviews them in depth for a study she is doing on race relations in the city. Is this a probability sample? What type of sample has she chosen?

4. An upper-level sociology class at a large urban university has 120 students, including 34 seniors, 57 juniors, 22 sophomores, and 7 freshmen.

a. Imagine that you choose one random student from the classroom (perhaps by using a random number table). What is the probability that the student will be a junior?

b. What is the probability that the student will be a freshman?

c. If you are asked to select a proportionate stratified sample of size 30 from the classroom, stratified by class level (senior, junior, and so on), how many students from each group will be in the sample?

d. If instead you are to select a disproportionate sample of size 20 from the classroom, with equal numbers of students from each class level in the sample, how many freshmen will be in the sample?

5. Can the standard error of a variable ever be larger than, or even equal in size to, the standard deviation for the same variable? Justify your answer by means of both a formula and a discussion of the relationship between these two concepts.

6. When taking a random sample from a very large population, how does the standard error of the mean change when

a. the sample size is increased from 100 to 1,600?

b. the sample size is decreased from 300 to 150?

c. the sample size is multiplied by 4?

7. Many television news shows conduct "instant" polls by providing an 800 number and asking an interesting question of the day for viewers to call and answer.

a. Is this poll a probability sample? Why or why not?

b. Specify the population from which the sample of calls is drawn.

8. The following table shows the number of active military personnel in 2002, by state (including the District of Columbia).

Alabama	11,354	Montana	3,512
Alaska	15,906	Nebraska	7,793
Arizona	22,448	Nevada	8,461
Arkansas	4,855	New Hampshire	326

(Continued)

California	123,948	New Jersey	6,306
Colorado	29,733	New Mexico	11,254
Connecticut	4,239	New York	20,882
Delaware	3,899	North Carolina	94,296
D.C.	12,767	North Dakota	7,465
Florida	55,815	Ohio	6,899
Georgia	64,392	Oklahoma	23,664
Hawaii	34,608	Oregon	705
Idaho	4,251	Pennsylvania	3,098
Illinois	25,036	Rhode Island	2,974
Indiana	1,041	South Carolina	37,943
Iowa	447	South Dakota	3,350
Kansas	15,819	Tennessee	2,554
Kentucky	34,081	Texas	115,100
Louisiana	16,541	Utah	5,447
Maine	2,689	Vermont	61
Maryland	30,928	Virginia	90,851
Massachusetts	2,427	Washington	38,521
Michigan	1,173	West Virginia	558
Minnesota	702	Wisconsin	532
Mississippi	14,005	Wyoming	3,292
Missouri	16,119		

Source: U.S. Bureau of the Census, *Statistical Abstract of the United States*, 2003, Table 518.

a. Calculate the mean and standard deviation for the population.
b. Now take 10 samples of size 5 from the population. Use either simple random sampling or systematic sampling with the help of the table of random numbers in Appendix A. Calculate the mean for each sample.
c. Once you have calculated the mean for each sample, calculate the mean of means (i.e., add up your 10 sample means and divide by 10). How does this mean compare with the mean for all states?
d. How does the value of the standard deviation that you calculated in 8a compare with the value of the standard error (i.e., the standard deviation of the sampling distribution)?
e. Construct two histograms, one for the distribution of values in the population and the other for the various sample means taken from 8b. Describe and explain any differences you observe among the two distributions.
f. It is important that you have a clear sense of the population that we are working with in this exercise. What is the population?

9. You've been asked to determine the percentage of students who would support increased ethnic diversity on your campus. You want to take a random sample of fellow students to make the estimate. Explain whether each of the following scenarios describes a random sample.
   a. You ask all students eating lunch in the cafeteria on a Tuesday at 12:30 P.M.
   b. You ask every 10th student from the list of enrolled students.
   c. You ask every 10th student passing by the student union.
   d. What sampling procedure would you recommend to complete your study?

10. For the total population of a large southern city, mean family income is $34,000, with a standard deviation (for the population) of $5,000.
    a. Imagine that you take a subsample of 200 city residents. What is the probability that your sample mean is between $33,000 and $34,000?
    b. For this same sample size, what is the probability that the sample mean exceeds $37,000?

11. A small population of $N = 10$ has values of 4, 7, 2, 11, 5, 3, 4, 6, 10, and 1.
    a. Calculate the mean and standard deviation for the population.
    b. Take 10 simple random samples of size 3, and calculate the mean for each.
    c. Calculate the mean and standard deviation of all these sample means. How closely does the mean of all sample means match the population mean? How is the standard deviation of the means related to the standard deviation for the population?

12. Imagine that you are working with a total population of 21,473 respondents. You assign each respondent a value from 0 to 21,473 and proceed to select your sample using the random number table in Appendix A. Staring at column 7, line 1 in Appendix A, which are the first five respondents that will be included in your sample?

13. The 2000 ISSP asked international respondents whether they thought that some sort of nuclear incident would take place during the next 5 years. The percentage of respondents from seven countries indicating that a nuclear incident was "likely" or "very likely" are shown below.

Country	% of Respondents
Great Britain	87.8
United States	60.5
Netherlands	40.3
Russia	81.1
Canada	51.1
Israel	60.5
Japan	71.2

    a. Assume that $\sigma_Y = 16.4$. Calculate the standard error and interpret.
    b. Write a report wherein you discuss the following: the standard error compared to the standard deviation of the population, the shape of the sampling distribution, and suggestions for reducing the standard error.

Estimation

I n this chapter we discuss the procedures involved in estimating population means and proportions. These procedures are based on the principles of sampling and statistical inference discussed in Chapter 10. Knowledge about the sampling distribution allows us to estimate population means and proportions from sample outcomes and to assess the accuracy of these estimates.

Example 1 A telephone survey was conducted during August 5–10, 2004, by the Pew Research Center. Based on a sample of 1,512 adults, this poll estimated that 60 percent of Americans oppose allowing gays and lesbians to legally marry; however, only 45 percent oppose allowing gays and lesbians to enter into legal agreements that confer many of the same legal rights as marriage.

Example 2 Each month the Bureau of Labor Statistics interviews a sample of about 50,000 adult Americans to determine job-related activities. Based on these interviews, monthly estimates are made of statistics such as the unemployment rate (the proportion who are unemployed), average earnings, the percentage of the work force working part-time, and the percentage collecting unemployment benefits. These estimates are considered so vital that they cause fluctuations in the stock market and influence economic policies of the federal government.

Example 3 The Pew National Survey of Religion and Politics conducted a telephone poll of 2,188 Protestants and 880 Catholics in 2004. Individuals were asked about their opinions of current social and political issues. When asked whether they supported a ban on stem-cell research, 37 percent of Protestants and 32 percent of Catholics agreed.

Each year the National Opinion Research Center (NORC) conducts the General Social Survey (GSS) on a representative sample of about 1,500 respondents. The GSS, from which many of the examples in this book are selected, is designed to provide social science researchers with a readily accessible database of socially relevant attitudes, behaviors, and attributes of a cross section of the U.S. adult population. For example, in analyzing the responses to the 2002 GSS, researchers found the average respondent's education was about 13.4 years. This average probably differs from the average of the population from which the GSS sample was drawn. However, we can establish that in most cases the sample mean (in this case, 13.4 years) is fairly close to the actual true average in the population.

As you read these examples, you may have questioned the reliability of some of the numbers. Based on a sample of about 1,000 respondents, is it possible to determine American attitudes toward gay marriage? What is the actual percentage of unemployed Americans? What is the actual percentage of Protestants and Catholics who think that stem-cell research ought to be banned? What is the actual average level of education in the United States?

▣ ESTIMATION DEFINED

American attitudes toward gay marriage, the actual percentage of unemployed Americans, the actual percentage of Protestants and Catholics in support of a ban on stem-cell research, and the actual average level of education in the United States are all population parameters. American attitudes toward gay marriage as taken from the Pew Research Center, the percentage of unemployed Americans as estimated by the Bureau of Labor and Statistics, the percentage of Protestants and Catholics in support of a ban on stem-cell research as taken from the Pew National Survey of Religion and Politics, and the average level of education in the United States as calculated from the GSS are all sample estimates of population parameters. Sample estimates are used to estimate population parameters; the mean number of years of education of 13.4 calculated from the GSS sample can be used to estimate the mean education of all adults in the United States. Similarly, based on a national sample of adult Americans, the Pew Research Center estimated the percentage of Americans opposed to gay marriage.

These are all illustrations of estimation. Estimation is a process whereby we select a random sample from a population and use a sample statistic to estimate a population parameter. We can use sample proportions as estimates of population proportions, sample means as estimates of population means, or sample variances as estimates of population variances.

Estimation   A process whereby we select a random sample from a population and use a sample statistic to estimate a population parameter.

Reasons for Estimation

Why estimate? The goal of most research is to find the population parameter. Yet we hardly ever have enough resources to collect information about the entire population. We rarely know the value of the population parameter. On the other hand, we can learn a lot about a population by randomly selecting a sample from that population and obtaining an estimate of the population parameter. The major objective of sampling theory and statistical inference is to provide estimates of unknown population parameters from sample statistics.

Point and Interval Estimation

Estimates of population characteristics can be divided into two types: point estimates and interval estimates. **Point estimates** are sample statistics used to estimate the exact value of a population parameter. When the Pew Research Center reports that 60 percent of Americans are opposed to allowing gay marriage, they are using a point estimate. Similarly, if we reported the average level of education of the population of adult Americans to be exactly 13.4 years, we would be using a point estimate.

The problem with point estimates is that sample estimates usually vary, and most result in some sort of sampling error. As a result, when we use a sample statistic to estimate the exact value of a population parameter, we never really know how accurate it is.

One method of showing accuracy is to use an interval estimate rather than a point estimate. In interval estimation, we identify a range of values within which the population parameter may fall. This range of values is called a **confidence interval**. Instead of using a single value, 13.4 years, as an estimate of the mean education of adult Americans, we could say that the population mean is somewhere between 12 and 14 years.

When we use confidence intervals to estimate population parameters, such as mean educational levels, we can also evaluate the accuracy of this estimate by assessing the likelihood that any given interval will contain the mean. This likelihood, expressed as a percentage or a probability, is called a **confidence level**. Confidence intervals are defined in terms of confidence levels. Thus, by selecting a 95 percent confidence level, we are saying that there is a 0.95 probability—or 95 chances out of 100—that a specified interval will contain the population mean. Confidence intervals can be constructed for any level of confidence, but the most common ones are the 90 percent, 95 percent, and 99 percent levels.

Confidence intervals can be constructed for many different parameters based on their corresponding sample statistics. In this chapter we describe the rationale and the procedure for the construction of confidence intervals for means and proportions.

Point estimate   A sample statistic used to estimate the exact value of a population parameter.

Confidence interval (interval estimate)   A range of values defined by the confidence level within which the population parameter is estimated to fall.

Confidence level   The likelihood, expressed as a percentage or a probability, that a specified interval will contain the population parameter.

✔ *Learning Check.*   *What is the difference between a point estimate and a confidence interval?*

▣ PROCEDURES FOR ESTIMATING CONFIDENCE INTERVALS FOR MEANS

To illustrate the procedure of establishing confidence intervals for means, we'll reintroduce one of the research examples mentioned in Chapter 10—assessing the needs of commuting students on our campus.

Recall that we have been given enough money to survey a random sample of 500 students. One of our tasks is to estimate the average commuting time of all 15,000 commuters on our campus—the population parameter. To obtain this estimate we calculate the average commuting time for the sample. Suppose the sample average is $\bar{Y} = 7.5$ hours per week, and we want to use it as an estimate of the true average commuting time for the entire population of commuting students.

Because it is based on a sample, this estimate is subject to sampling error. We do not know how close it is to the true population mean. However, based on what the Central Limit Theorem tells us about the properties of the sampling distribution of the mean, we know that with a large enough sample size, most sample means will tend to be close to the true population mean. Therefore, it is unlikely that our sample mean, $\bar{Y} = 7.5$, deviates much from the true population mean.

▣ A Closer Look 11.1
Estimation as a Type of Inference

The goal of inferential statistics is to say something meaningful about the population, based entirely on information from a sample of that population. A confidence interval attempts to do just that: By knowing a sample mean, sample size, and sample standard deviation, we are able to say something about the population from which that sample was drawn.

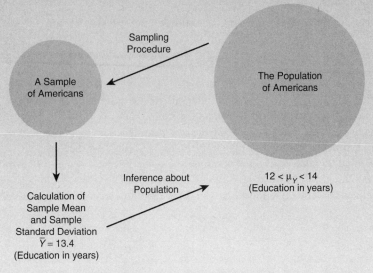

We know exactly what our sample mean is. Combining this information with the sample standard deviation and sample size gives us a range within which we can confidently say that the population mean falls.

We know that the sampling distribution of the means is approximately normal with a mean $\mu_{\bar{Y}}$ equal to the population mean μ_Y, and a standard error $\sigma_{\bar{Y}}$ (standard deviation of the sampling distribution) as follows:

$$\sigma_{\bar{Y}} = \frac{\sigma_Y}{\sqrt{N}} \tag{11.1}$$

This information allows us to use the normal distribution to determine the probability that a sample mean will fall within a certain distance—measured in standard deviation (standard error) units, or Z scores—of μ_Y or $\mu_{\bar{Y}}$. We can make the following assumptions:

- 68 percent of all random sample means will fall within ±1 standard error of the true population mean

- 95 percent of all random sample means will fall within ±1.96 standard errors of the true population mean
- 99 percent of all random sample means will fall within ±2.58 standard errors of the true population mean

Based on these assumptions and the value of the standard error, we can establish a range of values—a confidence interval—that is likely to contain the actual population mean. We can also evaluate the accuracy of this estimate by assessing the likelihood that this range of values will actually contain the population mean.

The general formula for constructing a confidence interval (CI) for any level is

$$CI = \bar{Y} \pm Z(\sigma_{\bar{Y}}) \tag{11.2}$$

Notice that to calculate a confidence interval, we take the sample mean and add to or subtract from it the product of a Z value and the standard error.

The Z score we choose depends on the desired confidence level. For example, to obtain a 95 percent confidence interval we would choose a Z of 1.96 because we know (from Appendix B) that 95 percent of the area under the curve lies between ±1.96. Similarly, for a 99 percent confidence level we would choose a Z of 2.58. The relationship between the confidence level and Z is illustrated in Figure 11.1 for the 95 percent and 99 percent confidence levels.

✔ **Learning Check.**   *If you don't understand the relationship between the confidence level and Z, review the material in Chapter 9. What would be the appropriate Z value for a 98 percent confidence interval?*

Figure 11.1   Relationship Between Confidence Level and Z for 95 and 99 Percent Confidence Intervals

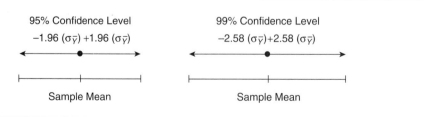

Source: Adapted from David Freedman, Robert Pisani, Roger Purves, and Ani Adhikari, *Statistics*, 2nd ed., p. 348. Copyright © 1991 by W. W. Norton & Co., Inc. Reprinted by permission of W. W. Norton & Co., Inc.

Determining the Confidence Interval

To determine the confidence interval for means, follow these steps:

1. Calculate the standard error of the mean.
2. Decide on the level of confidence, and find the corresponding Z value.
3. Calculate the confidence interval.
4. Interpret the results.

Let's return to the problem of estimating the mean commuting time of the population of students on our campus. How would you find the 95 percent confidence interval?

Calculating the Standard Error of the Mean   Let's suppose that the standard deviation for our population of commuters is $\sigma_Y = 1.5$. We calculate the standard error for the sampling distribution of the mean:

$$\sigma_{\overline{Y}} = \frac{\sigma_Y}{\sqrt{N}} = \frac{1.5}{\sqrt{500}} = 0.07$$

Deciding on the Level of Confidence and Finding the Corresponding Z Value   We decide on a 95 percent confidence level. The Z value corresponding to a 95 percent confidence level is 1.96.

Calculating the Confidence Interval   The confidence interval is calculated by adding and subtracting from the observed sample mean the product of the standard error and Z:

$$95\% \text{ CI} = 7.5 \pm 1.96(0.07)$$
$$= 7.5 \pm 0.14$$
$$= 7.36 \text{ to } 7.64$$

The 95 percent CI for the mean commuting time is illustrated in Figure 11.2.

Interpreting the Results   We can be 95 percent confident that the actual mean commuting time—the true population mean—is not less than 7.36 hours and not greater than 7.64 hours. In other words, if we collected a large number of samples ($N = 500$) from the population of commuting students, 95 times out of 100, the true population mean would be included within our computed interval. With a 95 percent confidence level, there is a 5 percent risk that we are wrong. Five times out of 100, the true population mean will not be included in the specified interval.

Figure 11.2   95 Percent Confidence Interval for the Mean Commuting Time ($N = 500$)

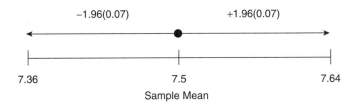

Remember that we can never be sure whether the population mean is actually contained within the confidence interval. Once the sample is selected and the confidence interval defined, the population mean either does or does not contain the population mean—but we will never be sure.

✔ **Learning Check.**   *What is the 90 percent confidence interval for the mean commuting time? (Hint: First, find the Z value associated with a 90 percent confidence level.)*

To further illustrate the concept of confidence intervals, let's suppose that we draw 10 different samples ($N = 500$) from the population of commuting students. For each sample mean, we construct a 95 percent confidence interval. Figure 11.3 displays these confidence intervals. Each horizontal line represents a 95 percent confidence interval constructed around a sample mean (marked with a circle).

The vertical line represents the population mean. Note that the horizontal lines that intersect the vertical line are the intervals that contain the true population mean. Only one out of the 10 confidence intervals does not intersect the vertical line, meaning it does not contain the population mean. What would happen if we continued to draw samples of the same size from this population and constructed a 95 percent confidence interval for each sample? For about 95 percent of all samples the specified interval would contain the true population mean, but for 5 percent of all samples it would not.

Figure 11.3   95 Percent Confidence Intervals for 10 Samples

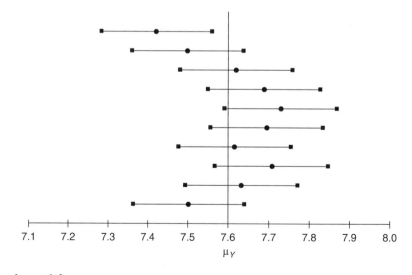

Reducing Risk

One way to reduce the risk of being incorrect is by increasing the level of confidence. For instance, we can increase our confidence level from 95 to 99 percent. The 99 percent confidence interval for our commuting example is

$$99\% \text{ CI} = 7.5 \pm 2.58(0.07)$$
$$= 7.5 \pm 0.18$$
$$= 7.32 \text{ to } 7.68$$

When using the 99 percent confidence interval, there is only a 1 percent risk that we are wrong and the specified interval does not contain the true population mean. We can be almost certain that the true population mean is included in the interval ranging from 7.32 to 7.68 hours per week. Notice that by increasing the confidence level we have also increased the width of the confidence interval from 0.28 (7.36–7.64) to 0.36 hours (7.32–7.68), thereby making our estimate less precise.

You can see that there is a trade-off between achieving greater confidence in an estimate and the precision of that estimate. Although using a higher level of confidence (such as 99 percent) increases our confidence that the true population mean is included in our confidence interval, the estimate becomes less precise as the width of the interval increases. Although we are only 95 percent confident that the interval ranging between 7.36 and 7.64 hours includes the true population mean, it is a more precise estimate than the 99 percent interval ranging from 7.32 to 7.68 hours. The relationship between the confidence level and the precision of the confidence interval is illustrated in Figure 11.4. Table 11.1 lists three commonly used confidence levels along with their corresponding Z values.

Figure 11.4 95 Percent Versus 99 Percent Confidence Intervals (Mean Commuting Time)

A 95% confidence interval

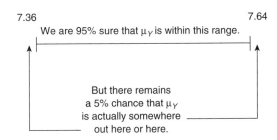

A 99% confidence interval

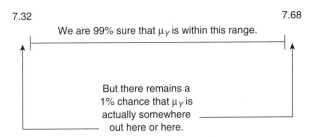

Table 11.1 Confidence Levels and
Corresponding Z Values

Confidence Level	Z Value
90%	1.65
95%	1.96
99%	2.58

Estimating Sigma

To calculate confidence intervals, we need to know the standard error of the sampling distribution, $\sigma_{\bar{Y}}$. The standard error is a function of the population standard deviation and the sample size:

$$\sigma_{\bar{Y}} = \frac{\sigma_Y}{\sqrt{N}}$$

In our commuting example we have been using a hypothetical value, $\sigma_Y = 1.5$, for the population standard deviation. Typically, both the mean (μ_Y) and the standard deviation (σ_Y) of the population are unknown to us. When $N \geq 50$, however, the sample standard deviation S_Y is a good estimate of σ_Y. The standard error is then calculated as follows:

$$S_{\bar{Y}} = \frac{S_Y}{\sqrt{N}} \qquad (11.3)$$

As an example, we'll estimate the mean hours per day that Americans spend watching television based on the 2002 GSS survey. The mean hours per day spent watching television for a sample of $N = 501$ is $\bar{Y} = 3.06$ hours, and the standard deviation is $S_Y = 2.39$ hours. Let's determine the 95 percent confidence interval for these data.

Calculating the Estimated Standard Error of the Mean The estimated standard error for the sampling distribution of the mean is

$$S_{\bar{Y}} = \frac{S_Y}{\sqrt{N}} = \frac{2.39}{\sqrt{501}} = 0.11$$

Deciding on the Level of Confidence and Finding the Corresponding Z Value We decide on a 95 percent confidence level. The Z value corresponding to a 95 percent confidence level is 1.96.

Calculating the Confidence Interval The confidence interval is calculated by adding to and subtracting from the observed sample mean the product of the standard error and Z:

$$95\% \text{ CI} = 3.06 \pm 1.96(0.11)$$
$$= 3.06 \pm 0.22$$
$$= 2.84 \text{ to } 3.28$$

Interpreting the Results We can be 95 percent confident that the actual mean hours spent watching television by Americans from which the GSS sample was taken is not less than 2.84 hours and not greater than 3.28 hours. In other words, if we drew a large number of samples ($N = 501$) from this population, then 95 times out of 100 the true population mean would be included within our computed interval.

Sample Size and Confidence Intervals

Researchers can increase the precision of their estimate by increasing the sample size. In Chapter 10 we learned that larger samples result in smaller standard errors and, therefore, in sampling distributions that are more clustered around the population mean (Figure 10.6). A more tightly clustered sampling distribution means that our confidence intervals will be narrower and more precise. To illustrate the relationship between sample size and the standard error, and thus the confidence interval, let's calculate the 95 percent confidence interval for our GSS data with (1) a sample of $N = 108$ and (2) a sample of $N = 1,890$.

With a sample size $N = 108$, the estimated standard error for the sampling distribution is

$$S_{\overline{Y}} = \frac{S_Y}{\sqrt{N}} = \frac{2.39}{\sqrt{108}} = 0.23$$

The 95 percent confidence interval is

$$95\% \text{ CI} = 3.06 \pm 1.96(0.23)$$
$$= 3.06 \pm 0.45$$
$$= 2.61 \text{ to } 3.51$$

With a sample size $N = 1,890$, the estimated standard error for the sampling distribution is

$$S_{\overline{Y}} = \frac{S_Y}{\sqrt{N}} = \frac{2.39}{\sqrt{1,890}} = 0.05$$

The 95 percent confidence interval is

$$95\% \text{ CI} = 3.06 \pm 1.96(0.05)$$
$$= 3.06 \pm 0.10$$
$$= 2.96 \text{ to } 3.16$$

In Table 11.2 we summarize the 95 percent confidence intervals for the mean number of hours watching television for these three sample sizes: $N = 108$, $N = 501$, and $N = 1,890$.

Notice that there is an inverse relationship between sample size and the width of the confidence interval. The increase in sample size is linked with increased precision of the confidence interval. The 95 percent confidence interval for the GSS sample of 108 cases is 0.90 hours. But the interval widths decrease to 0.44 hours and 0.20 hours, respectively, as the sample sizes increase to $N = 501$ and then to $N = 1,890$. We had to nearly quadruple the size of the sample (from 501 to 1,890) to reduce the confidence interval by about one half[1] (from 0.44 hours to 0.20 hours). In general, although the precision of estimates increases steadily

---

[1]The slight variation is due to rounding.

Table 11.2   95 Percent Confidence Interval and Width for Mean
Income for Three Different Sample Sizes

Sample Size	Confidence Interval	Interval Width	S_Y	$S_{\bar{Y}}$
$N = 108$	2.61–3.51	0.90	2.39	0.23
$N = 501$	2.84–3.28	0.44	2.39	0.11
$N = 1,890$	2.96–3.16	0.20	2.39	0.05

with sample size, the gains would appear to be rather modest after N reaches 1,890. An important factor to keep in mind is the increased cost associated with a larger sample. Researchers have to consider at what point the increase in precision is too small to justify the additional cost associated with a larger sample.

✓ *Learning Check.* *Why do smaller sample sizes produce wider confidence intervals? (See Figure 11.5.) (Hint: Compare the standard errors of the mean for the three sample sizes.)*

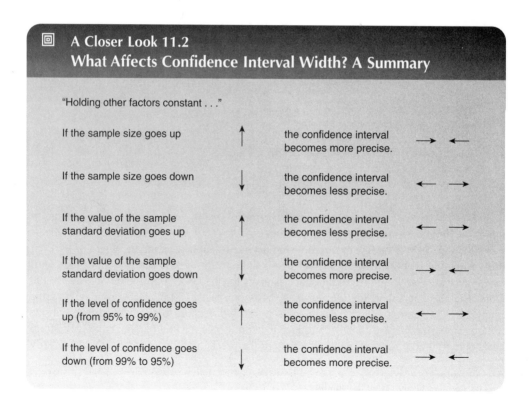

🔲 **A Closer Look 11.2**
What Affects Confidence Interval Width? A Summary

"Holding other factors constant . . ."

If the sample size goes up	↑	the confidence interval becomes more precise. → ←
If the sample size goes down	↓	the confidence interval becomes less precise. ← →
If the value of the sample standard deviation goes up	↑	the confidence interval becomes less precise. ← →
If the value of the sample standard deviation goes down	↓	the confidence interval becomes more precise. → ←
If the level of confidence goes up (from 95% to 99%)	↑	the confidence interval becomes less precise. ← →
If the level of confidence goes down (from 99% to 95%)	↓	the confidence interval becomes more precise. → ←

Figure 11.5   The Relationship Between Sample Size and Confidence Interval Width

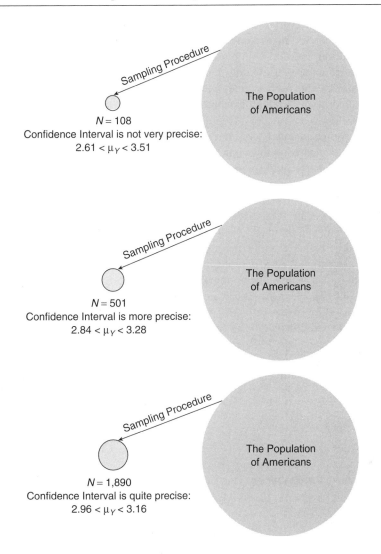

▣ STATISTICS IN PRACTICE:
HISPANIC MIGRATION AND EARNINGS

Tienda and Wilson investigated the relationship between migration and the earnings of Hispanic men.[2] Based on a sample of the 1980 census that included 5,726 Mexicans, 5,908

---

[2]Adapted from Marta Tienda and Franklin D. Wilson, "Migration and the Earnings of Hispanic Men," *American Sociological Review* 57 (1992): 661–678.

Puerto Ricans, and 3,895 Cubans, Tienda and Wilson argued that these three Hispanic groups varied markedly in socioeconomic characteristics because of differences in the timing and circumstances of their immigration to the United States. The authors claimed that the period of entry and the circumstances prompting migration affected the geographical distribution and the employment opportunities of each group. For example, Puerto Ricans were disproportionately located in the Northeast, where the labor market was characterized by the highest unemployment rates, whereas the majority of Cuban immigrants resided in the Southeast, where the unemployment rate was the lowest in the United States.

Tienda and Wilson also noted persistent differences in educational levels among Mexicans and Puerto Ricans compared with Cubans. About 60 percent of Mexicans and Puerto Ricans had not completed high school, compared with 42 percent of Cuban men. At the other extreme, 17 percent of Cuban men were college graduates, compared with about 4 percent of Mexican men and Puerto Rican men.

These differences in migrant status and educational level were likely to be reflected in disparities in earnings among the three groups. Tienda and Wilson anticipated that the earnings of Cubans would be higher than the earnings of Mexicans and Puerto Ricans. As hypothesized, with average earnings of $16,368 ($S_Y = \$3,069$) Cubans were at the top of the income hierarchy. Puerto Ricans were at the bottom of the income hierarchy, with earnings averaging $12,587 ($S_Y = \$8,647$). Mexican men were intermediate among the groups, with average annual earnings of $13,342 ($S_Y = \$9,414$).

Although Tienda and Wilson did not calculate confidence intervals for their estimates, we will use their data to calculate a 95 percent confidence interval for the mean income for the three groups of Hispanic men.

To find the 95 percent confidence interval for Cuban income, we first estimate the standard error:

$$S_{\overline{Y}} = \frac{3,069}{\sqrt{3,895}} = 49.17$$

Then we calculate the confidence interval:

$$95\% \ CI = 16,368 \pm 1.96(49.17)$$
$$= 16,368 \pm 96$$
$$= 16,272 \ to \ 16,464$$

For Puerto Rican income, the estimated standard error is

$$S_{\overline{Y}} = \frac{8,647}{\sqrt{5,908}} = 112.50$$

and the confidence interval is

$$95\% \ CI = 12,587 \pm 1.96(112.50)$$
$$= 12,587 \pm 220$$
$$= 12,367 \ to \ 12,807$$

Finally, for Mexican income the estimated standard error is

$$S_{\bar{Y}} = \frac{9,414}{\sqrt{5,726}} = 124.41$$

and the confidence interval is

$$95\% \; CI = 13,342 \pm 1.96(124.41)$$
$$= 13,342 \pm 244$$
$$= 13,098 \text{ to } 13,586$$

The confidence intervals for mean annual income of Cuban, Puerto Rican, and Mexican immigrants are illustrated in Figure 11.6. We can say with 95 percent confidence that the true income mean for each Hispanic group lies somewhere within the corresponding confidence interval. Notice that the confidence intervals do not overlap, revealing great disparities in earnings among the three groups. As noted earlier, highest interval estimates are for Cubans, followed by Mexicans and then Puerto Ricans.

Figure 11.6   The 95 Percent Confidence Intervals for the Mean Income of Puerto Ricans, Mexicans, and Cubans

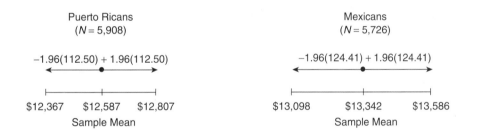

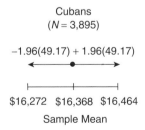

▣ CONFIDENCE INTERVALS FOR PROPORTIONS

Confidence intervals can also be computed for sample proportions or percentages in order to estimate population proportions or percentages. The procedures for estimating proportions and percentages are identical. Any of the formulas presented for proportions can be applied to percentages, and vice versa. We can obtain a confidence interval for a percentage by calculating the confidence interval for a proportion and then multiplying the result by 100.

The same conceptual foundations of sampling and statistical inference that are central to the estimation of population means—the selection of random samples and the special properties of the sampling distribution—are also central to the estimation of population proportions.

Earlier we saw that the sampling distribution of the means underlies the process of estimating population means from sample means. Similarly, the *sampling distribution of proportions* underlies the estimation of population proportions from sample proportions. Based on the Central Limit Theorem, we know that with sufficient sample size the sampling distribution of proportions is approximately normal, with mean μ_p equal to the population proportion π and with a standard error of proportions (the standard deviation of the sampling distribution of proportions) equal to

$$\sigma_p = \sqrt{\frac{(\pi)(1-\pi)}{N}} \tag{11.5}$$

where

σ_p = the standard error of proportions
π = the population proportion
N = the population size

However, since the population proportion, π, is unknown to us (that is what we are trying to estimate), we can use the sample proportion, p, as an estimate of π. The estimated standard error then becomes

$$S_p = \sqrt{\frac{(p)(1-p)}{N}} \tag{11.6}$$

where

S_p = the estimated standard error of proportions
p = the sample proportion
N = the sample size

As an example, let's calculate the estimated standard error for the survey by Pew Research Center. Based on a random sample of 1,512 adults, the percentage opposed to allowing gay marriage was estimated to be 60 percent. Based on Formula 11.5, with $p = .60$, $1 - p = (1 - 0.60) = 0.40$, and $N = 1,512$, the standard error is

$$S_p = \sqrt{\frac{(0.60)(1-0.60)}{1,512}} = 0.01$$

We will have to consider two factors to meet the assumption of normality with the sampling distribution of proportions: (1) the sample size N and (2) the sample proportions p and $1 - p$. When p and $1 - p$ are about 0.50, a sample size of at least 50 is sufficient. But when $p > 0.50$ (or $1 - p < 0.50$), a larger sample is required to meet the assumption of normality. Usually, a sample of 100 or more is adequate for any single estimate of a population proportion.

Procedures for Estimating Proportions

Because the sampling distribution of proportions is approximately normal, we can use the normal distribution to establish confidence intervals for proportions in the same manner that we used the normal distribution to establish confidence intervals for means.

The general formula for constructing confidence intervals for proportions for any level of confidence is

$$CI = p \pm Z(S_p)$$

where

CI = the confidence interval
p = the observed sample proportion
Z = the Z corresponding to the confidence level
S_p = the estimated standard error of proportions

Let's examine this formula in more detail. Notice that to obtain a confidence interval at a certain level we take the sample proportion and add to or subtract from it the product of a Z value and the standard error. The Z value we choose depends on the desired confidence level. We want the area between the mean and the selected $\pm Z$ to be equal to the confidence level.

For example, to obtain a 95 percent confidence interval we would choose a Z of 1.96 because we know (from Appendix B) that 95 percent of the area under the curve is included between ±1.96. Similarly, for a 99 percent confidence level we would choose a Z of 2.58. (The relationship between confidence level and Z values was illustrated in Figure 11.1.)

To determine the confidence interval for a proportion, we follow the same steps used to find confidence intervals for means:

1. Calculate the estimated standard error of the proportion.

2. Decide on the desired level of confidence, and find the corresponding Z value.

3. Calculate the confidence interval.

4. Interpret the results.

To illustrate these steps we use the results of the Pew Research Center survey on the percentage of American opposed to allowing gay marriage.

Calculating the Estimated Standard Error of the Proportion The standard error of the proportion .60 (60%) with a sample $N = 1,512$ is 0.01.

Deciding on the Desired Level of Confidence and Finding the Corresponding Z Value We choose the 95 percent confidence level. The Z corresponding to a 95 percent confidence level is 1.96.

Calculating the Confidence Interval   We calculate the confidence interval by adding to and subtracting from the observed sample proportion the product of the standard error and Z:

$$95\% \text{ CI} = 0.60 \pm 1.96(0.01)$$
$$= 0.60 \pm 0.02$$
$$= 0.58 \text{ to } 0.62$$

Interpreting the Results   We are 95 percent confident that the true population proportion is somewhere between 0.58 and 0.62. In other words, if we drew a large number of samples from the population of adults, then 95 times out of 100 the confidence interval we obtained would contain the true population proportion. We can also express this result in percentages and say that we are 95 percent confident that the true population percentage of Americans opposed to allowing gay marriage is included somewhere within our computed interval of 58 percent to 62 percent.

✔ *Learning Check.*   *Calculate the confidence interval for Pew Research Center survey using percentages rather than proportions. Your results should be identical with ours except that they are expressed in percentages.*

Note that with a 95 percent confidence level there is a 5 percent risk that we are wrong. If we continued to draw large samples from this population, in 5 out of 100 samples the true population proportion would not be included in the specified interval.

We can decrease our risk by increasing the confidence level from 95 to 99 percent.

$$99\% \text{ CI} = 0.60 \pm 2.58(0.01)$$
$$= 0.60 \pm 0.03$$
$$= 0.57 \text{ to } 0.63$$

When using the 99 percent confidence interval we can be almost certain (99 times out of 100) that the true population proportion is included in the interval ranging from 0.57 (57%) to 0.63 (63%). However, as we saw earlier, there is a trade-off between achieving greater confidence in making an estimate and the precision of that estimate. Although using a 99 percent level increased our confidence level from 95 percent to 99 percent (thereby reducing our risk of being wrong from 5% to 1%), the estimate became less precise as the width of the interval increased.*

---

* The relationship between sample size and interval width when estimating means also holds true for sample proportions. When the sample size increases, the standard error of the proportion decreases, and therefore the width of the confidence interval decreases as well.

▣ STATISTICS IN PRACTICE: RELIGIOSITY AND SUPPORT FOR STEM-CELL RESEARCH

Poll or survey results may be limited to a single estimate of a parameter. For instance, the Pew National Survey of Religion and Politics reported the estimated percentage of Protestants and Catholics in the United States in support of a ban on stem-cell research. Most survey studies, however, are not limited to single estimates for the overall population. Often, separate estimates are reported for subgroups within the overall population of interest. The 2004 Pew National Survey of Religion and Politics, for example, compared support for a ban on stem-cell research among Protestant and Catholic respondents.

When estimates are reported for subgroups, the confidence intervals are likely to vary from subgroup to subgroup. Each confidence interval is based on the confidence level, the standard error of the proportion (which can be estimated from p), and the sample size. Even when a confidence interval is reported only for the overall sample, we can easily compute separate confidence intervals for each of the subgroups if the confidence level and the size of each of the subgroups are included.

To illustrate this, let's calculate the 95 percent confidence intervals for the proportions of Protestants and Catholics who support a ban on stem-cell research. Out of 2,188 Protestants included in the overall sample, 0.37 (or 37%) favor a ban on stem-cell research. In contrast, 0.32 (or 32%) of the 880 Catholic respondents agree.

Calculating the Estimated Standard Error of the Proportion

The estimated standard error for the proportion of Protestants is

$$S_p = \sqrt{\frac{(0.37)(1 - 0.37)}{2,188}} = 0.01$$

The estimated standard error for the proportion of Catholics who support a ban on stem-cell research is

$$S_p = \sqrt{\frac{(0.32)(1 - 0.32)}{880}} = 0.02$$

Deciding on the Desired Level of Confidence and Finding the Corresponding *Z* Value

We choose the 95 percent confidence level, with a corresponding Z value of 1.96.

Calculating the Confidence Interval

For Protestants,

$$95\% \text{ CI} = 0.37 \pm 1.96(0.01)$$
$$= 0.37 \pm 0.02$$
$$= 0.35 \text{ to } 0.39$$

and for Catholics,

$$95\% \text{ CI} = 0.32 \pm 1.96(0.02)$$
$$= 0.32 \pm 0.04$$
$$= 0.28 \text{ to } 0.36$$

The 95 percent confidence interval for the proportion of Protestant and Catholic respondents who support a ban on stem-cell research is illustrated in Figure 11.7.

Interpreting the Results

We are 95 percent confident that the true population proportion supporting a ban on stem-cell research is somewhere between 0.35 and 0.39 (or between 35% and 39%) for Protestants, and somewhere between 0.28 and 0.36 (or between 28% and 36%) for Catholics. Based on the sample, it is clear that only a minority of Protestants and Catholics supports a ban on stem-cell research.

Figure 11.7   The 95 Percent Confidence Interval for the Proportion of Protestants and Catholics in Support of a Ban on Stem-Cell Research

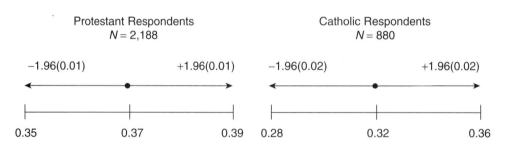

MAIN POINTS

• The goal of most research is to find population parameters. The major objective of sampling theory and statistical inference is to provide estimates of unknown parameters from sample statistics.

• Researchers make point estimates and interval estimates. Point estimates are sample statistics used to estimate the exact value of a population parameter. Interval estimates are ranges of values within which the population parameter may fall.

• Confidence intervals can be used to estimate population parameters such as means or proportions. Their accuracy is defined with the confidence level. The most common confidence levels are 90, 95, and 99 percent.

- To establish a confidence interval for a mean or a proportion, add or subtract from

the mean or the proportion the product of the standard error and the Z value corresponding to the confidence level.

KEY TERMS

confidence interval (interval estimate)
confidence level

estimation
point estimate

SPSS DEMONSTRATION

[GSS02PFP-A]

*Producing Confidence Intervals
Around a Mean*

SPSS calculates confidence intervals around a sample mean or proportion with the Explore procedure. Let's investigate the average ideal number of children for men and women.

Activate the Explore procedure by selecting the *Analyze* menu, *Descriptive Statistics*, then *Explore*. The opening dialog box has spaces for both dependent and independent variables. Place CHLDIDEL in the Dependent List box, and SEX in the Factor List box, as shown in Figure 11.8. [For this analysis, we've eliminated response categories 8 (as many as I want) and 9 (don't know or no answer).]

Click on the *Statistics* button. Notice that the Descriptives choice also includes the confidence interval for the mean, which by default is calculated at the 95 percent confidence level. Let's change that to the 99 percent level by erasing the "95" and substituting "99" (Figure 11.9).

Click on *Continue* to return to the main dialog box. Recall that Explore produces several statistics and plots by default. For this example, we don't need to view the graphics, so click on the *Statistics* button in the *Display* section. Your screen should now look like Figure 11.10.

Click on *OK* to run the procedure.

Figure 11.8

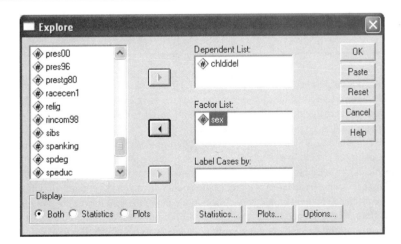

Figure 11.9

Figure 11.10

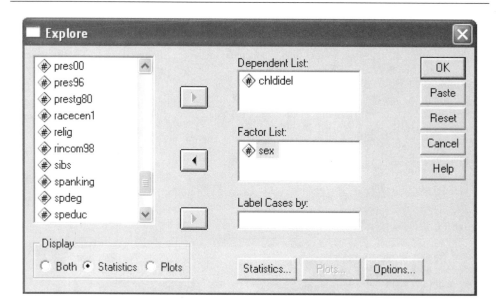

The output from the Explore procedure (Figure 11.11) is broken into two parts, one for males and one for females. The mean ideal number of children for males is 2.70; for females, it's 3.09. Our data indicate that, on average, the ideal number of children is greater for women than it is for men.

The 99 percent confidence interval for males runs from about 2.43 to 2.98 children. One way to interpret this result is to state that, in 100 samples of size 220 of males (the number of males in this sample) from the U.S. adult population, we would expect the confidence interval to include the true population value for the mean ideal number of children 99 times out of those 100. We can never be sure that in this particular sample the confidence interval includes the population mean. As explained in this chapter, any one sample's confidence interval either does or does not contain the (unknown) population mean, so no probability value can be associated with a particular confidence interval. Still, our best estimate for the mean ideal number of children falls within a narrow range of only about .55. For females, the 99 percent confidence interval is slightly wider, varying from about 2.80 to 3.38 children, or only .58.

Figure 11.11

Case Processing Summary

| | | | Cases | | | | | |
| | | Valid | | Missing | | Total | | |
	SEX	N	Percent	N	Percent	N	Percent
CHLDIDEL	1 MALE	220	32.9%	448	67.1%	668	100.0%
	2 FEMALE	277	33.3%	555	66.7%	832	100.0%

Descriptives

SEX			Statistic	Std. Error
CHLDIDEL 1 MALE	Mean		2.70	.106
	99% Confidence Lower Bound		2.43	
	Interval for Mean Upper Bound		2.98	
	5% Trimmed Mean		2.53	
	Median		2.00	
	Variance		2.474	
	Std. Deviation		1.573	
	Minimum		0	
	Maximum		8	
	Range		8	
	Interquartile Range		1.00	
	Skewness		2.154	.164
	Kurtosis		5.158	.327
2 FEMALE	Mean		3.09	.111
	99% Confidence Lower Bound		2.80	
	Interval for Mean Upper Bound		3.38	
	5% Trimmed Mean		2.93	
	Median		2.00	
	Variance		3.394	
	Std. Deviation		1.842	
	Minimum		0	
	Maximum		8	
	Range		8	
	Interquartile Range		1.00	
	Skewness		1.803	.146
	Kurtosis		2.416	.292

SPSS PROBLEMS

[GSS02PFP-A]

1. Recall that the GSS sample includes men and women from 18 to 89 years of age. Does it matter that we may have responses from men and women of diverse ages? Would our results change if we selected a younger sample of men and women?
    a. To take the SPSS demonstration one step further, use the Select Cases procedure to select respondents based on the variable AGE who are less than or equal to 35 years old. Do this

by selecting *Data* and then *Select Cases*. Next, select "*If Condition is Satisfied*" and then click on *If*. Find and highlight the variable AGE in the scroll down box on the left of your screen. Click the arrow next to the scroll down box. AGE will now appear in the box on the right. Now, tell SPSS that you want to select respondents who are 35 years of age or less. The box on the right should now read: AGE <= 35. Click *Continue* and then *OK*.

b. Using this younger sample, repeat the Explore procedure that we just completed in the demonstration. What differences exist between men and women in this younger sample on the ideal number of children? How do these results compare with those based on the entire sample?

2. Calculate the 90 percent confidence interval for the following variables, comparing lower, working, middle, and upper classes (CLASS) in the GSS sample. First, tell SPSS that we want to select all cases in the sample by selecting *Data* and then *All Cases*, and then *OK*. Then use the Explore procedure using CLASS as your factor variable (*Analyze, Descriptive Statistics, Explore*) Make a summary statement of your findings.
   a. CHILDS (Number of children in the household)
   b. EDUC (Respondent's highest year of school completed)
   c. PAEDUC (Father's highest year of school completed)
   d. PRESG80 (Respondent's occupational prestige)
   e. MAEDUC (Mother's highest year of school completed)

CHAPTER EXERCISES

1. In a study of crime, the FBI found that 13.2 percent of all Americans had been victims of crime during a 1-year period. This result was based on a sample of 1,105 adults.
   a. Estimate the percentage of U.S. adults who were victims at the 90 percent confidence level. State in words the meaning of the result.
   b. Estimate the percentage of victims at the 99 percent confidence level.
   c. Imagine that the FBI doubles the sample size in a new sample but finds the same value of 13.2 percent for the percentage of victims in the second sample. By how much would the 90 percent confidence interval shrink? By how much would the 99 percent confidence interval shrink?
   d. Considering your answers to (a), (b), and (c), can you suggest why national surveys, such as those by Gallup, Roper, or the *New York Times*, typically take samples of size 1,000 to 1,500?

2. Use the data on income from Chapter 9, Exercise 5.

	Mean	Standard Deviation	N
Lower class	10.34	3.18	89
Working class	12.64	2.50	675
Middle class	14.19	2.39	675
Upper class	15.00	3.27	49

   a. Construct the 95 percent confidence interval for the mean number of years of education for lower class and middle class respondents.
   b. Construct the 99 percent confidence interval for the mean number of years of education for lower class and middle class respondents.

c. As our confidence in the result increases, how does the size of the confidence interval change? Explain why this is true.

3. The United States has often adopted an isolationist foreign policy designed to stay out of foreign entanglements (as George Washington advised his fellow citizens more than two hundred years ago). In the 2000 ISSP, 1,132 respondents from the United States were asked whether they agreed that international agreements were a national priority. The data show that 897 of the 1,132 either strongly agreed or agreed that international agreements were an important national priority.

a. Estimate the proportion of all adult Americans who strongly agreed or agreed at the 95 percent confidence level.

b. Estimate the proportion of all adult Americans who strongly agreed or agreed at the 99 percent confidence level.

c. If you were going to write a report on this poll result, would you prefer to use the 99 percent or 95 percent confidence interval? Explain why.

4. Use the data in Chapter 5, Exercise 6, about occupational prestige and education.

PRESTG80			*Statistic*
OCCUPATIONAL PRESTIGE SCORE	High school diploma	Mean	40.39
		Median	40.00
		Std. deviation	11.44
		Minimum	17
		Maximum	86
		Range	69
		Interquartile range	16
	Bachelor's degree	Mean	53.19
		Median	52.00
		Std. deviation	12.48
		Minimum	19
		Maximum	75
		Range	56
		Interquartile range	17

a. Construct the 90 percent confidence interval for occupational prestige for respondents with only a high school diploma ($N = 1,412$).

b. Construct the 90 percent confidence interval for occupational prestige for respondents with a bachelor's degree ($N = 437$). State in words the meaning of the result.

c. Use these statistics to discuss differences in occupational prestige scores by educational attainment.

5. A newspaper does a poll to determine the likely vote for the incumbent mayor, Ann Johnson, in the upcoming election. They find that 52 percent of the voters favor Johnson in a sample of 500 registered voters. The newspaper asks you, their statistical consultant, to tell them whether they should declare Johnson the likely winner of the election. What is your advice? Why?

6. The police department in your city was asked by the mayor's office to estimate the cost of crime. The police began their study with burglary records, taking a random sample of 500 files since there were too many crime records to calculate statistics for all the crimes committed.
   a. If the average dollar loss in a burglary, for this sample of size 500, is $678, with a standard deviation of $560, construct the 95 percent confidence interval for the true average dollar loss in burglaries.
   b. An assistant to the mayor, who claims to understand statistics, complains about your confidence interval calculation. She asserts that the dollar losses from burglaries are not normally distributed, which in turn makes the confidence interval calculation meaningless. Assume that she is correct about the distribution of money loss. Does that imply that the calculation of a confidence interval is not appropriate? Why or why not?

7. From the 2002 GSS subsample, we find that 79.6 percent of respondents believe in some form of life after death ($N = 666$).
   a. What is the 95 percent confidence interval for the percent of the U.S. population that believe in life after death?
   b. Without doing any calculations, make an educated guess at the lower and upper bounds of 90 percent and 99 percent confidence intervals.

8. A social service agency plans to conduct a survey to determine the mean income of its clients. The director of the agency prefers that you measure the mean income very accurately, to within ± $500. From a sample taken 2 years ago, you estimate that the standard deviation of income for this population is about $5,000. Your job is to figure out the necessary sample size to reduce sampling error to ± $500.
   a. Do you need to have an estimate of the current mean income to answer this question? Why or why not?
   b. What sample size should be drawn to meet the director's requirement at the 95 percent level of confidence? (Hint: Use the formula for a confidence interval and solve for N, the sample size.)
   c. What sample size should be drawn to meet the director's requirement at the 99 percent level of confidence?

9. Data from a 2002 General Social Survey subsample show that the mean number of children per respondent was 1.82, with a standard deviation of 1.69. A total of 1,498 people answered this question. Estimate the population mean number of children per adult using a 90 percent confidence interval.

10. In their November 1999 report, "Kids and Media at the New Millennium," the Kaiser Family Foundation released information on children's media use.[3] Based on a nationally representative sample of 3,155 children ages 2 to 18, the Kaiser Family Foundation report documents the amount and type of media use by American children. In their study, they found that 64 percent of children spend more than 1 hour a day watching television and 20 percent of children spend more than 1 hour a day reading. Estimate at the 99 percent confidence level the proportion of all children who watch television for more than 1 hour a day. Estimate at the 99 percent confidence level the proportion of children who read more than 1 hour per day.

---

[3]Data from this publication were reprinted with permission of the Henry J. Kaiser Family Foundation of Menlo Park, California. The Kaiser Family Foundation is an independent health care philanthropy and is not associated with Kaiser Permanente or Kaiser Industries.

11. According to the Gallup Organization, there has been little change in public opinion in favor of the death penalty in cases of murder. For their survey on the death penalty, the Gallup Poll surveyed 1,012 adults during August 29–September 5, 2000. In response to the question, "Are you in favor of the death penalty for a person convicted of murder?" 67 percent were in favor. What is the 95 percent confidence interval for the percentage of adult Americans who support the death penalty in cases of murder?

12. The Social Security system in the United States may encounter serious financial difficulties as baby boomers begin to retire in the future. Several polls have asked Americans their opinion about the financial condition of Social Security. In one poll, taken in August 2000 by the Gallup Organization (Gallup Poll Topics: A–Z, Social Security), 42 percent of a sample of 798 nonretired persons said they did not think "the Social Security system will be able to pay you a benefit when you retire."
    a. Calculate the 95 percent confidence interval to estimate the percentage of Americans who don't think that Social Security will be able to provide for them.
    b. Calculate the 99 percent confidence interval.
    c. Are both of these results compatible with the conclusion that fewer than 50 percent of Americans believe that the Social Security system will not be able to pay their benefits after retirement?

13. Whether one views homosexual relations as wrong is closely related to whether one views homosexuality as a biological trait or based on one's environment and socialization. Thus, it is not surprising that several religious groups that condemn homosexual relations have proclaimed their ability to "cure" gays of their sexual orientation. After all, their assumption is that homosexuality is not a trait that a person is born with. In 2002, GSS respondents ($N = 475$) were asked what they thought about homosexual relations. The data show that 56 percent believed that homosexual relations were always wrong, while 33 percent believed that homosexual relations were not wrong at all.
    a. For each reported percentage, calculate the 95 percent confidence interval.
    b. About 10 percent of GSS respondents were in the middle, some saying that homosexual relations were almost always wrong or sometimes wrong. Calculate the 95 percent confidence interval.
    c. What conclusions can you draw about the public's opinions of homosexual behavior based on your calculations?

14. What should be the top priority of the president of the United States? Should it be to improve education? Health care? Or race relations? The Gallup Organization asked these questions of 1,018 adults. (Data from Jeffrey J. Jones, January 10, 2001. Used by permission of the Gallup Poll.) Among the respondents, 50 percent indicated that education should be a top priority, 43 percent thought that improving health care should be a top priority, and 28 percent said that improving race relations should be a top priority of the administration.
    a. Calculate the 95 percent confidence interval for each proportion reported.
    b. Write a summary based on your calculations. Does any issue have overwhelming support as a top priority (over 50%)?

15. In recent decades there has been increasing concern about the state of the environment. Aside from the environment as a political issue, some environmentalists have begun to address how people view nature and the environment. Is nature simply a commodity—something to be owned, manipulated, abused, and eventually profitable? Or is nature something sacred—because it has always existed without human intervention, it is a gift from God, etc.?
    a. How do you think the Western world views nature? How about the United States, in particular?

b. In 2000, approximately 26 percent of ISSP respondents ($N = 326$) from the United States said that nature was important, but not sacred. This compares with 15 percent of Chilean respondents ($N = 221$) who said that nature was important, but not sacred. Construct confidence intervals for the percentage of American and Chilean respondents in the population who said that nature was important, but not sacred, at the 90 percent confidence level.

c. Approximately 75 percent of Filipino respondents ($N = 850$) said that nature is sacred because it is God's creation. With 99 percent confidence, construct an interval that contains the population percentage of Filipinos who think that nature is sacred because it is God's creation.

d. Only 8 percent of respondents from Denmark thought that nature is sacred because it is God's creation. Construct the narrowest possible confidence interval. Choose from one of three confidence levels and one of three sample sizes: 90 percent, 95 percent, or 99 percent and $N = 50$, $N = 100$, $N = 105$. Interpret your results.

Testing Hypotheses

M ore than a quarter of a century after Congress enacted major legislation aimed at equalizing opportunity in the workplace, African Americans continue to experience considerable earnings disadvantages relative to other workers in the labor market. For example, whites are almost twice as likely to be employed in managerial and professional fields, whereas blacks are much more concentrated in semiskilled labor and service occupations. Moreover, unemployment rates are considerably higher for blacks. These differences in labor market experiences are reflected in persistent gaps in income. Thus, we would expect the average income of black Americans to be lower than the average earnings of all Americans.

We drew a random sample of African Americans ($N = 100$) working full-time from the General Social Survey (GSS) and calculated their mean income for 1998. Based on census information[1] we also know the mean income of all Americans in 1998. We can thus compare the mean earnings of blacks with the mean earnings of all employed Americans in 1998. By comparing these means we are asking whether it is reasonable to consider the sample of black Americans a random sample that is representative of the population of workers in the United States. We expect to find the sample of blacks to be unrepresentative of the population of workers because we assume that blacks experience a considerable earnings disadvantage relative to other workers in the labor market.

The mean earnings for our sample of blacks is $24,100. This figure is considerably lower than $28,985, the mean earnings of the population of full-time workers obtained from the census. But is the observed gap of $4,885 ($28,985 – $24,100) large enough to convince us that the sample of blacks is not representative of the population?

There is no easy answer to this question. The sample mean of $24,100 is considerably lower than $28,985, but it is an estimate based on a single sample. Thus, it could mean one of two things: (1) the average earnings of the black population are indeed lower than the national average, or (2) the average earnings of the black population are about the same as the national average, and this sample happens to show a particularly low mean.

How can we decide which of these explanations makes more sense? Because most estimates are based on single samples and different samples may result in different estimates, sampling results cannot be used directly to make statements about a population. We need a procedure that allows us to evaluate hypotheses about population parameters based on sample statistics. In Chapter 11 we saw that population parameters can be estimated from sample statistics. In this chapter we will learn how to use sample statistics to make decisions about population parameters. This procedure is called **statistical hypothesis testing**.

Statistical hypothesis testing   A procedure that allows us to evaluate hypotheses about population parameters based on sample statistics.

▣ ASSUMPTIONS OF STATISTICAL HYPOTHESIS TESTING

Statistical hypothesis testing requires several assumptions. These assumptions include considerations of the level of measurement of the variable, the method of sampling, the shape of

---

[1]*Current Population Survey,* March 1998, P-60 Series.

the population distribution, and the sample size. The specific assumptions may vary, depending on the test or the conditions of testing. However, without exception, *all* statistical tests assume random sampling. Tests of hypotheses about means also assume interval-ratio level of measurement and require that the population under consideration be normally distributed or that the sample size be larger than 50.

Based on our data, we can test the hypothesis that average earnings of black Americans are lower than the average national earnings. The test we are considering meets these conditions:

1. The sample is a subgroup in the GSS sample, which is a national probability sample, randomly selected.

2. The variable *income* is measured at the interval-ratio level.

3. We cannot assume that the population is normally distributed. However, because our sample size is sufficiently large ($N > 50$), we know, based on the Central Limit Theorem, that the sampling distribution of the mean will be approximately normal.

▣ STATING THE RESEARCH AND NULL HYPOTHESES

Hypotheses are usually defined in terms of interrelations between variables and are often based on a substantive theory. Earlier, we defined *hypotheses* as tentative answers to research questions. They are tentative because they can be verified only after being empirically tested. The testing of hypotheses is an important step in this verification process.

The Research Hypothesis (H_1)

Our first step is to formally express the hypothesis in a way that makes it amenable to a statistical test. The substantive hypothesis is called the **research hypothesis** and is symbolized by H_1. Research hypotheses are always expressed in terms of population parameters because we are interested in making statements about population parameters based on our sample statistics.

In our research hypothesis (H_1), we state that the average wages of blacks are lower than the average wages of all Americans. We are stating a hypothesis about the relationship between race and wages in the general population by comparing the mean earnings of the black population to the mean national earnings of \$28,985. Symbolically, we use μ_Y to represent the black population mean; our hypothesis can be expressed as

$$H_1: \mu_Y < \$28,985$$

In general, the research hypothesis (H_1) specifies that the population parameter is one of the following:

1. Not equal to some specified value: $\mu_Y \neq$ some specified value

2. Greater than some specified value: $\mu_Y >$ some specified value

3. Less than some specified value: $\mu_Y <$ some specified value

> ***Research hypothesis (H₁)***   A statement reflecting the substantive hypothesis. It is always expressed in terms of population parameters, but its specific form varies from test to test.

The Null Hypothesis (H_0)

Is it possible that in the population there is no real difference between the mean wages of blacks and the mean wages of all Americans and that the observed difference of \$4,885 (\$28,985 – \$24,100) is actually due to the fact that this particular sample happened to contain blacks with lower earnings? Since statistical inference is based on probability theory, it is not possible to prove or disprove the research hypothesis directly. We can, at best, estimate the *likelihood* that it is true or false.

To assess this likelihood, statisticians set up a hypothesis that is counter to the research hypothesis. The **null hypothesis**, symbolized as H_0, contradicts the research hypothesis and usually states that there is no difference between the population mean and some specified value. It is also referred to as the hypothesis of "no difference." Our null hypothesis can be stated symbolically as

$$H_0: \mu_Y = \$28,985$$

Rather than directly testing the substantive hypothesis (H_1) that there is a difference between the mean wages of blacks and the wages nationally, we test the null hypothesis (H_0) that there is no difference in earnings. In hypothesis testing, we hope to reject the null hypothesis in order to provide support for the research hypothesis. Rejection of the null hypothesis will strengthen our belief in the research hypothesis and increase our confidence in the importance and utility of the broader theory from which the research hypothesis was derived.

> ***Null hypothesis (H₀)***   A statement of "no difference" that contradicts the research hypothesis and is always expressed in terms of population parameters.

More About Research Hypotheses: One- and Two-Tailed Tests

In a **one-tailed test,** the research hypothesis is directional; that is, it specifies that a population mean is either less than (<) or greater than (>) some specified value. We can express our research hypothesis as either

$H_1: \mu_Y <$ some specified value

or

$H_1: \mu_Y >$ some specified value

The research hypothesis we've stated for the average earnings of the black population is a one-tailed test.

When a one-tailed test specifies that the population mean is *greater than* some specified value, we call it a **right-tailed test** because we will evaluate the outcome at the right tail of the sampling distribution. If the research hypothesis specifies that the population mean is *less than* some specified value, it is called a **left-tailed test** because the outcome will be evaluated at the left tail of the sampling distribution. Our example is a left-tailed test because the research hypothesis states that the mean earnings of the black population are less than $28,985.

Sometimes, we have some theoretical basis to believe there is a difference between groups, but we cannot anticipate the direction of that difference. For example, we may have reason to believe that the average income of African Americans is *different* from that of the general population, but we may not have enough research or support to predict whether it is *higher* or *lower*. When we have no theoretical reason for specifying a direction in the research hypothesis, we conduct a **two-tailed test**. The research hypothesis specifies that the population mean is not equal to some specified value. For example, we can express the research hypothesis about the mean earnings of blacks as

$$H_1: \mu_Y \neq \$28,985$$

With both one-tailed and two-tailed tests, our null hypothesis of no difference remains the same. It can be expressed as

$$H_0: \mu_Y = \text{some specified value}$$

One-tailed test   A type of hypothesis test that involves a directional hypothesis. It specifies that the values of one group are either larger or smaller than some specified population value.

Right-tailed test   A one-tailed test in which the sample outcome is hypothesized to be at the right tail of the sampling distribution.

Left-tailed test   A one-tailed test in which the sample outcome is hypothesized to be at the left tail of the sampling distribution.

Two-tailed test   A type of hypothesis test that involves a nondirectional research hypothesis. We are equally interested in whether the values are less than or greater than one another. The sample outcome may be located at both the low and high ends of the sampling distribution.

▣ DETERMINING WHAT IS SUFFICIENTLY IMPROBABLE: PROBABILITY VALUES AND ALPHA

Now let's put all our information together. We're assuming that our null hypothesis ($\mu_Y = \$28,985$) is true, and we want to determine whether our sample evidence casts doubt on that assumption, suggesting that actually our research hypothesis, $\mu_Y < \$28,985$, is correct. What are the chances that we would have randomly selected a sample of African Americans such that the average earnings are this much lower than $28,985, the average for the general population? We can determine the chances or probability because of what we know about the sampling distribution and its properties. We know, based on the Central Limit Theorem, that if our sample size is larger than 50, the sampling distribution of the mean is approximately normal, with a mean $\mu_{\bar{Y}} = \mu_Y$ and a standard deviation (standard error) of

$$\sigma_{\bar{Y}} = \frac{\sigma_Y}{\sqrt{N}}$$

We are going to assume that the null hypothesis is true and then see if our sample evidence casts doubt on that assumption. We have a population mean $\mu_Y = \$28,985$ and a standard deviation $\sigma_Y = \$23,335^2$. Our sample size is $N = 100$, and the sample mean is $24,100. We can assume that the distribution of means of all possible samples of size $N = 100$ drawn from this distribution would be approximately normal, with a mean of $28,985 and a standard deviation of

$$\sigma_{\bar{Y}} = \frac{\sigma_Y}{\sqrt{N}} = \frac{23,335}{\sqrt{100}} = 2,333.5$$

This sampling distribution is shown in Figure 12.1. Also shown in Figure 12.1 is the mean earnings we observed for our sample of African Americans.

Because this distribution of sample means is normal, we can use Appendix B to determine the probability of drawing a sample mean of $24,100 or smaller from this population. We will translate our sample mean into a Z score so we can determine its location relative to the population mean. In Chapter 9, we learned how to translate a raw score into a Z score by using Formula 9.1:

$$Z = \frac{Y - \bar{Y}}{S_Y}$$

Figure 12.1   Sampling Distribution of Sample Means Assuming H_0 Is True for a Sample $N = 100$

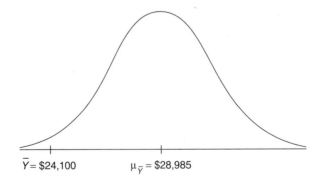

$\bar{Y} = \$24,100$ $\mu_{\bar{Y}} = \$28,985$

²Ibid.

Because we are dealing with a sampling distribution in which our raw score is $\overline{Y}$, the mean is μ_Y, and the standard deviation (standard error) is $\frac{\sigma_Y}{\sqrt{N}}$, we need to modify the formula somewhat:

$$Z = \frac{\overline{Y} - \mu_Y}{\frac{\sigma_Y}{\sqrt{N}}} \tag{12.1}$$

Converting the sample mean to a Z score equivalent is called computing the *test statistic*. The Z value we obtain is called the **Z statistic (obtained)**. The obtained Z gives us the number of standard deviations (standard errors) that our sample is from the hypothesized value (μ_Y or $\mu_{\overline{Y}}$), assuming the null hypothesis is true. For our example, the obtained Z is

$$Z = \frac{24,100 - 28,985}{\frac{23,335}{\sqrt{100}}} = -2.09$$

Z statistic (obtained)   The test statistic computed by converting a sample statistic (such as the mean) to a Z score. The formula for obtaining Z varies from test to test.

Before we determine the probability of our obtained Z statistic, let's determine whether it is consistent with our research hypothesis. Recall that we defined our research hypothesis as a left-tailed test ($\mu_Y < \$28,985$), predicting that the difference would be assessed on the left tail of the sampling distribution. The negative value of our obtained Z statistic confirms that we will be evaluating the difference on the left tail. (If we had a positive obtained Z, it would mean the difference would have to be evaluated at the right tail of the distribution, contrary to our research hypothesis.)

To determine the probability of observing a Z value of -2.09, assuming that the null hypothesis is true, look up the value in Appendix B to find the area to the left of (below) the negative Z of 2.09. Recall from Chapter 9, where we calculated Z scores and their probability, that the Z values are located in column A. The P value is the probability to the left of the obtained Z, or the "area beyond Z" in column C. This area includes the proportion of all sample means that are \$24,100 or lower. The proportion is .0183 (Figure 12.2). This value is the probability of getting a result as extreme as the sample result if the null hypothesis is true; it is symbolized as P. Thus, for our example, $P \leq .0183$.

A **P value** can be defined as the actual probability associated with the obtained value of Z. It is a measure of how unusual or rare our obtained statistic is, compared to what is stated in our null hypothesis. The smaller the P value, the more evidence we have that the null hypothesis is not true.

P value   The probability associated with the obtained value of Z.

Figure 12.2   The Probability (P) Associated with $Z \leq -2.09$

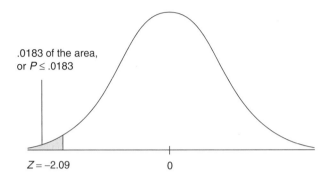

.0183 of the area, or $P \leq .0183$

$Z = -2.09$ 0

Researchers usually define in advance what is a sufficiently improbable Z value by specifying a cutoff point below which P must fall to reject the null hypothesis. This cutoff point, called **alpha** and denoted by the Greek letter α, is customarily set at the .05, .01, or .001 level. Let's say that we decide to reject the null hypothesis if $P \leq .05$. The value .05 is referred to as alpha (α); it defines for us what result is sufficiently improbable to allow us to take the risk and reject the null hypothesis. An alpha (α) of .05 means that even if the obtained Z statistic is due to sampling error, so the null hypothesis is true, we would allow a 5 percent risk of rejecting it. Alpha values of .01 and .001 are more cautionary levels of risk. The difference between P and alpha is that P is the *actual probability* associated with the obtained value of Z, whereas alpha is the level of probability *determined in advance* at which the null hypothesis is rejected. The null hypothesis is rejected when $P \leq \alpha$.

Alpha (α) The level of probability at which the null hypothesis is rejected. It is customary to set alpha at the .05, .01, or .001 level.

We have already determined that our obtained Z has a probability value of .0183. Since our observed P is less than .05 ($P = .0183 < \alpha = .05$), we can reject the null hypothesis. The value of .0183 means that fewer than 183 out of 10,000 samples drawn from this population are likely to have a mean that is 2.09 Z scores below the hypothesized mean of $28,985. Another way to say it is: There are only 183 chances out of 10,000 (or 1.83%) that we would draw a random sample with a $Z \leq -2.09$ if the mean earnings of blacks were equal to mean earnings of all Americans.

Based on the P value(s), we can also make a statement regarding the "significance" of the results. If the P value is equal to or less than our alpha level, our obtained Z statistic is considered *statistically significant*—that is to say, it is very unlikely to have occurred by random chance or sampling error. We can state that the difference between the average earnings of all Americans and the average earnings of African Americans is significant at the .05 level, or we can specify the actual level of significance by saying that the level of significance is .0183.

Recall that our hypothesis was a one-tailed test ($\mu_Y < \$28,985$). In a two-tailed test, sample outcomes may be located at both the high and the low ends of the sampling distribution. Thus, the null hypothesis will be rejected if our sample outcome falls either at the left or right tail of the sampling distribution. For instance, a .05 alpha or P level means that H_0 will be rejected if our sample outcome falls among either the lowest or the highest 5 percent of the sampling distribution.

Suppose we had expressed our research hypothesis about the mean income of blacks as

$$H_1: \mu_Y \neq \$28,985$$

The null hypothesis to be directly tested still takes the form $H_0: \mu_Y = \$28,985$ and our obtained Z is calculated using the same formula (12.1) as was used with a one-tailed test. To find P for a two-tailed test, look up the area in column C of Appendix B that corresponds to your obtained Z (as we did earlier) and then multiply it by 2 to obtain the two-tailed probability. Thus the two-tailed P value for $Z = -2.09$ is $.0183 \times 2 = .0366$. This probability is less than our stated alpha ($< .05$) and thus we reject the null hypothesis.

◙ THE FIVE STEPS IN HYPOTHESIS TESTING: A SUMMARY

Regardless of the particular application or problem, statistical hypothesis testing can be organized into five basic steps. Let's summarize these steps:

1. Making assumptions

2. Stating the research and null hypotheses and selecting alpha

3. Selecting the sampling distribution and specifying the test statistic

4. Computing the test statistic

5. Making a decision and interpreting the results

Making Assumptions   Statistical hypothesis testing involves making several assumptions regarding the level of measurement of the variable, the method of sampling, the shape of the population distribution, and the sample size. In our example we made the following assumptions:

1. A random sample was used.

2. The variable *income* is measured on an interval-ratio level of measurement.

3. Because $N > 50$, the assumption of normal population is not required.

Stating the Research and Null Hypotheses and Selecting Alpha   The substantive hypothesis is called the *research hypothesis* and is symbolized by H_1. Research hypotheses are always expressed in terms of population parameters because we are interested in making statements about population parameters based on sample statistics. Our research hypothesis was

$$H_1: \mu_Y < \$28,985$$

The *null hypothesis,* symbolized as H_0, contradicts the research hypothesis in a statement of no difference between the population mean and some specified value. For our example, the null hypothesis was stated symbolically as

$$H_0: \mu_Y = \$28,985$$

We set alpha at .05, meaning that we would reject the null hypothesis if the probability of our obtained Z was less than or equal to .05.

Selecting the Sampling Distribution and Specifying the Test Statistic   The normal distribution and the Z statistic are used to test the null hypothesis.

Computing the Test Statistic   Based on Formula 12.1, our Z statistic is −2.09.

Making a Decision and Interpreting the Results   We confirm that our obtained Z is on the left tail of the distribution, consistent with our research hypothesis. Based on our obtained Z statistic of −2.09, we determine that its P value is .0183, less than our .05 alpha level. We can reject the null hypothesis of no difference between the mean income of African Americans and the mean income of all Americans. We thus conclude that the income of black Americans is, on average, significantly lower than the national average.

◙ ERRORS IN HYPOTHESIS TESTING

We should emphasize that because our conclusion is based on sample data, we will never really know if the null hypothesis is true or false. In fact, as we have seen, there is a 1.83 percent chance that the null hypothesis is true and that we are making an error by rejecting it.

The null hypothesis can be either true or false, and in either case it can be rejected or not rejected. If the null hypothesis is true and we reject it nonetheless, we are making an incorrect decision. This type of error is called a **Type I error**. Conversely, if the null hypothesis is false but we fail to reject it, this incorrect decision is a **Type II error**.

In Table 12.1 we show the relationship between the two types of errors and the decisions we make regarding the null hypothesis. The probability of a Type I error—rejecting a true hypothesis—is equal to the chosen alpha level. For example, when we set alpha at the .05 level, we know the probability that the null hypothesis is in fact true is .05 (or 5%).

We can control the risk of rejecting a true hypothesis by manipulating alpha. For example, by setting alpha at .01, we are reducing the risk of making a Type I error to 1 percent. Unfortunately, however, Type I errors and Type II errors are inversely related; thus, by reducing alpha and lowering the risk of making a Type I error, we are increasing the risk of making a Type II error.

Table 12.1   Type I and Type II Errors

	True State of Affairs	
Decision Made	*H_0 is True*	*H_0 is False*
Reject H_0	Type I error (α)	Correct decision
Do not reject H_0	Correct decision	Type II error

As long as we base our decisions on sample statistics and not population parameters, we have to accept a degree of uncertainty as part of the process of statistical inference.

Type I error   The probability associated with rejecting a null hypothesis when it is true.

Type II error   The probability associated with failing to reject a null hypothesis when it is false.

✓ *Learning Check.*   *The implications of research findings are not created equal. For example, researchers might hypothesize that eating spinach increases the strength of weight lifters. Little harm will be done if the null hypothesis that eating spinach has no effect on the strength of weight lifters is rejected in error. The researchers would most likely be willing to risk a high probability of a Type I error, and all weight lifters would eat spinach. However, when the implications of research have important consequences, the balancing act between Type I and Type II errors becomes more important. Can you think of some examples where researchers would want to minimize Type I errors? When might they want to minimize Type II errors?*

The *t* Statistic and Estimating the Standard Error

The Z statistic we have calculated (Formula 12.1) to test the hypothesis involving a sample of black Americans assumes that the population standard deviation σ_Y is known. The value of σ_Y is required in order to calculate the standard error

$$\sigma_Y/\sqrt{N}$$

In most situations σ_Y will not be known and we will need to estimate it using the sample standard deviation S_Y. We then use the *t* statistic instead of the Z statistic to test the null hypothesis. The formula for computing the *t* statistic is

$$t = \frac{\overline{Y} - \mu_Y}{S_Y/\sqrt{N}} \tag{12.2}$$

The *t* value we calculate is called the **t statistic (obtained)**. The obtained *t* represents the number of standard deviation units (or standard error units) that our sample mean is from the hypothesized value of μ_Y, assuming the null hypothesis is true.

---

t statistic (obtained)   The test statistic computed to test the null hypothesis about a population mean when the population standard deviation is unknown and is estimated using the sample standard deviation.

The *t* Distribution and Degrees of Freedom

To understand the *t* statistic, we should first be familiar with its distribution. The **t distribution** is actually a family of curves, each determined by its *degrees of freedom.* The concept of degrees of freedom is used in calculating several statistics, including the *t* statistic. The **degrees of freedom (df)** represent the number of scores that are free to vary in calculating each statistic.

To calculate the degrees of freedom, we must know the sample size and whether there are any restrictions in calculating that statistic. The number of restrictions is then subtracted from the sample size to determine the degrees of freedom. When calculating the *t* statistic for a one-sample test, we start with the sample size *N* and lose 1 degree of freedom for the population standard deviation we estimate.[3] Notice that degrees of freedom will increase as the sample size increases. In the case of a single-sample mean, the df is calculated as follows:

$$df = N - 1 \qquad\qquad (12.3)$$

Comparing the *t* and *Z* Statistics

Notice the similarities between the formulas for the *t* and *Z* statistics. The only apparent difference is in the denominator. The denominator of *Z* is $\sigma_Y/\sqrt{N}$, the standard error based on the population standard deviation σ_Y. For the denominator of *t* we replace $\sigma_Y/\sqrt{N}$ with $S_Y/\sqrt{N}$, the estimated standard error based on the sample standard deviation.

However, there is another important difference between the *Z* and *t* statistics: Because it is estimated from sample data, the denominator of the *t* statistic is subject to sampling error. The sampling distribution of the test statistic is not normal, and the standard normal distribution cannot be used to determine probabilities associated with it.

In Figure 12.3 we present the *t* distribution for several dfs. Like the standard normal distribution, the *t* distribution is bell shaped. The *t* statistic, similar to the *Z* statistic, can have positive and negative values. A positive *t* statistic corresponds to the right tail of the distribution; a negative value corresponds to the left tail. Notice that when the df is small, the *t* distribution is much flatter than the normal curve. But as the degrees of freedom increase, the shape of the *t* distribution gets closer to the normal distribution, until the two are almost identical when df is greater than 120.

---

[3]To compute the sample variance for any particular sample, we must first compute the sample mean. Since the sum of the deviations about the mean must equal zero, only *N* − 1 of the deviation scores are free to vary with each variance estimate.

Figure 12.3 The Normal Distribution and *t* Distributions for 1, 5, 20, and ∞ Degrees of Freedom

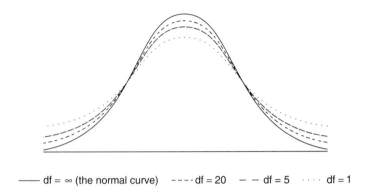

―― df = ∞ (the normal curve)   ---- df = 20   ― ― df = 5   · · · · df = 1

---

t distribution A family of curves, each determined by its degrees of freedom (df). It is used when the population standard deviation is unknown and the standard error is estimated from the sample standard deviation.

Degrees of freedom (df) The number of scores that are free to vary in calculating a statistic.

Appendix C summarizes the *t* distribution. We have reproduced a small part of this appendix in Table 12.2. Note that the *t* table differs from the normal (*Z*) table in several ways. First, the column on the left side of the table shows the degrees of freedom. The *t* statistic will vary depending on the degrees of freedom, which must first be computed (df = $N - 1$). Second, the probabilities or alpha, denoted as significance levels, are arrayed across the top of the table in two rows, the first for a one-tailed and the second for a two-tailed test. Finally, the value of *t*, listed as the entries of this table, are a function of (1) the degrees of freedom, (2) the level of significance, and (3) whether the test is a one- or a two-tailed test.

To illustrate the use of this table, let's determine the probability of observing a *t* value of 2.021 with 40 degrees of freedom and a two-tailed test. Locating the proper row (df = 40) and column (two-tailed test), we find the *t* statistic of 2.021 corresponding to the .05 level of significance. Restated, we can say that the probability of obtaining a *t* statistic of 2.021 is .05, or that there are less than 5 chances out of 100 that we would have drawn a random sample with an obtained *t* of 2.021 if the null hypothesis were correct.

Statistics in Practice: The Earnings of White Women

To illustrate the application of the *t* statistic, let's test a two-tailed hypothesis about a population mean μ_Y. We drew a 2002 GSS sample ($N = 371$) of white females who work

Table 12.2   Values of the t Distribution

	Level of Significance for One-Tailed Test					
	.10	.05	.025	.01	.005	.0005
	Level of Significance for Two-Tailed Test					
df	.20	.10	.05	.02	.01	.001
1	3.078	6.314	12.706	31.821	63.657	636.619
2	1.886	2.920	4.303	6.965	9.925	31.598
3	1.638	2.353	3.182	4.541	5.841	12.941
4	1.533	2.132	2.776	3.747	4.604	8.610
5	1.476	2.015	2.571	3.365	4.032	6.859
10	1.372	1.812	2.228	2.764	3.169	4.587
15	1.341	1.753	2.131	2.602	2.947	4.073
20	1.325	1.725	2.086	2.528	2.845	3.850
25	1.316	1.708	2.060	2.485	2.787	3.725
30	1.310	1.697	2.042	2.457	2.750	3.646
40	1.303	1.684	2.021	2.423	2.704	3.551
60	1.296	1.671	2.000	2.390	2.660	3.460
80	1.289	1.658	1.980	2.358	2.617	3.373
∞	1.282	1.645	1.960	2.326	2.576	3.291

Source: Abridged from R. A. Fisher and F. Yates, *Statistical Tables for Biological, Agricultural and Medical Research*, Table 111. Copyright © R. A. Fisher and F. Yates 1963. Reprinted by permission of Pearson Education Limited.

full-time. We found their mean earnings to be $28,889 with a standard deviation $S_Y = \$21,071$. Based on the *Current Population Survey*,[4] we also know that the 2002 mean earnings nationally for all women is $\mu_Y = \$24,146$. However, we do not know the value of the population standard deviation. We want to determine whether the sample of white women is representative of the population of all full-time women workers. Although we suspect that white American women experience a relative advantage in earnings, we are not sure enough to predict that their earnings are indeed higher than the earnings of all women nationally. Therefore, the statistical test is two-tailed.

Let's apply the five-step model to test the hypothesis that the average earnings of white women differ from the average earnings of all women working full-time in the United States.

Making Assumptions   Our assumptions are as follows:

1. A random sample is selected.

2. Because $N > 50$, the assumption of normal population is not required.

3. The level of measurement of the variable *income* is interval-ratio.

---

[4]*Current Population Survey,* March 2003, P-60 Series.

Stating the Research and the Null Hypotheses and Selecting Alpha   The research hypothesis is

$$H_1: \mu_Y \neq \$24,146$$

and the null hypothesis is

$$H_0: \mu_Y = \$24,146$$

We'll set alpha at .05, meaning that we will reject the null hypothesis if the probability of our obtained statistic is less than or equal to .05.

Selecting the Sampling Distribution and Specifying the Test Statistic   We use the t distribution and the t statistic to test the null hypothesis

Computing the Test Statistic   We first calculate the df associated with our test

$$df = (N - 1) = (371 - 1) = 370$$

To evaluate the probability of obtaining a sample mean of $28,889, assuming the average earnings of white women are equal to the national average of $24,146, we need to calculate the obtained t statistic using Formula 12.2:

$$t = \frac{\overline{Y} - \mu_Y}{S_Y / \sqrt{N}} = \frac{28,889 - 24,146}{21,071 / \sqrt{371}} = 4.33$$

Making a Decision and Interpreting the Results   Given our research hypothesis, we will conduct a two-tailed test. To determine the probability of observing a t value of 4.33 with 370 degrees of freedom, let's refer to Table 12.2. From the first column, we can see that 370 degrees of freedom is not listed, so we'll have to use the last row, df = ∞, to locate our obtained t statistic.

Though our obtained t statistic of 4.33 is not listed in the last row of t statistics, in fact, it is greater than the last value listed in the row, 3.291. The t statistic of 3.291 corresponds to the .001 level of significance for two-tailed tests. Restated, we can say that our obtained t statistic of 4.33 is greater than 3.291 or the probability of obtaining a t statistic of 4.33 is less than .001 ($P < .001$). This P value is below our .05 alpha level. The probability of obtaining the difference of $4,743 ($28,889 − $24,146) between the income of white women and the national average for all women, if the null hypothesis were true, is extremely low. We have sufficient evidence to reject the null hypothesis and conclude that the average earnings of white women are significantly different from the average earnings of all women. The difference of $4,743 is significant at the .05 level. We can also say that the level of significance is less than .001.

▣ TESTING HYPOTHESES ABOUT TWO SAMPLES

At the beginning of this chapter we illustrated the process of hypothesis testing with an example involving a single sample of African Americans. We were interested in testing the statistical significance of the difference between a sample value, the average earnings of

Table 12.3   Years of Education for White and Black Men, GSS 2002

	White Men Sample 1	Black Men Sample 2
Mean	13.47	12.51
Standard deviation	3.15	2.89
Variance	9.92	8.35
N	554	76

African Americans, and a population value, the average earnings of all Americans. In practice, social scientists are much more interested in situations involving two parameters than those involving one. We may be interested in finding out, for example, whether the average earnings of black women in the United States are lower than the average earnings of white men. Or we may wish to know whether black women earn the same, less, or more than black men.

For example, in the 2002 GSS, white men reported an average of 13.47 years of education and black men, an average of 12.51 years. Data for years of education are shown in Table 12.3. These sample averages could mean either (1) the average number of years of education for white men is higher than the average for black men or (2) the average for white men is actually about the same as for black men, but our sample just happens to indicate a higher average for white men.

The statistical procedures discussed in the following sections allow us to test whether the differences we observe between two samples are large enough for us to conclude that the populations from which these samples are drawn are different as well. We present tests for the significance of the differences between two groups. Primarily, we consider differences between sample means and differences between sample proportions.

Hypothesis testing with two samples follows the same structure as for single sample tests. The assumptions of the test are stated; the research and null hypotheses are formulated and the alpha level selected; the sampling distribution and the test statistic are specified; the test statistic is computed; and a decision is made whether or not to reject the null hypothesis.

The Assumption of Independent Samples

One important difference between one- and two-sample hypothesis testing involves sampling procedures. With a two-sample case we assume that the samples are independent of each other. The choice of sample members from one population has no effect on the choice of sample members from the second population. In our comparison of white and black men, we are assuming that the selection of white men is independent of the selection of black men. (The requirement of independence is also satisfied by selecting one sample randomly, then dividing the sample into appropriate subgroups. For example, we could randomly select a sample and then divide it into groups based on gender, religion, income, or any other attribute we are interested in.)

Stating the Research and Null Hypotheses

The second difference between one- and two-sample tests is in the form taken by the research and the null hypotheses. In one-sample tests both the null and the research hypotheses are statements about a single population parameter, μ_Y. In contrast, with two-sample tests we compare two population parameters.

Our research hypothesis (H_1) is that the average years of education is higher for white men than for black men. We are stating a hypothesis about the relationship between race and education in the general population by comparing the mean educational attainment of white men with the mean educational attainment of black men. Symbolically, we use μ to represent the population mean; the subscript 1 refers to our first sample (white men) and 2 to our second sample (black men). Our hypothesis can then be expressed as

$$H_1: \mu_1 > \mu_2$$

Because H_1 specifies that the mean education for white men is larger than the mean education for black men, it is a directional hypothesis. Thus, our test will be a one-tailed test. Alternatively, if there were not enough basis for deciding which population mean score is larger, the research hypothesis would state that the two population means are not equal and our test would be a two-tailed test:

$$H_1: \mu_1 \neq \mu_2$$

In either case the null hypothesis states that there are no differences between the two population means:

$$H_0: \mu_1 = \mu_2$$

We are interested in rejecting the null hypothesis of no difference so that we have sufficient support for our research hypothesis.

✔ *Learning Check.* *For the following research situations, state your research and null hypotheses:*

- *There is a difference between the mean statistics grades of social science majors and the mean statistics grades of business majors.*
- *The average number of children in two-parent black families is lower than the average number of children in two-parent nonblack families.*
- *Grade point averages are higher among girls who participate in organized sports than among girls who do not.*

▣ THE SAMPLING DISTRIBUTION OF THE DIFFERENCE BETWEEN MEANS

The sampling distribution allows us to compare our sample results with all possible sample outcomes and estimate the likelihood of their occurrence. Tests about differences between two sample means are based on the **sampling distribution of the difference between means.** The sampling distribution of the difference between two sample means is a theoretical probability distribution that would be obtained by calculating all the possible mean differences $(\bar{Y}_1 - \bar{Y}_2)$ by drawing all possible independent random samples of size N_1 and N_2 from two populations.

The properties of the sampling distribution of the difference between two sample means are determined by a corollary to the Central Limit Theorem. This theorem assumes that our samples are independently drawn from normal populations, but that with sufficient sample size ($N_1 > 50$, $N_2 > 50$) the sampling distribution of the difference between means will be approximately normal, even if the original populations are not normal. This sampling distribution has a mean $\mu_{\bar{Y}_1} - \mu_{\bar{Y}_2}$ and a standard deviation (standard error)

$$\sigma_{\bar{Y}_1 - \bar{Y}_2} = \sqrt{\frac{\sigma^2_{Y_1}}{N_1} + \frac{\sigma^2_{Y_2}}{N_2}} \tag{12.4}$$

which is based on the variances in each of the two populations ($\sigma^2_{Y_1}$ and $\sigma^2_{Y_2}$).

Sampling distribution of the difference between means   A theoretical probability distribution that would be obtained by calculating all the possible mean differences $(\bar{Y}_1 - \bar{Y}_2)$ that would be obtained by drawing all the possible independent random samples of size N_1 and N_2 from two populations where N_1 and N_2 are both greater than 50.

Estimating the Standard Error

Formula 12.4 assumes that the population variances are known and we can calculate the standard error $\sigma_{\bar{Y}_1 - \bar{Y}_2}$ (the standard deviation of the sampling distribution). However, in most situations the only data we have are based on sample data, and we do not know the true value of the population variances, $\sigma^2_{Y_1}$ and $\sigma^2_{Y_2}$. Thus, we need to estimate the standard error from the sample variances, $S^2_{Y_1}$ and $S^2_{Y_2}$. The estimated standard error of the difference between means is symbolized by $S_{\bar{Y}_1 - \bar{Y}_2}$ (instead of $\sigma_{\bar{Y}_1 - \bar{Y}_2}$).

Calculating the Estimated Standard Error

When we can assume that the two population variances are equal, we combine information from the two sample variances to calculate the estimated standard error.

$$S_{\bar{Y}_1 - \bar{Y}_2} = \sqrt{\frac{(N_1 - 1)S^2_{Y_1} + (N_2 - 1)S^2_{Y_2}}{(N_1 + N_2) - 2}} \sqrt{\frac{N_1 + N_2}{N_1 N_2}} \tag{12.5}$$

where $S_{\bar{Y}_1-\bar{Y}_2}$ is the estimated standard error of the difference between means, and $S_{Y_1}^2$ and $S_{Y_2}^2$ are the variances of the two samples. As a rule of thumb, when either sample variance is more than *twice* as large as the other, we can no longer assume that the two population variances are equal and would need to use Formula 12.8 in A Closer Look 12.1.

The *t* Statistic

As with single-sample means, we use the *t* distribution and the **t statistic** whenever we estimate the standard error for a difference between means test. The *t* value we calculate is the **obtained *t*.** It represents the number of standard deviation units (or standard error units) that our mean difference ($\bar{Y}_1 - \bar{Y}_2$) is from the hypothesized value of $\mu_1 - \mu_2$, assuming the null hypothesis is true.

The formula for computing the *t* statistic for a difference between means test is

$$t = \frac{\bar{Y}_1 - \bar{Y}_2}{S_{\bar{Y}_1-\bar{Y}_2}} \tag{12.6}$$

where $S_{\bar{Y}_1-\bar{Y}_2}$ is the estimated standard error.

Calculating the Degrees of Freedom for a Difference Between Means Test

To use the *t* distribution for testing the difference between two sample means, we need to calculate degrees of freedom. As we saw earlier, the degrees of freedom (df) represent the number of scores that are free to vary in calculating each statistic. When calculating the *t* statistic for the two-sample test, we lose 2 degrees of freedom, one for every population variance we estimate. When population variances are assumed equal or the size of both samples is greater than 50, the df is calculated as follows:

$$df = (N_1 + N_2) - 2 \tag{12.7}$$

When we cannot assume that the population variances are equal and when the size of one or both samples is equal to or less than 50, we use Formula 12.9 in A Closer Look 12.1 to calculate the degrees of freedom.

▣ THE FIVE STEPS IN HYPOTHESIS TESTING ABOUT DIFFERENCE BETWEEN MEANS: A SUMMARY

As with single-sample tests, statistical hypothesis testing involving two sample means can be organized into five basic steps. Let's summarize these steps:

1. Making assumptions

2. Stating the research and null hypotheses and selecting alpha

---

▣ A Closer Look 12.1
Calculating the Estimated Standard Error
and the Degrees of Freedom (df) When
the Population Variances Are Assumed Unequal

If the variances of the two samples ($S_{Y_1}^2$ and $S_{Y_2}^2$) are very different (one variance is twice as large as the other), the formula for the estimated standard error becomes

$$S_{\bar{Y}_1-\bar{Y}_2} = \sqrt{\frac{S_{Y_1}^2}{N_1} + \frac{S_{Y_2}^2}{N_2}} \qquad (12.8)$$

When the population variances are unequal and the size of one or both samples is equal to or less than 50, we use another formula to calculate the degrees of freedom associated with the t statistic:[4]

$$df = \frac{(S_{Y_1}^2/N_1 + S_{Y_2}^2/N_2)^2}{(S_{Y_1}^2/N_1)^2/(N_1-1) + (S_{Y_2}^2/N_2)^2/(N_2-1)} \qquad (12.9)$$

---

3. Selecting the sampling distribution and specifying the test statistic

4. Computing the test statistic

5. Making a decision and interpreting the results

Making Assumptions In our example we made the following assumptions:

1. Independent random samples are used.

2. The variable *years of education* is measured at an interval-ratio level of measurement.

3. Because $N_1 > 50$ and $N_2 > 50$, the assumption of normal population is not required.

4. The population variances are assumed equal.

Stating the Research and Null Hypotheses and Selecting Alpha The research hypothesis we are testing is that the mean education of white men is higher than the mean education of black men. Symbolically the research hypothesis is expressed as

$$H_1: \mu_1 > \mu_2$$

---

[4]Degrees of freedom formula based on Dennis Hinkle, William Wiersma, and Stephen Jurs, *Applied Statistics for the Behavioral Sciences* (Boston: Houghton Mifflin, 1998), p. 268.

with μ_1 representing the mean education of white men and μ_2 the mean education of black men. Note that because H_1 specifies that the mean education for white men is greater than the mean education for black men, it is a directional hypothesis. Thus, our test is a right-tailed test.

The null hypothesis states that there are no differences between the two population means, or

$$H_0: \mu_1 = \mu_2$$

We are interested in rejecting the null hypothesis of no difference so that we have sufficient support for our research hypothesis. We will reject the null hypothesis if the probability of t (obtained) is less than or equal to .05.

Selecting the Sampling Distribution and Specifying the Test Statistic   The t distribution and the t statistic are used to test the significance of the difference between the two sample means.

Computing the Test Statistic   To test the null hypothesis about the differences between the mean education of white and black men, we need to translate the ratio of the observed differences to its standard error into a t statistic (based on data presented in Table 12.3). The obtained t statistic is calculated using Formula 12.6:

$$t = \frac{\overline{Y}_1 - \overline{Y}_2}{S_{\overline{Y}_1 - \overline{Y}_2}}$$

where $S_{\overline{Y}_1 - \overline{Y}_2}$ is the estimated standard error of the sampling distribution. Because the population variances are assumed equal, df is $(N_1 + N_2) - 2 = (554 + 76) - 2 = 628$ and we can combine information from the two sample variances to estimate the standard error (Formula 12.5):

$$S_{\overline{Y}_1 - \overline{Y}_2} = \sqrt{\frac{(554 - 1)3.15^2 + (76 - 1)2.89^2}{(554 + 76) - 2}} \sqrt{\frac{554 + 76}{554(76)}} = .38$$

We substitute this value into the denominator for the t statistic (Formula 12.6):

$$t = \frac{13.47 - 12.51}{.38} = 2.53$$

Making a Decision and Interpreting the Results   We confirm that our obtained t is on the right tail of the distribution, consistent with our research hypothesis. Based on our obtained t statistic of 2.53, we determine that its P value is less than .01. This is less than our .05 alpha level, and we can reject the null hypothesis of no difference between white and black men's mean education. We conclude that white men, on average, have significantly higher years of education than black men do.

Statistics in Practice: The Earnings of Asian-American Men

Because of their socioeconomic achievement in American society, Asian Americans have been cited as a "model minority."[5] According to the 2000 census, the level of education and the

---

[5]Roger Daniels, *Asian Americans: Chinese and Japanese in the United States Since 1850* (Seattle: University of Washington Press, 1988).

average family income of Asian Americans are the highest among minority groups in the United States. Among Asian Americans, the achievements of Chinese and Japanese have been particularly impressive. The 2003 census reports that 44.7 percent of Asian Americans between the ages of 15 and 64 completed a bachelor's degree or higher compared with 25.1 percent of whites and 14.7 percent of blacks.[6] Similar trends have been observed in the relative earnings of Asian Americans. For example, in 2003, the median income of Asian Americans was the highest at $55,500, 117 percent of the median income of non-Hispanic white households.[7]

The success of Chinese and Japanese Americans and their image as a "model minority" challenge the predominant view that being nonwhite is an inherent liability to achievement in American society. It seems to reinforce the notion advanced by human capital theorists that the economic success of immigrant group members is determined solely by individual human capital (credentials and skills) and not by race or national origin. Thus, according to the human capital perspective, we would expect to find earnings parity between Chinese and Japanese Americans and non-Hispanic whites who have similar credentials.

Other researchers argue that the "model minority" image diverts attention from problems such as employment discrimination and economic marginality confronting Asian Americans. They suggest that the earnings parity between Asian Americans and white workers may be due to overachievement in educational attainment, longer working hours, and regional concentration of Asians in states such as California, where earnings are generally higher than in other states.[8] Thus, when factors such as education, work experience, and job training are controlled for, the earnings of Chinese and Japanese are lower than the earnings of white Americans. It is suggested that racial discrimination is the most likely explanation for these earning differentials.

To examine these competing explanations for the socioeconomic status of Chinese and Japanese Americans, Zhou and Kamo analyzed the earning patterns of Chinese and Japanese Americans relative to the earnings of non-Hispanic whites. Zhou and Kamo argue that, as a group, Chinese and Japanese Americans have not achieved earnings parity with whites with identical credentials. They based their analysis on random samples from the 1980 census.[9] Their sample is limited to male workers between the ages of 25 and 64 who were in the labor force.[10]

Table 12.4 shows the mean earnings and standard deviations for U.S.-born Chinese, Japanese, and non-Hispanic whites who reside in California and who have a college degree.[11]

---

[6]Nicole Stoops, *Educational Attainment in the United States: 2003* (U.S. Census Bureau, Current Population Report, P20-550).

[7]DeNavas-Walt, Carmen, Bernadette Proctor and Robert J. Mills. 2004. *Income in the United States: 2003*. Current Population Reports, P60-226. Washington, DC: U.S. Government Printing Office.

[8]Zhou Min and Yoshinori Kamo; William P. O'Hare and Judy C. Felt, *Asian Americans: America's Fastest Growing Minority Group* (Washington, DC: Population Reference Bureau, 1991).

[9]U.S. Bureau of the Census, *Public-Use Microdata Samples*, 1983.

[10]To justify the exclusion of females from the analysis, Zhou and Kamo argue that the nature of female employment differs from that of male employment and that patterns of female labor participation vary among racial and ethnic groups.

[11]Zhou and Kamo analyzed the earnings of additional subsamples of Chinese, Japanese, and non-Hispanic whites. We are focusing on U.S.-born persons who reside in California and who have a college degree.

Table 12.4   Means and Standard Deviations for Earnings of College-Educated Chinese, Japanese, and Non-Hispanic White Americans (in California)

	Chinese	*Japanese*	*Non-Hispanic White*
Mean	$21,439	$22,907	$24,891
Standard deviation	10,289	11,120	14,225
N	471	758	2,123

Source: Adapted from M. Zhou and Y. Kamo, *The Sociological Quarterly* 35, no. 4 (November 1994), Table 2, p. 591. © 1994 by JAI Press. Reprinted by permission of the University of California Press.

The table shows that Chinese and Japanese Americans have not achieved earnings parity with whites, despite similar credentials (college degree) and employment in the same labor market (California). The mean earnings for whites ($\overline{Y}$ = $24,891) were higher than the earnings of either Japanese ($\overline{Y}$ = $22,907) or Chinese American ($\overline{Y}$ = $21,439) workers.

Does the gap of $3,452 ($24,891 − $21,439) or $1,984 ($24,891 − $22,907) between the earnings of Chinese and Japanese Americans, respectively, and the earnings of whites provide support for Zhou and Kamo's argument? We can use the data shown in Table 12.4 and the procedure of hypothesis testing about differences between means to answer this question. We limit our discussion to a test of the difference in mean earnings between Chinese Americans and non-Hispanic whites. We calculate the *t* statistic to test whether the observed difference in earnings between these two groups is large enough for us to conclude that the populations from which these samples are drawn are different as well.

Making Assumptions   Our assumptions are as follows:

1. Independent random samples are selected.

2. The level of measurement of the variable *income* is interval-ratio.

3. Because $N_1 > 50$ and $N_2 > 50$, the assumption of normal population is not required.

4. The population variances are assumed equal.

Stating the Research and Null Hypotheses and Selecting Alpha   If Chinese Americans as a group have not achieved earnings parity with whites with identical credentials, as suggested by Zhou and Kamo, we would expect the earnings of non-Hispanic white men to be higher than the earnings of Chinese American men. The research hypothesis we will test is that the mean earnings for the population of non-Hispanic white men are greater than the mean earnings of the population of Chinese American men.

Symbolically, the hypothesis is expressed as

$$H_1: \mu_1 > \mu_2$$

with μ_1 representing the mean earnings of white men and μ_2 the mean earnings of Chinese American men. Note that because H_1 specifies that the mean earnings for white men are greater than the mean earnings for Chinese men, it is a directional hypothesis. Thus, our test is a right-tailed test.

The null hypothesis states that there are no differences between the two population means, or

$$H_0: \mu_1 = \mu_2$$

We are interested in rejecting the null hypothesis of no difference so that we have sufficient support for our research hypothesis that white men's earnings are higher than Chinese men's. We will reject the null hypothesis if the probability of t (obtained) is less than or equal to .05.

Selecting the Sampling Distribution and Specifying the Test Statistic   We use the t distribution and the t statistic to test the significance of the difference between the two sample means.

Computing the Test Statistic   To test the null hypothesis about the difference between the mean earnings of Chinese Americans and non-Hispanic whites, we need to translate the ratio of the observed difference to its standard error into a t statistic. The obtained t statistic is calculated using Formula 12.6

$$t = \frac{\overline{Y}_1 - \overline{Y}_2}{S_{\overline{Y}_1 - \overline{Y}_2}}$$

where $S_{\overline{Y}_1 - \overline{Y}_2}$ is the estimated standard error of the sampling distribution. Because the population variances are assumed equal, df is $N_1 + N_2 - 2 = 2{,}123 + 471 - 2 = 2{,}592$.

Because we assume that the population variances are equal, we can combine information from the two sample variances to estimate the standard error (Formula 12.5):

$$S_{\overline{Y}_1 - \overline{Y}_2} = \sqrt{\frac{(N_1 - 1)S_{Y_1}^2 + (N_2 - 1)S_{Y_2}^2}{(N_1 + N_2) - 2}} \sqrt{\frac{N_1 + N_2}{N_1 N_2}}$$

For our example, the estimate for $S_{\overline{Y}_1 - \overline{Y}_2}$ is

$$S_{\overline{Y}_1 - \overline{Y}_2} = \sqrt{\frac{2{,}122(14{,}225)^2 + 470(10{,}289)^2}{(2{,}123 + 471) - 2}} \sqrt{\frac{2{,}123 + 471}{(2{,}123)(471)}} = 692.48$$

We substitute this value into the denominator for the t statistic (Formula 12.6):

$$t = \frac{24{,}891 - 21{,}439}{692.48} = 4.98$$

Making a Decision and Interpreting the Results   Consistent with our research hypothesis, the obtained t is positive, on the right tail of the distribution. Since the obtained t of 4.98 is greater than $t = 3.291$ (df $= \infty$, one-tailed test; see Appendix C), we can state that its probability is less than .0005. This P value is below our .05 alpha level. The probability of obtaining this difference between non-Hispanic white and Chinese American men, if the null hypothesis were true, is extremely low. We have sufficient evidence to reject the null hypothesis and conclude that the average earnings of non-Hispanic white males are significantly higher than the average earnings of Chinese American men. The difference of $3,462 is significant at the .0005 level.

✓ *Learning Check.* *Would you change your decision in the previous example if we selected an alpha of .01?*

Using the data presented in Table 12.4, test the null hypothesis that the mean earnings of Japanese Americans are equal to the mean earnings of non-Hispanic whites. Set alpha at .05. What is your conclusion?

⊡ TESTING THE SIGNIFICANCE OF THE DIFFERENCE BETWEEN TWO SAMPLE PROPORTIONS

In the preceding sections, we have learned how to test for the significance of the difference between two population means when the variable is measured at an interval-ratio level. Yet numerous variables in the social sciences are measured at a nominal or an ordinal level. These variables are often described in terms of proportions. For example, we might be interested in comparing the proportion of blacks and whites who support gun control or the proportion of men and women who support federal funding for abortion. In this section we present statistical inference techniques to test for significant differences between two sample proportions.

Hypothesis testing with two sample proportions follows the same structure as the statistical tests presented earlier. The assumptions of the test are stated; the research and null hypotheses are formulated; the sampling distribution and the test statistic are specified; the test statistic is calculated; and a decision is made whether or not to reject the null hypothesis.

Statistics in Practice: Equalizing Income

Do middle and working class individuals feel the same way about the government's responsibility to equalize income? We can use the 2000 International Social Survey Programme (ISSP) data to test the null hypothesis that the proportion of middle and working class respondents from different countries who believe that it is the government's responsibility to reduce income differences is equal. The proportion of middle class respondents who reported that they agree that it is the government's responsibility to reduce income differences was 0.18 (p_1); the proportion of lower class with the same response was 0.30 (p_2). Four hundred fifty middle class (N_1) and 300 lower class respondents (N_2) answered this question.

Making Assumptions   Our assumptions are as follows:

1. Independent random samples of $N_1 > 50$ and $N_2 > 50$ are used.

2. The level of measurement of the variable is nominal.

Stating the Research and Null Hypotheses and Selecting Alpha   We propose a two-tailed test, that the population proportions for middle and working class are not equal.

$$H_1: \pi_1 \neq \pi_2$$

$$H_0: \pi_1 = \pi_2$$

We decide to set alpha at .05.

Selecting the Sampling Distribution and Specifying the Test Statistic   The population distributions of dichotomies are not normal. However, based on the central limit theorem, we know that the sampling distribution of the difference between sample proportions is normally distributed when the sample size is large (when $N_1 > 50$ and $N_2 > 50$), with mean μ_{p1-p2} and the estimated standard error S_{p1-p2}. Therefore, we can use the normal distribution as the sampling distribution and calculate Z as the test statistic.[12]

The formula for computing the Z statistic for a difference between proportions test is

$$Z = \frac{p_1 - p_2}{S_{p1-p2}} \tag{12.10}$$

where p_1 and p_2 are the sample proportions for middle and working class, and S_{p1-p2} is the estimated standard error of the sampling distribution of the difference between sample proportions.

The estimated standard error is calculated using the following formula:

$$S_{p1-p2} = \sqrt{\frac{p_1(1 - p_1)}{N_1} + \frac{p_2(1 - p_2)}{N_2}} \tag{12.11}$$

Calculating the Test Statistic   We calculate the standard error using Formula 12.11:

$$S_{p1-p2} = \sqrt{\frac{.18(1 - .18)}{450} + \frac{.30(1 - .30)}{300}} = .03$$

Substituting this value into the denominator of Formula 12.10. we get

$$Z = \frac{.18 - .30}{.03} = -4.00$$

---

[12]The sample proportions are unbiased estimates of the corresponding population proportions. Therefore, we can use the Z statistic, although our standard error is estimated from the sample proportions.

Making a Decision and Interpreting the Results   Our obtained Z of -4.00 indicates that the difference between the two proportions will be evaluated at the left tail of the Z distribution. To determine the probability of observing a Z value of $|-4.00|$ if the null hypothesis is true, look up the value in Appendix B (column C) to find the area to the right of (above) the obtained Z.

We can state that the P value of -4.00 is less than .0001 for a one-tailed test. However, for a two-tailed test we'll have to multiply P by 2 $(.0001 \times 2 = .0002)$. The probability of -4.00 for a two-tailed test is less than our alpha level of .05 $(.0002 < .05)$.

Thus, we reject the null hypothesis of no difference and conclude that there is a significant class difference in the opinion that it is the government's responsibility to reduce income differences. Working-class respondents are more likely than middle class respondents to believe that it is the government's responsibility to reduce income differences.

✓ *Learning Check.*   *Would you change your decision if we selected an alpha of .01? Why or why not?*

Statistics in Practice: Equalizing Income and Educational Attainment

Would our results be different if we compared educational attainment—those with a secondary degree and those with a college degree? We have randomly selected a sample from the 2000 ISSP survey and compared the proportion of secondary degree and university degree respondents who agreed that the government should reduce income differences.

Secondary Degree	University Degree
$p_1 = 0.21$	$p_2 = 0.20$
$N_1 = 356$	$N_2 = 197$

Making Assumptions Our assumptions are as follows:

1. Independent random samples of $N_1 > 50$ and $N_2 > 50$ are used.

2. The level of measurement of the variable is nominal.

Stating the Research and Null Hypotheses and Selecting Alpha   We assume that the population proportions are not equal.

$$H_1: \pi_1 \neq \pi_2$$
$$H_0: \pi_1 = \pi_2$$

We will set alpha at .05.

Selecting the Sampling Distribution and Specifying the Test Statistic   The sampling distribution is the normal distribution, and the test statistic is Z.

Calculating the Test Statistic   We calculate the standard error using Formula 12.11:

$$S_{p_1-p_2} = \sqrt{\frac{.21(1-.21)}{356} + \frac{.20(1-.20)}{197}} = .04$$

Substituting this value into the denominator of Formula 12.10, we get

$$Z = \frac{.21 - .20}{.04} = .25$$

Making a Decision and Interpreting the Results   The positive Z obtained of .25 indicates that the difference between the proportions will be evaluated at the right tail of the distribution. The two-tailed probability of .25 is $.4013 \times 2 = .8026$. This is a very large P value—much larger than our alpha of .05. We cannot reject the null hypothesis of no difference. We conclude that the observed differences between ISSP respondents with a secondary degree and those with a university degree regarding government's responsibility to reduce income differences probably do not reflect a difference that would have been seen had the entire population been measured.

▣ READING THE RESEARCH LITERATURE: REPORTING THE RESULTS OF STATISTICAL HYPOTHESIS TESTING

Let's conclude with an example of how the results of statistical hypothesis testing are presented in the social science research literature. Keep in mind that the research literature does not follow the same format or the degree of detail that we've presented in this chapter. For example, most research articles do not include a formal discussion of the null hypothesis or the sampling distribution. The presentation of statistical analyses and detail will vary according to the journal's editorial policy or the standard format for the discipline.

It is not uncommon for a single research article to include the results of 10 to 20 statistical tests. Results have to be presented succinctly and in summary form. An author's findings are usually presented in a summary table that may include the sample statistics (for instance, the sample means), the obtained test statistics (t or Z), the alpha level, and an indication of whether or not the results are statistically significant.[13]

Kenneth Ferraro (1996) examined the link between fear and the perceived risk of violent crime among men and women.[14] According to Ferraro, gender is the most important predictor

---

[13]A similar discussion is presented in Joseph F. Healey, *Statistics: A Tool for Social Research,* 3rd ed. (Belmont, CA: Wadsworth, 1999), pp. 216–217.

[14]Kenneth F. Ferraro, "Women's Fear of Victimization: Shadow of Sexual Assault?" *Social Forces* 75, no. 2 (1996): 667–690.

Table 12.5   Means and Standard Deviations for Victimization Fear and Perceived
Risk by Sex

	Men	*Women*	*Mean Difference*
Type of Fear			
Sexual assault	2.21+2.47	6.09+3.36	3.88**
Murder	3.48+3.05	5.30+3.67	1.82**
Robbery	3.66+2.54	5.05+3.16	1.39**
Assault	4.31+2.92	5.69+3.40	1.38**
Burglary/home	3.85+2.97	5.90+3.42	2.05**
Car theft	4.25+2.79	4.76+2.90	.51**
Burglary/away	5.18+2.82	6.18+2.98	1.00**
Cheat/con	3.40+2.56	3.89+2.90	.49**
Vandalism	4.31+2.58	4.89+2.95	.58**
Panhandler	2.36+1.86	3.36+2.57	1.00**
Fear (total)	37.02+18.41	51.22+23.25	14.20**
Type of Risk			
Sexual assault	1.38+1.03	2.98+2.33	1.60**
Murder	1.80+1.64	2.27+2.19	.47**
Robbery	2.66+1.98	3.21+2.51	.55**
Assault	2.38+1.85	2.76+2.34	.38**
Burglary/home	2.07+1.77	2.79+2.21	.72**
Car theft	3.60+2.43	3.79+2.66	.19
Burglary/away	3.55+2.29	4.20+2.57	.65**
Cheat/con	4.08+2.90	3.60+2.78	−.48**
Vandalism	3.37+2.34	3.72+2.75	.35*
Panhandler	4.54+3.48	3.83+3.31	−.71**
Risk (total)	29.34+13.84	33.26+17.98	3.92**

*p < .05 **p < .01 *Note:* Differences assessed by *t*-test of means.

Source: Adapted from K. Ferraro, "Women's Fear of Victimization: Shadow of Sexual Assault?" *Social Forces* 72, no. 2 (1996), Table 2, p. 676. Copyright © The University of North Carolina Press. Used by permission.

of the fear of crime. Women are more likely to report fear of crime than men. His sample is based on the Fear of Crime in America Survey 1990, with 1,101 respondents. Given a list of 10 types of crime, respondents were asked to rate their fear of being a victim on a 10-point scale; 1 indicates they were "not afraid at all" and 10 means that they were "very afraid." They were also asked to rate the chance that each of the 10 crimes would happen to them in the coming year on a 10-point scale; 1 means that they consider "it's not at all likely" and 10 indicates that "it's very likely" to occur. In addition, Ferraro calculated the total scores for all fear and risk items. Ferraro's results are summarized in Table 12.5.

Let's examine this table carefully. Each row represents the mean responses and standard deviation for men and women for a particular type of fear or risk. In the last column of the table, Ferraro presents the mean difference between the scores for men and women (subtracting men from women). Obtained *t*-test statistics are not presented. Yet the footnote informs us the differences were assessed by *t*-test of means (there is no indication whether the tests were one- or two-tailed). The asterisks in the last column indicate which differences are statistically significant at the .05 (*) and .01 (**) levels. Note that with the exception of one risk comparison (car theft), all mean comparisons are significant.

Ferraro concludes:

> For each battery of victimization, the personal offenses are listed first. Consistent with most previous research, women in the Fear of Crime in America survey are more fearful of all the ten offenses considered. The differences are greater for the personal crimes including rape, burglary, and robbery, but the difference in fear of sexual assault is particularly dramatic. Women in this sample were more afraid of rape than murder. . . . While the difference in total victimization fear is over 14, the difference in the total across types of perceived risk is less than 4. Men are more likely than women to perceive high risk of being approached by a panhandler and cheated or conned out of money. There is no significant gender difference in perceived risk of car theft.[15]

MAIN POINTS

• Statistical hypothesis testing is a decision-making process that enables us to determine whether a particular sample result falls within a range that can occur by an acceptable level of chance. The process of statistical hypothesis testing consists of five steps: making assumptions; stating the research and null hypotheses and selecting alpha; selecting a sampling distribution and a test statistic; computing the test statistic; and making a decision and interpreting the results.

• Statistical hypothesis testing may involve a comparison between a sample mean and a population mean or a comparison between two sample means. If we know the population variance(s) when testing for differences between means, we can use the Z statistic and the normal distribution. However, in practice, we are unlikely to have this information.

• When testing for differences between means when the population variance(s) are unknown, we use the t statistic and the t distribution.

• Tests involving differences between proportions follow the same procedure as tests for differences between means when population variances are known. The test statistic is Z, and the sampling distribution is approximated by the normal distribution.

KEY TERMS

alpha (α)
degrees of freedom (df)
left-tailed test
null hypothesis (H_0)
one-tailed test
P value
research hypothesis (H_1)
right-tailed test

sampling distribution of the difference between means
statistical hypothesis testing
t distribution
t statistic (obtained)
two-tailed test
Type I error
Type II error
Z statistic (obtained)

---

[15]Ibid, p. 675.

ON YOUR OWN

Log on to the web-based student study site at http://www.pineforge.com/frankfort-nachmiasstudy4 for additional study questions, quizzes, web resources, and links to social science journal articles reflecting the statistics used in this chapter.

SPSS DEMONSTRATIONS

[GSS02PFP-B]

Demonstration 1: Producing a One-Sample T Test

In this chapter we discussed methods of testing differences in means between a sample and a population value. SPSS includes a One-Sample T Test procedure to do this test. SPSS does not do the test with the Z statistic; instead, it uses the *t* statistic to test for all mean differences. The One-Sample T Test procedure can be found under the *Analyze* menu choice, then under *Compare Means*, where it is labeled *One-Sample T Test*. The opening dialog box (Figure 12.4) requires that you place at least one variable in the Test Variable(s) box. Then a test value must be specified.

We'll use the 2002 GSS data for this demonstration. The standard workweek is thought to be 40 hours, so let's test to see whether American adults work that many hours each week. In this example, place HRS1 in the Test Variable(s) box and "40" in the Test Value box. Then click on *OK* to run the procedure.

The output from the One-Sample T Test procedure is not very extensive (see Figure 12.5). A total of 955 people answered the question about number of hours worked per week. The mean number of hours worked is 41.90, with a standard deviation of 14.13. Below this, SPSS lists the test value, 40. It includes the two-tailed significance, or probability, for the one-sample test. This value is .000, given the calculated *t* statistic of 4.145, with 954 degrees of freedom. Thus, at the .01 significance level, we would reject the null hypothesis and conclude that American adults work more than 40 hours a week.

SPSS also supplies a 95 percent confidence interval for the mean difference between the test value and the sample mean. Here, the confidence interval runs from 1.00 to 2.79, providing estimates of how much more than 40 hours per week Americans work.

Figure 12.4

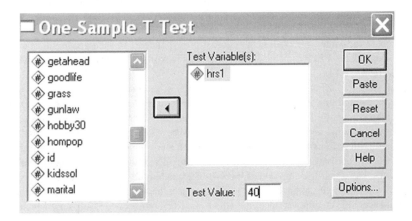

Figure 12.5

One-Sample Statistics

	N	Mean	Std. Deviation	Std. Error Mean
NUMBER OF HOURS WORKED LAST WEEK	955	41.90	14.131	.457

One-Sample Test

	Test Value = 40					
					95% Confidence Interval of the Difference	
	t	df	Sig. (2-tailed)	Mean Difference	Lower	Upper
NUMBER OF HOURS WORKED LAST WEEK	4.145	954	.000	1.895	1.00	2.79

Demonstration 2: Producing a Test of Mean Differences

In this chapter we have also discussed methods of testing differences in means or proportions between two samples (or groups). The Two-Sample T Test procedure can be found under the *Analyze* menu choice, then under *Compare Means*, where it is labeled *Independent-Samples T Test*.

The opening dialog box requires that you specify various test variables (the dependent variable) and one independent or grouping variable (Figure 12.6). We'll test the null hypothesis that whites and blacks work the same number of hours each week by using the variable HRS1. Place that variable in the Test Variable(s) box and RACECEN1 in the Grouping Variable box. When you do so, question marks appear next to RACECEN1, indicating that you must supply two values to define the two groups (independent samples). Click on *Define Groups*. Then put "1" in the first box and "2" in the second box (1 = white and 2 = black), as shown in Figure 12.7. Then click on *Continue* and *OK* to run the procedure.

The output from Independent Samples T Test (Figure 12.8) is detailed and contains more information than we have reviewed in this chapter. The first part of the output displays the mean number of hours worked for whites and blacks, the number of respondents in each group, the standard deviation, and the standard error of the mean. We see that whites worked 1.14 hours more per week than blacks $(42.10 - 40.96 = 1.14)$.

Recall from the chapter that an important decision of the *t* statistic calculation is whether the variances of the two groups are equal. If the variances are assumed equal, you can use a simple formula to calculate the number of degrees of freedom. However, SPSS can take account of those times when the variances are unequal and still calculate an appropriate *t* statistic and degrees of freedom. To do so, SPSS does a direct test of whether or not the variances for hours worked are identical for blacks and whites (we did not discuss this test in the chapter). This is the Levene Test in the second part of the output. The test has a null hypothesis that the variances are equal. SPSS reports a probability of .255 for this test, so, for example, at the .05 level, we fail to reject the null hypothesis that the variances are equal. We would conclude that the variances are equal. Based on our decision, we know which of the *t* tests to

Figure 12.6

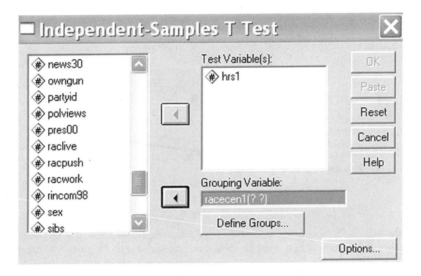

Figure 12.7

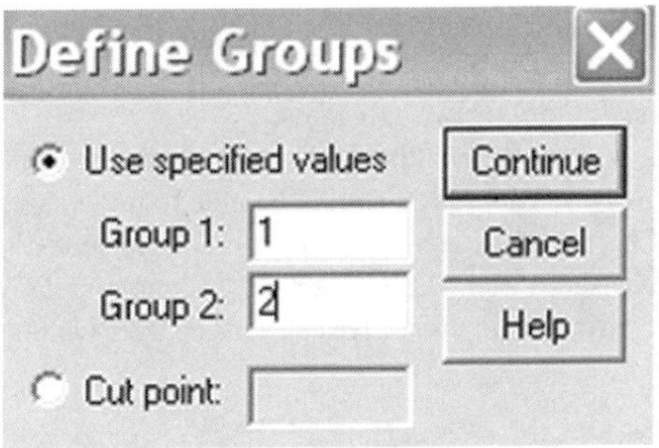

Figure 12.8

Group Statistics

		N	Mean	Std. Deviation	Std. Error Mean
NUMBER OF HOURS WORKED LAST WEEK	WHAT IS RS RACE 1ST MENTION	761	42.10	14.169	.514
	WHITE BLACK OR AFRICAN AMERICAN	131	40.96	13.178	1.151

(Continued)

Independent Samples Test

		Levene's Test for Equality of Variances		t-test for Equality of Means						
									95% Confidence Interval of the Difference	
		F	Sig.	t	df	Sig. (2-tailed)	Mean Difference	Std. Error Difference	Lower	Upper
NUMBER OF HOURS WORKED LAST WEEK	Equal variances assumed	1.299	.255	.855	890	.393	1.134	1.327	−1.470	3.739
	Equal variances not assumed			.900	185.630	.370	1.134	1.261	−1.353	3.621

use in the bottom section of the output. Because the Levene Test suggests that the variances are *equal*, we use the t-test result reported in the first line. The actual *t* value is .855, with 890 degrees of freedom. The two-tailed exact significance [Sig. (2-tailed)] is listed as .393. SPSS calculates the exact significance, so there is no need to look in Appendix C. This probability is much larger than the .05 or .01 level, so we conclude that there is no significant difference in hours worked between whites and blacks. We cannot reject the null hypothesis.

What if we wanted to do a one-tailed test instead? SPSS does not directly list the probability for a one-tailed test, but it is easy to calculate. If we had specified a directional research hypothesis—such as that men work more hours than women—we would simply take the probability reported by SPSS and divide it in half for a one-tailed test. Because the probability is so large in this case, our conclusion will be the same whether we do a one- or two-tailed test.

The last bit of output on each line is the 95 percent confidence interval for the mean difference in hours worked between the two groups. You should be able to understand this based on the discussion of confidence intervals in Chapter 11. It is helpful information when testing mean differences because the actual mean difference measured (here 1.134 hours) is a sample mean, which will vary from sample to sample. The 95 percent confidence interval gives us a range over which the sample mean differences are likely to vary.

SPSS PROBLEMS

[GSS02PFP-B]

1. Use the 2002 GSS file to investigate whether or not Americans have at least two children per person. Use the One-Sample T Test procedure to do this test with the variable CHILDS. Do the test at the .01 significance level. What did you find? Do Americans have two children, more, or less?

2. How are individuals who support legalization of marijuana different from those who do not? Use the variable GRASS as your independent or grouping variable. Investigate whether there is a significant difference between these two groups in terms of their age (AGE), number of children (CHILDS), and education (EDUC). Assume that alpha is .05 for a two-tailed test. Based on your analysis, write three Step 5 type statements summarizing your findings.

3. Extend your analysis in Exercise 2, this time by comparing individuals who support/do not support the death penalty (CAPPUN). Use the same dependent variables, AGE, CHILDS, and

EDUC, to estimate t-tests. Assume alpha is .05 for a two-tailed test. Prepare a statement to summarize your findings.

4. The GSS includes a measure of political party affiliation (PARTYID). Test whether there is a significant difference between strong Democrats (coded 0) and strong Republicans (coded 6) in their age (AGE), number of children (CHILDS), and educational attainment (EDUC). Assume alpha is .05 for a two-tailed test. Prepare a statement to summarize your findings.

5. In this exercise, we will use data from the ISSP 2000, comparing average family/household income between two countries. First, you'll need to run frequencies of the variable COUNTRY. Take note of what each country is coded; for example, the United States is coded as "6." Also note that the United States sample is smaller than 50 and cannot be used for t-test comparisons. [The t-test examples we calculated in the chapter included sample sizes 50 or larger. If not, the assumption of normal population is required.] Based on the countries that have sample sizes larger than 50, select three pairs for comparison, e.g., Mexico vs. Chile, Denmark vs. Switzerland, Norway vs. Netherlands. Using INCOME as the dependent variable and the countries as your independent (factor) variable, estimate the t-test values.

CHAPTER EXERCISES

1. It is known that, nationally, doctors working for HMOs (health maintenance organizations) average 13.5 years of experience in their specialties, with a standard deviation of 7.6 years. The executive director of an HMO in a western state is interested in determining whether or not its doctors have less experience than the national average. A random sample of 150 doctors from the HMO shows a mean of only 10.9 years of experience.
   a. State the research and the null hypotheses to test whether or not doctors in this HMO have less experience than the national average.
   b. Using an alpha level of .01, make this test.

2. Consider the problem facing security personnel at a military facility in the Southwest. Their job is to detect infiltrators (spies trying to break in). The facility has an alarm system to assist the security officers. However, sometimes the alarm doesn't work properly, and sometimes the officers don't notice a real alarm. In general, the security personnel must decide between these two alternatives at any given time:

   H_0: Everything is fine; no one is attempting an illegal entry.

   H_1: There are problems; someone is trying to break into the facility.

   Based on this information, fill in the blanks in these statements:
   a. A "missed alarm" is a Type ____ error, and its probability of occurrence is denoted by ____.
   b. A "false alarm" is a Type ____ error.

3. For each of the following situations determine whether a one- or a two-tailed test is appropriate. Also, state the research and the null hypotheses.
   a. You are interested in finding out if the average household income of residents in your state is different from the national average household. According to the U.S. census, for 2003, the national average household income is $43,318.
   b. You believe that students in small liberal arts colleges attend more parties per month than students nationwide. It is known that nationally, undergraduate students attend an average of 3.2 parties per month. The average number of parties per month will be calculated from a random sample of students from small liberal arts colleges.
   c. A sociologist believes that the average income of elderly women is lower than the average income of elderly men.

    d. Is there a difference in the amount of study time on-campus and off-campus students devote to their schoolwork during an average week? You prepare a survey to determine the average number of study hours for each group of students.

    e. Reading scores for a group of third-graders enrolled in an accelerated reading program are predicted to be higher than the scores for nonenrolled third-graders.

    f. Stress (measured on an ordinal scale) is predicted to be lower for adults who own dogs (or other pets) than for non–pet owners.

4. a. For each situation in Exercise 3, describe the Type I and Type II errors that could occur.

    b. What are the general implications of making a Type I error? Of making a Type II error?

    c. When would you want to minimize Type I error? Type II error?

5. One way to check on how representative a survey is of the population from which it was drawn is to compare various characteristics of the sample with the population characteristics. A typical variable used for this purpose is age. The 2002 GSS survey of the American adult population found a mean age of 46.21 and a standard deviation of 17.45 for its sample of 1,496 adults. Assume that we know from census data that the mean age of all American adults is 43.10. Use this information to answer these questions.

    a. State the research and the null hypotheses for a two-tailed test.

    b. Calculate the t statistic and test the null hypothesis at the .05 significance level. What did you find?

    c. What is your decision about the null hypothesis? What does this tell us about how representative the sample is of the American adult population?

6. Sociologists Bellas, Ritchey, and Parmer (2001)[16] conducted an analysis of the gender gap in faculty salaries. Data consistently indicate how women faculty members earn less than their male counterparts. Differences have been attributed to individual difference, as well as employer and institutional discrimination. For their analysis the sociologists obtained salary data on 158 men and 148 women from a large public research university. In 1985, men were estimated to earn an average of $42,340.62 ($S_Y = 9,639.19$), while women earned $33,865.98 ($S_Y = 8,298.82$).

    a. Use the t test to conduct a one-tailed test of the null hypothesis, $\alpha = .05$, comparing male salaries to female salaries. What can you conclude?

    b. Would your conclusions have been different if you had used a two-tailed test?

7. In this exercise, we will examine the attitudes of liberals and conservatives toward affirmative action policies in the workplace. Data from the 2002 GSS reveal that 15% of conservatives ($N = 185$) and 18% of liberals ($N = 128$) indicate that they "strongly support" or "support" affirmative action policies for African Americans in the workplace.

    a. What is the appropriate test statistic? Why?

    b. Test the null hypothesis with a one-tailed test (conservatives are less likely to support affirmative action policies than liberals); $\alpha = .05$. What do you conclude about the difference in attitudes between conservatives and liberals?

    c. If you conducted a two-tailed test with $\alpha = .05$, would your decision have been different?

8. Let's continue our analysis of liberals and conservatives, taking a look this time at differences in their educational attainment. We obtain the following information from the 2002 GSS—the

---

[16]Data from Bellas, Marcia, P. Neal Ritchey, and Penelope Parmer. 2001. "Gender Differences in Salaries and Salary Growth Rates of University Faculty: An Exploratory Study." *Sociological Perspectives* 44 (2): 163-187.

average educational attainment for conservatives is 13.40 years ($S_Y = 3.08$) and the average educational attainment for liberals is 13.72 years ($S_Y = 3.29$).

   a. Do you have enough information to test the null hypothesis that there is no difference in level of education between liberals and conservatives? Why or why not?

   b. What if you learned that there were 269 conservatives and 188 liberals? Do you now have enough information? Why or why not?

   c. Test the hypothesis at the .01 alpha level.

9. The gender gap—differences between men and women in their political attitudes and behavior—was one of the central issues in the 2000 presidential election. The gender gap is evident in the tendency of women to hold liberal views and to vote Democratic more often than men. The GSS asked respondents who they voted for in the 2000 election and found that among 524 female voters, 259 voted for Democratic candidate Al Gore. On the other hand, among 421 male voters, 149 voted for Gore. Do these differences reflect a real gender gap in the population of voters?

   a. What proportion of males voted for Gore? What proportion of females?

   b. If you wanted to test the research hypothesis that the proportion of male voters is less than female voters, would you conduct a one- or a two-tailed test?

   c. Test the null hypothesis at the .01 alpha level. What do you conclude?

10. Data from the 2002 GSS show that 84.3 percent (198 out of 235) of females and 75 percent (198 out of 264) of males are in favor of requiring gun permits. You wonder whether there is any difference between males and females in the population in their support for gun permits. Use a test of the difference between proportions when answering these questions.

   a. What is the research hypothesis? Should you conduct a one- or a two-tailed test? Why?

   b. Test your hypothesis at the .05 level. What do you conclude?

11. We compare educational attainment between men and women for the GSS 2002 and the ISSP 2000. Data from each data set are presented in Table 12.6.

   a. Use the GSS 2002 information to determine whether males have significantly higher educational attainment than women. Test at the .05 alpha level.

   b. Use the ISSP 2000 data to test whether there is a significant difference in educational attainment between men and women. Test at the .01 alpha level.

Table 12.6   Years of Education for Men and Women, GSS 2002 and ISSP 2000

	GSS 2002	*ISSP 2000*
Males	$\overline{Y} = 13.34$ $S_Y = 3.13$ $N = 666$	$\overline{Y} = 11.85$ $S_Y = 3.98$ $N = 618$
Females	$\overline{Y} = 13.22$ $S_Y = 2.88$ $N = 829$	$\overline{Y} = 11.34$ $S_Y = 3.74$ $N = 732$

12. Do men and women have different beliefs on the ideal number of children in a family? Based on the following GSS 2002 data and obtained t statistic, what would you conclude? (Assume a two-tailed test; $\alpha = .05$.)

	Men	Women
Mean ideal number of children	2.70	3.09
Standard deviation	1.57	1.84
N	220	227
Obtained t statistic	**−2.71**	

13. Does trade union membership vary by social class? Data from the ISSP 2000 reveal that among those who identified themselves as working class ($N = 257$) 33% were currently union members. However, among the middle class ($N = 411$), only 28% were union members. Is there a significant difference in proportion of union membership between these two social classes? Set alpha at .05. What can you conclude?

14. We recalculated our comparison of ideal number of children, this time only for men and women 45 years of age or younger. Our results are presented below.

	Men	Women
Mean ideal number of children	2.58	2.92
Standard deviation	1.35	1.72
N	113	153
Obtained t statistic	**−1.70**	

What conclusions can you draw, based on the same alpha of .05? How do these results compare with those in Exercise 12?

The Chi-Square Test

F igures collected by the U.S. Department of Justice suggest that violent crime is not an equal opportunity offender. Your chances of being a victim of a violent crime are strongly influenced by your age, race, gender, and neighborhood. For example, you are far more likely to be a victim of crime if you live in a city rather than in a suburb or in the country; if you are a young black male rather than a middle-aged white male; or if you are a black woman between the ages of 16 and 24 rather than a white woman of the same age.

As we learned at the end of the previous chapter, the fear of being a crime victim—regardless of actual victimization—is greater for women at every age and of every race than for men.[1] We now extend Kenneth Ferraro's analysis with an analysis of two General Social Survey (GSS) variables: *fear of walking alone at night* and *gender*. Based on a random sample

---

[1]Kenneth F. Ferraro, "Women's Fear of Victimization: Shadow of Sexual Assault?" *Social Forces* 75, no. 2 (1996): 667–690.

Table 13.1   Percentage of Men and Women Afraid to Walk Alone in
Their Neighborhood at Night: GSS 2002

Afraid	Men	Women	Total
No	83.3%	57.2%	71.1%
	(150)	(91)	(241)
Yes	16.7%	42.8%	28.9%
	(30)	(68)	(98)
Total (N)	100.0%	100.0%	100.0%
	(180)	(159)	(339)

taken from the 2002 GSS data set, these data confirm the observation that fear of crime differs according to gender: 42.8 percent of the women surveyed compared with only 16.7 percent of the men are afraid to walk alone in their neighborhoods at night.

The percentage differences between males and females in perception of safety, shown in Table 13.1, suggest that there is a relationship between gender and fear in our sample. In inferential statistics we base our statements about the larger population on what we observe in our sample. How do we know whether the gender differences in Table 13.1 reflect a real difference in the perception of safety among the larger population? How can we be sure that these differences are not just a quirk of sampling? If we took another sample, would these differences be wiped out or even reversed?

Let's assume that men and women are equally likely to be afraid to walk alone at night—that in the population from which this sample was drawn there are no real differences between them. What would be the expected percentages of men and women who would be afraid to walk alone at night?

If gender and fear were not associated, we would expect the same percentage of men and women to be fearful. Similarly, we would expect to see the same percentage of men and women who are not fearful. These percentages should be equal to the percentage of "fearful" and "not fearful" respondents in the sample as a whole. The last column of Table 13.1—the row marginals—displays these percentages: 28.9 percent of all respondents were afraid to walk alone at night, whereas 71.1 percent were not afraid. Therefore, if there were no association between gender and fear, we would expect to see 28.9 percent of the men and 28.9 percent of the women in the sample afraid to walk alone at night. Similarly, 71.1 percent of the men and 71.1 percent of the women would not be afraid to do so.

Table 13.2 shows these hypothetical expected percentages. Because the percentage distributions of the variable *fear* are identical for men and women, we can say that Table 13.2 demonstrates a perfect model of "no association" between the variable *fear* and the variable *gender.*

If there is an association between gender and fear, then at least some of the observed percentages in Table 13.1 should differ from the hypothetical expected percentages shown in Table 13.2. On the other hand, if gender and fear are not associated, the observed percentages should approximate the expected percentages shown in Table 13.2. In a cell-by-cell comparison of Tables 13.1 and 13.2, you can see that there is quite a disparity between the observed

Table 13.2   Percentage of Men and Women Afraid to Walk Alone in
Their Neighborhood at Night: Hypothetical Data
Showing No Association

Afraid	Men	Women	Total
No	71.1%	71.1%	71.1% (241)
Yes	28.9%	28.9%	28.9% (98)
Total (N)	100.0% (180)	100.0% (159)	100.0% (339)

percentages and the hypothetical percentages. For example, in Table 13.1, 83.3 percent of the men reported that they were not afraid, whereas the corresponding cell for Table 13.2 shows that only 71.1 percent of the men report no fear. The remaining three cells reveal similar discrepancies.

Are the disparities between the observed and expected percentages large enough to convince us that there is a genuine pattern in the population? The *chi-square* statistic helps answer this question. It is obtained by comparing the actual observed frequencies in a bivariate table with the frequencies that are generated under an assumption that the two variables in the cross-tabulation are not associated with each other. If the observed and expected values are very close, the chi-square statistic will be small. If the disparities between the observed and expected values are large, the chi-square statistic will be large. In the following sections, we will learn how to compute the chi-square statistic in order to determine whether the differences between men's and women's fear of walking alone in their neighborhood at night could have occurred simply by chance.

◙ THE CONCEPT OF CHI-SQUARE AS A STATISTICAL TEST

The **chi-square test** (pronounced kai-square and written as χ^2) is an inferential statistics technique designed to test for significant relationships between two variables organized in a bivariate table. The test has a variety of research applications and is one of the most widely used tests in the social sciences. Chi-square requires no assumptions about the shape of the population distribution from which a sample is drawn. It can be applied to nominally or ordinally measured variables.

Chi-square test   An inferential statistics technique designed to test for significant relationships between two variables organized in a bivariate table.

◙ THE CONCEPT OF STATISTICAL INDEPENDENCE

When two variables are not associated (as in Table 13.2), one can say that they are **statistically independent**. That is, an individual's score on one variable is independent of his/her score on the second variable. We identify statistical independence in a bivariate table by comparing the distribution of the dependent variable in each category of the independent variable. When two variables are statistically independent, the percentage distributions of the dependent variable within each category of the independent variable are identical. The hypothetical data presented in Table 13.2 illustrate the notion of statistical independence. The distributions of the dependent variable *fear* are identical within each category of the independent variable *gender*: 28.9 percent of each group are afraid to walk alone at night, and 71.1 percent of each group are not afraid. Based on Table 13.2, we would say that level of fear is independent of one's gender.[2]

Independence (statistical) The absence of association between two cross-tabulated variables. The percentage distributions of the dependent variable within each category of the independent variable are identical.

✔ *Learning Check.* *The data we will use to practice calculating chi-square are taken from the International Social Survey Programme (ISSP) 2000. They examine the relationship between sex (independent variable) and respondent's belief in whether a nuclear accident is likely to happen in the next 5 years (the dependent variable), as shown in the following bivariate table:*

Sex and Likelihood of a Nuclear Accident in Next 5 Years

	Sex		
Nuclear Accident	*Men*	*Women*	*Total*
Likely	62.9%	69.8%	66.5%
	(373)	(452)	(825)
Unlikely	37.1%	30.2%	33.5%
	(220)	(196)	(416)
Total	100%	100%	100%
(*N*)	(593)	(648)	(1241)

Construct a bivariate table (in percentages) showing no association between sex and belief in the likelihood of a nuclear accident.

---

[2]Because statistical independence is a symmetrical property, the distribution of the independent variable within each category of the dependent variable will also be identical. That is, if gender and fear were statistically independent, we would also expect to see the distribution of gender identical in each category of the variable fear.

▣ THE STRUCTURE OF HYPOTHESIS TESTING WITH CHI-SQUARE

The chi-square test follows the same five basic steps as the statistical tests presented in Chapter 12: (1) making assumptions; (2) stating the research and null hypotheses and selecting alpha; (3) selecting the sampling distribution and specifying the test statistic; (4) computing the test statistic; and (5) making a decision and interpreting the results. Before we apply the five-step model to a specific example, let's discuss some of the elements that are specific to the chi-square test.

The Assumptions

The chi-square test requires no assumptions about the shape of the population distribution from which the sample was drawn. However, like all inferential techniques it assumes random sampling. It can be applied to variables measured at a nominal and/or an ordinal level of measurement.

Stating the Research and the Null Hypotheses

The research hypothesis (H_1) proposes that the two variables are related in the population.

H_1: The two variables are related in the population. (Gender and fear of walking alone at night are statistically dependent.)

Like all other tests of statistical significance, the chi-square is a test of the null hypothesis. The null hypothesis (H_0) states that no association exists between two cross-tabulated variables in the population, and therefore the variables are statistically independent.

H_0: There is no association between the two variables. (Gender and fear of walking alone at night are statistically independent.)

✓ **Learning Check.** *Refer to the data in the previous Learning Check. Are the variables* sex *and* belief in the likelihood of a nuclear accident *statistically independent? Write out the research and the null hypotheses for your practice data.*

The Concept of Expected Frequencies

Assuming that the null hypothesis is true, we compute the cell frequencies we would expect to find if the variables are statistically independent. These frequencies are called **expected frequencies** (and are symbolized as f_e). The chi-square test is based on cell-by-cell comparisons between the expected frequencies (f_e) and the frequencies actually observed (**observed frequencies** are symbolized as f_o).

Expected frequencies (f_e)   The cell frequencies that would be expected in a bivariate table if the two variables were statistically independent.

Observed frequencies (f_o)   The cell frequencies actually observed in a bivariate table.

Calculating the Expected Frequencies

The difference between f_o and f_e will determine the likelihood that the null hypothesis is true and that the variables are, in fact, statistically independent. When there is a large difference between f_o and f_e, it is unlikely that the two variables are independent, and we will probably reject the null hypothesis. On the other hand, if there is little difference between f_o and f_e, the variables are probably independent of each other, as stated by the null hypothesis (and therefore we will not reject the null hypothesis).

The most important element in using chi-square to test for the statistical significance of cross-tabulated data is the determination of the expected frequencies. Because chi-square is computed on actual frequencies instead of on percentages, we need to calculate the expected frequencies based on the null hypothesis.

In practice, expected frequencies are more easily computed directly from the row and column frequencies than from percentages. We can calculate the expected frequencies using this formula:

$$f_e = \frac{(\text{column marginal})(\text{row marginal})}{N} \tag{13.1}$$

To obtain the expected frequencies for any cell in any cross-tabulation in which the two variables are assumed independent, multiply the row and column totals for that cell and divide the product by the total number of cases in the table.

Let's use this formula to recalculate the expected frequencies for our data on gender and fear as displayed in Table 13.1. Consider the men who were not afraid to walk alone at night (the upper left cell). The expected frequency for this cell is the product of the row total (241) and the column total (180) divided by all the cases in the table (339):

$$f_e = \frac{241 \times 180}{339} = 127.96$$

For men who are afraid to walk alone at night (the lower left cell) the expected frequency is

$$f_e = \frac{98 \times 180}{339} = 52.04$$

Next, let's compute the expected frequencies for women who are not afraid to walk alone at night (the upper right cell):

$$f_e = \frac{241 \times 159}{339} = 113.04$$

Finally, the expected frequency for women who are afraid to walk alone at night (the lower right cell) is

$$f_e = \frac{98 \times 159}{339} = 45.96$$

These expected frequencies are displayed in Table 13.3.

Note that the table of expected frequencies contains the identical row and column marginals as the original table (Table 13.1). Although the expected frequencies usually differ from the observed frequencies (depending on the degree of relationship between the variables), the row and column marginals must always be identical with the marginals in the original table.

Table 13.3 Expected Frequencies of Men and Women Afraid to Walk Alone in Their Neighborhood at Night

Afraid	Men	Women	Total
No	127.96	113.04	241
Yes	52.04	45.96	98
Total (*N*)	180	159	339

✓ **Learning Check.** *Calculate the expected frequencies for* sex *and* belief in the likelihood of a nuclear accident *and construct a bivariate table. Are your column and row marginals the same as in the original table?*

Calculating the Obtained Chi-Square

The next step in calculating chi-square is to compare the differences between the expected and observed frequencies across all cells in the table. In Table 13.4, the expected frequencies are shown in the shaded area in each cell below the corresponding observed frequencies. Note that the difference between the observed and expected frequencies in each cell is quite large. Is it large enough to be significant? The way we decide is by calculating the **obtained chi-square** statistic:

$$\chi^2 = \sum \frac{(f_o - f_e)^2}{f_e} \tag{13.2}$$

where

f_o = observed frequencies
f_e = expected frequencies

Table 13.4   Observed and Expected Frequencies of Men and Women Afraid to Walk Alone in Their Neighborhood at Night

Afraid	Men (f_o) (f_e)	Women (f_o) (f_e)	Total
No	150	91	241
	127.96	113.04	
Yes	30	68	98
	52.04	45.96	
Total (N)	180	159	339

According to this formula, for each cell subtract the expected frequency from the observed frequency, square the difference, and divide by the expected frequency. After performing this operation for every cell, sum the results to obtain the chi-square statistic.

Let's follow these procedures using the observed and expected frequencies from Table 13.4. Our calculations are displayed in Table 13.5. The obtained χ^2 statistic, 28.00, summarizes the differences between the observed frequencies and the frequencies we would expect to see if the null hypothesis were true and the variables—gender and fear of walking alone at night—were not associated. Next, we need to interpret our obtained chi-square statistic and decide whether it is large enough to allow us to reject the null hypothesis.

✔ **Learning Check.**   *Using the format of Table 13.5, construct a table to calculate chi-square for sex and belief in the likelihood of a nuclear accident.*

Chi-square (obtained)   The test statistic that summarizes the differences between the observed (f_o) and the expected (f_e) frequencies in a bivariate table.

The Sampling Distribution of Chi-Square

In Chapters 10 through 12, we've learned that test statistics such as Z and t have characteristic sampling distributions that tell us the probability of obtaining a statistic, assuming the

Table 13.5   Calculating Chi-Square

Fear and Gender	f_o	f_e	$f_o - f_e$	$(f_o - f_e)^2$	$\dfrac{(f_o - f_e)^2}{f_e}$
Men not afraid	150	127.96	22.04	485.76	3.80
Men afraid	30	52.04	−22.04	485.76	9.33
Women not afraid	91	113.04	−22.04	485.76	4.30
Women afraid	68	45.96	22.04	485.76	10.57

$$\chi^2 = \sum \frac{(fo - fe)^2}{fe} = 28.00$$

null hypothesis is true. In the same way, the sampling distribution of chi-square tells the probability of getting values of chi-square, assuming no relationship exists in the population.

Like other sampling distributions, the chi-square sampling distributions depend on the degrees of freedom. In fact, the χ^2 sampling distribution is not one distribution, but—like the t distribution—is a family of distributions. The shape of a particular chi-square distribution depends on the number of degrees of freedom. This is illustrated in Figure 13.1, which shows chi-square distributions for 1, 5, and 9 degrees of freedom. Here are some of the main properties of the chi-square distributions that can be observed in this figure:

Figure 13.1   Chi-Square Distributions for 1, 5, and 9 Degrees of Freedom

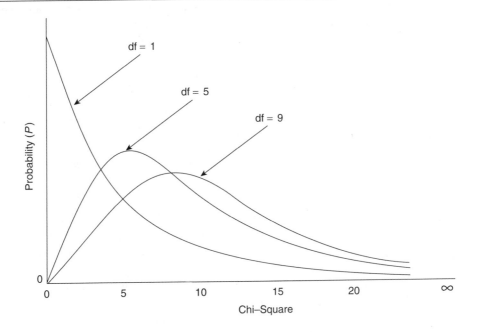

- The distributions are positively skewed. The research hypothesis for the chi-square is always a one-tailed test.
- Chi-square values are always positive. The minimum possible value is zero, with no upper limit to its maximum value. A χ^2 of zero means that the variables are completely independent; the observed frequencies in every cell are equal to the corresponding expected frequencies.
- As the number of degrees of freedom increases, the χ^2 distribution becomes more symmetrical and, with df greater than 30, begins to resemble the normal curve.

Determining the Degrees of Freedom

In Chapter 12 we defined degrees of freedom (df) as the number of values that are free to vary. With cross-tabulation data, we find the degrees of freedom using the following formula:

$$df = (r - 1)(c - 1) \tag{13.3}$$

where

r = the number of rows
c = the number of columns

Thus, Table 13.1 with 2 rows and 2 columns has $(2 - 1)(2 - 1)$ or 1 degree of freedom. If the table had 3 rows and 2 columns it would have $(3 - 1)(2 - 1)$ or 2 degrees of freedom.

The degrees of freedom in a bivariate table can be interpreted as the number of cells in the table for which the expected frequencies are free to vary, given that the marginal totals are already set. Based on our data in Table 13.3, suppose we first calculate the expected frequencies for men who are not afraid to walk alone at night ($f_e = 127.96$). Because the sum of the expected frequencies in the first column is set at 180, the expected frequency of men who are afraid has to be 52.04($180 - 127.96$). Similarly, all other cells are predetermined by the marginal totals and are not free to vary. Therefore, this table has only 1 degree of freedom.

> ✓ *Learning Check.*   *Degrees of freedom is sometimes a difficult concept to grasp. Review this section and if you don't understand the concept of degrees of freedom, ask your instructor for further explanation. How many degrees of freedom are there in your practice example?*

Appendix D shows values of the chi-square distribution for various dfs. Notice how the table is arranged with the degrees of freedom listed down the first column and the level of significance (or P values) arrayed across the top. For example, with 5 df, the probability associated with a chi-square as large as 15.086 is .01. An obtained chi-square as large as 15.086 would occur only once in 100 samples.

✔ **Learning Check.** *Based on Appendix D, identify the probability for each chi-square value (df in parentheses):*

- *12.307 (15)*

- *20.337 (21)*

- *54.052 (24)*

Let's get back to our example and find the probability of the chi-square we obtained. We can see that 28.00 does not appear on the first row (df = 1); in fact, it exceeds the largest chi-square value of 10.827 ($P = .001$). We can establish that the probability of obtaining a χ^2 of 28.00 is less than .001 if the null hypothesis were true. If our alpha was preset at .05, the probability of 10.827 would be well below this. Therefore, we can reject the null hypothesis that gender and fear are not associated in the population from which our sample was drawn. Remember, the larger the χ^2 statistic, the smaller the P value, providing us with more evidence to reject the null hypothesis. We can be very confident of our conclusion that there is a relationship between gender and fear in the population because the probability of this result occurring due to sampling error is less than .001, a very rare occurrence.

Review

To summarize our discussion, let's apply the five-step process of hypothesis testing.
Making Assumptions   Our assumptions are as follows:

1. A random sample of $N = 339$ was selected.

2. The level of measurement of the variable *gender* is nominal.

3. The level of measurement of the variable *fear* is nominal.

Stating the Research and Null Hypotheses and Selecting Alpha   The research hypothesis, H_1, is that there is a relationship between gender and fear (that is, gender and fear are statistically dependent). The null hypothesis, H_0, is that there is no relationship between gender and fear in the population (that is, gender and fear are statistically independent). Alpha is set at .05.
Selecting the Sampling Distribution and Specifying the Test Statistic   Both the sampling distribution and the test statistic are chi-square.
Computing the Test Statistic   We should first determine the degrees of freedom associated with our test statistic:

$$df = (r - 1)(c - 1) = (2 - 1)(2 - 1) = (1)(1) = 1$$

Next, in order to calculate chi-square, we calculate the expected frequencies under the assumption of statistical independence. To obtain the expected frequencies for each cell, we multiply its row and column marginal totals and divide the product by N. The expected frequencies are displayed in Table 13.3.

Are these expected frequencies different enough from the observed frequencies presented in Table 13.1 to justify rejection of the null hypothesis? To find out, we calculate the chi-square statistic of 28.00. The calculations are shown in Table 13.5.

Making a Decision and Interpreting the Results   To determine the probability of obtaining our chi-square of 28.00, we refer to Appendix D. With 1 degree of freedom, the probability of obtaining 28.00 is less than .001 (less than our alpha of .05). We reject the null hypothesis that there is no difference in the level of fear among men and women. Thus, we can conclude that in the population from which our sample was drawn, fear does vary by gender. Based on our sample data, we know that women are more likely to report being afraid to walk alone in their neighborhood at night than men.

> ✔ ***Learning Check.***   *What decision can you make about the association between* sex *and belief in the likelihood of a nuclear accident? Should you reject the null hypothesis at the .05 alpha level? At the .01 level?*

▣ THE LIMITATIONS OF THE CHI-SQUARE TEST: SAMPLE SIZE AND STATISTICAL SIGNIFICANCE

Although we found the relationship between gender and fear to be statistically significant, this in itself does not give us much information about the *strength* of the relationship or its *substantive significance* in the population. Statistical significance only helps us to evaluate whether the argument (the null hypothesis) that the observed relationship occurred by chance is reasonable. It does not tell us anything about the relationship's theoretical importance or even if it is worth further investigation.

The distinction between statistical and substantive significance is an important one in applying any of the statistical tests discussed in Chapter 12. However, this distinction is of particular relevance for the chi-square test because of its sensitivity to sample size. The size of the calculated chi-square is directly proportional to the size of the sample, independent of the strength of the relationship between the variables.

For instance, suppose that we cut the observed frequencies for every cell in Table 13.1 exactly in half—which is equivalent to reducing the sample size by one-half. This change will not affect the percentage distribution of fear among men and women; therefore, the size of the percentage difference and the strength of the association between gender and fear will remain the same. However, reducing the observed frequencies by half will cut down our calculated chi-square by exactly half, from 28.00 to 14.00. (Can you verify this calculation?) Conversely, had we doubled the frequencies in each cell, the size of the calculated chi-square would have doubled, thereby making it easier to reject the null hypothesis.

This sensitivity of the chi-square test to the size of the sample means that a relatively strong association between the variables may not be significant when the sample size is small. Similarly, even when the association between variables is very weak, a large sample may result in a statistically significant relationship. However, just because the calculated chi-square is large and we are able to reject the null hypothesis by a large margin does not imply that the relationship between the variables is strong and substantively important.

Another limitation of the chi-square test is that it is sensitive to small expected frequencies in one or more of the cells in the table. Generally, when the expected frequency in one or more of the cells is below 5, the chi-square statistic may be unstable and lead to erroneous conclusions. There is no hard-and-fast rule regarding the size of the expected frequencies. Most researchers limit the use of chi-square to tables that either (1) have no f_e values below 5 in value *or* (2) have no more than 20 percent of the f_e values below 5 in value.

Testing the statistical significance of a bivariate relationship is only a small step, albeit an important one, in examining a relationship between two variables. A significant chi-square suggests that a relationship, weak or strong, probably exists in the population and is not due to sampling fluctuation. However, to establish the strength of the association, we need to employ measures of association such as gamma, lambda, or Pearson's *r* (refer to Chapters 7 and 8). Used in conjunction, statistical tests of significance and measures of association can help determine the importance of the relationship and whether it is worth additional investigation.

▣ A Closer Look 13.1
Decision: Fail to Reject the Null Hypothesis

In Chapter 12 we learned how to test for differences between proportions. Tests between proportions can always be expressed as 2×2 (a bivariate table with 2 rows and 2 columns) chi-square tests. Let's take a look at the difference in educational attainment among men and women, in a 2×2 table.

We've collapsed educational level into two categories: high school or less versus some college or more. Results are presented in the following table.

Percentage of Men and Women by Educational Level

Educational Level	Men	Women	Total
High school or less	44.6%	46.7%	45.8%
	(262)	(344)	(606)
Some college or more	55.4%	53.3%	54.2%
	(325)	(392)	(717)
Total	100%	100%	100%
(N)	(587)	(736)	(1323)

The research hypothesis is

H_1: There is a relationship between gender and educational level. (Gender and educational level are statistically dependent.)

Our null hypothesis is

H_0: There is no relationship between gender and educational level. (Gender and educational level are statistically independent.)

We set alpha at .05. The sampling distribution is chi-square; the test statistic is chi-square. The df for the table is

$$df = (r - 1)(c - 1)$$
$$= (2 - 1)(2 - 1)$$
$$= (1)(1)$$
$$= 1$$

To compute the obtained chi-square, we first determine the expected frequencies under the assumption of statistical independence. These expected frequencies are shown in the following table.

Expected Frequencies of Men and Women and Educational Level

Educational Level	Men (f_e)	Women (f_e)	Total (f_e)
High school or less	268.88	337.12	606
Some college or more	318.12	398.88	717
Total (N)	587	736	1323

Next we calculate the obtained chi-square, as shown in the following table.

Calculating Chi-Square for Educational Level

Educational Level	f_o	f_e	$f_o - f_e$	$(f_o - f_e)^2$	$\dfrac{(f_o - f_e)^2}{f_e}$
Men/High school	262	268.88	−6.88	47.33	0.18
Men/College	325	318.12	6.88	47.33	0.15
Women/High school	344	337.12	6.88	47.33	0.14
Women/College	392	398.88	−6.88	47.33	0.12

$$\chi^2 = \sum \frac{(f_o - f_e)^2}{f_e} = .59$$

To determine if the observed frequencies are significantly different from the expected frequencies, we determine the P value of our calculated chi-square, .59. With 1 degree of freedom, our chi-square falls between critical chi-squares .455 ($P = .50$) and 1.074 ($P = .30$). We can say that the probability of .59 is between .50 and .30. Both P values indicate non-rare occurrences and are greater than our alpha level. Therefore, we fail to reject our null hypothesis that there is no difference in educational level between men and women. We have inconclusive evidence that there is a relationship between gender and educational attainment.

▣ STATISTICS IN PRACTICE: EDUCATION AND ATTENDANCE OF RELIGIOUS SERVICES

In Chapter 6 (Table 6.9) we examined the relationship between educational level and attendance of religious services based on a sample of ISSP respondents. We repeat the analysis here using chi-square as an inferential test. Individuals were asked to identify their own college degree attainment (none, secondary degree, and university degree) and their level of church attendance (never, infrequently, or two to three times per month or more). These data are shown in Table 13.6. This bivariate table shows a clear pattern of negative association between education (the independent variable) and church attendance (the dependent variable). For instance, whereas 27.8 percent of individuals with university degrees attended church at least two to three times per month, 66.2 percent of those with no degree reported the same level of attendance. Similarly, only 5.2 percent of respondents with no degree reported never attending church, whereas 37.3 percent of the respondents with a university degree fell into that category.

The differences in the levels of church attendance among the three educational groups seem sizable. However, it is not clear whether these differences are due to chance or sampling fluctuations, or whether they reflect a real pattern of association in the population. To answer these questions we perform a chi-square test following the five-step model of testing hypotheses.

Making Assumptions Our assumptions are as follows:

1. A random sample of $N = 440$ is selected.

2. The level of measurement of the variable *education* is ordinal.

3. The level of measurement of the variable *church attendance* is ordinal.

Table 13.6 Attendance of Religious Services by Educational Level: ISSP 2000

ATTENDANCE OF RELIGIOUS SERVICES	EDUCATIONAL LEVEL			
	None	*Secondary Degree*	*University Degree*	*Total*
Never	4 (5.2%)	77 (32.5%)	47 (37.3%)	128 (29.1%)
Infrequently	22 (28.6%)	83 (35.0%)	44 (34.9%)	149 (33.9%)
2–3 × per month or more	51 (66.2%)	77 (32.5%)	35 (27.8%)	163 (37.0%)
	77 (100%)	237 (100%)	126 (100%)	440 (100%)

Stating the Research and Null Hypotheses and Selecting Alpha Our hypotheses are:

H_1: There is a relationship between education and church attendance. (Education and church attendance are statistically dependent.)

H_0: There is no relationship between education and church attendance in the population. (Education and church attendance are statistically independent.)

For this test, we'll select an alpha of .01.

Selecting the Sampling Distribution and Specifying the Test Statistic The sampling distribution is chi-square; the test statistic is chi-square.

Computing the Test Statistic The degrees of freedom for Table 13.6 is

$$df = (r - 1)(c - 1) = (3 - 1)(3 - 1) = (2)(2) = 4$$

To calculate chi-square, first we determine the expected frequencies under the assumption of statistical independence. To obtain the expected frequency for each cell, we multiply its row and column marginal totals and divide the product by N. Following are the calculations for all cells in Table 13.6.

For no degree/never attending

$$f_e = \frac{128 \times 77}{440} = 22.40$$

For no degree/infrequently

$$f_e = \frac{149 \times 77}{440} = 26.08$$

For no degree/two to three times per month or more

$$f_e = \frac{163 \times 77}{440} = 28.52$$

For secondary degree/never attending

$$f_e = \frac{128 \times 237}{440} = 68.94$$

For secondary degree/infrequently

$$f_e = \frac{149 \times 237}{440} = 80.26$$

For secondary degree/two to three times per month or more

$$f_e = \frac{163 \times 237}{440} = 87.80$$

For university degree/never attending

$$f_e = \frac{128 \times 126}{440} = 36.65$$

For university degree/infrequently

$$f_e = \frac{149 \times 126}{440} = 42.67$$

Finally, for university degree/two to three times per month or more

$$f_e = \frac{163 \times 126}{440} = 46.68$$

We next calculate the chi-square to determine whether these expected frequencies are different enough from the observed frequencies to justify rejection of the null hypothesis. These calculations are shown in Table 13.7.

Making a Decision and Interpreting the Results   To determine if the observed frequencies are significantly different from the expected frequencies, we compare our calculated chi-square with Appendix D. With 4 degrees of freedom, our chi-square of 41.71 exceeds the largest listed chi-square value of 18.645 ($P = .001$). We determine that the probability of observing our obtained chi-square of 41.71 is less than .001, and less than our alpha of .01. We can reject the null hypothesis that there are no differences in the level of attendance among the different educational groups. Thus, we conclude that in the population from which our sample was drawn, church attendance does vary by educational attainment.

Table 13.7   Calculating Chi-Square for Education and Church Attendance

Education/Church Attendance	f_o	f_e	$f_o - f_e$	$(f_o - f_e)^2$	$\dfrac{(f_o - f_e)^2}{f_e}$
None/never	4	22.40	−18.40	338.56	15.11
None/infrequently	22	26.08	−4.08	16.65	.64
None/2–3×	51	28.52	22.48	505.35	17.72
Secondary/never	77	68.94	8.06	64.96	.94
Secondary/infrequently	83	80.26	2.74	7.51	.09
Secondary/2–3×	77	87.80	−10.80	116.64	1.33
University/never	47	36.65	10.35	107.12	2.92
University/infrequently	44	42.67	1.33	1.77	.04
University/2–3×	35	46.68	−11.68	136.42	2.92

$$\chi^2 = \sum \frac{(f_o - f_e)^2}{f_e} = 41.71$$

▣ READING THE RESEARCH LITERATURE: SIBLING COOPERATION AND ACADEMIC ACHIEVEMENT

In earlier chapters we have examined a number of examples showing how the results of statistical analyses are presented in the professional literature. Rarely do research articles go through the detailed steps of reasoning and calculation that are presented in this chapter. In most applications, the calculated chi-square is presented together with the results of a bivariate analysis. Occasionally, an appropriate measure of association summarizes the strength of the relationship between the variables.

Such an application is illustrated in Table 13.8, which is taken from an article by Carl Bankston III (1998)[3] about the relationship between academic achievement and cooperative family relations among Vietnamese American high school students. In his study, the author examined sibling cooperation among Vietnamese American families: How does help given to or received from other siblings relate to grade achievement? Bankston hypothesized that among Vietnamese American families, there is a strong emphasis on the family as a "cooperative unit." He used 1994 survey data to test whether "cooperation among siblings is significantly associated with academic success."

Data were collected from two regular public high schools and one honors public high school near Vietnamese communities in New Orleans. Surveys were distributed during English classes to a total of 402 students.

Table 13.8 Average Letter Grade Received by Reported Sibling Cooperation on Schoolwork—Percentage Reported

a. Help Received from Siblings

	Never Help	Sometimes	Often or Always	Row Total (N)
C or less	35.0	13.7	6.8	19.6 (63)
B	50.5	58.3	52.3	55.0 (177)
A	13.6	28.0	40.9	25.5 (82)
Column total (N)	32.0 (103)	54.3 (175)	13.7 (44)	100 (322)

Note: $\chi^2 = 29.33$; $p < .001$

(Continued)

[3]Carl L. Bankston III, "Sibling Cooperation and Scholastic Performance Among Vietnamese American Secondary Students: An Ethnic Social Relations Theory," *Sociological Perspectives* 41, no. 1 (1998): 167–184.

b. Help Given to Siblings

	Never Help	Sometimes	Often or Always	Row Total (N)
C or less	54.3	18.3	8.7	19.9 (67)
B	32.6	57.7	60.0	55.1 (185)
A	13.0	24.0	31.3	25.0 (84)
Column total (N)	13.7 (46)	52.1 (175)	34.2 (115)	100 (336)

Note: $\chi^2 = 44.32$; $p < .001$

Source: Adapted from Carl L. Bankston III, "Sibling Cooperation and Scholastic Performance Among Vietnamese American Secondary Students: An Ethnic Social Relations Theory," *Sociological Perspectives* 41, no. 1 (1998), pp. 174, 175. Used by permission of JAI Press, Inc.

Bankston relied on two primary measures for his analysis. Grade performance was based on self-reported letter grades: A, B, or C or less. He used an ordinal measure to determine how often the respondent provided his/her siblings with assistance and how often he/she received assistance from siblings: "never," "sometimes," or "often or always."

Table 13.8 shows the results cross-tabulated for the dependent variable, *average letter grade,* by the two independent variables: (a) reported *help received from siblings* and (b) reported *help given to siblings*. The table displays the percentage of cases in each cell (the observed frequencies) and row and column marginals.

Below each table Bankston reports the obtained chi-square and the actual significance (the *P* value) of the obtained statistic. This information indicates a very significant relationship between sibling cooperation and grade performance. A level of significance less than .001 means that a chi-square as high as 29.33 or 44.32 would have occurred less than once in 1,000 samples if the two variables were not related. In other words, the probability of the relationship occurring due to sampling fluctuations is less than 1 out of 1,000.

Based on these tables, Bankston presents the following analysis:

Most of these students are good students, regardless of the amount of help they report receiving from siblings. 55% of all students in the survey who had siblings or who answered the question reported grades averaging to a "B." Among those whose brothers and sisters never helped, 51% received "B" averages; among those whose brothers and sisters sometimes helped, 58% achieved "B" averages; among those whose brothers and sisters often or always helped, 52% achieved "B" averages.

Despite this clustering at the level of the "B," however, those who reported a great deal of help from siblings were much more likely than others to make "A"s and those who reported no help from siblings were much more likely to receive "C"s or less. 35% of those who reported never receiving help from siblings had grades averaging "C" or less, while only 15% of those who reported never receiving help from siblings had grades averaging to "A." By contrast, only 7% of those who reported that their siblings often or always helped them had grades averaging to a "C" or less, and 41% of them had grades averaging to "A."

Table 2 [Table 13.8b] presents a cross tabulation of averaged grade received by the amount of help reported given to siblings. Comparing Tables 1 [Table 13.8a] and 2, it appears that there is a slight tendency among respondents to over report help given to siblings and underreport help received. Despite this apparent slight bias in reporting, the very fact that help given and help received are systematically related to academic performance supports the view that these reported data are meaningful, if not necessarily precise, measures of amounts of sibling cooperation. Those who report more sibling cooperation do better in school than those who report less.

The greater the amount of help that respondents report giving to their siblings, the better the respondents tend to do in school. Over half of those who say they never help their siblings have grades averaging to "C" or less; fewer than one out of ten who say they often or always help their siblings have grades averaging to "C" or less.[4]

✔ *Learning Check.*   *For the ISSP table for sex and likelihood of a nuclear accident in 5 years, the value of the obtained chi-square is 6.52 with 1 degree of freedom. Based on Appendix D, we determine that its probability is less than .02. This probability is less than our alpha level of .05. We reject the null hypothesis of no relationship between sex and likelihood of a nuclear accident. If we reduce our sample size by half, the obtained chi-square is 3.26. Determine the P value for 3.26. What decision can you make about the null hypothesis?*

MAIN POINTS

- The chi-square test is an inferential statistics technique designed to test for a significant relationship between variables organized in a bivariate table. The test is conducted by testing the null hypothesis that no association exists between two cross-tabulated variables in the population and, therefore, the variables are statistically independent.

- The obtained chi-square (χ^2) statistic summarizes the differences between the observed frequencies (f_o) and the expected frequencies (f_e)—the frequencies we would have expected to see if the null hypothesis were true and the variables were not associated.

- The sampling distribution of chi-square tells the probability of getting values of chi-square, assuming no relationship exists in the population. The shape of a particular chi-square sampling distribution depends on the number of degrees of freedom.

[4]Ibid, p. 175.

KEY TERMS

chi-square (obtained)
chi-square test
expected frequencies (f_e)

independence (statistical)
observed frequencies (f_o)

ON YOUR OWN

Log on to the web-based student study site at http://www.pineforge.com/frankfort-nachmiasstudy4 for additional study questions, quizzes, web resources, and links to social science journal articles reflecting the statistics used in this chapter.

SPSS DEMONSTRATION

[GSS02PFP-A]

Producing the Chi-Square Statistic for Cross-Tabulations

The SPSS Crosstabs procedure was previously demonstrated in Chapters 6 and 7. This procedure can also be used to calculate a chi-square value for a bivariate table.

Click on *Analyze, Descriptive Statistics*, and *Crosstabs*, then on the *Statistics* button. You will see the dialog box shown in Figure 13.2.

To request the chi-square statistic click on the Chi-square box in the upper left corner. Notice that the chi-square choice is not grouped with the nominal or ordinal measures of association that we discussed in Chapter 7. SPSS separates it because the chi-square test is not a measure of association, but a test of independence of the row and column variables.

Click on *Continue*. In this demonstration we will look at the relationship between social class (CLASS) and happiness (HAPPY). Place HAPPY in the Row(s) box and CLASS in the Column(s) box. Then click on *OK* to run the procedure.

Figure 13.2

Crosstabs: Statistics

☑ Chi-square   ☐ Correlations   [Continue]

Nominal
  ☐ Contingency coefficient
  ☐ Phi and Cramér's V
  ☐ Lambda
  ☐ Uncertainty coefficient

Ordinal
  ☐ Gamma
  ☐ Somers' d
  ☐ Kendall's tau-b
  ☐ Kendall's tau-c

[Cancel]

[Help]

Nominal by Interval
  ☐ Eta

☐ Kappa
☐ Risk
☐ McNemar

☐ Cochran's and Mantel-Haenszel statistics
  Test common odds ratio equals: [1]

Figure 13.3

Chi-Square Tests

	Value	df	Asymp. Sig. (2-sided)
Pearson Chi-Square	46.030[a]	6	.000
Likelihood Ratio	35.760	6	.000
Linear-by-Linear Association	20.411	1	.000
N of Valid Cases	743		

a. 1 cells (8.3%) have expected count less than 5. The minimum expected count is 2.37.

The resulting output includes the chi-square statistics shown in Figure 13.3. SPSS produces quite a bit of output, perhaps more than expected. We will concentrate on the first row of information, the Pearson chi-square.

The Pearson chi-square has a value of 46.030 with 6 degrees of freedom. SPSS calculates the significance of this chi-square to be less than .000. The interpretation is that happiness and social class are related. Specifically, as social class increases, it appears that men and women are more likely to report being "very happy."

The last portion of the output from SPSS allows us to check for the assumption that all expected values in each cell of the table are 5 or greater. The output indicates that only one cell or 8.3 percent has values less than 5. This is lower than our threshold of 20 percent.

The gamma for this table is –.225 (the output is not shown here, but can be selected in the Crosstabs Statistics window). This indicates a very weak negative relationship between social class and happiness, yet the significance associated with the chi-square value for the table indicates there is little chance that the two variables are independent. These two statements are not contradictory. The magnitude of a relationship is not necessarily related to the statistical significance of that same relationship.

SPSS PROBLEMS

[GSS02PFP-A]

1. The GSS 2002 contains a series of questions about the role of women at home and at work. It is very likely that the responses to these questions vary by sex—or do they?
   a. Use SPSS to investigate the relationship between SEX and FECHLD (a working mother does not hurt her children). Create a bivariate table and ask for appropriate percentages and expected values. Does the table have a large number of cells with expected values less than 5? Are there any surprises in the data?
   b. Have SPSS calculate chi-square for each table.
   c. Test the null hypothesis at the .05 significance level in each table. What do you conclude?
   d. Select another demographic variable (DEGREE or CLASS) and investigate its relationship with FECHLD.

2. Is it better for a man to work and a woman to stay at home? Women and men were asked this question in the GSS 2002. Investigate the relationship between marital status (MARITAL) and responses to this question (FEMFAM). Have SPSS calculate the cross-tabulation of both variables, along with chi-square (set alpha at .05). What can you conclude?

3. Investigate the relationship between SEX and FEPRESCH, do preschool children suffer when a mother works.

a. Have SPSS calculate the cross-tabulation of the variables. What percent of women disagree or strongly disagree with the statement? What percent of men feel the same?

b. Test the null hypothesis at the .05 significance level. What do you conclude?

4. Throughout the textbook we have been illustrating the use of SPSS with items measuring support for women's employment and women's rights. Many variables are predictors of abortion attitudes (whether they are a *cause* of the attitudes is a question that can't be answered directly by statistics). In this exercise we want you to explore the relationship of several variables to some of the abortion items (ABPOOR, ABNORMORE, ABRAPE). Good predictors to use are a general measure of political position (POLVIEWS), attitudes toward homosexuality (HOMOSEX), religious preference (RELIG), and religiosity (ATTEND).

a. Create bivariate tables with some or all of these predictors and some of the GSS abortion items.

b. Have SPSS calculate the appropriate percentages and chi-squares (set alpha at .05). You may have to recode some tables or drop some categories to complete the analysis.

c. Summarize which variables are good predictors of which abortion attitudes. Did you find any general pattern?

d. Add the demographic variables RACE, CLASS, and SEX as control variables to a couple of these tables to see whether differences emerge between categories of respondents.

CHAPTER EXERCISES

1. In previous exercises we have examined the relationship between race and the fear of walking alone at night. In Chapter 6 Exercise 1, we created a bivariate table with these two variables to investigate their joint relationship. Now we can extend the analysis by calculating chi-square for the same table.

a. Use the data from Chapter 6, Exercise 1, to calculate chi-square for the bivariate table of *race* and *fear of walking alone at night.*

b. What is the number of degrees of freedom for this table?

c. Test the null hypothesis that race and fear of walking alone are independent (alpha = .05). What do you conclude? Is your conclusion consistent with your description of the percentage differences in Chapter 6, Exercise 1?

d. It's always important to test the assumption that the expected value in each cell is at least 5. Does any cell fail to meet this criterion?

2. In Chapter 6, Exercise 5, we investigated the relationship between respondent's gender/race and his/her confidence in medicine. Now we can calculate chi-square for this table to determine whether our previous advice to the neighborhood clinic was correct.

		Race and Sex of Respondent				
		White Males	*Black Males*	*White Females*	*Black Females*	*Total*
Confidence in Medicine	A great deal	135	17	136	25	313
	Only some	162	23	214	47	446
	Hardly any	33	6	41	11	91
Total		330	46	391	83	850

Source: General Social Survey 2002.

a. How many degrees of freedom does the table have?
b. Calculate chi-square for the table. What is the expected number of black females who have "only some" confidence in medical care?
c. Test the hypothesis that confidence in medicine and gender/race are independent. What is the P value of your obtained chi-square? What do you conclude (alpha = .05)? Is this consistent with your findings in Chapter 6?

3. The issue of how much should be spent to solve particular U.S. social problems is a complex matter, and people have diverse and conflicting ideas on these issues. Not surprisingly, race and social class have an impact on how people perceive the extent of government spending. The 2002 GSS contains several questions on these topics. The bivariate tables present race and the variable NATFARE, which asked whether we were spending too much, too little, or the right amount of money to address welfare, and race and the variable NATEDUC, which asked about the amount of spending for education. To make your task easier, the expected value (f_e) is also included as the second number in each cell.

		Whites	*Blacks*
Spending on welfare	Too little	107	36
		122.2	20.8
	About right	214	34
		211.9	36.1
	Too much	249	27
		235.9	40.1
Total		570	97

		Whites	*Blacks*
Spending on education	Too little	413	83
		422.8	73.2
	About right	130	16
		124.5	21.5
	Too much	35	1
		30.7	5.3
Total		578	100

a. What is the number of degrees of freedom for each table?
b. Calculate chi-square for each table.
c. Test whether RACE and NATFARE are independent (alpha = .01). What do you conclude?
d. Test whether RACE and NATEDUC are independent (alpha = .01). What do you conclude?
e. To further specify the relationship, calculate an appropriate measure of association for each table. Refer to Chapter 7 if necessary.

4. In Chapter 6, Exercise 4, we studied whether there was a relationship between church attendance and views about homosexual relations.
   a. Calculate chi-square for the table.
   b. Test the null hypothesis that the two variables are independent at the .05 alpha level. What did you find? Is this consistent with what you decided in Chapter 6?
   c. Which cell has the greatest difference between the expected value (f_e) and the actual value (f_o)? What did you discover when you tried to answer this question?

		Church Attendance			
		Never	*Several Times a Year*	*Every Week*	*Total*
Homosexual	Always wrong	43	23	53	119
Relations	Not wrong at all	42	22	10	74
Total		85	45	63	193

5. Use the following GSS 2002 data to investigate the relationship between educational attainment and attitudes toward premarital sex. In that earlier exercise, you used percentage differences and an appropriate measure of association for the table to study the relationships.
   a. Calculate chi-square for the following table.
   b. Based on an alpha of .01, test whether educational degree is independent of attitudes toward premarital sex.

Attitudes Toward Premarital Sex	*Less Than High School*	*High School*	*Bachelor's Degree or Higher*	*Total*
Always wrong	24	49	22	95
Almost always wrong	4	11	2	17
Sometimes wrong	2	42	21	65
Not wrong at all	14	92	39	145
Total	44	194	84	322

6. We continue our analysis from Exercise 3, this time examining the relationship between social class (CLASS) and spending on welfare (NATFARE) and education (NATEDUC).
   a. Calculate the value of chi-square for each table. What is the number of degrees of freedom for each table?
   b. Based on an alpha of .01, do you reject the null hypothesis?

		Lower Class	Working Class	Middle Class	Upper Class
Spending on welfare	Too little	18	59	73	4
	About right	12	109	132	10
	Too much	16	142	121	12
Total		46	310	326	26

		Lower Class	Working Class	Middle Class	Upper Class
Spending on education	Too little	36	232	244	19
	About right	10	71	70	4
	Too much	2	13	18	3
Total		48	316	332	26

7. In Chapter 7, Exercise 1, we studied the relationship between the race of violent offenders and the race of their victims.

	Offender Race		
Victim Race	White	Black	Other
White	3000	483	58
Black	227	2852	11
Other	51	28	109

   a. Complete the five-step model for these data, selecting an alpha of .01 to test whether these two variables are independent.
   b. Describe the relationship you found, using all available information.

8. In Chapter 7, Exercise 4, we first examined the relationship between social class and health. In Chapter 7, we calculated gamma as a measure of association for the table.
   a. Note that the table is so large that several cells have an expected value of less than 5. Is this a serious violation of the expected value assumption?
   b. Check this table for the smallest expected frequency. What is its value?
   c. In order to reduce the number of small cell sizes, we can group the cells in logical subsets for each variable to create a table with at least three or four categories for each variable. Calculate chi-square for your new, collapsed table. How many degrees of freedom does this table have?
   d. Based on alpha = .05, test the null hypothesis that social class and respondent's health are independent. What do you conclude?
   e. Would your decision change if alpha were set at .01?

Health	Lower Class	Working Class	Middle Class	Upper Class	Total
Poor	9	17	11	2	39
Fair	11	58	37	6	112
Good	14	133	150	5	302
Excellent	5	86	110	14	215
Total	39	294	308	27	668

Social Class (spanning header over Lower Class, Working Class, Middle Class, Upper Class)

9. Women and men were asked in the ISSP 2000 whether they agreed that it was the government's responsibility to reduce differences among citizens (GOVDIFF). The following bivariate table presents their educational attainment along with their level of agreement.

		Primary Degree	Secondary Degree	University Degree
Government's	Strongly agree	69	76	39
responsibility	Agree	94	124	69
to reduce	Neither agree nor disagree	54	60	29
differences	Disagree	29	66	42
	Strongly disagree	11	30	18
Total		257	356	197

a. Calculate the value of the chi-square for this table. What is the number of degrees of freedom of the table?
b. Based on alpha =.05, do you reject the null hypothesis?
c. Would your decision change if alpha = .01?

10. We'll continue examining the variable GOVDIFF using social class (CLASS) as the independent variable. Complete the five-step model for these data, setting alpha at .05.

GOVDIFF	Lower	Working	Lower Middle	Middle	Upper Middle	Upper
Strongly agree	30	90	49	81	20	7
Agree	50	125	72	178	44	4
Neither	21	45	26	64	13	2
Disagree	10	35	20	92	25	6
Strongly disagree	2	5	12	35	20	5
Total	113	300	179	450	122	24

11. In our SPSS illustration, we investigated the relationship between social class (CLASS) and happiness (HAPPY). Let's extend our investigation, this time examining the relationship between marital status (MARITAL) and happiness. Note: Not all marital categories are represented in the table.

Happiness Rating	Married	Divorced	Never Married	Total
Very happy	90	16	38	144
Pretty happy	103	50	89	242
Not too happy	17	20	23	60
Total	210	86	150	446

a. Complete the five-step model for these data (alpha = .05). What do you conclude?
b. Where are the greatest differences between the observed and expected frequencies? Which cells contribute the most to the chi-square value?

12. Are women happier in marriage than men? Let's examine the relationship between sex and marital happiness for this sample of 313 adults from the GSS 2002.

Happiness Rating	Men	Women	Total
Very happy	116	75	191
Pretty happy	51	60	111
Not too happy	6	5	11
Total	173	140	313

a. What percent of women report being "very happy" in marriage? What percent of men?
b. Complete the five-step model for these data (alpha = .05). What do you conclude?
c. Would your Step 5 final decision change if alpha were set at .01? Why or why not?

Analysis of Variance

M any research questions require us to look at multiple samples or groups, at least more than two at a time. We may be interested in studying the influence of ethnic identity (white, African American, Asian American, Latino/a) on church attendance, the influence of social class (lower, working, middle, upper) on opinions about the 2004 presidential campaign, or the effect of educational attainment (less than high school, high school graduate, some college, college graduate) on household income. Notice that each of these examples requires a comparison between multiple demographic or ethnic groups, more than the two group comparisons we reviewed in Chapter 12. While it would be easy to confine our analyses between two groups, our social world is much more complex and diverse.

Table 14.1 Educational Attainment (Measured in Years) for Four
GSS 2002 Groups

White Males $n_1 = 6$	Black Males $n_2 = 4$	White Females $n_3 = 6$	Black Females $n_4 = 5$
16	16	16	14
18	12	12	10
14	11	14	12
14	14	14	13
16		11	11
16		11	

Source: GSS 2002.

Let's say that we're interested in examining educational attainment—on average, how many years of education do Americans achieve? Recent 2003 census data indicated that over 85 percent of Americans had completed at least high school.[1] Despite these educational gains, do disparities still exist between different social groups?

In Chapter 12, Testing Hypotheses, we introduced statistical techniques to assess the difference between two sample means or proportions. For our example in Table 12.3, we compared the difference in educational attainment for white men and black men. But what if we wanted to extend our analysis to include females? Is there significant variation in educational attainment among black females, white females, black men, and white men?

Based on the 2002 GSS, we've taken a random sample of 21 men and women, grouped them into four demographic categories, and included their educational attainment in Table 14.1. With the *t*-test statistic we covered in Chapter 12, we could analyze only two samples at a time. We would have to analyze the mean educational attainment of black females versus white females' educational attainment, black females versus black men, and black females versus white men, and so on. (Confirm that we would have to analyze six different pairs!) In the end, we would have a tedious series of *t*-test statistic calculations and we still wouldn't be able to answer our original question: Is there a difference in educational attainment among all *four* demographic groups?

There is a statistical technique that will allow us to examine all four samples simultaneously. This technique is called analysis of variance (ANOVA). ANOVA follows the same five-step model of hypothesis testing that we used with *t*-test and Z-test for proportions (in Chapter 12) and chi-square (in Chapter 13). In this chapter, we review the calculations for ANOVA, discuss how we can test the significance of r^2 and R^2 using ANOVA, and discuss two applications of ANOVA from the research literature.

UNDERSTANDING ANALYSIS OF VARIANCE

Recall that the *t*-test examines the difference between two means ($\bar{Y}_1$, $\bar{Y}_2$), while the null hypothesis assumed there was no difference between them: $\mu_1 = \mu_2$. Rejecting the null

---

[1]Nicole Stoops, "Educational Attainment in the United States, 2003." *Current Population Reports*, 2004, P20-550, p. 1.

hypothesis meant that there was a significant difference between the two mean scores (or the populations from which the samples were drawn). In our Chapter 12 example, we analyzed the difference between mean years of education for white men and black men. Based on our *t*-test statistic, we rejected the null hypothesis, concluding that white men, on average, have significantly more years of education than black men do.

The logic of ANOVA is the same, but extending to two or more samples. For the data presented in Table 14.1, ANOVA will allow us to examine the variation among four means ($\overline{Y}_1, \overline{Y}_2, \overline{Y}_3, \overline{Y}_4$) and the null hypothesis can be stated as: $\mu_1 = \mu_2 = \mu_3 = \mu_4$. Rejecting the null hypothesis for ANOVA indicates that there is a significant variation among the four samples (or the four populations from which the samples were drawn) and that at least one sample mean is significantly different from the others. In our example, it suggests that years of education (dependent variable) do vary by group membership (independent variable). When ANOVA procedures are applied to data with one dependent and one independent variable, it is called a **one-way ANOVA**.

The means, standard deviations, and variances for the samples have been calculated and are shown in Table 14.2. Notice that the four mean educational years are not identical, with white males having the highest educational attainment. Also, based on the standard deviations, we can tell that the samples are relatively homogeneous with deviations within 1.51 to 2.59 years of the mean. We already know that there is a difference between the samples, but the question remains: Is this difference significant? Do the samples reflect a relationship between demographic group membership and educational attainment in the general population?

> ✓ *Learning Check.* *We've calculated the mean and standard deviation scores for each sample in Table 14.2. Compute each mean (Chapter 4) and standard deviation (Chapter 5) and confirm that our statistics are correct.*

Table 14.2   Means, Variances, and Standard Deviations for Four GSS Groups

White Males $n_1 = 6$	*Black Males* $n_2 = 4$	*White Females* $n_3 = 6$	*Black Females* $n_4 = 5$
16	16	16	14
18	12	12	10
14	11	14	12
14	14	14	13
16		11	11
16		11	
$\overline{Y}_1 = 15.67$	$\overline{Y}_2 = 13.25$	$\overline{Y}_3 = 13.00$	$\overline{Y}_4 = 12.00$
$s_1 = 1.51$	$s_2 = 2.22$	$s_3 = 2.00$	$s_4 = 1.58$
$s_1^2 = 2.27$	$s_2^2 = 4.92$	$s_3^2 = 4.00$	$s_4^2 = 2.50$
	$\overline{Y} = 13.57$		

To determine whether the differences are significant, analysis of variance examines the differences **between** our four samples, as well as the differences **within** a single sample. The differences can also be referred to as variance or variation, which is why ANOVA is the analysis of *variance*. What is the difference between one sample's mean score and the overall mean? What is the variation of individual scores within one sample? Are all the scores alike (no variation), or is there a broad variation in scores? ANOVA allows us to determine whether the variance between samples is larger than the variance within the samples. If the variance is larger between samples than the variance within samples, we know that educational attainment varies significantly across the samples. It would support the notion that group membership explains the variation in educational attainment.

Analysis of variance or ANOVA   An inferential statistics technique designed to test for a significant relationship between two variables in two or more samples.

▣ THE STRUCTURE OF HYPOTHESIS TESTING WITH ANOVA

The Assumptions

ANOVA requires several assumptions regarding the method of sampling, the level of measurement, the shape of the population distribution, and the homogeneity of variance.

1. Independent random samples are used. Our choice of sample members from one population has no effect on the choice of sample members from the second, third, or fourth population. For example, the selection of white males has no effect on the selection of any other sample.

2. The dependent variable, years of education, is an interval-ratio level of measurement. Some researchers also apply ANOVA to ordinal level measurements.

3. The population is normally distributed. Though we cannot confirm whether the populations are normal, given that our N is so small, we must assume that the population is normally distributed in order to proceed with our analysis.

4. The population variances are equal. Based on our calculations in Table 14.2, we see that the sample variances, though not identical, are relatively homogeneous.[2]

---

[2]Since the N in our computational example is small ($N = 21$), the assumptions of normality and homogeneity of variance are required. We've selected a small N in order to demonstrate the calculations for F and have proceeded with Assumptions 3 and 4. If a researcher is not comfortable with making these assumptions for a small sample, s/he can increase the size of N. In general, the F test is known to be robust with respect to moderate violations of these assumptions. A larger N increases the F test's robustness to severe departures from the normality and homogeneity of variance assumptions.

Stating the Research and the Null Hypotheses and Setting Alpha

The research hypothesis (H_1) proposes that at least one of the means is different. We do not identify which one(s) will be different, or larger or smaller, we only predict that a difference does exist.

$$H_1: \text{At least one mean is different from the others.}$$

ANOVA is a test of the null hypothesis of no difference between any of the means. Since we're working with four samples, we include four μs in our null hypothesis.

$$H_0: \mu_1 = \mu_2 = \mu_3 = \mu_4$$

As we did in other models of hypothesis testing, we'll have to set our alpha. Alpha is the level of probability at which we'll reject our null hypothesis. For this example, we'll set alpha at .05.

The Concepts of Between and Within Total Variance

A caution before we proceed: since we're working with four different samples and a total of 21 respondents, we'll have a lot of calculations. It's important to be consistent with your notations (don't mix up numbers for sample 1 with sample 4) and be careful with your calculations.

Our primary set of calculations has to do with the two types of variance, between-group and within-group. The estimate of each variance has two parts, the **sum of squares** and **degrees of freedom.**

The **between-group sum of squares** or **SSB** measures the difference in average years of education between our four groups. Sum of squares is short for "sum of squared deviations." For SSB, what we're measuring is the sum of squared deviations between each sample mean to the overall mean score. The formula for the between-group sum of squares can be presented as:

$$SSB = \sum n_k (\bar{Y}_k - \bar{Y})^2 \tag{14.1}$$

where n_k = the number of cases in a sample (k represents the number of different samples),

$\bar{Y}_k$ = the mean of a sample, and

$\bar{Y}$ = the overall mean.

SSB can also be understood as the amount of variation in the dependent variable (years of education) that can be attributed to or explained by the independent variable (the four demographic groups).

Within-group sum of squares or **SSW** measures the variation of scores within a single sample or as in our example, the variation in years of education within one group. SSW is also referred to as the amount of unexplained variance, since it is what is left after we consider the effect of the specified independent variable. The formula for SSW measures the sum of squared deviations within each group, between each individual score with its sample mean.

$$SSW = \sum (Y_i - \bar{Y}_k)^2 \tag{14.2}$$

where Y_i = each individual score in a sample, and

$\overline{Y}_k$ = the mean of a sample.

Even with our small sample size, if we were to use Formula 14.2, we'd have a tedious and cumbersome set of calculations. Instead, we suggest using the following computational formula for within-group variation or SSW:

$$SSW = \sum Y_i^2 - \sum \frac{\left(\sum Y_k \right)^2}{n_k} \tag{14.3}$$

where Y_i^2 = the squared scores from each sample,

$\sum Y_k$ = the sum of the scores of each sample, and

n_k = the total of each sample.

Together, the explained (SSB) and unexplained (SSW) variances compose the amount of total variation in scores. The **total sum of squares** or **SST** can be represented by

$$SST = \sum (Y_i - \overline{Y}_k)^2 = SSB + SSW \tag{14.4}$$

🔲 A Closer Look 14.1
Decomposition of SST

According to Formula 14.4, sum of squares total (SST) is equal to

$$SST = \sum (Y_i - \overline{Y}_k)^2 = SSB + SSW$$

You can see that the between sum of squares (explained variance) and within sum of squares (unexplained variance) account for the total variance (SST) in a particular dependent variable. How does that apply to a single case in our educational attainment example? Let's take the fifth white male in Table 14.1 with 16 years of education.

His total deviation (corresponding to SST) is based on the difference between his score from the overall mean (Formula 14.4). His score is quite a bit higher than the overall mean education of 13.57 years. The difference of his score from the overall mean is 2.43 years (16 – 13.57). Between-group deviation (corresponding to SSB) can be determined by measuring the difference between his group average from the overall mean (Formula 14.1). We've already commented on the higher educational attainment for white males (average of 15.67 years) when compared to the other three demographic groups. The deviation between the group average for white males and overall average is 2.10 years (15.67 – 13.57). Finally, the within-group deviation (corresponding to SSW, Formula 14.2) is based on the difference between the fifth white male's years of education and the group average for white males: .33 years (16 – 15.67).

So for the fifth white male in our sample, SSB + SSW = SST or 2.10 + .33 = 2.43.

In a complete ANOVA problem, we're computing these two sources of deviation (SSB and SSW) to obtain SST (Formula 14.4).

where Y_i = each individual score, and

$\overline{Y}$ = the overall mean.

The second part of estimating the between-group and within-group variances is calculating the degrees of freedom. Degrees of freedom were also discussed in Chapters 12 and 13. For ANOVA, we have to calculate two degrees of freedom. For SSB, the degrees of freedom are determined by

$$\text{dfb} = k - 1 \tag{14.5}$$

where k = number of samples.

For SSW, the degrees of freedom are determined by

$$\text{dfw} = N - k \tag{14.6}$$

where N = total number of cases and k = number of samples.

Finally, we can estimate the between-group variance by calculating **mean square between.** Simply stated, mean squares are averages computed by dividing each sum of squares by its corresponding degrees of freedom. Mean square between can be represented by

$$\text{Mean square between} = \text{SSB/dfb} \tag{14.7}$$

and the within-group variance or **mean square within** can be represented by

$$\text{Mean square within} = \text{SSW/dfw} \tag{14.8}$$

The F Statistic

Together the mean square between (Formula 14.7) and mean square within (Formula 14.8) compose the obtained F **ratio** or F **statistic**. Developed by R.A. Fisher, the F statistic is the ratio of between-group variance to within-group variance and is determined by Formula 14.9:

$$F = \frac{\textit{Mean square between}}{\textit{Mean square within}} = \frac{SSB/dfb}{SSW/dfw} \tag{14.9}$$

We know that a larger obtained F statistic means that there is more between-group variance than within-group variance, increasing the chances of rejecting our null hypothesis. In Table 14.3, we present additional calculations to compute F.

Let's calculate between-group sum of squares and degrees of freedom based on Formulas 14.1 and 14.5. The calculation for SSB is

$$\sum n_k (\overline{Y}_k - \overline{Y})^2$$
$$= 6(15.67 - 13.57)^2 + 4(13.25 - 13.57)^2 + 6(13.00 - 13.57)^2 + 5(12.00 - 13.57)^2$$
$$= 41.14$$

Table 14.3   Computational Worksheet for ANOVA

White Males $n_1 = 6$		Black Males $n_2 = 4$		White Females $n_3 = 6$		Black Females $n_4 = 5$	
Y_1	Y_1^2	Y_2	Y_2^2	Y_3	Y_3^2	Y_4	Y_4^2
16	256	16	256	16	256	14	196
18	324	12	144	12	144	10	100
14	196	11	121	14	196	12	144
14	196	14	196	14	196	13	169
16	256			11	121	11	121
16	256			11	121		

$$\bar{Y}_1 = 15.67 \qquad \bar{Y}_2 = 13.25 \qquad \bar{Y}_3 = 13.00 \qquad \bar{Y}_4 = 12.00$$
$$s_1 = 1.51 \qquad s_2 = 2.22 \qquad s_3 = 2.00 \qquad s_4 = 1.58$$
$$\sum Y_1 = 94 \qquad \sum Y_2 = 53 \qquad \sum Y_3 = 78 \qquad \sum Y_4 = 60$$
$$\sum Y_1^2 = 1484 \qquad \sum Y_2^2 = 717 \qquad \sum Y_3^2 = 1034 \qquad \sum Y_4^2 = 730$$

$$\bar{Y} = 13.57$$

The degrees of freedom for SSB is $k - 1$ or $4 - 1 = 3$. Based on Formula 14.7, the mean square between is

$$\frac{41.14}{3} = 13.71$$

The within-group sum of squares and degrees of freedom are based on formulas 14.3 and 14.6. The calculation for SSW is

$$\sum Y_i^2 - \sum \frac{\left(\sum Y_k \right)^2}{n_k}$$

$$= (1484 + 717 + 1034 + 730) - \left(\frac{94^2}{6} + \frac{53^2}{4} + \frac{78^2}{6} + \frac{60^2}{5} \right)$$

$$= 3965 - 3908.92 = 56.08$$

The degrees of freedom for SSW is $N - k = 21 - 4 = 17$. Based on Formula 14.8, the mean square within is

$$\frac{56.08}{17} = 3.30$$

Finally, our calculation of F is based on formula 14.9:

$$F = \frac{13.71}{3.30} = 4.15$$

Making a Decision

To determine the probability of calculating an F statistic of 4.15, we rely on Appendix E, the distribution of the F statistic. Appendix E lists the corresponding values of the F distribution for various degrees of freedom and two levels of significance, .05 and .01. Table 14.4 displays the distribution of F for .05 level of significance.

Since we set alpha at .05, we'll refer to the table marked "$p = .05$." Notice that Appendix F includes two df's. These refer to our degrees of freedom, df1 = dfb and df2 = dfw.

Because of the two degrees of freedom, we'll have to determine the probability of our obtained F differently than we did with t-test or chi-square. For this ANOVA example, we'll have to determine the corresponding F, also called the **critical F**, when dfb = 3 and dfw = 17 and alpha = .05.

Based on Appendix E, the critical F is 3.20, while our obtained F (the one that we calculated) is 4.15. Since our obtained F is greater than the critical F (4.15 > 3.20), we know that its probability is less than .05, extending into the shaded area. (If our obtained F was less than 3.20, we could determine that its probability was greater than our alpha of .05, in the unshaded area of the F distribution curve. Refer to Figure 14.1.) We can reject the null hypothesis of no difference and conclude that there is a significant difference in educational attainment between the four groups.

Figure 14.1

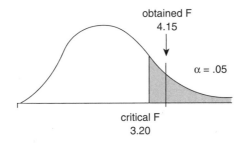

obtained F
4.15

$\alpha = .05$

critical F
3.20

◙ THE FIVE STEPS IN HYPOTHESIS TESTING: A SUMMARY

To summarize, we've calculated an analysis of variance test, examining the difference between four demographic groups and their average years of education.

Making assumptions:

1. Independent random samples are used.

2. The dependent variable, years of education, is an interval-ratio level of measurement.

Table 14.4

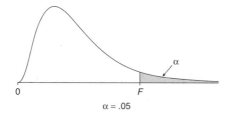

$\alpha = .05$

| | | | | | df_1 | | | | | |
df_2	1	2	3	4	5	6	8	12	24	∞
1	161.40	199.50	215.70	224.60	230.20	234.00	238.90	243.90	249.00	254.30
2	18.51	19.00	19.16	19.25	19.30	19.33	19.37	19.41	19.45	19.50
3	10.13	9.55	9.28	9.12	9.01	8.94	8.84	8.74	8.64	8.53
4	7.71	6.94	6.59	6.39	6.26	6.16	6.04	5.91	5.77	5.63
5	6.61	5.79	5.41	5.19	5.05	4.95	4.82	4.68	4.53	4.36
6	5.99	5.14	4.76	4.53	4.39	4.28	4.15	4.00	3.84	3.67
7	5.59	4.74	4.35	4.12	3.97	3.87	3.73	3.57	3.41	3.23
8	5.32	4.46	4.07	3.84	3.69	3.58	3.44	3.28	3.12	2.93
9	5.12	4.26	3.86	3.63	3.48	3.37	3.23	3.07	2.90	2.71
10	4.96	4.10	3.71	3.48	3.33	3.22	3.07	2.91	2.74	2.54
11	4.84	3.98	3.59	3.36	3.20	3.09	2.95	2.79	2.61	2.40
12	4.75	3.88	3.49	3.26	3.11	3.00	2.85	2.69	2.50	2.30
13	4.67	3.80	3.41	3.18	3.02	2.92	2.77	2.60	2.42	2.21
14	4.60	3.74	3.34	3.11	2.96	2.85	2.70	2.53	2.35	2.13
15	4.54	3.68	3.29	3.06	2.90	2.79	2.64	2.48	2.29	2.07
16	4.49	3.63	3.24	3.01	2.85	2.74	2.59	2.42	2.24	2.01
17	4.45	3.59	3.20	2.96	2.81	2.70	2.55	2.38	2.19	1.96
18	4.41	3.55	3.16	2.93	2.77	2.66	2.51	2.34	2.15	1.92
19	4.38	3.52	3.13	2.90	2.74	2.63	2.48	2.31	2.11	1.88
20	4.35	3.49	3.10	2.87	2.71	2.60	2.45	2.28	2.08	1.84
21	4.32	3.47	3.07	2.84	2.68	2.57	2.42	2.25	2.05	1.81
22	4.30	3.44	3.05	2.82	2.66	2.55	2.40	2.23	2.03	1.78
23	4.28	3.42	3.03	2.80	2.64	2.53	2.38	2.20	2.00	1.76
24	4.26	3.40	3.01	2.78	2.62	2.51	2.36	2.18	1.98	1.73
25	4.24	3.38	2.99	2.76	2.60	2.49	2.34	2.16	1.96	1.71
26	4.22	3.37	2.98	2.74	2.59	2.47	2.32	2.15	1.95	1.69
27	4.21	3.35	2.96	2.73	2.57	2.46	2.30	2.13	1.93	1.67
28	4.20	3.34	2.95	2.71	2.56	2.44	2.29	2.12	1.91	1.65
29	4.18	3.33	2.93	2.70	2.54	2.43	2.28	2.10	1.90	1.64
30	4.17	3.32	2.92	2.69	2.53	2.42	2.27	2.09	1.89	1.62
40	4.08	3.23	2.84	2.61	2.45	2.34	2.18	2.00	1.79	1.51
60	4.00	3.15	2.76	2.52	2.37	2.25	2.10	1.92	1.70	1.39
120	3.92	3.07	2.68	2.45	2.29	2.17	2.02	1.83	1.61	1.25
∞	3.84	2.99	2.60	2.37	2.21	2.09	1.94	1.75	1.52	1.00

Source: R. A. Fisher and F. Yates, *Statistical Tables for Biological, Agricultural and Medical Research,* 6th ed. Copyright © R. A. Fisher and F. Yates 1963. Reprinted by permission of Pearson Education Limited.

3. The population is normally distributed.

4. The population variances are equal.

Stating the research and null hypothesis and selecting alpha:

H_1: At least one mean is different from the others.
H_0: $\mu_1 = \mu_2 = \mu_3 = \mu_4$

$\alpha = .05$

Selecting the sampling distribution and specifying the test statistic:
The F distribution and F statistic are used to test the significance of the difference between the four sample means.

Computing the test statistic:
We need to calculate the between-group and within-group variation (sum of squares and degrees of freedom).
We estimate SSB = 13.71 (dfb = 3) and SSW = 3.30 (dfw = 17). Based on formula 14.9,

$$F = \frac{13.71}{3.30} = 4.15$$

Making a decision and interpreting the results:
We reject the null hypothesis of no difference and conclude that the groups are different in their educational attainment. Our F obtained of 4.15 is greater than the critical F of 3.20. The probability of 4.15 is less than .05. F doesn't advise us about which groups are different, only that educational attainment does differ significantly by demographic group members. However, based on the sample data, we know that the only group to achieve a college education average was white males (15.67 years). If we were to rank the remaining means, second highest educational attainment was among black males (13.25), then white females (13.00), and finally, the lowest education attainment was for black females (12.00). The mean education years for all three groups were at least 2 years lower than the mean score for white males. Educational attainment does differ significantly by race and gender group membership.

▣ TESTING THE SIGNIFICANCE OF R^2 USING ANOVA

Analysis of variance (ANOVA) can easily be applied to determine the statistical significance of the regression model as expressed in r^2 (see Chapter 8, pp. 284–287). In fact, when you look closely, ANOVA and regression analysis can look very much the same. In both methods we attempt to account for variation in the dependent variable in terms of the independent variable, except that in ANOVA the independent variable is a categorical variable (nominal or ordinal, e.g., gender or social class) and with regression it is an interval-ratio variable (e.g., income measured in dollars).

In Chapter 8 we developed a regression model that helped us predict the dependent variable, "willingness to pay higher prices to protect the environment," based on the independent

▣ **A Closer Look 14.2**
Assessing the Relationship Between Variables

Based on our five-step model of F we've determined that there is a significant difference between the four demographic groups in their educational attainment. We rejected the null hypothesis and concluded that years of education (our dependent variable) do vary by group membership (our independent variable). But can we say anything about how strong the relationship is between the variables?

The correlation ratio or eta-squared (eta^2) allows us to make a statement about the strength of the relationship. Eta-squared is determined by the following:

$$\text{Eta}^2 = \frac{SSB}{SST} \qquad (14.10)$$

The ratio of SSB to SST (SSB + SSW) represents the proportion of variance that is explained by the group (or independent) variable. Eta^2 indicates the strength of the relationship between the independent and dependent variables, ranging in value from 0 to 1.0. As eta-squared approaches 0, the relationship between the variables is weaker, and as eta-squared approaches 1, the relationship between the variables is stronger.

Based on our ANOVA example,

$$\text{Eta}^2 = \frac{13.71}{13.71 + 3.30} = \frac{13.71}{17.01} = 0.80$$

We can state that 80% of the variation in educational attainment can be attributed to demographic group membership. Or, phrased another way, 80% of the variation in the dependent variable (educational attainment) can be explained by the independent variable (group membership). So how strong is this relationship? We can base our determination of strength on A Closer Look 7.1 from Chapter 7, the same scale we used to assess gamma. Based on A Closer Look 7.1, we can conclude that there is a weak to moderate relationship between group membership and educational attainment.

variable, the "country's GNP per capita." We found that the coefficient of determination, r^2, reflects the proportion of total variation in the dependent variable explained by the independent variable. An r^2 of 0.361 means that the independent variable (GNP per capita) explains 36.1 percent of the variation in the dependent variable (the percentage willing to pay higher prices).

Like other descriptive statistics, r^2 is an estimate based on sample data. Once r^2 is obtained, we should assess the probability that the linear relationship between GNP per capita and the percentage willing to pay higher prices, as expressed in r^2, is really zero in the population (given the observed sample coefficient). In other words, we must test r^2 for statistical significance.

With ANOVA we decomposed the total variation in the dependent variable into portions explained (SSB) and unexplained (SSW) by the independent variable. Next, we calculated the

▣ A Closer Look 14.3
Educational Attainment and the
International Social Survey Programme 2000

We continue our investigation on educational attainment, focusing on the relationship between educational attainment (measured in years) and country of origin. We've selected a random sample of 15 respondents, five each from the United States, Canada, and Mexico.

United States $(n_1 = 5)$		Canada $(n_2 = 5)$		Mexico $(n_3 = 5)$	
Y_1	Y_1^2	Y_2	Y_2^2	Y_3	Y_3^2
14	196	15	225	8	64
15	225	16	256	9	81
12	144	16	256	11	121
16	256	18	324	12	144
14	196	18	324	14	196

$$\bar{Y}_1 = 14.20 \qquad \bar{Y}_2 = 16.60 \qquad \bar{Y}_2 = 10.80$$
$$s_1 = 1.48 \qquad s_2 = 1.34 \qquad s_3 = 2.39$$
$$\sum Y_1 = 71 \qquad \sum Y_2 = 83 \qquad \sum Y_3 = 54$$
$$\sum Y_1^2 = 1017 \qquad \sum Y_2^2 = 1385 \qquad \sum Y_3^2 = 606$$

$$\bar{Y} = 13.87$$

Our null hypothesis is that all educational means are equal. If we fail to reject the null hypothesis, this would indicate that there is no difference in educational attainment between respondents from the U.S., Canada, and Mexico.

Our calculations are:

SSB: $5(14.2 - 13.87)^2 + 5(16.60 - 13.87)^2 + 5(10.80 - 13.87)^2$
= .5445 + 37.2645 + 47.1245
= 84.93

Mean square between: 84.93/2 = 42.47

$$\text{SSW: } (1017 + 1385 + 606) - \left(\frac{71^2}{5} + \frac{83^2}{5} + \frac{54^2}{5} \right)$$

= 3008 − (1008.2 + 1377.8 + 583.2)
= 3008 − 2969.2
= 38.8

Mean square within: 38.8/12 = 3.23

F: 42.47/3.23 = 13.15

Decision: If we set alpha at .05, F critical would be 3.88 (df$_1$ = 2 and df$_2$ = 12). Based on our F obtained of 13.15, we would reject the null hypothesis and conclude that at least one of the means is significantly different from the others. Canadian respondents had the highest years of school completed (16.60), followed by U.S.(14.20) and Mexico (10.80) respondents.

mean squares between (SSB/dfb) and mean squares within (SSW/dfw). The statistical test, F, is the ratio of the mean square between to the mean square within (Formula 14.9).

$$F = \frac{Mean\ square\ between}{Mean\ square\ within} = \frac{SSB/dfb}{SSW/dfw}$$

With regression analysis we decomposed the total variation in the dependent variable into portions explained, SSR (*regression sum of squares*), and unexplained, SSE (*residual sum of squares*). Similar to ANOVA, the *mean squares regression* and the *mean squares residual* are calculated by dividing each sum of squares by its corresponding degrees of freedom (*df*). The degrees of freedom associated with SSR (df_r) are equal to K, which refers to the number of independent variables in the regression equation.

$$Mean\ squares\ Regression = \frac{SSR}{df_r} = \frac{SSR}{K} \qquad (14.11)$$

For SSE, the degrees of freedom (df_e) is equal to $[N - (K + 1)]$, with N equal to the sample size.

$$Mean\ squares\ Residual = \frac{SSE}{df_e} = \frac{SSE}{[N - (K + 1)]} \qquad (14.12)$$

In Table 14.5, for example, we present the analysis of variance summary table for willingness to pay higher prices to protect the environment and the country's GNP per capita.

In the table under the heading *Source of Variation* are displayed the regression, residual, and total sums of squares. The column marked *df* shows the degrees of freedom associated with both the *regression* and *residual* sum of squares. In the bivariate case, SSR has 1 degree of freedom associated with it. The *df* associated with SSE is $[N - (K + 1)]$, where K refers to the number of independent variables in the regression equation. In the bivariate case, with one independent variable—GNP per capita—SSE has $N - 2$ *df* associated with it $[N - (1 + 1)]$. Finally, the mean squares regression (MSR) and the mean squares residual (MSE) are calculated by dividing each sum of squares by its corresponding degrees of freedom. For our example,

Table 14.5   Analysis of Variance Summary Table for GNP per Capita and the Percentage Willing to Pay Higher Prices

Source of Variation	Sum of Squares	df	Mean Squares	F
Regression	844.97	1	844.97	7.91
Residual	1495.81	14	106.84	
Total	2340.78	15		

$$Mean\ squares\ Regression = \frac{SSR}{1} = \frac{844.97}{1} = 844.97$$

$$Mean\ squares\ Residual = \frac{SSE}{14} = \frac{1495.81}{14} = 106.84$$

The* F *Statistic   Together the mean squares regression and the mean squares residual compose the obtained ***F* ratio or *F* statistic**. The *F* statistic is the ratio of the mean squares regression to the mean squares residual:

$$F = \frac{Mean\ square\ Regression}{Mean\ square\ Residual} = \frac{SSR/df_r}{SSE/df_e} \tag{14.13}$$

The *F* ratio thus represents the size of the mean squares regression relative to the size of the mean squares residual. The larger the mean squares regression relative to the mean squares residual, the larger the *F* ratio and the more likely r^2 is significantly larger than zero in the population. We are testing the null hypothesis that r^2 is zero in the population.

Let's calculate the *F* ratio for our example of GNP per capita and the percentage willing to pay higher prices:

$$F = \frac{Mean\ square\ Regression}{Mean\ square\ Residual} = \frac{844.97}{106.84} = 7.91$$

Making a Decision   To determine the probability of calculating an *F* statistic of 7.91, we rely on Appendix E, Distribution of *F*. Appendix E lists the corresponding values of the *F* distribution for various degrees of freedom and two levels of significance, .05 and .01. We will set alpha at .05 and thus we will refer to the table marked "$p = .05$." Notice that Appendix E includes two *df's*. For the numerator, df_1 refers to the df_r associated with the mean squares regression; for the denominator, df_2 refers to the df_e associated with the means squares residual. For our example, we compare our obtained *F* (7.91) to the *critical F*. When the *df's* are 1 (numerator) and 14 (denominator), and alpha = .05, the critical *F* is 4.60. Since our obtained *F* is larger than the critical *F* (7.91 > 4.60), we can reject the null hypothesis that r^2 is zero in the population and conclude that the linear relationship between GNP per capita and percentage willing to pay higher prices as expressed in r^2 is probably greater than zero in the population (given our observed sample coefficient).

✔ ***Learning Check.***   *Test the null hypothesis that the linear relationship between GNP per capita and the percentage that view nature or the environment as sacred in the population. The mean squares regression is 885.34 with 1 degree of freedom. The means squares residual is 186.01, with 12 degrees of freedom. Calculate the F statistic and assess its significance.*

▣ ANOVA FOR MULTIPLE LINEAR REGRESSION

The ANOVA summary table for multiple regression is nearly identical to the one for simple linear regression, except that the degrees of freedom are adjusted to reflect the number of independent variables in the model.

We conducted an ANOVA test to assess the probability that the linear relationship between teen pregnancy and the combined effect of expenditure per pupil and the unemployment rate (see Chapter 8), as expressed by R^2, is really zero. The results of this test are summarized in Table 14.6. The obtained F statistic of 7.85 is shown in this table. With 2 and 43 df, we would need an F of 5.18 to reject the null hypothesis that $R^2 = 0$ at the .01 level. Since our obtained F exceeds that value (7.85 > 5.18) we can reject the null hypothesis with $p < .01$.

Table 14.6 Analysis of Variance Summary Table for Teen Pregnancy Rate with Expenditure per Pupil and Unemployment Rate

Source of Variation	Sum of Squares	df	Mean Squares	F
Regression	11,733.76	2	5,866.88	7.85
Error	32,143.58	43	747.52	
Total	43,877.34	45		

F statistic for multiple regression   The ratio of the mean squares regression to the mean squares residual. The larger the F ratio, the more likely it is that r^2 is greater than zero in the population.

▣ READING THE RESEARCH LITERATURE: SELF-IMAGE AND ETHNIC IDENTIFICATION

Like bivariate, t-test, or chi-square analyses, ANOVA can help us understand how the categories of experience—race, age, class, and/or gender—shape our social lives. As we did in our first example, comparing the impact of race and gender on educational attainment, ANOVA allows us to investigate a variety of social categories by comparing the differences between them. We conclude our ANOVA discussion with two examples of how ANOVA is presented and interpreted in the social science literature. As a statistical method, ANOVA is most commonly used in the field of social psychology. Both of our research examples come from social psychology journals.

Our first example is based on Elirea Bornman's 1999 research on ethnic identification in South Africa.[3] Politics, language, ethnicity, race, and culture all play an important role in defining self-image and ethnic identity in South Africa. But more importantly according to Bronman, the relationship between ethnic identity and self-image is highly relevant in a society characterized by "complex pluralization on the basis of racial and ethnic differences."[4] Bronman hypothesized that "each group's attitude toward their ingroup as well as the socio-economic and political changes in South Africa since the 1990s may have influenced the relationship between ethnic identification and self-image."[5]

Bornman examined the differences in self-image and ethnic identity, comparing Afrikaans-speaking whites, blacks, and English-speaking whites. Based on a survey conducted by the Human Sciences Research Council, Bornman analyzed a sample of South Africans' responses to several scales measuring self-image and ethnic identity. The Rosenberg Self-Esteem Scale (1965) measured both negative and positive self-image. For negative self-image, a higher score indicated a more negative self-image; for the positive self-image scale, a higher score indicated a more positive self-image.[6] Bornman created three measures of ethnic identification: ethnic identification (higher score indicates stronger identification, loyalty, respect, and pride with one's own ethnic group); exploration, identity achievement, and involvement (high score reflected higher levels of exploration, achievement of a well-defined identity, and involvement with and participation in cultural and other activities); and ambivalence versus protection (a higher score indicated a greater willingness to protect and preserve the identity of the group and less uncertainty about the membership in one's ethnic group).[7] Table 14.7 shows the means and standard deviations for Bornman's three South African groups. Degrees of freedom, along with the *F* statistic, are reported in the last two columns of the table. All *F* statistics are significant at the .01 level, indicated by the asterisks.

Based on her data analysis, Bornman writes:

On average, the Afrikaans-speaking Whites had the most positive self-image. Their mean score for negative self-image was the lowest, and they also had the highest mean score for positive self-image. In contrast, the Black respondents had the highest mean score for negative self-image and the lowest mean score for positive self-image, an indication that they had, on average, a more negative self-image than members of the two White groups. The results of the ANOVAs indicated that statistically significant differences existed between the mean scores of the various groups; for negative self-image, $F(2, 291) = 99.47$, $p = .0001$; for positive self-image, $F(2, 921) = 32.47$,

[3]Elirea Bornman, "Self Image and Ethnic Identification in South Africa," *Journal of Social Psychology,* 139 no. 4 (1999): 411–425.

[4]Ibid., p. 412.

[5]Ibid., p. 414.

[6]M. Rosenberg, *Society and the Adolescent Self-Image*. Princeton, NJ: Princeton University Press, 1965.

[7]Bornman, pp. 415–416. Bornman based the three measures of ethnic identification on responses to two scales, Phinney's (1992) Multigroup Measure of Ethnic Identity and Bornman's own 1988 ethnic identity scale.

Table 14.7 Mean Scores and the Results of Analyses of Variance for the Various Scales

	Afrikaans-speaking Whites	Blacks	English-speaking Whites	df	F
SELF-IMAGE SCALES					
Negative self-image	6.66 (2.39)	9.21 (2.89)	7.35 (2.89)	2, 921	99.47**
Positive self-image	17.49 (2.08)	15.97 (3.31)	17.30 (2.29)	2, 921	32.47**
SCALES ASSOCIATED WITH ETHNIC IDENTITY					
Ethnic identification	37.92 (6.45)	36.71 (6.97)	35.01 (6.67)	2, 922	8.52**
Exploration, identity achievement, and involvement	22.84 (4.76)	20.73 (5.64)	21.13 (4.43)	2, 921	16.79**
Ambivalence versus protection	15.36 (3.33)	12.64 (3.65)	13.42 (3.14)	2, 922	61.71**

Note: Standard deviation in parentheses.

**$p < .01$

$p = .0001$. . . . The Afrikaans-speaking Whites had the highest mean score for ethnic identification; the mean score for the English-speaking Whites was the lowest. The results of an ANOVA indicated the existence of significant differences, $F(2, 922) = 8.52$, $p = .0002$. . . . The Afrikaans-speaking Whites had the highest mean scores for the other two variables associated with ethnic identification (i.e., exploration, identity achievement, and involvement and ambivalence versus protection). However, the Black respondents had the lowest mean scores for these variables. The results of the ANOVAs were as follows: $F(2, 921) = 16.79$, $p = .0001$, for exploration, identity achievement, and involvement; $F(2, 922) = 61.71$, $p = .0001$, for ambivalence versus protection.[8]

Notice that in her discussion, Bornman reports each significant F statistic, along with its degrees of freedom and probability. While she does not specifically mention the mean scores, she does identify which groups have the highest or lowest scores on the measures for self-image and ethnic identification.

[8]Ibid., pp. 417–418.

🔲 READING THE RESEARCH LITERATURE: EFFECTS OF AUTHORITY STRUCTURES AND GENDER ON INTERACTION IN TASK GROUPS

Our second example comes from Cathryn Johnson, Jody Clay Warner, and Stephanie J. Funk from the *Social Psychology Quarterly* (1996).[9] The authors compared the impact of gender on the type of interaction and behavior of men and women in task groups. Their research tested predictions based on two theories: status characteristic theory and the socialization/normative approach. Status characteristic theory explains how gender is a status characteristic with two states, male and female, with males being valued more highly than females. As a status characteristic, gender influences the development and maintenance of status hierarchies in small groups by specifically defining expectations of male/female behavior. When men and women work together, the theory expects men to participate and communicate at a higher level than women because both males and females will have higher performance expectations of men. In all-female or all-male groups, gender status should not affect an individual's performance expectation. Participation would be based on individual differences, not just because one is female or male. The second theory tested by Johnson, Warner, and Funk is called the socialization/normative approach. The approach explains how women and men learn specific interaction norms. For example, women are more supportive, while men are more direct in their interactions. The theory predicts that regardless of the group structure, men and women will communicate based on their socialized interaction.

Their research is based on observations of three types of four-person groups: all-female groups in a women's college, all-male groups in a coeducational setting, and all-female groups in a coeducational setting. The student groups were asked to discuss and resolve a jury case involving an individual seeking action against another person's insurance company for injuries resulting from a fall down a flight of stairs. The group had to decide on an award ranging from $0 to $50,000.

Based on transcripts of the group's discussion, several dependent variables were measured: active task behaviors (giving opinions or information), directive behaviors (suggesting what direction the group should take), counterarguments (clear contradiction to a point made by another group member), agreements (concurring with a statement), positive socioemotional behaviors (giving a compliment or showing consideration for others in the group), and passive task behaviors (asking for opinions or information).[10] The results of their study are presented in Table 14.8.

Based on their findings, the authors offer the following interpretation:

> We first examined the effect of group type (women's college female groups, coeducational college female groups, and coeducational college male groups) on each task and

---

[9]Cathryn Johnson, Jody Clay Warner, and Stephanie J. Funk, "Effects of Authority Structures and Gender on Interaction in Same-Sex Task Groups," *Social Psychology Quarterly*, 59 no. 3 (1996): 221–236.

[10]Ibid., p. 228.

Table 14.8   Means, Standard Deviations, Univariate *F*-Tests

| | *Condition Means*[a] | | | *ANOVA Results*[b] |
	Women's College Females (n = 40)	*Coed College Males (n = 40)*	*Coed College Females (n = 40)*	*Group Type*
Active Task	8.88	9.56	8.30	*F* = .968
Behaviors	(5.66)	(3.61)	(2.00)	*p* = .383
Directive	.15	.27	.31	*F* = 2.263
Behaviors	(.22)	(.40)	(.37)	*p* = .077
Counter	.36	.74	.46	*F* = 4.868
Arguments	(.46)	(.72)	(.47)	*p* = .009
Agreements	2.13	1.33	2.06	*F* = 2.177
	(1.78)	(1.22)	(1.84)	*p* = .059
Positive Socio-	.15	.15	.10	*F* = .343
Emotional Behaviors	(.24)	(.33)	(.27)	*p* = .710
Passive Task	.87	1.03	.73	*F* = .700
Behaviors	(.84)	(1.36)	(1.21)	*p* = .499

a. For each member, frequency for each behavior was divided by length of time talked (measured in seconds) and multiplied by 60 to represent the rate of behavior per minute. Numbers in parentheses are standard deviations.
b. df = 2, 117.

socio-emotional behavior, using analysis of variance. . . . As shown in Table 2 [Table 14.8], we found no significant differences between the three types of groups in their rates of active task behaviors, directive behaviors, and passive task behaviors. In addition, we found no significant differences in rates of directive questions and rates of questions as opinions or counterarguments. This set of results for these behaviors substantially supports the original formulation of status characteristics theory. Analysis of variance also reveals, however, that group type has a significant effect on two variables: counter arguments ($F(2,117) = 4.868$, $p = .009$) and agreements ($F(2,117) = 2.900$, $p = .059$). Inspection of the means shows that men have the highest rates of counterarguments ($M = .74$ for males, $M = .46$ for coeducational females, $M = .36$ for women's college females) and the lowest rates of agreements ($M = 1.33$ for males, $M = 2.06$ for coeducational females, and $M = 2.13$ for women's college females).[11]

In reporting the *F* statistic for each of the behaviors, the authors also include the degrees of freedom, along with the probability of *F*. In their discussion, the notation "*M*" represents the "mean." While clearly the *F* statistic for "counterarguments" is significant at .009, they identify the *F* statistic for "agreements" as significant at $p = .059$. However, with an alpha level at .05, we would not have agreed with their interpretation. If alpha were set lower at .10, "counterarguments" and "agreements" along with "directive behaviors" would have been significant.

---

[11]Ibid, pp. 230–231.

MAIN POINTS

- Analysis of variance (ANOVA) procedures allow us to examine the variation in means in more than two samples. To determine whether the difference in mean scores is significant, ANOVA examines the differences between multiple samples, as well as the differences within a single sample.

- One-way ANOVA is a procedure using one dependent variable and one independent variable. The five-step hypothesis-testing model is applied to one-way ANOVA.

- The test statistic for ANOVA is F. The F statistic is the ratio of between-group variance to within-group variance.

- ANOVA can also be applied to determine the significance of the coefficient of determination (r^2) for a regression model and the multiple coefficient of determination (R^2) for a multiple regression model.

KEY TERMS

between-group sum of squares (SSB)
F critical
F obtained
F ratio or F statistic (F)
mean square between
mean square regression
mean square residual

mean square within
one-way ANOVA
regression sum of squares (SSR)
residual sum of squares (SSE)
total sum of square (SST)
within-group sum of squares (SSW)

ON YOUR OWN

Log on to the web-based student study site at http://www.pineforge.com/frankfort-nachmiasstudy4 for additional study questions, quizzes, web resources, and links to social science journal articles reflecting the statistics used in this chapter.

SPSS DEMONSTRATIONS

[GSS02PFP-A]

Demonstration 1: Computing Analysis of Variance Models

Social scientists have examined the association between a woman's fertility decisions (deciding whether and/or when to have a child) and her wages, employment status, and education, along with other socioeconomic and demographic factors. Research has indicated that different social groups may have different norms and values about fertility.[12]

In this first example, we'll investigate the relationship between a woman's educational attainment and the age at which her first child was born. Using education as the independent variable and age at which her first child was born as the dependent variable, we can assess whether there is a relationship between educational attainment and age at first childbirth.

---

[12]Sandra Hofferth, "Childbearing Decision Making and Family Well-being: A Dynamic, Sequential Model," *American Sociological Review* 48, no. 4 (1983): 533.

Figure 14.2

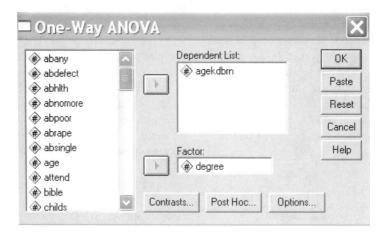

We'll use two variables for our analysis, the variable DEGREE (five categories of educational attainment) and AGEKDBRN (respondent's age when her first child was born). But first, we'll restrict our analysis to women in the GSS sample (using *Data – Select Cases* command. Note that you will have to select the option "If the condition is satisfied," then type SEX = 2 to restrict your analysis to women.).

We can compute the ANOVA model by clicking on *Analyze, Compare Means,* then *One-Way Anova.* The opening dialog box requires that we insert AGEKDBRN in the box labeled "Dependent List" and in the box labeled "Factor" insert DEGREE (see Figure 14.2).

Click on the Options button at the lower right. Click on "Descriptive" in the Statistics box. This will produce a table of means and standard deviations along with the ANOVA statistics (Figure 14.3). Click on "Continue" in the Options box, then "OK" in the One-Way ANOVA box.

Figure 14.3

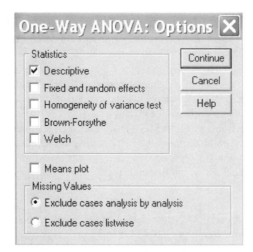

Figure 14.4

Descriptives

agekdbrn R'S AGE WHEN 1ST CHILD BORN

| | N | Mean | Std. Deviation | Std. Error | 95% Confidence Interval for Mean | | Minimum | Maximum |
					Lower Bound	Upper Bound		
0 LT HIGH SCHOOL	103	20.14	4.236	.417	19.31	20.96	13	35
1 HIGH SCHOOL	368	22.19	4.666	.243	21.71	22.67	14	42
2 JUNIOR COLLEGE	51	22.57	3.251	.455	21.65	23.48	15	33
3 BACHELOR	82	24.96	4.232	.467	24.03	25.89	13	35
4 GRADUATE	40	28.20	5.698	.901	26.38	30.02	18	41
Total	644	22.62	4.904	.193	22.24	23.00	13	42

ANOVA

agekdbrn R'S AGE WHEN 1ST CHILD BORN

	Sum of Squares	df	Mean Squares	F	Sig.
Between Groups	2400.306	4	600.076	29.347	.000
Within Groups	13065.960	639	20.448		
Total	15466.266	643			

SPSS produces two tables, Descriptives and ANOVA (Figure 14.4). We're interested in the *F* statistic and significance in the ANOVA table (a partial table is presented here). Based on the data, *F* is 29.347 with a significance of .000. We know that the difference between the educational groups is significantly different. The higher one's educational attainment, the age of first childbirth increases. The oldest average age at first childbirth is for women with graduate degrees (28.20 years of age), followed by the group of women with bachelor's degrees (24.96 years of age). The youngest group of first mothers is among women with less than a high school diploma. On average, women with less than a high school diploma had their first child at 20.14 years of age. When compared with graduate degree women, this is a difference of 8.06 years.

Demonstration 2: Producing an ANOVA Table—Regression-Bivariate [ISSP00PFP]

Do people with more education work longer hours at their jobs? This question was explored using SPSS in Chapter 8. We used SPSS to calculate the best-fitting regression line and the coefficient of determination. This procedure is located by clicking on *Analyze*, *Regression*, then *Linear*. The Linear Regression dialog box is shown in Figure 14.5. There are several boxes in which to enter the dependent variable, WRKHRS, and one or more independent variables, EDUCYRS. The Linear Regression dialog box offers many other choices, but the default output from the procedure contains all that we need.

SPSS produces a great deal of output. We've selected two portions of the output to review here (Figure 14.6). Under the Model Summary, the coefficient of determination is labeled "R Square." Its value is .001, which is very weak. Educational attainment explains little of the variation in hours worked, only about 0.1 percent. This is probably not too surprising; for example, people who own a small business may have no more than a high school education but work very long hours in their business.

Figure 14.5

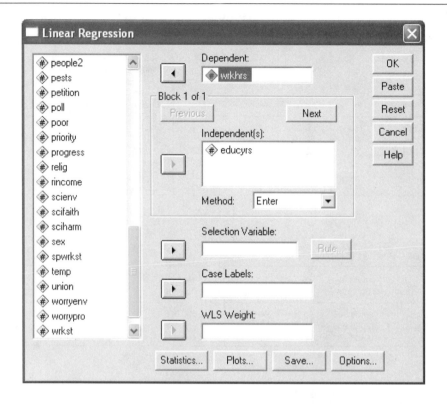

Figure 14.6

Model Summary

Model	R	R Square	Adjusted R Square	Std. Error of the Estimate
1	.028[a]	.001	−.001	13.229

a. Predictors: (Constant), EDUCYRS

ANOVA[b]

Model		Sum of Squares	df	Mean Square	F	Sig.
1	Regression	100.205	1	100.205	.573	.449[a]
	Residual	127395.9	728	174.994		
	Total	127496.1	729			

a. Predictors: (Constant), EDUCYRS

b. Dependent Variable: WRKHRS

Figure 14.7

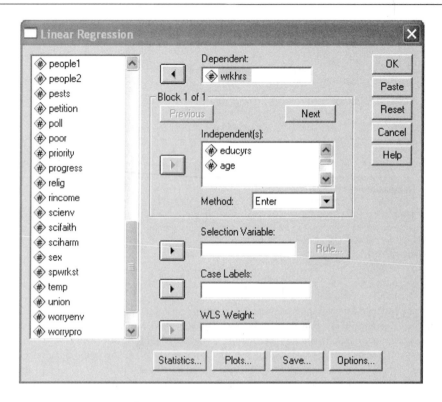

The ANOVA table provides the results of the analysis of variance test. The table includes Regression and Residual Sum of Squares, as well as Mean Squares. To test the null hypothesis that r^2 is zero, you will only need the statistic shown in the last column labeled "Sig." This is the p value associated with the F ratio listed in the column headed "F." The F statistic is .573 and its associated p value is .449. This means that there is a high probability (.449) that r^2 is really zero in the population, given the observed r^2 of .001. We therefore cannot reject the null hypothesis at the .05 level.

Demonstration 3: Producing an ANOVA Table: Regression-Multiple

What other variables, in addition to education, affect the number of hours worked? One answer to this question is that age (AGE) has something to do with the number of hours worked per week. This question was also explored in Chapter 8. To answer this question, we will use SPSS to calculate a multiple regression equation and a multiple coefficient of determination. This procedure is similar to the one used to generate the bivariate regression equation. Click on *Analyze*, *Regression*, then *Linear*. The Linear Regression dialog box (Figure 14.7) provides boxes in which to enter the dependent variable, WRKHRS, and the independent variables, EDUCYRS and AGE. We place EDUCYRS (number of years of education) and AGE (age in years) in the box for the independent variables and WRKHRS (the number of hours worked per week) in the box for the dependent variable, and click on *OK.*

SPSS produces a great deal of output. We've selected two portions of the output to review here (Figure 14.8).

Figure 14.8

Model Summary

Model	R	R Square	Adjusted R Square	Std. Error of the Estimate
1	.028[a]	.001	−.002	13.220

a. Predictors: (Constant), AGE, EDUCYRS

ANOVA[b]

Model		Sum of Squares	df	Mean Square	F	Sig.
1	Regression	101.486	2	50.743	.290	.748[a]
	Residual	126523.5	724	174.756		
	Total	126625.0	726			

a. Predictors: (Constant), AGE, EDUCYRS

b. Dependent Variable: WRKHRS

Under the Model Summary, the multiple correlation coefficient labeled "R" is .028. This tells us that education and age are weakly associated with hours worked. The coefficient of determination is labeled "R Square." Its value is .001. In addition, SPSS provides an "Adjusted R Square" which is −.002 (i.e., 0.0). The "Adjusted R Square" adjusts the R^2 coefficient for the number of predictors in the equation. Generally, the adjusted R^2 will be lower, relative to R^2, the larger the number of predictors. An R^2 of .001 means that educational attainment and age jointly explain about 0.1 percent of the variation in hours worked.

The ANOVA table tells us that the significance of F (.290) is .748. In other words, since $p > .05$ we cannot reject the null hypothesis that R^2 is zero at the .05 level.

SPSS PROBLEMS

1. Based on GSS02PFP-A, let's continue to examine the relationship between fertility decisions and education. But this time, we'll analyze the relationship for men.
   a. Run a Select Cases, selecting only men for the analysis.
   b. Compute an ANOVA model for men, using age at first-born child (AGEKDBRN) as the dependent variable and educational degree (DEGREE) as the independent variable.
   c. Based on the SPSS output, what can you conclude about the relationship between degree attainment and AGEKDBRN for men? How do these results compare to the results for women in the SPSS demonstration?
   d. Compute a second ANOVA model for men, using number of children (CHILDS) as the dependent variable and educational degree (DEGREE) as the independent variable. Based on your results, what conclusions can you draw?

2. Repeat Exercise 1b, substituting respondent's social class (CLASS) as the independent variable in separate models for men and women. What can you conclude about the relationship between CLASS and AGEKDBRN?

3. We'll continue our analysis of fertility decisions, examining responses to the question, What is the ideal number of children a family should have? (variable CHLDIDEL). Based on GSS02PFP-A, use CHLDIDEL as your dependent variable and DEGREE as your independent

variable. Is there a significant difference in number of ideal children among different educational groups? (Option: You can run three sets of analyses—first, for all GSS respondents; second, an ANOVA model for women only; and finally, a model for men.)

4. As we did in Chapter 8, use GSS02PFP-A to investigate the relationship between the respondent's education (EDUC) and the education received by his or her father and mother (PAEDUC and MAEDUC, respectively).

   a. What is the *F* statistic, and what is its significance? Can you reject the null hypothesis that R^2 is equal to zero?

   b. Did taking into account the respondent's mother's education improve our prediction? Compare your results from 4a to those when only a respondent's father's education is considered as the independent variable.

5. Based on GSS02PFP-B, examine attitudes toward affirmative action based on two variables: AFFRMACT and DISCAFF. AFFRMACT measures respondent support of preferential hiring and promotion of blacks (a higher score indicates strong opposition). For the variable DISCAFF, individuals reported how likely is it that a white person won't get a job or promotion while an equally or less qualified black person gets one (a higher score indicates "not very likely"). Using AFFRMACT and DISCAFF as your dependent variables, determine whether there are significant differences in attitudes by social class (CLASS)? (You should have two ANOVA models, with CLASS as the independent variable in both models.)

6. In Chapter 8 we used the 2000 ISSP data file to study the relationship between the number of persons in a respondent's household (HOMPOP) and number of hours worked per week (WRKHRS). Run the same analysis with HOMPOP as the independent variable and WRKHRS as the dependent variable (using data from ISSP00PFP). This time, however, focus on the ANOVA summary table.

   What is the *F* statistic? What is its *p* value? Is the relationship between the number of persons in a respondent's household and the number of hours worked per week significant? Interpret the results.

7. As we did in Chapter 8, use the same variables as in Exercise 1, but do the analysis separately for Americans and Russians. Begin by locating the variable COUNTRY. Click *Data*, *Split File*, and then select *Organize Output by Groups*. Insert COUNTRY into the box and click *OK*. Now SPSS will split your results by country.

   Locate the ANOVA summary tables for Americans and Russians. Interpret the *F* statistic from each table. Is each model statistically significant?

8. As we did in Chapter 8, use the same variables as in Exercise 1, but do the analysis separately for women and men. Begin by locating the variable SEX. Click *Data*, *Split File*, and then select *Organize Output by Groups*. Insert SEX into the box and click *OK*. (Note: Be sure to remove COUNTRY from the box if it is still there from the previous exercise.) Now SPSS will split your results by sex.

   Is there any difference between the regression equations for men and women? Compare the *F* statistics and the statistical significance.

9. As we did in Chapter 8, use the same variables as in Exercise 1, but do the analysis separately for married and divorced respondents. Begin by locating the variable MARITAL. Click *Data*, *Split File*, and then select *Organize Output by Groups*. Insert MARITAL into the box and click *OK*. (Note: Be sure to remove SEX and/or COUNTRY from the box if either is still there from the previous exercises.) Now SPSS will split your results by marital status.

   Find the *F* ratios and the significance level for married and divorced respondents. On the basis of these values, which model appears to be suited to make predictions in terms of the number of hours worked per week?

CHAPTER EXERCISES

1. In Chapter 8 we examined the relationship between being concerned about the state of the environment and actually donating money to environmental groups for respondents from eight countries around the world.

Country	Percent Concerned	Percent Donating Money
United States	33.8	22.8
Austria	35.5	27.8
Netherlands	30.1	44.8
Slovenia	50.3	10.7
Russia	29.0	1.6
Philippines	50.1	6.8
Spain	35.9	7.4
Denmark	27.2	22.3

Source: International Social Survey Programme 2000.

Using the formulas discussed in this chapter, as well as those in Chapter 8, calculate the F statistic. What is its p value? Is the relationship between being concerned about the environment and donating money to environmental groups significant? Interpret the results.

2. In Chapter 8 we examined the relationship between a person's educational attainment and the number of children he or she has. Our hypothesis was that as one's educational level increases, he or she has fewer children. Now, we would like to know if the relationship between education and the number of children is significant. Investigate this conjecture with 25 cases drawn from the 2002 GSS file. The following table displays educational attainment, in years, and the number of children for each respondent.

EDUC	CHILDS	EDUC	CHILDS
16	0	12	2
12	1	12	3
12	3	11	1
6	6	12	2
14	2	11	2
14	2	12	0
16	2	12	2
12	2	12	3
17	2	12	4
12	3	12	1
14	4	14	0
13	0	12	3
12	1		

Calculate the F statistic and interpret. Assume that $\alpha = .05$ and test the null hypothesis that $r^2 = 0$.

3. In this exercise we examine the relationship between union membership and respondent's earning based on data from the ISSP 2000. SPSS ANOVA output is presented in Figure 14.9.
   a. Rank groups by average earnings, starting from the lowest salary to the highest.
   b. Based on the SPSS output, what can you conclude about the relationship between union membership and earnings? Assume $\alpha = .05$.

4. The SPSS output shown in Figure 14.10 displays the relationship between education (measured in years) and television viewing (measured in hours) based on 2002 GSS data. We can hypothesize that as educational attainment increases, hours of television viewing will decrease, indicating a negative relationship between the two variables.

   Discuss the significance of the overall model based on F and its p level. Is the relationship between education and television viewing significant?

Figure 14.9

Descriptives

rincome Earnings

	N	Mean	Std. Deviation	Std. Error	95% Confidence Interval for Mean Lower Bound	95% Confidence Interval for Mean Upper Bound	Minimum	Maximum
1. Member, currently	278	99812.78	150022.230	8997.735	82100.15	117525.40	85	999996
2. Once member, former member	20	31046.40	42129.192	9420.374	11329.33	50763.47	30	130000
3. No member	531	44614.56	89818.279	3897.781	36957.56	52271.55	8	999996
Total	829	62797.60	115889.592	4025.012	54897.17	70698.03	8	999996

ANOVA

rincome Earnings

	Sum of Squares	df	Mean Squares	F	Sig.
Between Groups	6E+011	2	2.883E+011	22.586	.000
Within Groups	1E+013	826	1.276E+010		
Total	1E+013	828			

5. In several bivariate tables in Chapter 6, we examined the relationship between support for abortion and preferred family size. We found that there was a relationship between larger preferred family size and no support for abortion. We extend this analysis to an ANOVA model based on GSS2002 data using support for abortion under any circumstance (1, yes; 2, no) as our independent grouping variable. For our dependent variable, we use CHILDS (number of actual children). SPSS output tables (both Descriptives and ANOVA) are presented in Figure 14.11.
   a. Which group had the highest average number of children?
   b. Is there a significant difference in the number of children between the two groups? (Assume that $\alpha = .05$.) Provide evidence to support your answer.

Figure 14.10

Model Summary

Model	R	R Square	Adjusted R Square	Std. Error of the Estimate
1	.177[a]	.031	.029	2.351

a. Predictors: (Constant), EDUC

ANOVA[b]

Model		Sum of Squares	df	Mean Square	F	Sig.
1	Regression	89.369	1	89.369	16.164	.000[a]
	Residual	2753.453	498	5.529		
	Total	2842.822	499			

a. Predictors: (Constant), EDUC

b. Dependent Variable: TVHOURS

Figure 14.11

Descriptives

childs NUMBER OF CHILDREN

	N	Mean	Std. Deviation	Std. Error	95% Confidence Interval for Mean		Minimum	Maximum
					Lower Bound	Upper Bound		
1 YES	208	1.66	1.640	.114	1.44	1.89	0	8
2 NO	282	2.04	1.903	.113	1.82	2.27	0	8
Total	490	1.88	1.804	.081	1.72	2.04	0	8

ANOVA

childs NUMBER OF CHILDREN

	Sum of Squares	df	Mean Square	F	Sig.
Between Groups	17.203	1	17.203	5.334	.021
Within Groups	1573.932	488	3.225		
Total	1591.135	489			

6. In earlier chapters, we analyzed the relationship between social class and health assessment. We repeat the analysis here for a random sample of 32 cases from the GSS 2002. Health is measured according to a four-point scale, 1, excellent; 2, good; 3, fair; 4, poor. Four social classes are

Lower Class	Working Class	Middle Class	Upper Class
3	2	2	2
2	1	3	1
2	3	1	1
2	2	1	2
3	2	2	1
3	2	3	1
3	2	3	1
3	3	2	3

reported here: lower, working, middle, and upper. Complete the five-step model for these data, using $\alpha = .05$.

7. We extend our analysis in Exercise 5 to include CHLDIDEL as our dependent variable. The grouping variable remains the same. SPSS output tables (both Descriptives and ANOVA) are presented in Figure 14.12.
   a. Which group had the highest average of ideal number of children?
   b. Is there a significant difference in the ideal number of children between the two groups? (Assume $\alpha = .05$.) Explain the reason for your answer.
   c. Would your answer change if alpha were set at .01?

Figure 14.12

Descriptives

chldidel IDEAL NUMBER OF CHILDREN

	N	Mean	Std. Deviation	Std. Error	95% Confidence Interval for Mean		Minimum	Maximum
					Lower Bound	Upper Bound		
1 YES	98	2.52	1.364	.138	2.25	2.79	0	8
2 NO	155	2.97	1.696	.136	2.70	3.24	0	8
Total	253	2.79	1.588	.100	2.60	2.99	0	8

ANOVA

chldidel IDEAL NUMBER OF CHILDREN

	Sum of Squares	df	Mean Squares	F	Sig.
Between Groups	12.014	1	12.014	4.838	.029
Within Groups	623.298	251	2.483		
Total	635.312	252			

8. In Chapter 6, we used ISSP 2000 data to analyze the relationship between educational attainment and church attendance. We discovered that as educational attainment increased, church attendance decreased.

    We selected a sample of 30 ISSP respondents, noting their educational status (no degree, secondary degree, and university degree) and their level of church attendance [never (0), infrequently (1), or two to three times per month or more (2)].

    Complete the five-step model for these data, using $\alpha = .01$.

No Degree	Secondary Degree	University Degree
2	2	0
1	2	0
1	2	0
2	1	0
2	1	1
2	0	1
0	2	0
2	1	1
2	1	2
2	2	1

9. Based on a sample of 32 ISSP respondents, we present their social class and the number of people in their household. Complete the five-step model for these data, set α at .01.

Lower	Working	Middle	Upper
3	5	2	2
4	5	3	1
5	4	3	1
4	3	4	2
5	4	3	3
5	3	2	2
4	2	3	2
3	4	4	3

10. As examined in Chapter 8, social scientists have long been interested in the aspirations and achievements of people in the United States. The GSS 2002 data set has information on the educational level of respondents and their mothers. Use the following information to see whether those whose mothers had more education are more likely to have more education themselves.

    Calculate the F statistic. What is its p value? Is the relationship between a respondent's mother's education and a respondent's education significant? Interpret the results.

Mother's Highest School Year Completed	Respondent's Highest School Year Completed
0	12
15	13
6	9
9	12
16	16
12	12
6	16
18	14
12	13
14	12
14	18
7	12

11. In Chapter 8 Exercise 5, we investigated the relationship between infant mortality rate and GNP in South America. The birth rates (number of live births per 1,000 inhabitants) in these same countries are shown in the following table:

Country	Birth Rate in 1999
Argentina	19
Bolivia	32
Brazil	20
Chile	18
Colombia	24
Ecuador	24
Paraguay	30
Peru	25
Uruguay	17
Venezuela	25

Source: Data from World Bank 2000.

Calculate the F statistic with GNP per capita as the independent variable and birth rate as the dependent variable. What is its p value? Is the relationship between GNP per capita and birth rate significant? Interpret the results.

12. As discussed in Chapter 8, individuals and families living below the poverty line face many obstacles, the least of which is access to health care. In many cases, those living below the poverty line are without any form of health insurance. Using data from the U.S. Census Bureau, analyze the relationship between living below the poverty line and access to health care.

Assume that the percent living below the poverty line is the independent variable. Can you reject the null hypothesis that $r^2 = 0$ at the .01 level? At the .05 level? Why or why not?

State	% Below Poverty Line (Average from 2001–2003)	% Without Health Insurance (Average from 2001–2003)
Alabama	15.1	13.3
California	12.9	18.7
Idaho	11.0	17.5
Louisiana	16.9	19.4
New Jersey	8.2	13.4
New York	14.2	15.5
Pennsylvania	9.9	10.7
Rhode Island	10.7	9.3
South Carolina	14.0	13.1
Texas	15.8	24.6
Washington	11.4	14.3
Wisconsin	8.8	9.5

Source: U.S. Bureau of the Census. Current Population Reports P60-226, *Income, Poverty and Health Insurance Coverage in the United States*: 2003.

13. We selected a sample of 14 Irish respondents from the ISSP 2000. We present their marital status along with their years of education (marital status is considered as the independent variable). Complete the five-step model for these data, using $\alpha = .05$.

Married	Separated	Single
11	13	16
10	14	15
11	13	14
9	13	15
11		15

Appendix A
Table of Random Numbers

A Table of 14,000 Random Units

Line/Col.	(1)	(2)	(3)	(4)	(5)	(6)	(7)	(8)	(9)	(10)	(11)	(12)	(13)	(14)
1	10480	15011	01536	02011	81647	91646	69179	14194	62590	36207	20969	99570	91291	90700
2	22368	46573	25595	85393	30995	89198	27982	53402	93965	34095	52666	19174	39615	99505
3	24130	48360	22527	97265	76393	64809	15179	24830	49340	32081	30680	19655	63348	58629
4	42167	93093	06243	61680	07856	16376	39440	53537	71341	57004	00849	74917	97758	16379
5	37570	39975	81837	16656	06121	91782	60468	81305	49684	60672	14110	06927	01263	54613
6	77921	06907	11008	42751	27756	53498	18602	70659	90655	15053	21916	81825	44394	42880
7	99562	72905	56420	69994	98872	31016	71194	18738	44013	48840	63213	21069	10634	12952
8	96301	91977	05463	07972	18876	20922	94595	56869	69014	60045	18425	84903	42508	32307
9	89579	14342	63661	10281	17453	18103	57740	84378	25331	12566	58678	44947	05585	56941
10	85475	36857	43342	53988	53060	59533	38867	62300	08158	17983	16439	11458	18593	64952
11	28918	69578	88231	33276	70997	79936	56865	05859	90106	31595	01547	85590	91610	78188
12	63553	40961	48235	03427	49626	69445	18663	72695	52180	20847	12234	90511	33703	90322
13	09429	93969	52636	92737	88974	33488	36320	17617	30015	08272	84115	27156	30613	74952
14	10365	61129	87529	85689	48237	52267	67689	93394	01511	26358	85104	20285	29975	89868
15	07119	97336	71048	08178	77233	13916	47564	81056	97735	85977	29372	74461	28551	90707
16	51085	12765	51821	51259	77452	16308	60756	92144	49442	53900	70960	63990	75601	40719
17	02368	21382	52404	60268	89368	19885	55322	44819	01188	65255	64835	44919	05944	55157
18	01011	54092	33362	94904	31273	04146	18594	29852	71585	85030	51132	01915	92747	64951
19	52162	53916	46369	58586	23216	14513	83149	98736	23495	64350	94738	17752	35156	35749
20	07056	97628	33787	09998	42698	06691	76988	13602	51851	46104	88916	19509	25625	58104
21	48663	91245	85828	14346	09172	30168	90229	04734	59193	22178	30421	61666	99904	32812
22	54164	58492	22421	74103	47070	25306	76468	26384	58151	06646	21524	15227	96909	44592
23	32639	32363	05597	24200	13363	38005	94342	28728	35806	06912	17012	64161	18296	22851
24	29334	27001	87637	87308	58731	00256	45834	15398	46557	41135	10367	07684	36188	18510
25	02488	33062	28834	07351	19731	92420	60952	61280	50001	67658	32586	86679	50720	94953

(Continued)

Line/Col.	(1)	(2)	(3)	(4)	(5)	(6)	(7)	(8)	(9)	(10)	(11)	(12)	(13)	(14)
26	81525	72295	04839	96423	24878	82651	66566	14778	76797	14780	13300	87074	79666	95725
27	29676	20591	68086	26432	46901	20849	89768	81536	86645	12659	92259	57102	80428	25280
28	00742	57392	39064	66432	84673	40027	32832	61362	98947	96067	64760	64584	96096	98253
29	05366	04213	25669	26422	44407	44048	37937	63904	45766	66134	75470	66520	34693	90449
30	91921	26418	64117	94305	26766	25940	39972	22209	71500	64568	91402	42416	07844	69618
31	00582	04711	87917	77341	42206	35126	74087	99547	81817	42607	43808	76655	62028	76630
32	00725	69884	62797	56170	86324	88072	76222	36086	84637	93161	76038	65855	77919	88006
33	69011	65797	95876	55293	18988	27354	26575	08625	40801	59920	29841	80150	12777	48501
34	25976	57948	29888	88604	67917	48708	18912	82271	65424	69774	33611	54262	85963	03547
35	09763	83473	73577	12908	30883	18317	28290	35797	05998	41688	34952	37888	38917	88050
36	91567	42595	27958	30134	04024	86385	29880	99730	55536	84855	29080	09250	79656	73211
37	17955	56349	90999	49127	20044	59931	06115	20542	18059	02008	73708	83317	36103	42791
38	46503	18584	18845	49618	02304	51038	20655	58727	28168	15475	56942	53389	20562	87338
39	92157	89634	94824	78171	84610	82834	09922	25417	44137	48413	25555	21246	35509	20468
40	14577	62765	35605	81263	39667	47358	56873	56307	61607	49518	89656	20103	77490	18062
41	98427	07523	33362	64270	01638	92477	66969	98420	04880	45585	46565	04102	46880	45709
42	34914	63976	88720	82765	34476	17032	87589	40836	32427	70002	70663	88863	77775	69348
43	70060	28277	39475	46473	23219	53416	94970	25832	69975	94884	19661	72828	00102	66794
44	53976	54914	06990	67245	68350	82948	11398	42878	80287	88267	47363	46634	06541	97809
45	76072	29515	40980	07391	58745	25774	22987	80059	39911	96189	41151	14222	60697	59583
46	90725	52210	83974	29992	65831	38857	50490	83765	55657	14361	31720	57375	56228	41546
47	64364	67412	33339	31926	14883	24413	59744	92351	97473	89286	35931	04110	23726	51900
48	08962	00358	31662	25388	61642	34072	81249	35648	56891	69352	48373	45578	78547	81788
49	95012	68379	93526	70765	10593	04542	76463	54328	02349	17247	28865	14777	62730	92277
50	15664	10493	20492	38391	91132	21999	59516	81652	27195	48223	46751	22923	32261	85653
51	16408	81899	04153	53381	79401	21438	83035	92350	36693	31238	59649	91754	72772	02338
52	18629	81953	05520	91962	04739	13092	97662	24822	94730	06496	35090	04822	86772	98289
53	73115	35101	47498	87637	99016	71060	88824	71013	18735	20286	23153	72924	35165	43040
54	57491	16703	23167	49323	45021	33132	12544	41035	80780	45393	44812	12515	98931	91202
55	30405	83946	23792	14422	15059	45799	22716	19792	09983	74353	68668	30429	70735	25499
56	16631	35006	85900	98275	32388	52390	16815	69298	82732	38480	73817	32523	41961	44437
57	96773	20206	42559	78985	05300	22164	24369	54224	35083	19687	11052	91491	60383	19746
58	38935	64202	14349	82674	66523	44133	00697	35552	35970	19124	63318	29686	03387	59846
59	31624	76384	17403	53363	44167	64486	64758	75366	76554	31601	12614	33072	60332	92325
60	78919	19474	23632	27889	47914	02584	37680	20801	72152	39339	34806	08930	85001	87820
61	03931	33309	57047	74211	63445	17361	62825	39908	05607	91284	68833	25570	38818	46920
62	74426	33278	43972	10119	89917	15665	52872	73823	73144	88662	88970	74492	51805	99378
63	09066	00903	20795	95452	92648	45454	09552	88815	16553	51125	79375	97596	16296	66092
64	42238	12426	87025	14267	20979	04508	64535	31355	86064	29472	47689	05974	52468	16834
65	16153	08002	26504	41744	81959	65642	74240	56302	00033	67107	77510	70625	28725	34191

(Continued)

Line/ Col.	(1)	(2)	(3)	(4)	(5)	(6)	(7)	(8)	(9)	(10)	(11)	(12)	(13)	(14)
66	21457	40742	29820	96783	29400	21840	15035	34537	33310	06116	95240	15957	16572	06004
67	21581	57802	02050	89728	17937	37621	47075	42080	97403	48626	68995	43805	33386	21597
68	55612	78095	83197	33732	05810	24813	86902	60397	16489	03264	88525	42786	05269	92532
69	44657	66999	99324	51281	84463	60563	79312	93454	68876	25471	93911	25650	12682	73572
70	91340	84979	46949	81973	37949	61023	43997	15263	80644	43942	89203	71795	99533	50501
71	91227	21199	31935	27022	84067	05462	35216	14486	29891	68607	41867	14951	91696	85065
72	50001	38140	66321	19924	72163	09538	12151	06878	91903	18749	34405	56087	82790	70925
73	65390	05224	72958	28609	81406	39147	25549	48542	42627	45233	57202	94617	23772	07896
74	27504	96131	83944	41575	10573	08619	64482	73923	36152	05184	94142	25299	84387	34925
75	37169	94851	39117	89632	00959	16487	65536	49071	39782	17095	02330	74301	00275	48280
76	11508	70225	51111	38351	19444	66499	71945	05422	13442	78675	84081	66938	93654	59894
77	37449	30362	06694	54690	04052	53115	62757	95348	78662	11163	81651	50245	34971	52924
78	46515	70331	85922	38329	57015	15765	97161	17869	45349	61796	66345	81073	49106	79860
79	30986	81223	42416	58353	21532	30502	32305	86482	05174	07901	54339	58861	74818	46942
80	63798	64995	46583	09765	44160	78128	83991	42865	92520	83531	80377	35909	81250	54238
81	82486	84846	99254	67632	43218	50076	21361	64816	51202	88124	41870	52689	51275	83556
82	21885	32906	92431	09060	64297	51674	64126	62570	26123	05155	59194	52799	28225	85762
83	60336	98782	07408	53458	13564	59089	26445	29789	85205	41001	12535	12133	14645	23541
84	43937	46891	24010	25560	86355	33941	25786	54990	71899	15475	95434	98227	21824	19585
85	97656	63175	89303	16275	07100	92063	21942	18611	47348	20203	18534	03862	78095	50136
86	03299	01221	05418	38982	55758	92237	26759	86367	21216	98442	08303	56613	91511	75928
87	79626	06486	03574	17668	07785	76020	79924	25651	83325	88428	85076	72811	22717	50585
88	85636	68335	47539	03129	65651	11977	02510	26113	99447	68645	34327	15152	55230	93448
89	18039	14367	61337	06177	12143	46609	32989	74014	64708	00533	35398	58408	13261	47908
90	08362	15656	60627	36478	65648	16764	53412	09013	07832	41574	17639	82163	60859	75567
91	79556	29068	04142	16268	15387	12856	66227	38358	22478	73373	88732	09443	82558	05250
92	92608	82674	27072	32534	17075	27698	98204	63863	11951	34648	88022	56148	34925	57031
93	23982	25835	40055	67006	12293	02753	14827	22235	35071	99704	37543	11601	35503	85171
94	09915	96306	05908	97901	28395	14186	00821	80703	70426	75647	76310	88717	37890	40129
95	50937	33300	26695	62247	69927	76123	50842	43834	86654	70959	79725	93872	28117	19233
96	42488	78077	69882	61657	34136	79180	97526	43092	04098	73571	80799	76536	71255	64239
97	46764	86273	63003	93017	31204	36692	40202	35275	57306	55543	53203	18098	47625	88684
98	03237	45430	55417	63282	90816	17349	88298	90183	36600	78406	06216	95787	42579	90730
99	86591	81482	52667	61583	14972	90053	89534	76036	49199	43716	97548	04379	46370	28672
100	38534	01715	94964	87288	65680	43772	39560	12918	86537	62738	19636	51132	25739	56947

Source: William H. Beyer, ed., *Handbook for Probability and Statistics,* 2nd ed. Copyright © 1966 CRC Press, Boca Raton, Florida. Used by permission.

Appendix B
The Standard Normal Table

The values in column A are Z scores. Column B lists the proportion of area between the mean and a given Z. Column C lists the proportion of area beyond a given Z. Only positive Z scores are listed. Because the normal curve is symmetrical, the areas for negative Z scores will be exactly the same as the areas for positive Z scores.

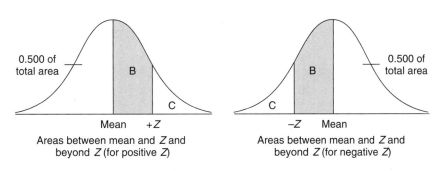

Areas between mean and Z and beyond Z (for positive Z)

Areas between mean and Z and beyond Z (for negative Z)

A	B	C	A	B	C	A	B	C
	Area Between	*Area Beyond*		*Area Between*	*Area Beyond*		*Area Between*	*Area Beyond*
Z	*Mean and Z*	*Z*	*Z*	*Mean and Z*	*Z*	*Z*	*Mean and Z*	*Z*
0.00	0.0000	0.5000	0.11	0.0438	0.4562	0.21	0.0832	0.4168
0.01	0.0040	0.4960	0.12	0.0478	0.4522	0.22	0.0871	0.4129
0.02	0.0080	0.4920	0.13	0.0517	0.4483	0.23	0.0910	0.4090
0.03	0.0120	0.4880	0.14	0.0557	0.4443	0.24	0.0948	0.4052
0.04	0.0160	0.4840	0.15	0.0596	0.4404	0.25	0.0987	0.4013
0.05	0.0199	0.4801	0.16	0.0636	0.4364	0.26	0.1026	0.3974
0.06	0.0239	0.4761	0.17	0.0675	0.4325	0.27	0.1064	0.3936
0.07	0.0279	0.4721	0.18	0.0714	0.4286	0.28	0.1103	0.3897
0.08	0.0319	0.4681	0.19	0.0753	0.4247	0.29	0.1141	0.3859
0.09	0.0359	0.4641	0.20	0.0793	0.4207	0.30	0.1179	0.3821
0.10	0.0398	0.4602						

(Continued)

A	B	C	A	B	C	A	B	C
Z	Area Between Mean and Z	Area Beyond Z	Z	Area Between Mean and Z	Area Beyond Z	Z	Area Between Mean and Z	Area Beyond Z
0.31	0.1217	0.3783	0.71	0.2611	0.2389	1.11	0.3665	0.1335
0.32	0.1255	0.3745	0.72	0.2642	0.2358	1.12	0.3686	0.1314
0.33	0.1293	0.3707	0.73	0.2673	0.2327	1.13	0.3708	0.1292
0.34	0.1331	0.3669	0.74	0.2703	0.2297	1.14	0.3749	0.1271
0.35	0.1368	0.3632	0.75	0.2734	0.2266	1.55	0.3749	0.1251
0.36	0.1406	0.3594	0.76	0.2764	0.2236	1.16	0.3770	0.1230
0.37	0.1443	0.3557	0.77	0.2794	0.2206	1.17	0.3790	0.1210
0.38	0.1480	0.3520	0.78	0.2823	0.2177	1.18	0.3810	0.1190
0.39	0.1517	0.3483	0.79	0.2852	0.2148	1.19	0.3830	0.1170
0.40	0.1554	0.3446	0.80	0.2881	0.2119	1.20	0.3849	0.1151
0.41	0.1591	0.3409	0.81	0.2910	0.2090	1.21	0.3869	0.1131
0.42	0.1628	0.3372	0.82	0.2939	0.2061	1.22	0.3888	0.1112
0.43	0.1664	0.3336	0.83	0.2967	0.2033	1.23	0.3907	0.1093
0.44	0.1700	0.3300	0.84	0.2995	0.2005	1.24	0.3925	0.1075
0.45	0.1736	0.3264	0.85	0.3023	0.1977	1.25	0.3944	0.1056
0.46	0.1772	0.3228	0.86	0.3051	0.1949	1.26	0.3962	0.1038
0.47	0.1808	0.3192	0.87	0.3078	0.1992	1.27	0.3980	0.1020
0.48	0.1844	0.3156	0.88	0.3106	0.1894	1.28	0.3997	0.1003
0.49	0.1879	0.3121	0.89	0.3133	0.1867	1.29	0.4015	0.0985
0.50	0.1915	0.3085	0.90	0.3159	0.1841	1.30	0.4032	0.0968
0.51	0.1950	0.3050	0.91	0.3186	0.1814	1.31	0.4049	0.0951
0.52	0.1985	0.3015	0.92	0.3212	0.1788	1.32	0.4066	0.0934
0.53	0.2019	0.2981	0.93	0.3238	0.1762	1.33	0.4082	0.0918
0.54	0.2054	0.2946	0.94	0.3264	0.1736	1.34	0.4099	0.0901
0.55	0.2088	0.2912	0.95	0.3289	0.1711	1.35	0.4115	0.0885
0.56	0.2123	0.2877	0.96	0.3315	0.1685	1.36	0.4131	0.0869
0.57	0.2157	0.2843	0.97	0.3340	0.1660	1.37	0.4147	0.0853
0.58	0.2190	0.2810	0.98	0.3365	0.1635	1.38	0.4612	0.0838
0.59	0.2224	0.2776	0.99	0.3389	0.1611	1.39	0.4177	0.0823
0.60	0.2257	0.2743	1.00	0.3413	0.1587	1.40	0.4192	0.0808
0.61	0.2291	0.2709	1.01	0.3438	0.1562	1.41	0.4207	0.0793
0.62	0.2324	0.2676	1.02	0.3461	0.1539	1.42	0.4222	0.0778
0.63	0.2357	0.2643	1.03	0.3485	0.1515	1.43	0.4236	0.0764
0.64	0.2389	0.2611	1.04	0.3508	0.1492	1.44	0.4251	0.0749
0.65	0.2422	0.2578	1.05	0.3531	0.1469	1.45	0.4265	0.0735
0.66	0.2454	0.2546	1.06	0.3554	0.1446	1.46	0.4279	0.0721
0.67	0.2486	0.2514	1.07	0.3577	0.1423	1.47	0.4292	0.0708
0.68	0.2517	0.2483	1.08	0.3599	0.1401	1.48	0.4306	0.0694
0.69	0.2549	0.2451	1.09	0.3621	0.1379	1.49	0.4319	0.0681
0.70	0.2580	0.2420	1.10	0.3643	0.1357	1.50	0.4332	0.0668

(Continued)

A	B	C	A	B	C	A	B	C
	Area	Area		Area	Area		Area	Area
	Between	Beyond		Between	Beyond		Between	Beyond
Z	Mean and Z	Z	Z	Mean and Z	Z	Z	Mean and Z	Z
1.51	0.4345	0.0655	1.91	0.4719	0.0281	2.31	0.4896	0.0104
1.52	0.4357	0.0643	1.92	0.4726	0.0274	2.32	0.4898	0.0102
1.53	0.4370	0.0630	1.93	0.4732	0.0268	2.33	0.4901	0.0099
1.54	0.4382	0.0618	1.94	0.4738	0.0262	2.34	0.4904	0.0096
1.55	0.4394	0.0606	1.95	0.4744	0.0256	2.35	0.4906	0.0094
1.56	0.4406	0.0594	1.96	0.4750	0.0250	2.36	0.4909	0.0091
1.57	0.4418	0.0582	1.97	0.4756	0.0244	2.37	0.4911	0.0089
1.58	0.4429	0.0571	1.98	0.4761	0.0239	2.38	0.4913	0.0087
1.59	0.4441	0.0559	1.99	0.4767	0.0233	2.39	0.4916	0.0084
1.60	0.4452	0.0548	2.00	0.4772	0.0228	2.40	0.4918	0.0082
1.61	0.4463	0.0537	2.01	0.4778	0.0222	2.41	0.4920	0.0080
1.62	0.4474	0.0526	2.02	0.4783	0.0217	2.42	0.4922	0.0078
1.63	0.4484	0.0516	2.03	0.4788	0.0212	2.43	0.4925	0.0075
1.64	0.4495	0.0505	2.04	0.4793	0.0207	2.44	0.4927	0.0073
1.65	0.4505	0.0495	2.05	0.4798	0.0202	2.45	0.4929	0.0071
1.66	0.4515	0.0485	2.06	0.4803	0.0197	2.46	0.4931	0.0069
1.67	0.4525	0.0475	2.07	0.4808	0.0192	2.47	0.4932	0.0068
1.68	0.4535	0.0465	2.08	0.4812	0.0188	2.48	0.4934	0.0066
1.69	0.4545	0.0455	2.09	0.4817	0.0183	2.49	0.4936	0.0064
1.70	0.4554	0.0466	2.10	0.4821	0.0179	2.50	0.4938	0.0062
1.71	0.4564	0.0436	2.11	0.4826	0.0174	2.51	0.4940	0.0060
1.72	0.4573	0.0427	2.12	0.4830	0.0170	2.52	0.4941	0.0059
1.73	0.4582	0.0418	2.13	0.4834	0.0166	2.53	0.4943	0.0057
1.74	0.4591	0.0409	2.14	0.4838	0.0162	2.54	0.4945	0.0055
1.75	0.4599	0.0401	2.15	0.4842	0.0158	2.55	0.4946	0.0054
1.76	0.4608	0.0392	2.16	0.4846	0.0154	2.56	0.4948	0.0052
1.77	0.4616	0.0384	2.17	0.4850	0.0150	2.57	0.4949	0.0051
1.78	0.4625	0.0375	2.18	0.4854	0.0146	2.58	0.4951	0.0049
1.79	0.4633	0.0367	2.19	0.4857	0.0143	2.59	0.4952	0.0048
1.80	0.4641	0.0359	2.20	0.4861	0.0139	2.60	0.4953	0.0047
1.81	0.4649	0.0351	2.21	0.4864	0.0136	2.61	0.4955	0.0045
1.82	0.4656	0.0344	2.22	0.4868	0.0132	2.62	0.4956	0.0044
1.83	0.4664	0.0336	2.23	0.4871	0.0129	2.63	0.4957	0.0043
1.84	0.4671	0.0329	2.24	0.4875	0.0125	2.64	0.4959	0.0041
1.85	0.4678	0.0322	2.25	0.4878	0.0122	2.65	0.4960	0.0040
1.86	0.4686	0.0314	2.26	0.4881	0.0119	2.66	0.4961	0.0039
1.87	0.4693	0.0307	2.27	0.4884	0.0116	2.67	0.4962	0.0038
1.88	0.4699	0.0301	2.28	0.4887	0.0113	2.68	0.4963	0.0037
1.89	0.4706	0.0294	2.29	0.4890	0.0110	2.69	0.4964	0.0036
1.90	0.4713	0.0287	2.30	0.4893	0.0107	2.70	0.4965	0.0035

(Continued)

A Z	B Area Between Mean and Z	C Area Beyond Z	A Z	B Area Between Mean and Z	C Area Beyond Z	A Z	B Area Between Mean and Z	C Area Beyond Z
2.71	0.4966	0.0034	3.01	0.4987	0.0013	3.31	0.4995	0.0005
2.72	0.4967	0.0033	3.02	0.4987	0.0013	3.32	0.4995	0.0005
2.73	0.4968	0.0032	3.03	0.4988	0.0012	3.33	0.4996	0.0004
2.74	0.4969	0.0031	3.04	0.4988	0.0012	3.34	0.4996	0.0004
2.75	0.4970	0.0030	3.05	0.4989	0.0011	3.35	0.4996	0.0004
2.76	0.4971	0.0029	3.06	0.4989	0.0011	3.36	0.4996	0.0004
2.77	0.4972	0.0028	3.07	0.4989	0.0011	3.37	0.4996	0.0004
2.78	0.4973	0.0027	3.08	0.4990	0.0010	3.38	0.4996	0.0004
2.79	0.4974	0.0026	3.09	0.4990	0.0010	3.39	0.4997	0.0003
2.80	0.4974	0.0026	3.10	0.4990	0.0010	3.40	0.4997	0.0003
2.81	0.4975	0.0025	3.11	0.4991	0.0009	3.41	0.4997	0.0003
2.82	0.4976	0.0024	3.12	0.4991	0.0009	3.42	0.4997	0.0003
2.83	0.4977	0.0023	3.13	0.4991	0.0009	3.43	0.4997	0.0003
2.84	0.4977	0.0023	3.14	0.4992	0.0008	3.44	0.4997	0.0003
2.85	0.4978	0.0022	3.15	0.4992	0.0008	3.45	0.4997	0.0003
2.86	0.4979	0.0021	3.16	0.4992	0.0008	3.46	0.4997	0.0003
2.87	0.4979	0.0021	3.17	0.4992	0.0008	3.47	0.4997	0.0003
2.88	0.4980	0.0020	3.18	0.4993	0.0007	3.48	0.4997	0.0003
2.89	0.4981	0.0019	3.19	0.4993	0.0007	3.49	0.4998	0.0002
2.90	0.4981	0.0019	3.20	0.4993	0.0007	3.50	0.4998	0.0002
2.91	0.4982	0.0018	3.21	0.4993	0.0007	3.60	0.4998	0.0002
2.92	0.4982	0.0018	3.22	0.4994	0.0006	3.70	0.4999	0.0001
2.93	0.4983	0.0017	3.23	0.4994	0.0006			
2.94	0.4984	0.0016	3.24	0.4994	0.0006	3.80	0.4999	0.0001
2.95	0.4984	0.0016	3.25	0.4994	0.0006			
2.96	0.4985	0.0015	3.26	0.4994	0.0006	3.90	0.4999	<0.0001
2.97	0.4985	0.0015	3.27	0.4995	0.0005	4.00	0.4999	<0.0001
2.98	0.4986	0.0014	3.28	0.4995	0.0005			
2.99	0.4986	0.0014	3.29	0.4995	0.0005			
3.00	0.4986	0.0014	3.30	0.4995	0.0005			

Appendix C
Distribution of t

df	Level of Significance for One-Tailed Test					
	.10	.05	.025	.01	.005	.0005
	Level of Significance for Two-Tailed Test					
	.20	.10	.05	.02	.01	.001
1	3.078	6.314	12.706	31.821	63.657	636.619
2	1.886	2.920	4.303	6.965	9.925	31.598
3	1.638	2.353	3.182	4.541	5.841	12.941
4	1.533	2.132	2.776	3.747	4.604	8.610
5	1.476	2.015	2.571	3.365	4.032	6.859
6	1.440	1.943	2.447	3.143	3.707	5.959
7	1.415	1.895	2.365	2.998	3.499	5.405
8	1.397	1.860	2.306	2.896	3.355	5.041
9	1.383	1.833	2.262	2.821	3.250	4.781
10	1.372	1.812	2.228	2.764	3.169	4.587
11	1.363	1.796	2.201	2.718	3.106	4.437
12	1.356	1.782	2.179	2.681	3.055	4.318
13	1.350	1.771	2.160	2.650	3.012	4.221
14	1.345	1.761	2.145	2.624	2.977	4.140
15	1.341	1.753	2.131	2.602	2.947	4.073
16	1.337	1.746	2.120	2.583	2.921	4.015
17	1.333	1.740	2.110	2.567	2.898	3.965
18	1.330	1.734	2.101	2.552	2.878	3.922
19	1.328	1.729	2.093	2.539	2.861	3.883
20	1.325	1.725	2.086	2.528	2.845	3.850
21	1.323	1.721	2.080	2.518	2.831	3.819
22	1.321	1.717	2.074	2.508	2.819	3.792
23	1.319	1.714	2.069	2.500	2.807	3.767
24	1.318	1.711	2.064	2.492	2.797	3.745
25	1.316	1.708	2.060	2.485	2.787	3.725

(Continued)

	Level of Significance for One-Tailed Test					
	.10	.05	.025	.01	.005	.0005
	Level of Significance for Two-Tailed Test					
df	.20	.10	.05	.02	.01	.001
26	1.315	1.706	2.056	2.479	2.779	3.707
27	1.314	1.703	2.052	2.473	2.771	3.690
28	1.313	1.701	2.048	2.467	2.763	3.674
29	1.311	1.699	2.045	2.462	2.756	3.659
30	1.310	1.697	2.042	2.457	2.750	3.646
40	1.303	1.684	2.021	2.423	2.704	3.551
60	1.296	1.671	2.000	2.390	2.660	3.460
120	1.289	1.658	1.980	2.358	2.617	3.373
∞	1.282	1.645	1.960	2.326	2.576	3.291

Source: Abridged from R. A. Fisher and F. Yates, *Statistical Tables for Biological, Agricultural and Medical Research*, 6th ed. Copyright © R. A. Fisher and F. Yates 1963. Reprinted by permission of Pearson Education Limited.

Appendix D
Distribution of Chi-Square

df	.99	.98	.95	.90	.80	.70	.50	.30	.20	.10	.05	.02	.01	.001
1	.03157	.03628	.00393	.0158	.0642	.148	.455	1.074	1.642	2.706	3.841	5.412	6.635	10.827
2	.0201	.0404	.103	.211	.446	.713	1.386	2.408	3.219	4.605	5.991	7.824	9.210	13.815
3	.115	.185	.352	.584	1.005	1.424	2.366	3.665	4.642	6.251	7.815	9.837	11.341	16.268
4	.297	.429	.711	1.064	1.649	2.195	3.357	4.878	5.989	7.779	9.488	11.668	13.277	18.465
5	.554	.752	1.145	1.610	2.343	3.000	4.351	6.064	7.289	9.236	11.070	13.388	15.086	20.517
6	.872	1.134	1.635	2.204	3.070	3.828	5.348	7.231	8.558	10.645	12.592	15.033	16.812	22.457
7	1.239	1.564	2.167	2.833	3.822	4.671	6.346	8.383	9.803	12.017	14.067	16.622	18.475	24.322
8	1.646	2.032	2.733	3.490	4.594	5.527	7.344	9.524	11.030	13.362	15.507	18.168	20.090	26.125
9	2.088	2.532	3.325	4.168	5.380	6.393	8.343	10.656	12.242	14.684	16.919	19.679	21.666	27.877
10	2.558	3.059	3.940	4.865	6.179	7.267	9.342	11.781	13.442	15.987	18.307	21.161	23.209	29.588
11	3.053	3.609	4.575	5.578	6.989	8.148	10.341	12.899	14.631	17.275	19.675	22.618	24.725	31.264
12	3.571	4.178	5.226	6.304	7.807	9.034	11.340	14.011	15.812	18.549	21.026	24.054	26.217	32.909
13	4.107	4.765	5.892	7.042	8.634	9.926	12.340	15.119	16.985	19.812	22.362	25.472	27.688	34.528
14	4.660	5.368	6.571	7.790	9.467	10.821	13.339	16.222	18.151	21.064	23.685	26.873	29.141	36.123
15	5.229	5.985	7.261	8.547	10.307	11.721	14.339	17.322	19.311	22.307	24.996	28.259	30.578	37.697
16	5.812	6.614	7.962	9.312	11.152	12.624	15.338	18.418	20.465	23.542	26.296	29.633	32.000	39.252
17	6.408	7.255	8.672	10.085	12.002	13.531	16.338	19.511	21.615	24.769	27.587	30.995	33.409	40.790
18	7.015	7.906	9.390	10.865	12.857	14.440	17.338	20.601	22.760	25.989	28.869	32.346	34.805	42.312
19	7.633	8.567	10.117	11.651	13.716	15.352	18.338	21.689	23.900	27.204	30.144	33.687	36.191	43.820
20	8.260	9.237	10.851	12.443	14.578	16.266	19.337	22.775	25.038	28.412	31.410	35.020	37.566	45.315
21	8.897	9.915	11.591	13.240	15.445	17.182	20.337	23.858	26.171	29.615	32.671	36.343	38.932	46.797
22	9.542	10.600	12.338	14.041	16.314	18.101	21.337	24.939	27.301	30.813	33.924	37.659	40.289	48.268
23	10.196	11.293	13.091	14.848	17.187	19.021	22.337	26.018	28.429	32.007	35.172	38.968	41.638	49.728
24	10.856	11.992	13.848	15.659	18.062	19.943	23.337	27.096	29.553	33.196	36.415	40.270	42.980	51.179
25	11.524	12.697	14.611	16.473	18.940	20.867	24.337	28.172	30.675	34.382	37.652	41.566	44.314	52.620
26	12.198	13.409	15.379	17.292	19.820	21.792	25.336	29.246	31.795	35.563	38.885	42.856	45.642	54.052
27	12.879	14.125	16.151	18.114	20.703	22.719	26.336	30.319	32.912	36.741	40.113	44.140	46.963	55.476
28	13.565	14.847	16.928	18.939	21.588	23.647	27.336	31.391	34.027	37.916	41.337	45.419	48.278	56.893
29	14.256	15.574	17.708	19.768	22.475	24.577	28.336	32.461	35.139	39.087	42.557	46.693	49.588	58.302
30	14.953	16.306	18.493	20.599	23.364	25.508	29.336	33.530	36.250	40.256	43.773	47.962	50.892	59.703

Source: R. A. Fisher & F. Yates, *Statistical Tables for Biological, Agricultural and Medical Research,* 6th ed. Copyright © R. A. Fisher and F. Yates 1963. Reprinted by permission of Pearson Education Limited.

Appendix E
Distribution of F

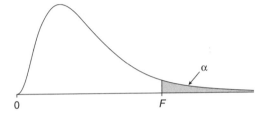

$$\alpha = .05$$

df_2	1	2	3	4	5	6	8	12	24	∞
					df_1					
1	161.4	199.5	215.7	224.6	230.2	234.0	238.9	243.9	249.0	254.3
2	18.51	19.00	19.16	19.25	19.30	19.33	19.37	19.41	19.45	19.50
3	10.13	9.55	9.28	9.12	9.01	8.94	8.84	8.74	8.64	8.53
4	7.71	6.94	6.59	6.39	6.26	6.16	6.04	5.91	5.77	5.63
5	6.61	5.79	5.41	5.19	5.05	4.95	4.82	4.68	4.53	4.36
6	5.99	5.14	4.76	4.53	4.39	4.28	4.15	4.00	3.84	3.67
7	5.59	4.74	4.35	4.12	3.97	3.87	3.73	3.57	3.41	3.23
8	5.32	4.46	4.07	3.84	3.69	3.58	3.44	3.28	3.12	2.93
9	5.12	4.26	3.86	3.63	3.48	3.37	3.23	3.07	2.90	2.71
10	4.96	4.10	3.71	3.48	3.33	3.22	3.07	2.91	2.74	2.54
11	4.84	3.98	3.59	3.36	3.20	3.09	2.95	2.79	2.61	2.40
12	4.75	3.88	3.49	3.26	3.11	3.00	2.85	2.69	2.50	2.30
13	4.67	3.80	3.41	3.18	3.02	2.92	2.77	2.60	2.42	2.21
14	4.60	3.74	3.34	3.11	2.96	2.85	2.70	2.53	2.35	2.13
15	4.54	3.68	3.29	3.06	2.90	2.79	2.64	2.48	2.29	2.07
16	4.49	3.63	3.24	3.01	2.85	2.74	2.59	2.42	2.24	2.01
17	4.45	3.59	3.20	2.96	2.81	2.70	2.55	2.38	2.19	1.96
18	4.41	3.55	3.16	2.93	2.77	2.66	2.51	2.34	2.15	1.92
19	4.38	3.52	3.13	2.90	2.74	2.63	2.48	2.31	2.11	1.88
20	4.35	3.49	3.10	2.87	2.71	2.60	2.45	2.28	2.08	1.84
21	4.32	3.47	3.07	2.84	2.68	2.57	2.42	2.25	2.05	1.81
22	4.30	3.44	3.05	2.82	2.66	2.55	2.40	2.23	2.03	1.78
23	4.28	3.42	3.03	2.80	2.64	2.53	2.38	2.20	2.00	1.76
24	4.26	3.40	3.01	2.78	2.62	2.51	2.36	2.18	1.98	1.73
25	4.24	3.38	2.99	2.76	2.60	2.49	2.34	2.16	1.96	1.71
26	4.22	3.37	2.98	2.74	2.59	2.47	2.32	2.15	1.95	1.69
27	4.21	3.35	2.96	2.73	2.57	2.46	2.30	2.13	1.93	1.67
28	4.20	3.34	2.95	2.71	2.56	2.44	2.29	2.12	1.91	1.65
29	4.18	3.33	2.93	2.70	2.54	2.43	2.28	2.10	1.90	1.64
30	4.17	3.32	2.92	2.69	2.53	2.42	2.27	2.09	1.89	1.62
40	4.08	3.23	2.84	2.61	2.45	2.34	2.18	2.00	1.79	1.51
60	4.00	3.15	2.76	2.52	2.37	2.25	2.10	1.92	1.70	1.39
120	3.92	3.07	2.68	2.45	2.29	2.17	2.02	1.83	1.61	1.25
∞	3.84	2.99	2.60	2.37	2.21	2.09	1.94	1.75	1.52	1.00

Source: R. A. Fisher and F. Yates, *Statistical Tables for Biological, Agricultural and Medical Research,* 6th ed. Copyright © R. A. Fisher and F. Yates 1963. Reprinted by permission of Pearson Education Limited.

$$\alpha = .01$$

df_2	df_1									
	1	2	3	4	5	6	8	12	24	∞
1	4052	4999	5403	5625	5764	5859	5981	6106	6234	6366
2	98.49	99.01	99.17	99.25	99.30	99.33	99.36	99.42	99.46	99.50
3	34.12	30.81	29.46	28.71	28.24	27.91	27.49	27.05	26.60	26.12
4	21.20	18.00	16.69	15.98	15.52	15.21	14.80	14.37	13.93	13.46
5	16.26	13.27	12.06	11.39	10.97	10.67	10.27	9.89	9.47	9.02
6	13.74	10.92	9.78	9.15	8.75	8.47	8.10	7.72	7.31	6.88
7	12.25	9.55	8.45	7.85	7.46	7.19	6.84	6.47	6.07	5.65
8	11.26	8.65	7.59	7.01	6.63	6.37	6.03	5.67	5.28	4.86
9	10.56	8.02	6.99	6.42	6.06	5.80	5.47	5.11	4.73	4.31
10	10.04	7.56	6.55	5.99	5.64	5.39	5.06	4.71	4.33	3.91
11	9.65	7.20	6.22	5.67	5.32	5.07	4.74	4.40	4.02	3.60
12	9.33	6.93	5.95	5.41	5.06	4.82	4.50	4.16	3.78	3.36
13	9.07	6.70	5.74	5.20	4.86	4.62	4.30	3.96	3.59	3.16
14	8.86	6.51	5.56	5.03	4.69	4.46	4.14	3.80	3.43	3.00
15	8.68	6.36	5.42	4.89	4.56	4.32	4.00	3.67	3.29	2.87
16	8.53	6.23	5.29	4.77	4.44	4.20	3.89	3.55	3.18	2.75
17	8.40	6.11	5.18	4.67	4.34	4.10	3.79	3.45	3.08	2.65
18	8.28	6.01	5.09	4.58	4.25	4.01	3.71	3.37	3.00	2.57
19	8.18	5.93	5.01	4.50	4.17	3.94	3.63	3.30	2.92	2.49
20	8.10	5.85	4.94	4.43	4.10	3.87	3.56	3.23	2.86	2.42
21	8.02	5.78	4.87	4.37	4.04	3.81	3.51	3.17	2.80	2.36
22	7.94	5.72	4.82	4.31	3.99	3.76	3.45	3.12	2.75	2.31
23	7.88	5.66	4.76	4.23	3.94	3.71	3.41	3.07	2.70	2.26
24	7.82	5.61	4.72	4.22	3.90	3.67	3.36	3.03	2.66	2.21
25	7.77	5.57	4.68	4.18	3.86	3.63	3.32	2.99	2.62	2.17
26	7.72	5.53	4.64	4.14	3.82	3.59	3.29	2.96	2.58	2.13
27	7.68	5.49	4.60	4.11	3.78	3.56	3.26	2.93	2.55	2.10
28	7.64	5.45	4.57	4.07	3.75	3.53	3.23	2.90	2.52	2.06
29	7.60	5.42	4.54	4.04	3.73	3.50	3.20	2.87	2.49	2.03
30	7.56	5.39	4.51	4.02	3.70	3.47	3.17	2.84	2.47	2.01
40	7.31	5.18	4.31	3.83	3.51	3.29	2.99	2.66	2.29	1.80
60	7.08	4.98	4.13	3.65	3.34	3.12	2.82	2.50	2.12	1.60
120	6.85	4.79	3.95	3.48	3.17	2.96	2.66	2.34	1.95	1.38
∞	6.64	4.60	3.78	3.32	3.02	2.80	2.51	2.18	1.79	1.00

Appendix F
A Basic Math Review

by James Harris

Y ou have probably already heard that there is a lot of math in statistics and for this reason you are somewhat anxious about taking a statistics course. Although it is true that courses in statistics can involve a great deal of mathematics, you should be relieved to hear that this course will stress interpretation rather than the ability to solve complex mathematical problems. With that said, however, you will still need to know how to perform some basic mathematical operations as well as understand the meanings of certain symbols used in statistics. Following is a review of the symbols and math you will need to know to successfully complete this course.

◉ SYMBOLS AND EXPRESSIONS USED IN STATISTICS

Statistics provides us with a set of tools for describing and analyzing *variables*. A variable is an attribute that can vary in some way. For example, a person's age is a variable because it can range from just born to over one hundred years old. "Race" and "gender" are also variables, though with fewer categories than the variable "age." In statistics, variables you are interested in measuring are often given a symbol. For example, if we wanted to know something about the age of students in our statistics class, we would use the symbol Y to represent the variable "age." Now let's say for simplicity we asked only the students sitting in the first row their ages—19, 21, 23, and 32. These four ages would be scores of the Y variable.

Another symbol that you will frequently encounter in statistics is Σ, or uppercase sigma. Sigma is a Greek letter that stands for summation in statistics. In other words, when you see the symbol Σ, it means you should sum all of the scores. An example will make this clear. Using our sample of students' ages represented by Y, the use of sigma as in the expression ΣY (read as: the sum of Y) tells us to sum all the scores of the variable Y. Using our example, we would find the sum of the set of scores from the variable "age" by adding the scores together:

$$19 + 21 + 23 + 32 = 95$$

So, for the variable "age," $\Sigma Y = 95$.

Sigma is also often used in expressions with an exponent, as in the expression $\sum Y^2$ (read as: the sum of squared scores). This means that we should first square all the scores of the Y variable and then sum the squared products. So using the same set of scores, we would solve the expression by squaring each score first and then adding them together:

$$19^2 + 21^2 + 23^2 + 32^2 = 361 + 441 + 529 + 1{,}024 = 2{,}355$$

So, for the variable "age," $\sum Y^2 = 2{,}355$.

A similar, but slightly different, expression, which illustrates the function of parentheses, is $(\sum Y)^2$ (read as: the sum of scores, squared). In this expression, the parentheses tell us to first sum all the scores and then square this summed total. Parentheses are often used in expressions in statistics, and they always tell us to perform the expression within the parentheses first and then the part of the problem that is outside of the parentheses. To solve this expression, we need to sum all the scores first. However, we already found that $\sum Y = 95$, so to solve the expression $(\sum Y)^2$, we simply square this summed total,

$$95^2 = 9{,}025$$

So, for the variable "age," $(\sum Y)^2 = 9{,}025$.

You should also be familiar with the different symbols that denote multiplication and division. Most students are familiar with the times sign ($\times$); however, there are several other ways to express multiplication. For example,

$$3(4) \qquad (5)6 \qquad (4)(2) \qquad 7 \bullet 8 \qquad 9 * 6$$

all symbolize the operation of multiplication. In this text, the first three are most often used to denote multiplication. There are also several ways division can be expressed. You are probably familiar with the conventional division sign ($\div$), but division can also be expressed in these other ways:

$$4/6 \qquad \frac{6}{3}$$

This text uses the latter two forms to express division.

In statistics you are likely to encounter greater than and less than signs ($>$, $<$), greater than or equal to and less than or equal to signs ($\geq$, $\leq$), and not equal to signs ($\neq$). It is important you understand what each sign means, though admittedly it is easy to confuse them. Use the following expressions for review. Notice that numerals and symbols are often used together:

$4 > 2$ means 4 is greater than 2

$H_1 > 10$ means H_1 is greater than 10

$7 < 9$ means 7 is less than 9

$a < b$ means a is less than b

$Y \geq 10$ means that the value for Y is a value greater than or equal to 10

$a \leq b$ means that the value for a is less than or equal to the value for b

$8 \neq 10$ means 8 does not equal 10

$H_1 \neq H_2$ means H_1 does not equal H_2

▣ PROPORTIONS AND PERCENTAGES

Proportions and percentages are commonly used in statistics and provide a quick way to express information about the relative frequency of some value. You should know how to find proportions and percentages.

Proportions are identified by P; to find a proportion apply this formula:

$$P = \frac{f}{N}$$

where f stands for the frequency of cases in a category and N the total number of cases in all categories. So, in our sample of four students, if we wanted to know the proportion of males in the front row, there would be a total of two categories, female and male. Because there are 3 females and 1 male in our sample, our N is 4; and the number of cases in our category "male" is 1. To get the proportion, divide 1 by 4:

$$P = \frac{f}{N} \qquad P = \frac{1}{4} = .25$$

So, the proportion of males in the front row is .25. To convert this to a percentage, simply multiply the proportion by 100 or use the formula for percentaging:

$$\% = \frac{f}{N} \times 100 \qquad \% = \frac{1}{4} \times 100 = 25\%$$

▣ WORKING WITH NEGATIVES

Addition, subtraction, multiplication, division, and squared numbers are not difficult for most people; however, there are some important rules to know when working with negatives that you may need to review.

1. When adding a number that is negative, it is the same as subtracting:
$$5 + (-2) = 5 - 2 = 3$$

2. When subtracting a negative number, the sign changes:
$$8 - (-4) = 8 + 4 = 12$$

3. When multiplying or dividing a negative number, the product or quotient is always negative:
$$6 \times -4 = -24, \; -10 \div 5 = -2$$

4. When multiplying or dividing two negative numbers, the product or quotient is always positive:

$$-3 \times -7 = 21, \, -12 \div -4 = 3$$

5. Squaring a number that is negative always gives a positive product because it is the same as multiplying two negative numbers:

$$-5^2 = 25 \text{ is the same as } -5 \times -5 = 25$$

▣ ORDER OF OPERATIONS AND COMPLEX EXPRESSIONS

In statistics you are likely to encounter some fairly lengthy equations that require several steps to solve. To know what part of the equation to work out first, follow two basic rules. The first is called the rules of precedence. They state that you should solve all squares and square roots first, then multiplication and division, and finally, all addition and subtraction from left to right. The second rule is to solve expressions in parentheses first. If there are brackets in the equation, solve the expression within parentheses first and then the expression within the brackets. This means that parentheses and brackets can override the rules of precedence. In statistics, it is common for parentheses to control the order of calculations. These rules may seem somewhat abstract here, but a brief review of their application should make them more clear.

To solve this problem,

$$4 + 6 \cdot 8 = 4 + 48 = 52$$

do the multiplication first and then the addition. Not following the rules of precedence will lead to a substantially different answer:

$$4 + 6 \cdot 8 = 10 \bullet 8 = 80$$

which is incorrect.

To solve this problem,

$$6 - 4(6)/3^2$$

first, find the square of 3,

$$6 - 4(6)/9$$

then do the multiplication and division from left to right,

$$6 - \frac{24}{9} = 6 - 2.67$$

and finally, work out the subtraction,

$$6 - 2.67 = 3.33$$

To work out the following equation, do the expressions within parentheses first:

$$(4 + 3) - 6(2)/(3 - 1)^2$$

First, solve the addition and subtraction in the parentheses,

$$(7) - 6(2)/(2)^2$$

Now that you have solved the expressions within parentheses, work out the rest of the equation based on the rules of precedence, first squaring the 2,

$$(7) - 6(2)/4$$

Then do the multiplication and division next:

$$(7) - \frac{12}{4} = (7) - 3$$

Finally, work out the subtraction to solve the equation:

$$7 - 3 = 4$$

The following equation may seem intimidating at first, but by solving it in steps and following the rules, even these complex equations should become manageable:

$$\sqrt{(8(4 - 2)^2)/(12/4)^2}$$

For this equation, work out the expressions within parentheses first; note that there are parentheses within parentheses. In this case, work out the inner parentheses first,

$$\sqrt{(8(2)^2)/3^2}$$

Now do the outer parentheses, making sure to follow the rules of precedence within the parentheses—square first and then multiply:

$$\sqrt{\frac{32}{3^2}}$$

Now, work out the square of 3 first and then divide:

$$\sqrt{\frac{32}{9}} = \sqrt{3.55}$$

Last, take the square root:

$$1.88$$

Answers to Odd-Numbered Exercises

Note to instructors: When calculating statistics, small rounding errors will change the exact value of final solutions. Small differences such as 0.02 or 0.03 between the values provided in the answers and students' calculations should not be viewed as an error.

CHAPTER 1

1. Once our research question, the hypothesis, and the study variables have been selected, we move on to the next stage in the research process–measuring and collecting the data. The choice of a particular data collection method or instrument depends on our study objective. After our data have been collected, we have to find a systematic way to organize and analyze our data and set up some set of procedures to decide what they mean.

3. a. Interval-ratio
    b. Nominal
    c. Interval-ratio
    d. Ordinal
    e. Nominal
    f. Ordinal
    g. Interval-ratio
    h. Nominal

5. There are many possible variables from which to choose. Some of the most common selections by students will probably be race, gender, family income, and the mother's and father's own educational attainment. Students can construct hypotheses such as "As family income increases, a person is likely to have more education," or "If the mother or father is a college graduate, their child is more likely to be a college graduate." Students may also choose such variables as IQ (those with more intelligence have more education) or personality traits (those who are self-motivated are more likely to have higher educational attainment).

7. In general, the difficulty with studying criminal acts (including hate crimes) is that the criminal act needs to first be reported. It is estimated that the majority of crimes are not reported to authorities. Data on reported crimes is routinely collected by the Federal Bureau of Investigation and the Bureau of Justice.

9. a. Age. This variable could be measured as an interval-ratio variable, with actual age in years reported. As discussed in the chapter, interval-ratio variables are the highest level of measurement and can also be measured at ordinal and nominal levels.

b. Annual Income could be measured as interval ratio if recorded in dollars or as ordinal if categorized in groupings such as $0 < 10,000$; $10,000 < 20,000$; $20,000 < 30,000$, etc.

c. Religiosity could be measured as an ordinal variable if it is defined as how strongly people feel about their religious beliefs, such as "not very strong;" "neutral;" and "very strong." It could also be measured as an interval-ratio level if measured as frequency of church attendance per week or month.

SPSS PROBLEMS

1. A reminder: the measurement levels for several variables are misidentified in SPSS. Students should confirm that the reported level of measurement is correct.

CHAPTER 2

1. a. Race is a nominal variable. Class is an ordinal variable, since the categories can be ordered from lower to higher status.

b. Frequency table for Race:

Race	Frequency
White	17
Nonwhite	13

Frequency table for Class:

Class	Frequency
Lower	3
Working	15
Middle	11
Upper	1

c. Proportion nonwhite is $13/30 = .43$. The percentage white is $(17/30) \times 100 = 56.7\%$.

d. The proportion middle class is $11/30 = .37$.

3. a. Frequency table for traumas experienced:

Number of Traumas	Percentage (%)
0	50.0%
1	36.7%
2	13.3%
Total $N = 30$	100.0

Trauma is an interval or ratio level variable, since it has a real zero point and a meaningful numeric scale.

b. People in this survey are more likely to have experienced no traumas last year (50% of the group).

c. The proportion who experienced one or more traumas is calculated by first adding 36.7% and 13.3% = 50%. Then divide that number by 100 to obtain .50, or half the group.

5. In this case, the independent variable is party affiliation and the dependent variable is views on gay marriage. By looking at the percentages, it is clear that a higher percentage of Republicans (68%) are in favor of the amendment than Democrats (38%) and Independents (48%). More Democrats (60%) are opposed to the amendment than Republicans (28%) and Independents (46%). Thus the data support the statement, showing that Republicans have more conservative views toward gay marriages as opposed to Democrats and Independents.

7. **Male**

	f	*%*	*Cf*	*C%*
Some high school	65	14.74	65	14.74
High school	233	52.83	298	67.57
Some college	27	6.12	325	73.70
College graduate	116	26.3	441	100
Total	441	100.00		

Female

	f	*%*	*Cf*	*C%*
Some high school	78	13.95	78	13.95
High school	311	55.64	389	69.59
Some college	43	7.69	432	77.28
College graduate	127	22.72	559	100.00
Total	559	100.00		

White

	f	*%*	*Cf*	*C%*
Some high school	95	11.83	95	11.83
High school	433	53.92	528	65.75
Some college	52	6.48	580	72.23
College graduate	223	27.77	803	100.00
Total	803	100.00		

Black

	f	%	Cf	C%
Some high school	36	29.03	36	29.03
High school	70	56.45	106	85.48
Some college	11	8.87	117	94.35
College graduate	7	5.65	124	100.00
Total	124	100.00		

- b. 32.42% of males; 30.41% of females
- c. 65.75% of whites; 85.48% of blacks
- d. The cumulative percentages are more similar for men and women (more than 67% in each group had at least a high school diploma: 67.57% for men and 69.59% for women) than the comparison of whites and blacks (more than 65% of whites had at least a high school diploma, while 89.07% of blacks reported the same). It appears that there is more educational inequality based on race than gender.

9. a. ordinal
   b. 13.8% of New Zealanders, 31.7% of Japanese, and 68.7% of Russians strongly agree or agree to the statement.
   c. These three countries have very different views on this subject. There is a high percentage of Russian respondents who agreed or strongly agreed (27.7 + 41.0 = 68.7) that the government should redistribute income among citizens. New Zealanders, on the other hand, had the highest percentage of those who disagree or strongly disagree (64.7%). The Japanese respondents did not show a great tendency in either direction. The most common answer for the Japanese sample was neither (34.1%).

11. Overall a higher percentage of minority respondents indicated that the government should play a "major role" in improving the economic and social position of minorities. Sixty-eight percent of blacks and 67% of Hispanics indicated this in their responses. On the other hand, 32% of whites agreed with this position (this is about 50% less than what was indicated by blacks and Hispanics). More whites, 51%, indicated that the government should play a "minor role."

13. a. The category with the highest percentages across all the reported survey years is salesperson. The percentage of respondents who said yes ranged from 68% (1977) to 90% (2001).
    b. There are two categories with the lowest percentages, elementary school teachers and clergy. The percentage of Americans who said agreed that homosexuals should be hired as elementary school teachers ranged from a low of 27% in 1997 to a high of 55% in 1996. For clergy, the percentage who said yes ranged from 36% (1977) to 54% (2001).
    c. Overall, Gallup Poll results indicate an increase in the level of support for hiring homosexuals in all listed occupations. Support has always been high for particular occupations (salesperson, doctor, and armed forces). But for the years measured, acceptance for each occupation increased: the acceptance of salespersons rose 22 percentage

points (68% to 90%); the acceptance of doctors rose 31 points (44% to 75%); the acceptance of the armed forces rose 19 points (51% to 70%). Even for the occupations with the lowest levels of support, elementary school teachers and clergy, in 2001 slightly over half of those surveyed agreed that homosexuals should be hired as both.

15. a. Republicans are more likely to be male, while Democrats are more likely to be female. The age distributions are approximately equal for both parties. In the category for race/ethnicity, there are more whites in the Republican Party (93%) versus the Democratic Party (75%). Blacks represent 19% of the Democratic Party, while blacks compose only 3% of the Republican Party. Adding both black and other nonwhite categories, the Democrat Party has 25% nonwhite membership, while the Republicans have only 7%. The distribution for education is approximately equal for both parties. The parties show little difference in the percentage who have some college or less: 76% (47 + 29) for Democrats and 75% (38 + 37) for Republicans. Regarding personal income, Republicans have higher levels of income, particularly in the category of "$50,000 or more." Slightly more Democrats, about 3%, reported being a member of a union.

   b. Gender—nominal
   Age—ordinal
   Race/ethnicity—nominal
   Education—ordinal
   Household income—ordinal
   Union membership—nominal

   c. There are several possible variables that could be added to this assessment of differences between Democrats and Republicans. Variables could include: marital status, sexual orientation, religious affiliation, number of children or attitudes toward specific political or social issues.

CHAPTER 3

1. Based on the graph, it is clear that there has been an increase in the nonmarital birth rates for each group from 1940 to 1990, with a decline around 1980 in each group except 15–19 years. The rate for ages 20–24 was the highest for each time period, except between 1957 and 1970 when the rates for 15–19 year olds were higher. In addition, the rate for 35–39 year olds remained below the rest except for the period 1962–1963. The oldest group of women had the lowest nonmarital birth rates and the two youngest age groups had the highest.

3. According to Figure 3.27, between the years 1990 and 2000 overall birthrates among unmarried teen women slightly decreased from 43 to 40. However, this trend does not hold across independent racial groups. When overall birthrate is partitioned by race, we see that there was no change among Whites(non-Hispanic), a large decrease among Blacks (from 106 to 77) and a moderate increase in birthrates among Hispanics (from 66 to 74).

5. a. Number of household members by sex (ISSP 2000)

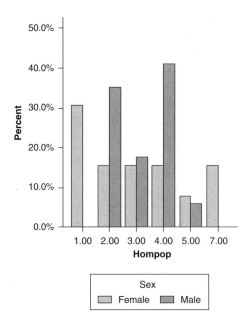

b. Females. More than 10% of females reported a household size of 7. The largest household size for males is 5.

c. Frequencies don't control or adjust for the total number of people in each group. There are more males (17) than females (13) in the survey, so percentages must be used to make the bars comparable.

7.   First, please note that you need to calculate the percentages for each group. We've presented the data in a bar graph (percentages). U.S. respondents were more similar to Canadian respondents: 41% of U.S. respondents and 35% of Canadian respondents selected "order in the nation." (Though more Canadian respondents indicated "give more people say," 40% versus 19% of Americans.) Both groups had similar support for "fight rising prices," 15% U.S. and 17% Canada. Among respondents from Mexico, "fight rising prices" was the highest category at 34%.

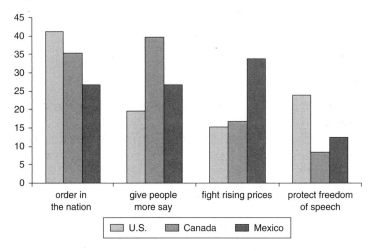

9. a. Years of education is an interval-ratio variable. Thus a histogram, which is suitable for graphing interval-ratio data, can be used for years of education.

    b. The following educational categories were created for this chart: 0–4, 5–8, 9–12, 13–16, 17–20 years (0 is included in the first category.) Category 1 should be 14 (1 + 1 + 8 + 1 + 3), category 2 should be 46 (6 + 8 + 7 + 25), category 3 should be 394 (22 + 29 + 49 + 294), category 4 should be 425 (97 + 135 + 56 + 137), and category 5 should be 118 (35 + 50 + 14 + 19).

Educational attainment, *N* = 997, GSS2002

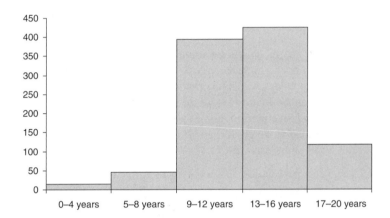

11. Bar graphs could be used to compare the percentages in different response categories among the three racial groups. Refer to Figure 3.5 for an example. Bar graphs can be used for nominal or ordinal variables.

13. a. If we wanted to compare percentages in different response categories between men and women, bar graphs would be most appropriate. In a single graph, we could compare both groups at once.

    b. Note: percentages should be calculated.

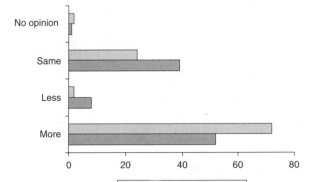

Attitudes toward gun laws, Gallup Poll 2000

CHAPTER 4

1. a. The mode can be found two ways, either by looking for the highest frequency (7570) or the highest percent (30.3%). The mode is the category that corresponds to these values, "Somewhat Dangerous"

    b. The median can be found two ways, either by using the frequencies column or by using the cumulative percentages. Either way, the median is "Somewhat Dangerous."

    c. The mode is simply the category with the highest frequency (or percentage) in the distribution. The median divides the distribution into two equal parts so that half the cases are below it and half above it.

    d. Because this variable is an ordinal level variable.

3. a. Interval-ratio. The mode can be found two ways, either by looking for the highest frequency (14) or the highest percent (18.7%). The mode is the category that corresponds to these values, "50 hours per week."

    The median can be found two ways, either by using the frequencies column or by using the cumulative percentages. The median is 42 hours per week.

    b. Since the median is merely a synonym for the 50th percentile, we already know that its value is 42 hours per week.

    $$25\text{th percentile} = (75 \times .25) = 18.75\text{th case} = 36 \text{ hours per week}$$
    $$75\text{th percentile} = (75 \times .75) = 56.25\text{th case} = 50 \text{ hours per week}$$

5. a. The mode can be found by looking for the highest frequency in each column; the mode for each group is listed below:

    "18–29": "Less than once a week"
    "30–39": "Once a day"
    "40–49": "Once a day"
    "50–59": "Once a day"

    The median can be found two ways, either by using the frequencies column or by using the cumulative percentages.

    "18–29": "Several times a week"
    "30–39": "Several times a week"
    "40–49": "Once a day"
    "50–59": "Once a day"

    b. It appears that the occurrence of prayer tends to increase with age. On average, 18–29 year old respondents pray the least (median = "Several times a week," mode = "Less than once a week"); whereas respondents in the upper categories (40–49 and 50–59) tend to pray frequently as the median and mode for both groups is "Once a day." On the surface this lends support for the idea of a "generation gap."

7.   We begin by multiplying each age by its frequency.

$$\overline{Y} = \frac{\sum fY}{N} = \frac{6427}{2765} = 2.32$$

So, the mean number of people per U.S. household is 2.32.

9. a. There appear to be a few outliers (i.e., extremely high values); this leads us to believe that the distribution is skewed in the positive direction.

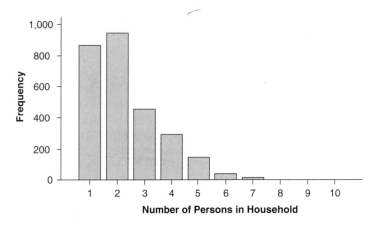

b. The median can be found two ways, either by using the frequencies column or by using the cumulative percentages. The median is 2.

Since the median (2) is less than the mean (2.32), we can conclude that the distribution is skewed in a positive direction. So our answer to question 9a is further supported.

11. a. Yes, both of these politicians can be correct, at least in a technical sense. One politician can be referring to the mean; the other could be using the median. It would be unusual if these two statistics were exactly equal. The average or mean income of Americans can be greater than the median if the distribution of income is positively skewed, which is certainly true.

13.   The median is simply the number in the middle. Since we have 10 cities in each group, we want the 5.5th case $(\frac{10+1}{2})$. For cities in the top 10, the 5.5th case falls between Houston and Phoenix. So, the median is 14.35 $(\frac{13.4+15.3}{2})$. For cities in the bottom 10, the 5.5th cases falls between Riverside and St. Petersburg. So, the median is 7.95 $(\frac{7.7+8.2}{2})$.

For cities in the top 10, the mean is 13.99 $(\frac{139.9}{10})$. For cities in the bottom 10, the mean is 10.39 $(\frac{103.9}{10})$. On average, cities with higher populations (i.e., top 10) had higher murder rates. The distribution of murder rates for cities in the bottom 10 is skewed in a positive direction because only a few cities (e.g., Newark) had very high murder rates. The pattern is not the same for cities in the top 10, which is skewed in a negative direction.

15. a. Number of marriages:

$$\overline{Y} = \frac{\sum Y}{N} = \frac{5470}{7} = 781.43$$

Number of divorces:

$$\overline{Y} = \frac{\sum Y}{N} = \frac{655}{7} = 93.57$$

b. Number of marriages:

$$\overline{Y} = \frac{\sum Y}{N} = \frac{4922}{6} = 820.33$$

Number of divorces:

$$\overline{Y} = \frac{\sum Y}{N} = \frac{455}{6} = 75.83$$

With the England removed, the average number of marriages increases to 820.33 With the United States removed, the average number of divorces drops to 75.83.

CHAPTER 5

1. a. The table reveals seven response categories for political views.
    b. The sum of the squared percentages is equal to 2311.36, the value of $\sum Pct^2$.
    c. Using the formula, we calculate the IQV as follows:

$$\text{IQV} = \frac{K(100^2 - \sum Pct^2)}{100^2(K-1)} = \frac{7(100^2 - 2311.36)}{100^2(7-1)} = \frac{53,820.48}{60,000} = 0.90$$

An IQV of 0.90 means that Americans are very diverse in their political opinions, thereby supporting our observations from the table.

3. a. The range of poverty rates in the South is from 10.5% to 19.6% or 9.1%. The range in the west is from 9.4% to 18.4% or 9.0%.
    b. For the South, the 25th percentile is 12.4% and the 75th percentile is 14.95%, So the IQR is 2.55%. The IQR for the West is larger (3.85%).
    c. Based on the range, the data do not support the idea that there is greater variability of poverty rates in southern states. The range in poverty levels in southern and western states is quite similar (9.1% and 9.0%). There is a difference in the IQRs between regions, though. The IQR is slightly larger for the western states.
    d.

$$\text{South: } S_Y^2 = \sqrt{S_Y} = \sqrt{\frac{\sum(Y - \overline{Y})^2}{N-1}} = \sqrt{\frac{57.44}{8}} = 2.68$$

$$\text{West: } S_Y^2 = \sqrt{S_Y} = \sqrt{\frac{\sum(Y - \overline{Y})^2}{N-1}} = \sqrt{\frac{71.83}{9}} = 2.82$$

e. On average, southern states deviated 2.68 percent from the mean; and western states deviated 2.83 percent from the mean. This implies slightly more variation in the distribution of western states.

The results are similar to one another (give or take a few hundredths). But, strictly speaking, the poverty rates in the south exhibit less variation when judged according to the standard deviation (and the IQR). Thus, the results are consistent with what we observed and discussed in 3b and 3c.

5. a. The range of percent increase in the elderly population for the western states is 59.5% (71.5% – 12.0%). The range of percent increase for the Midwestern states is 7.6% (10.0% – 2.4%). The western states have a much larger range.
    b. The IQR for the western states is 14.4% (29.3% – 14.9%). The IQR for the Midwestern states is 3.9% (7.9% – 4.0%). Again, the value for the western states is greater.
    c. There is great variability in the increase in the elderly population in the western states, chiefly caused by the large increases in Nevada, Alaska, and Arizona as measured by either the range or the IQR.

7. a. The mean, or the average, is 52.55 percent; and, on average, the percentage of respondents who agreed that modern life harms the environment deviated from the mean by 12.65 percent.
    b. The 25th percentile, 41.80, means that 25 percent of cases fall below 41.80 percent. Likewise, the 75th percentile, 58.95, means that 75 percent of all cases fall below 58.95 percent.

| 25th percentile | 10(.25) = 2.5th case | So, (40.7 + 42.9)/2 = 41.80 |
| 75th percentile | 10(.75) = 7.5th case | So, (55.0 + 62.9)/2 = 58.95 |

  c. The mean percentage that agree that modern life harms the environment is higher for western European countries ($\bar{Y} = 53.95$) than eastern European countries ($\bar{Y} = 50.45$). However, the distribution of western European countries exhibits more variability than the distribution of eastern European countries. On average, western European countries deviated 14.28 percent from the mean, whereas eastern European countries deviated 11.43 percent from the mean.

9. a. The mean number of crimes is 3470.81 and the mean amount of dollars spent on police protection per capita is $167.67.
    b. The standard deviation for the number of crimes is 675.62 and the standard deviation for the amount of dollars spent on police protection per capita is $48.18.
    c. The IQR for the number of crimes is 887 and the IQR for the amount of dollars spent on police protection per capita is $62.80.
    d. Because the number of crimes and police expenditures per capita are measured according to different scales, it isn't appropriate to directly compare the mean, standard deviation, or IQR for one variable with the other. But we can talk about each distribution separately. We know from examining the mean and standard deviation for the number of crimes that the standard deviation is large, 675.62, indicating a wide dispersion of scores from the mean. The IQR for the number of crimes is wide at

887 points. The IQR tells us that half of the 21 states are clustered between 3881.5 and 2994.5 (the difference between the lower and upper quartiles). For the number of crimes, states like New Hampshire, North Dakota, and Illinois contribute more to its variability because they have values far from the mean (both above and below). What can be said about the distribution for police expenditures per capita?

e. Among other considerations, we need to consider the economic conditions in each state. A downturn in the local and state economy may play a part in the number of crimes and police expenditures per capita.

11. a. Because Table 5.5 is so large, only the formula for the standard deviation is included here.

The standard deviation for the percentage increase in the elderly population from 1990 to 2000 is 13.89 percent.

$$S_Y^2 = \sqrt{S_Y} = \sqrt{\frac{\sum(Y - \overline{Y})^2}{N - 1}} = \sqrt{\frac{9642.64}{50}} = 13.89$$

b. The IQR is 14.8; the IQR is slightly larger.
c. Yes, the standard deviation would lead to about the same conclusion concerning the variability of the increase in elderly population as do the box plot and IQR. However, the box plot shows the actual range of values, which the standard deviation cannot (even though all values are used to calculate the statistic).

13. Since the variable is interval-ratio, according to our text we should use variance (or standard deviation), range, or IQR. Among these three measures, variance and/or standard deviation is preferred. For measurements of central tendency, as discussed in Chapter 4, if we are looking for the average percent of women's labor force participation for these 10 countries, we should rely on the mean.

On average, non-European respondents work more hours per week. Both the mean and median are higher for non-European respondents than for European respondents. Also, the distribution of European countries tended to exhibit more variability; the standard deviation for respondents from European countries is 6.29 hours, while for respondents from non-European countries it is 3.77 hours. The IQRs also attest to more variability in the distribution of European respondents (IQR = 6.04) compared to non-European respondents (IQR = 3.40).

15. a. In 2003, the difference between the 80th and 20th percentiles was $68,883 ($86,867 – $17,984). In 1993, that difference was $59,338 ($75,594 – $16,256)
b. We're referring to the IQR, which is difference between the 25th and 75th percentiles. In the table we see that the closest percentiles listed are the 20th and the 80th percentiles. Thus, our answer from 12a, gives us an idea as to the value of the IQR in 1993, 1998, and 2003.

CHAPTER 6

1. a. The independent variable is race; the dependent variable is fear of walking alone at night.

b. Approximately 69 percent of whites (69.52%) are not afraid to walk alone in their neighborhoods at night, whereas approximately 63 percent of blacks (62.5%) are not afraid to walk alone. This amounts to about a 7 percent difference (69.2% – 62.5%) between whites and blacks who are not afraid to walk alone at night, indicating a weak relationship. Also, although we compared percentages differences in this exercise, it is important to keep in mind that our sample size impairs our ability to make any statistically meaningful comparisons.

c. There is some difference in fears between homeowners and renters: 25.0 percent of homeowners and 38.5 percent or renters are afraid to walk in their neighborhood at night. The difference between the two groups is 13.5 percent. Thus, there is a weak to moderate relationship between homeownership and fear of walking in one's neighborhood at night.

3. a. DV: political views
   IV: attitudes about homosexual relations

   b. To calculate this we need to locate the column marginal (474) and divided it by the total (757) and the multiply this value by 100. We get 62.6 percent of those polled view homosexual relations as wrong.

$$\left(\frac{474}{757}\right) \times 100 = 62.6\%$$

   c. Looking at conservatives, 42.8 percent of those who view homosexual relations as always wrong identified as conservatives, whereas only 22.2 percent of those who think that homosexual relations are not wrong at all identified as conservatives. This makes for a 20.6 percent difference, indicating a moderate relationship. Similarly, we can make the same comparison for liberals: 17.7 percent of those who view homosexual relations as always wrong identified as liberals, whereas 38.9 percent of those who think that homosexual relations are not wrong at all identified as liberals. This makes for a 21.2 percent difference, indicating a moderate relationship.

5. a. The proportion of respondents who report having a great deal of confidence in medicine is 0.37. This was calculated by dividing the row marginal (313) by the total (850). The proportion of respondents who report having hardly any confidence in medicine is 0.11. This was calculated by dividing the row marginal (91) by the total (850).

   b. Overall, men more than women report that they have a great deal of confidence in medicine. If we combine black and white men, we see that approximately 40% report having a great deal of confidence in medicine. If we combine black and white women, we see that approximately 34% report having a great deal of confidence in medicine, indicating a 6 percent difference between men and women who report having a great deal of confidence in medicine.

   c. Reported confidence in medicine appears to be slightly higher among whites in the GSS sample. If this is reflective of their community, the clinic could work on improving confidence in their medical services among their black clients.

7. a. College plans for female seniors changed dramatically from 1980, when 45.0% stated that they definitely or probably would not attend college, to 1995, when that same response only represented 19.5% of the sample. In those fifteen years, the percentage of females definitely planning on going to college rose 26.2 percentage points (from 33.6% in 1980 to 59.8% in 1995). While these are dramatic changes, it is also interesting to observe the relative stagnancy of the percentage of females who responded that they would probably go to college: 21.3% in 1980, 20.5% in 1990, and 20.7% in 1995.

    b. Males exhibit a similar trend, though not nearly as pronounced. The change in percent of respondents definitely planning on attending college rose just 13 percentage points (from 48.6% in 1980 to 35.6% in 1995).

9. a. At all four educational levels, the trend is about the same. The elementary and college levels boast the greatest reductions in the percentage of white classmates from 1980 to 2001. Although these reductions are modest, it is important to note that the percentage reductions of white classmates have increased over the past decade relative to two decades ago. Among black college students, there was a 2.4 percent reduction in the percentage of white classmates from 1980 to 1990; between 1990 and 2001 this number was 5.3 percent. In other words, the percentage change more than doubled!

    b. Among black elementary school students, there was a 7.9 percent decrease in the percentage of white classmates from 1980 to 2001. For black high school students, this number was –7.3 percent. And among black college students, this number was –7.7 percent. Thus, the percentage decrease in white classmates is stronger at the elementary and college levels. One sociological explanation for this may be that more and more blacks are attending college these days.

    c. Among black elementary school students, there was a 6.2 percent decrease in the percentage of white classmates from 1990 to 2001. For black middle school students this number was –3.9 percent. For black high school students, this number was –4.1 percent. And among black college students, this number was –5.3 percent.

11. a. 70.9 percent of respondents who had seen an X-rated movie in the past year thought that extramarital sex was always wrong; 79.5 percent of respondents who had not seen an X-rated movie in the past year thought that extramarital sex was always wrong. Thus, there is only a percentage difference of 8.6 percent. Accordingly, it does not appear that the organization's assertion is supported by the evidence.

    b. Among males, 68.3 percent of respondents who had seen an X-rated movie in the past year thought that extramarital sex was always wrong; 82.5 percent of respondents who had not seen an X-rated movie in the past year thought that extramarital sex was always wrong. This makes for a percentage difference of 14.2 percent, a weak relationship, but more of a relationship than in 10a.

        Among females, 77.1 percent of respondents who had seen an X-rated movie in the past year thought that extramarital sex was always wrong; 76.8 percent of respondents who had not seen an X-rated movie in the past year thought that extramarital sex was always wrong. This makes for a percentage difference of 0.3 percent, no relationship.

c. This is an example of a conditional relationship because, for males, the relationship gets stronger and, for females, the relationship gets weaker. We can conclude that, given the information available in this problem, the relationship appears to be conditioned by gender.

13. a. It can be argued either way. The belief that science does more harm than good is a more general than a belief that science solves environmental problems. Thus, if we think about thing deductively, a belief that science solves environmental problems seems to follow from a belief that science does more harm than good.

   Alternatively, a belief that science solves (or fails to solve) environmental problems may shape a further belief that science does more harm that good.

   Ultimately, it is up to the researcher making a decision in light of the relevant research literature.

   b. There are two scenarios, depending on the assignment of variables.

		Science Does More Harm Than Good					
		Strongly agree	*Agree*	*Neither agree nor disagree*	*Disagree*	*Strongly disagree*	*Total*
Science Solves	Strongly agree	22.7%	4.4%	2.7%	2.8%	9.2%	5.1%
Environmental	Agree	20.8%	34.9%	18.1%	24.9%	22.2%	24.6%
Problems	Neither agree nor disagree	11.4%	17.7%	37.6%	20.4%	16.3%	22.6%
	Disagree	23.0%	33.2%	30.4%	40.7%	29.6%	34.4%
	Strongly disagree	22.1%	9.8%	11.1%	11.2%	22.7%	13.3%
Total		100.0%	100.0%	100.0%	100.0%	100.0%	100.0%

		Science Solves Environmental Problems					
		Strongly agree	*Agree*	*Neither agree nor disagree*	*Disagree*	*Strongly disagree*	*Total*
Science Does	Strongly agree	22.7%	4.3%	2.6%	3.4%	8.5%	5.1%
More	Agree	15.6%	25.9%	14.3%	17.6%	13.4%	18.2%

(Continued)

		Science Solves Environmental Problems					
		Strongly agree	*Agree*	*Neither agree nor disagree*	*Disagree*	*Strongly disagree*	*Total*
Harm Than Good	Neither agree nor disagree	11.7%	16.3%	36.8%	19.6%	18.5%	22.1%
	Disagree	20.9%	38.8%	34.6%	45.4%	32.1%	38.3%
	Strongly disagree	29.1%	14.7%	11.7%	14.0%	27.6%	16.2%
Total		100.0%	100.0%	100.0%	100.0%	100.0%	100.0%

c. Treating "science does more harm than good" as the independent variable, 22.7 percent of respondents who strongly agreed that science does more harm than good strongly agreed that science solves environmental problems, whereas only 2.7 percent of those who were indifferent to whether science does more harm than good strongly agree that science solves environmental problems. This amounts to a 20.0% difference, indicating a moderate relationship. What can be said if we order the variables differently?

15. a. 68.7 percent of Spanish respondents never avoid driving a car for environmental reasons, whereas 32 percent of Dutch respondents never avoid driving a car for environmental reasons. This amounts to a 36.7 percent difference, indicating a moderate relationship between the country of origin and whether respondents avoid driving for environmental reasons.

b. Among other considerations, Austria and the Netherlands are wealthier nations, and it would appear that concern about the environment, as measured by whether one avoids driving a car for environmental reasons, is more prevalent in more wealthy countries. Take another look at the percentaged table from 15a. Relative to respondents from the other countries listed (excluding Spain), Irish respondents overwhelmingly never abstain from driving a car for environmental reasons.

Accordingly, other variables that may be of interest here might include GNP, or each country's gross national product. We consider GNP in chapter 8.

CHAPTER 7

1. a. We will make 3278 errors, because we predict that all victims fall in the modal category (white). E1 = 6819 – 3541 = 3278.

b. For white offenders, we could make 278 errors; for black offenders, 511 errors; and for other offenders, we would make 69 errors. E2 = 858.

c. The proportional reduction in error is then (3278 – 858)/3278 = .7383 This indicates a very strong relationship between the two variables. We can reduce the error in predicting victim's race based upon race of offender by 73.83%.

3. a. Those with less than a secondary degree: $(115 + 40)/298 = .52$
    Those with a university degree: $(91 + 56)/201 = 0.73$

    b. Ns: 105121
    $14(57 + 20 + 85 + 28 + 151 + 91 + 57 + 56) + 55(85 + 28 + 151 + 91 + 57 + 56) +$
    $74(151 + 91 + 57 + 56) + 115(57 + 56) + 8(20 + 28 + 91 + 56) + 57(28 + 91 + 56) +$
    $85(91 + 56) + 151(56).$

    Nd: 67775
    $6(55 + 57 + 74 + 85 + 115 + 151 + 40 + 57) + 20(74 + 85 + 115 + 151 + 40 + 57) +$
    $28(115 + 151 + 40 + 57) + 91(40 + 57) + 8(55 + 74 + 115 + 40) + 57(74 + 115 + 40)$
    $+ 85(115 + 40) + 151(40)$

    c. Gamma = $(105121 – 67775)/(105121 + 67775) = .216$
    Gamma indicates a weak positive relationship for the table. As education increases, respondents are more likely to disagree with the statement. Knowing respondent education, we can decrease the error in predicting their position on the impact of science by 21.6%.

5. a. Lambda is appropriate when one or both of the variables is nominal. Happiness rating is ordinal, while marital status is nominal. You could estimate the lambda value will be close to zero since the majority of marital categories indicate "pretty happy".

    b. Lambda = 0. E1 = $505 – 281 = 224$; E2 = $107 + 15 + 36 + 5 + 61 = 224$

    The lambda indicates that there is no relationship between marital status and happiness. The mode for all marital status categories is "pretty happy," there is no variation in happiness ratings by marital status. Yet, if we examine the percentages, it appears that lambda might not be the best measure of association for the table. The percentages seem to indicate that a larger percentage of married respondents indicate being very happy (43% versus 16% widowed, 19% divorced, 21% separated and 25% never married). In addition, a larger percentage of widowed or separated respondents (67% and 64%) reported being "pretty happy" in contrast to the other groups (49% married, 58% divorced, and 59% never married).

7. a. Gamma is equal to $[(27)(87) – (63)(142)]/[(27)(87) + (63)(142)] = –.58$.

    b. Gamma is negative indicating that males (coded 1) are more likely to report not being afraid to walk at night (coded 2), whereas women (code 2) are more likely to reporting being afraid to walk at night (coded 1). The gamma of –.58 reveals a moderate negative relationship between the variables; 58% of the error is reduced in predicting fear based on respondent's sex.

9. The appropriate measure is gamma since both variables are ordinal measurements. The gamma is –.122 for the table.

    Ns = $47(151 + 97 + 70 + 25 + 50 + 25 + 29 + 8) + 136(70 + 25 + 50 + 25 + 29 + 8)$
    $+ 49(50 + 25 + 29 + 8) + 53(29 + 8) + 65(97 + 25 + 25 + 8) + 151(25 + 25 + 8) +$
    $70(25 + 8) + 50(8) = 78529$

    Nd = $46(136 + 151 + 49 + 70 + 53 + 50 + 24 + 29) + 97(49 + 70 + 53 + 50 + 24 +$
    $29) + 25(53 + 50 + 24 + 29) + 25(24 + 29) + 65(136 + 49 + 53 + 24) + 151(49 + 53$
    $+ 24) + 70(53 + 24) + 50(24) = 100398.$

Using education to predict attitudes toward animal testing results in a 12.2% reduction in error. This is a very weak positive relationship between the two variables. People with higher education were more likely to support animal testing. Over 70% of those with a university degree reported that they agreed or strongly agreed to the statement.

11.   Bible and PREMARSX: The calculated lambda = .20. E1 = 313 − 138 = 175; E2 = 40 + 90 + 10 = 140.

Bible and HOMOSEX: The calculated lambda = .10. E1 = 306 − 167 = 139; E2 = 17 + 91 + 17 = 125.

Bible and PORNLAW: The calculated lambda = .007. E1 = 320 − 181 = 139; E2 = 50 + 74 + 14 = 138.

The strongest association with belief about the Bible is attitudes about premarital sex. Twenty percent of the error in predicting responses to the statement can be reduced based on information about belief about the Bible. Only 10% of the error is reduced in predicting attitudes toward homosexuality based on one's belief about the Bible and less than 1% of the error is reduced in predicting attitudes toward pornography laws based on one's belief about the Bible.

13.   The appropriate measure of association for this table is lambda—with one ordinal variable (school performance) and a nominal variable (type of smoker). The lambda is .03, indicating a weak association between these two variables. If we had information about type of smoker, we would only improve our prediction about school performance by 3.0%.

CHAPTER 8

1.   a. On the scatterplot below the regression line has been plotted to make it easier to see the relationship between the two variables.

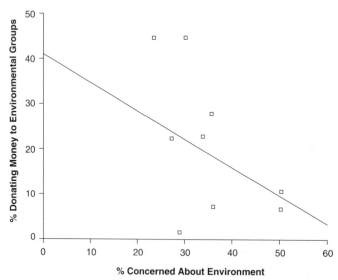

b. The scatterplot shows that there is a general linear relationship between the two variables. There is not a lot of scatter about the straight line describing the relationship. As the percentage of respondents concerned about the environment increases, the percentage of respondents donating money to environmental groups decreases.

c. The Pearson correlation coefficient between the two variables is −0.40. This seems reasonable because it is in line with the fact that the scatterplot indicated a negative relationship. A correlation coefficient of −0.40 indicates a moderate, negative relationship between being concerned about the environment and actually donating money to environmental groups.

3. a. The correlation coefficient is 0.91.

b. The correlation coefficient is very close to 1, which means that relatively high values of unwed births for whites are closely associated with relatively high values of unwed births for nonwhites, and vice versa. Hence, the two birth rates have changed together, increasing over time.

5. a. Yes, as indicated by the negative b of −.141, the relationship between the variables is negative.

b. Based on the coefficient output, we can predict that an individual with 16 years of education will watch television 2.662 hours per day. In contrast, someone with 12 years of education (high school degree) is predicted to watch 3.226 hours of television per day. This is .564 hours more than someone with a college degree. We used the regression equation $\hat{Y} = 4.918 + (-.141)X$ to make these predictions.

c. Looking at the scatterplot, a straight line does approximate the data; however, it does so rather poorly, leaving quite a bit of scatter on either side of the regression line.

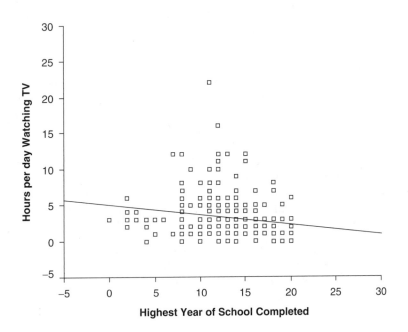

7. a. The scatterplot indicates a weak, positive relationship between a respondent's mother's education and a respondent's education.

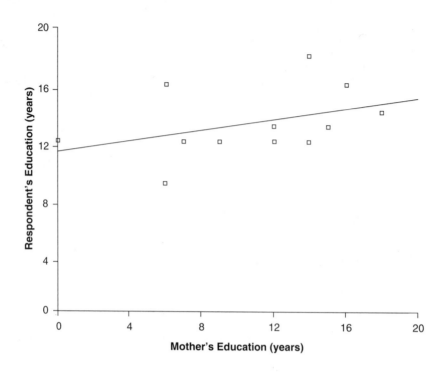

    b. See the regression line on the scatterplot above. The slope of the regression equation is .186. The intercept is 11.25. A straight line fits the data reasonably well. In general, the higher a respondent's mother's education, the higher a respondent's education.
    c. The error of prediction for the second case is about −1.04 years of education. That is, we predicted that he or she would have 14.04 years of education ($\hat{Y} = 11.25 + .186(15)$), but this person actually has 13 years of education. The error of prediction for a person whose mother had 18 years of education and had 14 years of education themselves is −.598 years of education ($\hat{Y} = 11.25 + .186(18)$). Here, we predicted a value of 14.598 but the person has less than that amount.
    d. For someone whose mother received 4 years of education, we predict about 11.99 years of education ($\hat{Y} = 11.25 + .186(4)$). For someone whose mother received 12 years of education, we predict about 13.48 years of education ($\hat{Y} = 11.25 + .186(12)$).
    e. On the least-squares line. Because both of the means are used to calculate the least-squares line, it makes sense that the point should fall on the line.

9. a. A straight line does seem to fit the data, as shown in the scatterplot.
    b. The equation, $\hat{Y} = 2.03 + 1.04X$, supports the assertion that a straight line best fits this data. In fact, b is 1.04 which indicates that for a 1 percent increase in those living

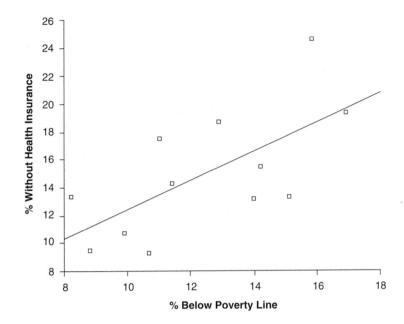

below the poverty line, there is a corresponding 1 percent (1.04%) increase in the percent of people without health insurance.

c. $5 = 2.03 + 1.04X$

$X = 2.86$ percent. So, about 2.86 percent need to be living below the poverty level in order to have only 5 percent without health insurance.

d. You cannot go outside the scope of your data. That said, it is interesting to see how closely poverty and lack of health insurance are related. Although we cannot generalize statistically, this example does help us to see just how closely the two are related. Thus, as we go about our research, this may be a consideration to keep in mind for future studies.

11. a. The b coefficient for education is $-.141$. As the amount of education increases by 1 year, the number of hours a person watches television decreases by .141 hours. The b coefficient for the number of children is .077. As the number of children in a family increases by 1, the number of hours a person watches television increases by .077. Yes, the relationship between education and hours of television viewing is as hypothesized.

b. For a person with 16 years of education and two children, the predicted hours of television viewed per day is about 2.50. In Exercise 5, the predicted number of television hours viewed for 16 years of education is 2.66 hours.

$\hat{Y} = 4.646 - .141(16) + .077(2) = 2.5$

c. The r^2 in Exercise 5 using only education is 0.031. Error is reduced by 3.1% when using education to explain the amount of television viewed. When adding the second predictor variable, number of children in a family the R^2 is 0.034. Error is reduced by 3.4% when using both education and number of children to explain television viewing.

13. a. False, both b and r will always have the same sign because both tell us the direction of the relationship.

   b. Both a and b refer to changes in the dependent variable.

   c. The coefficient of determination, r^2, is PRE measure. PRE stands for proportional reduction of error. This means that r^2 indicates the extent to which prediction error is reduced when we take into account the independent variable in our predictions

   d. False. The regression equation serves many functions, most notably to make predictions. Whether a regression equation models a causal relationship is a matter of meeting the causal requirements as they were discussed in Chapter 1.

   Regression equations are commonly used to model relationships wherein researchers look for how changes in one or more variables (referred to as the independent variables) correspond with changes in another variable (referred to as the dependent variable).

CHAPTER 9

1. a. For an applicant with a score of 115

$$Z = \frac{115 - 98}{13} = 1.31$$

From Appendix B, area beyond 1.31 is about 0.0951, or the 90.5th percentile. Applicant is accepted because this is above the 88th percentile.

   b. An area beyond Z of about 0.1210 corresponds to a Z value of 1.17. This is translated into a cutoff score by:

$$\frac{\text{cutoff score} - 98}{13} = 1.17$$

$$\text{cutoff score} \cong 113$$

Students should round off to the whole number that gives a value closest to a Z of 1.17.

   c. Z value is 1.17

3. a. California's equivalent Z score is

$$Z = \frac{35,116,000 - 5,755,980}{6,386,850} = 4.6$$

   b. For any normal distribution, about 15.87% of all cases should fall less than one standard deviation below the mean. For the distribution of population values, one standard deviation below the mean is less than zero (5,755,980 − 6,386,850), so clearly no states have population values lower than one standard deviation below the mean.

However, for a normal distribution, about 8 states $(0.1587 \times 50 = 7.94)$ should fall less than -1 standard deviations below the mean.

c. The distribution of population must not be normal, since a true normal distribution would have cases well below the mean. Therefore, the median would be a better measure of central tendency.

5. a. Among working class respondents:
   The Z score for a value of 12 is

$$Z = \frac{12.00 - 12.64}{2.50} = -0.26$$

The Z score for a value of 16 is

$$Z = \frac{16.00 - 12.64}{2.50} = 1.34$$

The area between a Z of -0.26 and the mean is 0.1026. The area between a Z of 1.34 and the mean is 0.4099, so the total area between the scores is

$$Area = 0.1026 + 0.4099 = 0.5125$$

So, the proportion of working class respondents with 12 to 16 years of education is 0.5125.

Among upper class respondents:

The Z score for a value of 12 is

$$Z = \frac{12.00 - 15.00}{3.27} = -0.92$$

The Z score for a value of 16 is

$$Z = \frac{16.00 - 15.00}{3.27} = 0.31$$

The area between a Z of -0.92 and the mean is 0.3212. The area between a Z of 0.31 and the mean is 0.1217, so the total area between the scores is

$$Area = 0.3212 + 0.1217 = 0.4429$$

So, the proportion of upper class respondents with 12 to 16 years of education is 0.4429.

b. Among working class respondents:
   The Z score for a value of 16 is

$$Z = \frac{16.00 - 12.64}{2.50} = 1.34$$

The area between a Z of 1.34 and the tail of the distribution (Column C) is 0.0901. So, the probability of a working class respondent having more than 16 years of education is 0.0901.

Among middle class respondents:
The Z score for a value of 16 is

$$Z = \frac{16.00 - 14.19}{2.39} = 0.76$$

The area between a Z of 0.76 and the tail of the distribution (Column C) is 0.2236. So, the probability of a working class respondent having more than 16 years of education is 0.2236.

c. Among lower class respondents:
The Z score for a value of 12 is

$$Z = \frac{12.00 - 10.34}{3.18} = 0.52$$

The area between a Z of 0.52 and the mean is 0.1985. To this we must add .50 (the lower half of the distribution) to 0.1985. So, the probability of a lower class respondent having less than 12 years of education is 0.6985 (0.1985 + 0.50).

Among upper class respondents:
The Z score for a value of 12 is

$$Z = \frac{12.00 - 15.00}{3.27} = -0.92$$

The area between a Z of –0.92 and the tail of the distribution is .1788. Remember, in this case, the fact that the Z score is a negative value tells us that we are working on the lower half of the distribution. Unlike our previous answer, we do not need to add .50. So the probability of an upper class respondent having less than 12 years of education is 0.1788.

d. First, we find the Z score that has 25%, or 0.25, of the area between it and the mean. This is a Z score of about = 0.68. The lower limit is

$$Y = \bar{Y} + Z(S_Y) = 12.64 + (-0.68)(2.50) = 10.94$$

And the upper limit is

$$Y = \bar{Y} + Z(S_Y) = 12.64 + 0.68(2.50) = 14.34$$

So the middle 50 percent of working class respondents falls between 10.94 and 14.34.

e. If years of education is positively skewed, then the proportion of cases with high levels of education will be less than for a normal distribution. This means, for example, that the probabilities associated with high levels of education will be smaller.

7. a. An occupational prestige score of 60 corresponds to a Z score of

$$Z = \frac{60 - 44.75}{13.75} = 1.11$$

The area between a Z of 1.11 and the tail of the distribution is 0.1335. So about 13 percent of whites should have occupational prestige scores above 60. This corresponds to about approximately 278 whites (0.1335×2082) in our sample who should have occupational prestige scores above 60.

b. An occupational prestige score of 60 corresponds to a Z score of

$$Z = \frac{60 - 36.81}{13.81} = 1.68$$

The area between a Z of 1.68 and the tail of the distribution is 0.0465. So about 5 percent of black should have occupational prestige scores above 60. This corresponds to about approximately 18 blacks (0.0465×383) in our sample who should have occupational prestige scores above 60.

c. An occupational prestige score of 30 corresponds to a Z score of

$$Z = \frac{30 - 44.75}{13.75} = -1.07$$

The area between a Z of –1.07 and the mean is 0.3577.

An occupational prestige score of 70 corresponds to a Z score of

$$Z = \frac{70 - 44.75}{13.75} = 1.84$$

The area between a Z of 1.84 and the mean is 0.4671. So the proportion of whites with occupational prestige scores between 30 and 70 is 0.8248 ($0.3577 + 0.4671$). Thus, approximately 1717 whites (0.8248×2082) in the sample should have occupational prestige scores between 30 and 70.

d. An occupational prestige score of 30 corresponds to a Z score of

$$Z = \frac{30 - 36.81}{13.81} = -0.49$$

The area between a Z of –0.49 and the mean is 0.1879.

An occupational prestige score of 60 corresponds to a Z score of

$$Z = \frac{60 - 36.81}{13.81} = 1.68$$

The area between a Z of 1.68 and the mean is 0.4535. So the proportion of blacks with occupational prestige scores between 30 and 60 is 0.6414 (0.1879 + 0.4535). Thus, approximately 246 blacks (0.6414 × 383) in the sample should have occupational prestige scores between 30 and 60.

9. a.

$$Z = \frac{1,000 - 678}{560} = 0.58$$

About 0.2810 of the distribution falls above the score, so that is the proportion of burglaries with dollar losses above $1,000.

b.

$$Z = \frac{400 - 678}{560} = -0.50$$

The area between the value and the mean is about 0.1915, so the total area above $400 is 0.50 + 0.1915 = 0.6915

c.

$$Z = \frac{500 - 678}{560} = -0.32$$

The area between the value and the lower tail of the distribution is 0.3745. About 37 percent of all burglaries had losses below $500.

11. a.

$$Z = \frac{8.0 - 3.0}{2.4} = 2.08$$

b.

$$Z = \frac{5.0 - 3.0}{2.4} = 0.83$$

The area between the value and the mean is about 0.2967, so the total area below this Z score is .50 + 0.2967 = 0.7967. The proportion of respondents that watch television less than 5 hours per day is 0.7967. This translates into approximately 721 respondents (905 × 0.7967) in the sample.

c. $Y = \bar{Y} + Z(S_Y) = 5.0 + 1.0(2.4) = 7.4$ hours per day

d.

$$Z = \frac{1.0 - 3.0}{2.4} = -0.83$$

The area between mean and the score is about 0.2967.

$$Z = \frac{6.0 - 3.0}{2.4} = 1.25$$

The area between mean and the score is about 0.3944. The percentage between these two scores is

$$(0.2967 + 0.3944)100 = 69.11\%$$

13. For any Z distribution, the value of the mean is 0. The standard deviation of a Z distribution is 1. Z distributions are based on the mean of a variable and are centered around that value, so they have a mean of 0 by definition. A Z score of 1 or -1 is equivalent to a score in the original distribution that is one standard deviation above or below the mean, respectively. This direct mapping from the original distribution to a Z score means that the standard deviation of a Z distribution must be equal to 1.

15. a. Make sure that you understand the properties of the normal distribution, as we will apply its properties to other statistical techniques and procedures in the following chapters. Refer to the first part of Chapter 9 for a discussion of the normal distribution.
  b. In general, standardized scores or Z scores provide a means to express the distance between the mean and a particular point or score. A Z score is the number of standard deviation units a raw score is from the mean. A negative Z score indicates that it is lower than the mean (or to the left of the mean); while a positive Z score indicates that it is higher than the mean (or to its right).

CHAPTER 10

1. a. Although there are problems with the collection of data from all Americans, the census is assumed to be complete, so the mean age would be a parameter.
  b. A statistic because it is estimated from a sample.
  c. A statistic because it is estimated from a sample.
  d. A parameter because the school has information on all employees.
  e. A parameter because the school would have information on all its students.

3. a. Assuming that the population is defined as all persons shopping at that shopping mall that day of the week, she is selecting a systematic random sample. A more precise definition might limit it to all persons passing by the department store at the mall that day.

   However, assuming that the population is defined as all Americans, this sort of sampling technique would qualify as nonprobability sampling.
  b. This is a stratified sample because voters were first grouped by county, and unless the counties have the same number of voters, it is a disproportionate stratified sample because the same number is chosen from each county. We can assume it was a probability sample, but we are not told exactly how the 50 voters were chosen from the lists.

    c. This is neither a simple random sample nor a systematic random sample. It might be thought of as a sample stratified on last name, but even then, choosing the first 20 names is not a random selection process.

    d. This is not a probability sample. Instead, it is a purposive sample chosen to represent a cross section of the population in New York City.

5.   The relationship between the standard error and the standard deviation is $\sigma_{\bar{Y}} = \dfrac{\sigma_Y}{\sqrt{N}}$ where $\sigma_{\bar{Y}} = \dfrac{\sigma_Y}{\sqrt{N}}$ is the standard error of the mean and σ_Y is the standard deviation. Since σ_Y is divided by $\sqrt{N}$, $\sigma_{\bar{Y}} = \dfrac{\sigma_Y}{\sqrt{N}}$ must always be smaller than σ_Y, except in the trivial case where $N = 1$.

Theoretically, the dispersion of the mean must be less than the dispersion of the raw scores. This implies that the standard error of the mean is less than the standard deviation.

7.   a. These polls are definitely not probability samples. No sampling is done by the television station to choose who calls the 800 number.

    b. The population is all those people who watch the television station and see the 800 number advertised.

9.   a. This is not a random sample. The students eating lunch on Tuesday are not necessarily representative of all students at the school, and you have no way of calculating the probability of inclusion of any student. Many students might, for example, rarely eat lunch at the cafeteria and therefore have no chance of being represented in your sample. The fact that you selected *all* the students eating lunch on Tuesday makes your selection appear to be a census of the population, but that isn't true either unless all the students ate at the cafeteria on Tuesday.

    b. This is a systematic random sample because names are drawn systematically from the list of all enrolled students.

    c. This would seem to be a systematic random sample as in (b), but it suffers from the same type of defect as the cafeteria sample. Unless all students pass by the student union, using that location as a selection criterion means that some students have no chance of being selected (but you don't know which ones). Samples are often drawn this way in shopping malls by choosing a central location from which to draw the sample. It is reasonable to assume that a sufficiently representative mix of shoppers will pass by a central location during any one period.

11.   a. Mean = 5.3; standard deviation = 3.27.

    b. Here are 10 means from random samples of size 3: 6.33, 5.67, 3.33, 5.00, 7.33, 2.33, 6.00, 6.33, 7.00, 3.00.

    c. The mean of these 10 sample means is 5.23. The standard deviation is 1.76. The mean of the sample means is very close to the mean for the population. The standard deviation of the sample means is much less than the standard deviation for the population. The standard deviation of the means from the samples is an estimate of the standard error of the mean we would find from one random sample of size 3.

13.   a. The standard error is calculated as follows

$$\sigma_{\bar{Y}} = \frac{\sigma_Y}{\sqrt{N}} = \frac{16.4}{\sqrt{7}} = 6.2$$

Pg 338-342

1,3,4,8,15

average deviation of any sample mean from the mean of
also be referred to as that standard deviation of the sampling

ses where $N = 1$, the standard error will always be less
d deviation of the population. This is expressed in the

$$\sigma_{\bar{Y}} = \frac{\sigma_Y}{\sqrt{N}}$$

distribution is normal; thus, even when working with a
w that the sampling distribution is normal. Suggestions
ror include increasing the sample size.

confidence level runs from 11.52 to 14.88 percent. This
es out of 100 that the confidence interval we calculate
percentage of victims in the American population. The
rticular sample either does or does not contain the

$$\sqrt{\frac{(13.2)(86.8)}{1,105}} = 1.018$$

5(1.018)
8
14.88%

b. Confidence interval $= 13.2 \pm 2.58(1.018)$
$= 13.2 \pm 2.63$
$= 10.57\%$ to 15.83%

c. If the sample size was doubled and the percentage remained at 13.2, the confidence
intervals will shrink by a factor of $\frac{1}{\sqrt{2}}$, or about 30%.

d. Sample sizes on the order of 1,000 to 1,500 are a trade-off between precision and cost.
Samples of that size yield confidence intervals for proportions with errors around ±3%, at
the 95% confidence level, which is sufficient for most purposes in applied social research.
Doubling sample size to 3,000 does reduce the errors but increases the costs by more.

3.   a.
$$S_p = \sqrt{\frac{(0.792)(1 - 0.792)}{1,132}} = 0.012$$

Confidence interval = 0.792 ± 1.96(0.012)
= 0.792 ± 0.024
= 0.768 to 0.816

b. Confidence interval = 0.792 ± 2.58(0.012)
= 0.792 ± 0.031
= 0.761 to 0.823

c. There is very little difference between the 95% or 99% confidence interval here because the sample size is reasonably large. The former interval is only ½ of a percentage point wide, the latter nearly 2/3 of a percentage point. Most large survey organizations use the 95% confidence interval routinely, and that seems like the best choice here. Our conclusions about Americans' opinions about foreign involvement will be the same in either case: Slightly more than a majority agrees that international agreements are a national priority. The intent of this problem is to get students to recognize that they always have a choice as to what confidence interval they choose for a particular problem.

5.
$$\sigma_{\bar{Y}} = \sqrt{\frac{(52)(48)}{500}} = 2.23\%$$

Confidence interval = 52 ± 1.96(2.23)
= 47.63% to 56.37%

We set the confidence interval at the 95% confidence level. However, no matter whether the 90%, 95% or 99% confidence level is chosen, the calculated interval includes values below 50% for the vote for Ann Johnson. Therefore, you should advise the newspaper to be cautious in declaring Johnson the likely winner because it is quite possible that less than a majority of voters support her.

7. a.
$$S_p = \sqrt{\frac{(0.796)(1 - 0.796)}{666}} = 0.016$$

Confidence interval = 0.796 ± 1.96(0.016)
= 0.796 ± 0.031
= 0.765 to 0.827 or 76.5% to 82.7%

b. Based on our answer in 7a, we know that a 90 percent confidence interval will be more precise than a 95 percent confidence interval that has a lower bound of 76.5 percent and an upper bound of 82.7 percent. Accordingly, a 90 percent confidence interval will have a lower bound that is greater than 76.5 percent and an upper bound that is less than 82.7 percent.

Additionally, we know that a 99 percent confidence interval will be less precise than what we calculated in 7a. Thus, the lower bound for a 99 percent confidence interval will be less than 76.5 percent and the upper bound will be greater than 82.7 percent.

9.
$$S_{\bar{Y}} = \frac{S_Y}{\sqrt{N}} = \frac{1.69}{\sqrt{1,498}} = 0.044$$

Confidence interval $= 1.82 \pm 1.65(0.044)$
$$= 1.82 \pm 0.073$$
$$= 1.75 \text{ to } 1.89$$

11.
$$\text{Standard error} = \sqrt{\frac{67(100 - 67)}{1012}} = 1.46$$

Confidence interval $= 67 \pm 1.96(1.46)$
$$= 67 \pm 2.90$$
$$= 64.1\% \text{ to } 69.9\%$$

13. a. For those who thought that homosexual relations were always wrong:

$$S_p = \sqrt{\frac{(56)(100 - 56)}{475}} = 2.28$$

Confidence interval $= 56 \pm 1.96(2.28)$
$$= 56 \pm 4.47$$
$$= \text{or } 51.53\% \text{ to } 60.47\%$$

For those who thought that homosexual relations were not wrong at all:

$$S_p = \sqrt{\frac{(33)(100 - 33)}{475}} = 2.16$$

Confidence interval $= 33 \pm 1.96(2.16)$
$$= 33 \pm 4.23$$
$$= 28.77\% \text{ to } 37.23\%$$

b.
$$S_p = \sqrt{\frac{(10)(100 - 10)}{475}} = 1.38$$

Confidence interval $= 10 \pm 1.96(1.38)$
$$= 10 \pm 2.70$$
$$= 7.30 \text{ to } 12.70$$

c. Because the 95% confidence interval for those who think that homosexual relations are always wrong does not include a value less than 50 percent, it would appear that one conclusion to be drawn is that the majority of the American public thinks that homosexual relations are always wrong.

15. a. Instructors should encourage students to think about this question and make some educated guesses.

   b. For American respondents:

$$S_p = \sqrt{\frac{(26)(100 - 26)}{326}} = 2.43$$

Confidence interval $= 26 \pm 1.65(2.43)$
$= 26 \pm 4.00$
$= 22\%$ to 30%

For Chilean respondents:

$$S_p = \sqrt{\frac{(15)(100 - 15)}{221}} = 2.40$$

Confidence interval $= 15 \pm 1.65(2.40)$
$= 15 \pm 3.96$
$= 11.04\%$ to 18.96%

   c.

$$S_p = \sqrt{\frac{(75)(100 - 75)}{850}} = 1.48$$

Confidence interval $= 75 \pm 2.58(1.48)$
$= 75 \pm 3.82$
$= 71.18\%$ to 78.82%

   d. We want to use a 90 percent confidence interval with $N = 105$.

$$S_p = \sqrt{\frac{(8)(100 - 8)}{105}} = 2.65$$

Confidence interval $= 8 \pm 1.65(2.65)$
$= 8 \pm 4.37$
$= 3.63\%$ to 12.37%

So, assuming a sample size of $N = 105$, we can be 90 percent confident that the percent of respondents from Denmark who felt that nature is sacred because it is God's creation is between 3.63% and 12.37% in the population.

CHAPTER 12

1. a. H_0: $\mu_Y = 13.5$ years; H_1: $\mu_Y < 13.5$ years.

b. The Z value obtained is –4.19. The p value for a Z of –4.19 is less than .001 for a one-tailed test. This is less than the alpha level of .01, so we reject the null hypothesis and conclude that the doctors at the HMO do have less experience than the population of doctors at all HMOs.

3. a. Two-tailed test, $\mu_1 \neq \$43,318$.
   b. One-tailed test, $\mu_1 > 3.2$.
   c. One-tailed test, $\mu_1 < \mu_2$
   d. Two-tailed test, $\mu_1 \neq \mu_2$.
   e. One-tailed test, $\mu_1 > \mu_2$.
   f. One-tailed test, $\mu_1 < \mu_2$.

5. a. $H_0: \mu_1 = 43.10$. $H_1:\mu_1 \neq 43.10$.
   b. Based on a two-tailed test with significance level of .05, the t is 6.91, and its p level
   ___ t the null hypothesis in favor of the research hypothesis. There
   ___ the mean age of the GSS sample and the mean age of all
   ___ clude that relative to age, the GSS sample is not representa-
   ___ lts.

$$t = \frac{46.21 - 43.10}{\frac{17.45}{\sqrt{1496}}} = 6.91$$

___ portions, the appropriate test statistic is Z.
= 0.2420. Since $p(0.2420) > $ alpha (0.05), we fail to reject the
___ ndicates that there is no statistical difference between conserv-
___ heir views toward affirmative action.

$$\sqrt{\frac{.15(1-.15)}{185} + \frac{.18(1-.18)}{128}} = .043$$

$$Z = \frac{.15 - .18}{.043} = -.698 = -.70$$

___, we would have to multiple p by 2, $0.2420 \times 2 = 0.4840$. Our deci-
___ the same, we would fail to reject the null hypothesis.

___ es: 0.35; for females: 0.49
___ ailed test. $H_0: \pi_1 = \pi_2$, $H_1: \pi_1 < \pi_2$

$$\sqrt{\frac{0.35(1-0.35)}{421} + \frac{0.49(1-0.49)}{524}} = 0.032$$

$$Z = \frac{0.35 - 0.49}{0.032} = -4.375$$

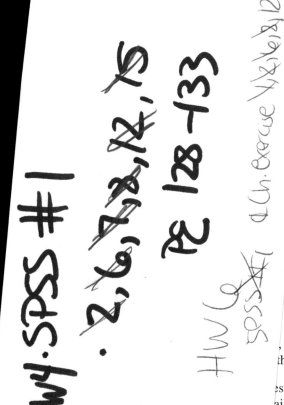

c. Since the probability of our obtained Z is less than .0001, we can reject the null hypothesis. We can conclude that there is a significant relationship between one's gender and tendency to vote Democrat. The proportion of women who voted for Gore is significantly greater than the proportion of men who voted for Gore (0.49 vs. 0.35).

11. a. One-tailed test, alpha = .05.

$$df = 666 + 829 - 2 = 1493$$

$$S_{\bar{Y}_1 - \bar{Y}_2} = \sqrt{\frac{(666-1)3.13^2 + (829-1)2.88^2}{(666+829) - 2}} \sqrt{\frac{666+829}{666(829)}} = \sqrt{8.96}\sqrt{.0027} = .155$$

$$t = \frac{13.34 - 13.22}{.155} = .774$$

Based on the t obtained of .774, we fail to reject the null hypothesis. For the GSS data set, there is no significant difference in educational attainment between male and female respondents.

b. $df = 618 + 732 - 2 = 1348$

$$S_{\bar{Y}_1 - \bar{Y}_2} = \sqrt{\frac{(618-1)3.98^2 + (732-1)3.74^2}{(618+732) - 2}} \sqrt{\frac{618+732-2}{618(732)}} = 0.19$$

$$t = \frac{11.85 - 11.34}{0.19} = 2.68$$

Based on t obtained of 2.68, we can reject the null hypothesis. The probability of 2.68 lies between .01 and .001 for a two-tailed test. Based on the ISSP data set, we conclude that there is a relationship between gender and educational attainment. Men have an average of .51 more years of education than women.

13.

$$S_{P_1 P_2} = \sqrt{\frac{.33(1-.33)}{257} + \frac{.28(1-.28)}{411}} = .037$$

$$Z = \frac{.33 - .28}{.037} = 1.35$$

The probability of obtaining $Z(1.35)$ is $.0885 \times 2 = .1770$. P is greater than our alpha of .01, we can not reject the null hypothesis. There is no significant relationship between social class and union membership.

CHAPTER 13

1. a.

Fear/Race	f_o	f_e	$\dfrac{(f_o - f_e)^2}{f_e}$
No/white	9	8.667	0.013
Yes/white	4	4.333	0.026
No/black	5	5.333	0.021
Yes/black	3	2.667	0.042
		$\chi^2 = 0.102$	

b. $(2 - 1)(2 - 1) = (1)(1) = 1$ degree of freedom

c. Based on Appendix D, we estimate that the probability of .102 is between .90 and .95, well above our .05 alpha level. Thus, we fail to reject the null hypothesis that race and fear of walking alone at night are statistically independent.

d. The cell for blacks who are afraid of walking alone has an expected value of 2.667. The cell for whites who are afraid of walking alone has an expected value of 4.333.

3. a. for RACE/NATFARE
$(3 - 1)(2 - 1) = (2)(1) = 2$ degrees of freedom

for RACE/NATEDUC
$(3 - 1)(2 - 1) = (2)(1) = 2$ degrees of freedom

b.

Natfare/Race	f_o	f_e	$\dfrac{(f_o - f_e)^2}{f_e}$
Too little/whites	107	122.2	1.891
About right/whites	214	211.9	0.021
Too much/whites	249	235.9	0.727
Too little/blacks	36	20.8	11.108
About right/blacks	34	36.1	0.122
Too much/blacks	27	40.1	4.280
		$\chi^2 = 18.149$	

Natfare/Race	f_o	f_e	$\dfrac{(f_o - f_e)^2}{f_e}$
Too little/whites	413	422.8	0.227
About right/whites	130	124.5	0.243
Too much/whites	35	30.7	0.602
Too little/blacks	83	73.2	1.312
About right/blacks	16	21.5	1.407
Too much/blacks	1	5.31	3.500
		$\chi^2 = 7.291$	

c. The estimated P value for the obtained chi-square for RACE and NATFARE is less than .001. Since this value is well below our alpha value of .05, we reject the null hypothesis. These two variables are statistically dependent.

d. The estimated P value for the obtained chi-square for RACE and NATEDUC is between .05 and .10. Since this value falls just above our alpha value of .05, we fail to reject the null hypothesis that these two variables are statistically independent.

e. The appropriate measure of association for these tables is lambda (λ).

for RACE/NATFARE
E1 = 667 – 276 = 391
E2 = (570 – 249) + (97 – 36) = 321 + 61 = 382
λ = (391 – 382) / 391 = 0.023

Given one's race, we can reduce the error in predicting his or her opinion on national spending on welfare by 2.3%. This indicates a weak relationship between the two variables.

for RACE/NATEDUC
E1 = 678 – 496 = 182
E2 = (578 – 413) + (100 – 83) = 165 + 17 = 182
λ = (182 – 182) / 182 = 0.000

There is no relationship between these two variables.

5. a.

Attitude Toward Premarital Sex/Educational Attainment	f_o	f_e	$\dfrac{(f_o - f_e)^2}{f_e}$
Always wrong/less than high school	24	12.981	9.353
Almost always wrong/less than high school	4	2.323	1.211
Sometimes wrong/less than high school	2	8.882	5.332

(Continued)

Attitude Toward Premarital Sex/Educational Attainment	f_o	f_e	$\frac{(f_o - f_e)^2}{f_e}$
Not wrong at all/less than high school	14	19.814	1.706
Always wrong/high school	49	57.236	1.185
Almost always wrong/high school	11	10.242	0.056
Sometimes wrong/high school	42	39.161	0.206
Not wrong at all/high school	92	87.360	0.246
Always wrong/bachelor's degree or higher	22	24.783	0.313
Almost always wrong/bachelor's degree or higher	2	4.435	1.337
Sometimes wrong/bachelor's degree or higher	21	16.957	0.964
Not wrong at all/bachelor's degree or higher	39	37.826	0.036
		$\chi^2 = 21.945$	

b. $(4 - 1)(3 - 1) = (3)(2) = 6$ degrees of freedom

Based on Appendix D, the probability of 21.945 is between .01 and .001. Since this P value is far below our alpha of .01, we reject the null hypothesis. One's educational degree and attitudes toward premarital sex are statistically dependent.

7. a.

Victim Race/Offender Race	f_o	f_e	$\frac{(f_o - f_e)^2}{f_e}$
White/white	3000	1702.214	989.446
Black/white	227	1485.411	1066.101
Other/white	51	90.374	17.154
White/black	483	1746.353	913.939
Black/black	2852	1523.929	1157.385
Other/black	28	92.718	45.174
White/other	58	92.433	12.827
Black/other	11	80.660	60.160
Other/other	109	4.907	2208.142
		$\chi^2 = 6470.328$	

$(3 - 1)(3 - 1) = (2)(2) = 4$ degrees of freedom

The probability of 6470.328 is less than .001. Since this value is much less than the given alpha of .01, we reject the null hypothesis that race of offender and race of victim are independent. These two variables are statistically dependent.

b. Offenders are most likely to choose victims of their own race. The proportion of "white" victims of "white" offenders is 0.915 (3000 out of 3278). "Black" victims of "black" offenders (2852 out of 3363) represent 0.848 of the category. Proportionally,

0.612 of the victims of "other" offenders (109 out of 178) are "other" as well. The calculated lambda value for this bivariate table is 0.738. This indicates a very strong relationship between the race of offender and the race of victim. Given the race of offender, we can reduce the error in predicting the race of victim by 73.8%.

9. a.

GOVDIFF/Educational Attainment	f_o	f_e	$\dfrac{(f_o - f_e)^2}{f_e}$
Strongly agree/primary degree	69	58.380	1.932
Agree/primary degree	94	91.060	0.095
Neither agree nor disagree/primary degree	54	45.372	1.641
Disagree/primary degree	29	43.468	4.816
Strongly disagree/primary degree	11	18.720	3.184
Strongly agree/secondary degree	76	80.869	0.293
Agree/secondary degree	124	126.138	0.036
Neither agree nor disagree/secondary degree	60	62.849	0.129
Disagree/secondary degree	66	60.212	0.556
Strongly disagree/secondary degree	30	25.931	0.638
Strongly agree/university degree	39	44.751	0.739
Agree/university degree	69	69.801	0.009
Neither agree nor disagree/university degree	29	34.779	0.960
Disagree/university degree	42	33.320	2.261
Strongly disagree/university degree	18	14.349	0.929
		$\chi^2 = 18.218$	

$(5 - 1)(3 - 1) = (4)(2) = 8$ degrees of freedom

b. According to Appendix D, the probability of 18.218 is between .01 and .02. Given the alpha of .05, we reject the null hypothesis that GOVDIFF and educational attainment are statistically independent since the probability of our chi-square value is less than the alpha.

c. Changing the alpha value to .01 would change the results because the probability of 18.218 is greater than the alpha. We would therefore fail to reject the null hypothesis.

11. a.

Happy/Marital	f_o	f_e	$\dfrac{(f_o - f_e)^2}{f_e}$
Very happy/married	90	67.803	7.267
Pretty happy/married	103	113.946	1.052
Not too happy/married	17	28.251	4.481

(Continued)

Happy/Marital	f_o	f_e	$\dfrac{(f_o - f_e)^2}{f_e}$
Very happy/divorced	16	27.767	4.987
Pretty happy/divorced	50	46.664	0.238
Not too happy/divorced	20	11.570	6.142
Very happy/never married	38	48.430	2.246
Pretty happy/never married	89	81.390	0.711
Not too happy/never married	23	20.179	0.394
		$\chi^2 = 27.518$	

$(3 - 1)(3 - 1) = (2)(2) = 4$ degrees of freedom

The probability of 27.518 is less than .001. Since this is less than the given alpha of .05, we reject the null hypothesis of independence between HAPPY and MARITAL. Marital status and happiness level are statistically dependent.

b.

Happy/Marital	$|f_o - f_e|$
Very happy/married	22.197
Pretty happy/married	10.946
Not too happy/married	11.251
Very happy/divorced	11.767
Pretty happy/divorced	3.336
Not too happy/divorced	8.430
Very happy/never married	10.430
Pretty happy/never married	7.610
Not too happy/never married	2.821

The greatest differences are in the cells of "Very happy/married" (22.197), "Very happy/divorced" (11.767), and "Not too happy/divorced" (11.251).

The greatest contributors to the chi-square value are "Pretty happy/married" (113.946), "Pretty happy/never married" (81.390), and "Very happy/married" (67.803).

CHAPTER 14

1.   We begin by calculating *SST, SSE,* and *SSR* using the table below.

$SST = \sum(Y - \bar{Y})^2 = 1415.858$
$SSE = \sum (Y - \hat{Y})^2 = 1190.158$
$SSR = SST - SSE = 1415.858 - 1190.158 = 225.700$

Y	$(Y-\bar{Y})$	$(Y-\bar{Y})^2$	$\hat{Y}$	$(Y-\hat{Y})$	$(Y-\hat{Y})^2$
22.8	4.775	22.801	19.70	3.098	9.596
27.8	9.775	95.551	18.63	9.171	84.098
44.8	26.775	716.901	22.04	22.763	518.159
10.7	−7.325	53.656	9.29	1.409	1.986
1.6	−16.425	269.781	22.73	−21.131	446.519
6.8	−11.225	126.001	9.42	−2.617	6.848
7.4	−10.625	112.891	18.38	−10.977	120.497
22.3	4.275	18.276	23.87	−1.567	2.455
		$\sum(Y-\bar{Y})^2 = 1415.858$			$\sum(Y-\hat{Y})^2 = 1190.158$

$\bar{Y} = 18.025$

* Note, answers may vary slightly due to rounding.
Instructions on how to calculate $\hat{Y}$ were introduced in Chapter 8 pp. 293-295.

Now, we plug in the values into the equations for the Mean Squares.

$$MSR = \frac{SSR}{df_r} = \frac{SSR}{K} = \frac{225.700}{1} = 225.700$$

$$MSE = \frac{SSE}{df_e} = \frac{SSE}{[N-(K+1)]} = \frac{1190.155}{[8-(1+1)]} = 198.359$$

Finally, we calculate the F ratio.

$$F = \frac{MSR}{MSE} = \frac{225.700}{198.359} = 1.138$$

Looking at the distribution of F with $\alpha = .05$ in Appendix E, with $df_1 = 1$ and $df_2 = 6$, it is clear that our F ratio of 1.138 is smaller than 5.99, the critical value of F needed to judge the relationship between being concerned about the environment and actually donating money to environmental groups as statistically significant. Thus, the p value is greater than .05; the relationship is not significant.

3. a. Once a member (Former member) = 31,046.40; Not a member = 44,614.56; Member, currently = 99,812.78.
   b. Based on F obtained of 22.586, we would reject the null hypothesis of no difference. We know from the SPSS output that probability of the F obtained is .000, less than our alpha of .05. The output indicates that current union members have the highest earnings, followed by those who are not members, then those who were formerly union members.

5. a. The group with the highest average number of children was the group that said "no" to abortion under any circumstance, 2.04 children.

   b. The SPSS results indicate that yes, there is a significant difference in the number of children between the two groups. F obtained is 5.334, significant at the .021 level (which is less than our .05 alpha).

7. a. Again the group with the highest average was the group of respondents who answered "no", 2.97 children.

   b. Yes, there is a significant difference in the number of ideal children for each group. Our F obtained is 4.838, significant at .029 (less than our alpha of .05). We would reject the null hypothesis and conclude that the means are significantly different. Those who would not support abortion report a higher number of ideal children (2.97) than those who do (2.52).

   c. If alpha were set at .01, we would fail to reject the null hypothesis. The probability of obtaining our F obtained is .029, greater than .01.

9.

$\bar{Y}_1 = 4.12$	$\bar{Y}_2 = 3.75$	$\bar{Y}_3 = 3$	$\bar{Y}_4 = 2$
$\sum Y_1 = 33$	$\sum Y_2 = 30$	$\sum Y_3 = 24$	$\sum Y_4 = 16$
$\sum Y_1^2 = 141$	$\sum Y_2^2 = 120$	$\sum Y_3^2 = 76$	$\sum Y_4^2 = 36$
$n_1 = 8$	$n_2 = 8$	$n_3 = 8$	$n_4 = 8$

$$\bar{Y} = 3.22$$
$$N = 32$$

$$\begin{aligned}
\text{SSB} &= 8(4.12 - 3.22)^2 + 8(3.75 - 3.22)^2 = 8(3 - 3.22)^2 + 8(2 - 3.22)^2 \\
&= 8(.81) + 8(.2809) + 8(.0484) + 8(1.4884) \\
&= 6.48 + 2.2472 + .3872 + 11.9072
\end{aligned}$$
SSB = 21.02

dfb = 4 − 1
dfb = 3
Mean square between = 21.02/3 = 7.01

$$\begin{aligned}
\text{SSW} &= (141 + 120 + 76 + 36) - [(33^2/8) + (30^2/8) + (24^2/8) + (16^2/8)] \\
&= 373 - (136.125 + 112.5 + 72 + 32) \\
&= 373 - 352.625
\end{aligned}$$
SSW = 20.375

dfw = 32 − 4
dfw = 28
Mean square within = 20.375/28 = 0.727679 = 0.73
F = 7.01/.73
F = 9.60

Decision: If we set alpha at .01, F critical would be 4.57 ($df_1 = 3$ and $df_2 = 28$). Based on our F obtained of 9.60, we would reject the null hypothesis and conclude that at least one of the means is significantly different than the others. Lower class respondents have the most people in their households (4.12), followed by working class respondents (3.75), middle class respondents (3), and lastly upper class respondents (2).

11.   We begin by calculating SST, SSE, and SSR using the table below.

$$SST = \sum(Y - \bar{Y})^2 = 224.40$$
$$SSE = \sum(Y - \hat{Y})^2 = 69.50$$
$$SSR = SST - SSE = 224.40 - 69.50 = 154.90$$

Y	$(Y-\bar{Y})$	$(Y-\bar{Y})^2$	$\hat{Y}$	$(Y-\hat{Y})$	$(Y-\hat{Y})^2$
19	−4.40	19.36	14.21	4.79	22.94
32	8.60	73.96	28.25	3.75	14.06
20	−3.40	11.56	21.01	−1.01	1.02
18	−5.40	29.16	20.29	−2.29	5.24
24	0.60	0.36	24.79	−.79	.62
24	0.60	0.36	27.23	−3.23	10.43
30	6.60	43.56	26.75	3.25	10.56
25	1.60	2.56	25.39	−0.39	0.15
17	−6.40	40.96	18.13	−1.13	1.28
25	1.60	2.56	23.21	1.79	3.20
		$\sum(Y-\bar{Y})^2 = 224.40$			$\sum(Y-\hat{Y})^2 = 69.50$
$\bar{Y} = 23.40$					

*Note, answers may vary slightly due to rounding.

Now, we plug in the values into the equations for the mean squares.

$$MSR = \frac{SSR}{df_r} = \frac{SSR}{K} = \frac{154.90}{1} = 154.90$$

$$MSE = \frac{SSE}{df_e} = \frac{SSE}{[N-(K+1)]} = \frac{69.50}{[10-(1+1)]} = 8.69$$

Finally, we calculate the F ratio.

$$F = \frac{MSR}{MSE} = \frac{154.90}{8.69} = 17.82$$

Looking at the distribution of F with $\alpha = .05$ in Appendix E, with $df_1 = 1$ and $df_2 = 8$, it is clear that our F ratio of 17.82 is larger than 4.96, the critical value of F needed to judge the relationship between GNP and birth rate as statistically significant. In fact, looking at the distribution of F with $\alpha = .01$ in Appendix E, 17.83 exceeds the critical value of 11.26. Thus, we know that the p value is less than .01; the relationship not significant.

13.

$\bar{Y}_1 = 10.4$	$\bar{Y}_2 = 13.25$	$\bar{Y}_3 = 15$
$\sum Y_1 = 52$	$\sum Y_2 = 53$	$\sum Y_3 = 75$
$\sum Y_1^2 = 544$	$\sum Y_2^2 = 703$	$\sum Y_3^2 = 1127$
$n_1 = 5$	$n_2 = 4$	$n_3 = 5$
	$\bar{Y} = 12.86$	
	$N = 14$	

$$SSB = 5(10.4 - 12.86)^2 + 4(13.25 - 12.86)^2 + 5(15 - 12.86)^2$$
$$= 5(6.0516) + 4(0.1521) + 5(4.5796)$$
$$= 30.258 + .6084 + 22.898$$
$$SSB = 53.76$$

$dfb = 3 - 1$
dfb = 2

Mean square between = 53.76/2 = 26.88

$$SSW = (544 + 703 + 1127) - [(52^2/5) + (53^2/4) + (75^2/5)]$$
$$= 2374 - (540.8 + 702.25 + 1125)$$
$$= 2374 - 2368.05$$
$$SSW = 5.95$$

$dfw = 14 - 3$
dfw = 11

Mean square within = 5.95/11 = 0.540909 = .54

$F = 26.88/.54$
F = 49.78

Decision: If we set alpha at 0.05, F critical would be 3.98 ($df_1 = 2$ and $df_2 = 11$). Based on our F obtained of 49.78, we would reject the null hypothesis and conclude that at least one of the means is significantly different than the others. Educational level is highest for single respondents (15 year – some college). Divorced respondents are next (13.25), followed lastly by married respondents (10.40).

INDEX/GLOSSARY